Lecture Notes in Computer Science

Edited by G. Goos and J. Hartmanis

486

J. van Leeuwen   N. Santoro (Eds.)

Distributed Algorithms

4th International Workshop
Bari, Italy, September 24–26, 1990
Proceedings

Springer-Verlag
Berlin Heidelberg New York London Paris
Tokyo Hong Kong Barcelona Budapest

Editorial Board
D. Barstow W. Brauer P. Brinch Hansen D. Gries D. Luckham
C. Moler A. Pnueli G. Seegmüller J. Stoer N. Wirth

Volume Editors

Jan van Leeuwen
Department of Computer Science, University of Utrecht
Padualaan 14, P.O. Box 80.089
3508 TB Utrecht, The Netherlands

Nicola Santoro
School of Computer Science, Carleton University
Ottawa, Canada K1S 5B6

CR Subject Classification (1991): C.2.2, C.2.4, D.4.1, D.4.4–5, F.1.1, F.2.2

ISBN 3-540-54099-7 Springer-Verlag Berlin Heidelberg New York
ISBN 0-387-54099-7 Springer-Verlag New York Berlin Heidelberg

This work is subject to copyright. All rights are reserved, whether the whole or part of the material is concerned, specifically the rights of translation, reprinting, re-use of illustrations, recitation, broadcasting, reproduction on microfilms or in other ways, and storage in data banks. Duplication of this publication or parts thereof is only permitted under the provisions of the German Copyright Law of September 9, 1965, in its current version, and a copyright fee must always be paid. Violations fall under the prosecution act of the German Copyright Law.

© Springer-Verlag Berlin Heidelberg 1991
Printed in Germany

Printing and binding: Druckhaus Beltz, Hemsbach/Bergstr.
2145/3140-543210 – Printed on acid-free paper

Preface

The 4th International Workshop on Distributed Algorithms (WDAG 90) was organized by the Istituto di Scienze dell' Informazione of the University of Bari (Italy) and held in Serra Alimini (near Otranto), some 200 kilometers southeast of Bari along the beautiful Adriatic coast, September 24-26, 1990. The workshop was intended as a forum for researchers, students and other interested persons to discuss the recent results and trends in the design and analysis of distributed algorithms for communication networks (i.e., on graphs) and decentralized systems. The earlier workshops on Distributed Algorithms were held in Ottawa (1985, proceedings published by Carleton University Press), Amsterdam (1987, see Lecture Notes in Computer Science, Vol. 312), and Nice (1989, see Lecture Notes in Computer Science, Vol. 392).

Papers were solicited describing original results in all areas of distributed algorithms and their applications including, e.g. distributed combinatorial algorithms, distributed optimization algorithms, distributed algorithms on graphs, distributed algorithms for decentralized systems, distributed data structures, distributed algorithms for control and communication, design of network protocols, routing algorithms, fail-safe and fault-tolerant distributed algorithms, distributed database techniques, algorithms for transaction management, replica-control algorithms, and other related fields. The program committee for the workshop consisted of

D. Dolev	(IBM Almaden Research Center and Hebrew University),
F. Mattern	(University of Kaiserslautern),
N. Santoro	(Carleton University, Ottawa), *co-chairman*
P. Spirakis	(Computer Technology Institute, Patras, and New York University),
R.B. Tan	(University of Oklahoma, Chickasha, and University of Utrecht),
S. Toueg	(Cornell University, Ithaca),
J. van Leeuwen	(University of Utrecht), *co-chairman*
P.M.B. Vitanyi	(CWI and University of Amsterdam),
S. Zaks	(Technion, Haifa).

Close to sixty papers were submitted out of which the program committee selected twenty-eight papers for presentation in the workshop. The selection reflects several current directions of research in the area of distributed algorithms, although certainly not all aspects of the field could be covered in the three-day workshop. The present volume contains the revised version of all papers presented in the workshop. The revised versions are based on the comments and suggestions received by the authors during and after the workshop. Several papers are in the form of preliminary reports on continuing research. The papers in this volume should give a good impression of the current work in distributed algorithms and stimulate further research. The sessions at the workshop were chaired by: Friedemann Mattern, Nicola Santoro, Paul Spirakis, Richard Tan, Sam Toueg/Gil Neiger, Jan van Leeuwen, Paul Vitanyi, and Shmuel Zaks.

We are grateful to the Centre for Parallel and Distributed Computing (*PARADISE*) of Carleton University (Ottawa, Canada) and the Department of Computer Science of the University of Utrecht (Utrecht, the Netherlands) for supporting the organization of the workshop, and to the Centro Internazionale Congressi S.r.l. and its staff for the superb local arrangements. The workshop was partly supported by IBM (Italy), Siemens (Italy) and by the Banca Vallone/Gruppo Ambrosiano. We also thank Annerie Deckers (Department of Computer Science, University of Utrecht) for her invaluable assistance during the preparation of the workshop.

Utrecht and Ottawa/Bari
March 1991

Jan van Leeuwen and Nicola Santoro
Co-Chairmen WDAG 90

CONTENTS

Self-Stabilizing Ring Orientation 1
    A. Israeli and M. Jalfon (Technion, Haifa)

Memory-Efficient Self-Stabilizing Protocols for General Networks 15
    Y. Afek (AT&T Bell Laboratories and Tel-Aviv University), S. Kutten and
    M. Yung (IBM T.J. Watson Research Center)

On the Computational Power Needed to Elect a Leader 29
    A. Itai (Technion, Haifa)

Spanning Tree Construction for Nameless Networks 41
    I. Lavallée and C. Lavault (INRIA, Rocquencourt)

A Linear Fault-Tolerant Naming Algorithm 57
    J. Beauquier (Université Paris 11), P. Gastin (Université Paris 6) and
    V. Villain (Université de Picardie, Amiens)

Distributed Data Structures: A Complexity-Oriented View 71
    D. Peleg (The Weizmann Institute, Rehovot)

An Improved Algorithm to Detect Communication Deadlocks in Distributed Systems 90
    B. Kröger (University of Osnabrück), R. Lüling, B. Monien (University of Paderborn)
    and O. Vornberger (University of Osnabrück)

On the Average Performance of Synchronized Programs in Distributed Networks 102
    S. Rajsbaum and M. Sidi (Technion, Haifa)

Distributed Algorithms for Reconstructing MST after Topology Change 122
    J. Park, T. Masuzawa, K. Hagihara, and N. Tokura (Osaka University)

Efficient Distributed Algorithms for Single-Source Shortest Paths and Related Problems
on Plane Networks 133
    R. Janardan and S. Wing Cheng (University of Minnesota, Minneapolis)

Stepwise Development of a Distributed Load Balancing Algorithm 151
    P. Grønning, T. Qvist Nielsen and H. H. Løvengreen (Technical University of Denmark,
    Lyngby)

Greedy Packet Scheduling 169
    I. Cidon, S. Kutten (IBM T.J. Watson Research Center), Y. Mansour (MIT) and
    D. Peleg (The Weizmann Institute, Rehovot)

Optimal Computation of Global Sensitive Functions in Fast Networks 185
    I. Cidon, I. Gopal and S. Kutten (IBM T.J. Watson Research Center)

Efficient Mechanism for Fairness and Deadlock-Avoidance in High-Speed Networks 192
    Y. Ofek and M. Yung (IBM T.J. Watson Research Center)

Strong Verifiable Secret Sharing 213
    C. Dwork (IBM Almaden Research Center)

Weak Consistency and Pessimistic Replica Control 228
    A. Sandoz and A. Schiper (Ecole Polytechnique Fédérale de Lausanne)

Localized-Access Protocols for Replicated Databases 245
    D. Agrawal and A. El Abbadi (University of California, Santa Barbara)

Weighted Voting for Operation Dependent Management of Replicated Data 263
    M. Obradovic and P. Berman (The Pennsylvania State University)

Wakeup under Read/Write Atomicity 277
    P. Jayanti and S. Toueg (Cornell University, Ithaca)

Time and Message Efficient Reliable Broadcasts 289
    T.D. Chandra and S. Toueg (Cornell University, Ithaca)

Early-Stopping Distributed Bidding and Applications 304
    N. Budhiraja, A. Gopal and S. Toueg (Cornell University, Ithaca)

Fast Consensus in Networks of Bounded Degree 321
    P. Berman (The Pennsylvania State University) and J. A. Garay (IBM T.J. Watson
    Research Center)

Common Knowledge and Consistent Simultaneous Coordination 334
    G. Neiger (Georgia Institute of Technology, Atlanta) and M. R. Tuttle (Cambridge Research
    Laboratory)

Agreement on the Group Membership in Synchronous Distributed Systems 353
    R. de Lemos and P. D. Ezhilchelvan (University of Newcastle upon Tyne)

Tight Bounds on the Round Complexity of Distributed 1-Solvable Tasks 373
    O. Biran, S. Moran and S. Zaks (Technion, Haifa)

A Time-Randomness Tradeoff for Communication Complexity 390
    R. Fleischer (University of the Saarland, Saarbrücken), H. Jung (Humboldt University,
    Berlin) and K. Mehlhorn (University of the Saarland, Saarbrücken)

Bounds on the Costs of Register Implementations 402
    S. Chaudhuri (University of Washington, Seattle) and J. Welch (University of
    North Carolina, Chapel Hill)

A Bounded First-In, First-Enabled Solution to the 1-Exclusion Problem 422
    Y. Afek (AT&T Bell Laboratories and Tel-Aviv University), D. Dolev (IBM Almaden
    Research Center and Hebrew University), E. Gafni (Tel-Aviv University and UCLA),
    M. Merritt (AT&T Bell Laboratories) and N. Shavit (IBM Almaden Research Center
    and Stanford University)

List of Participants 432

Author Index 433

Self-Stabilizing Ring Orientation

Amos Israeli [*]
Dept. of Electrical Engineering
Technion — Israel

Marc Jalfon [‡]
Dept. of Computer Science
Technion — Israel
and Int$_e$l — Haifa, Israel

Abstract

A *self-stabilizing* system is a distributed system which can be started in any *possible* global state. Once started the system regains its consistency by itself, without any kind of an outside intervention. A *ring* is a distributed system in which all processors are connected in a ring. A ring is *oriented* if all processors in the ring agree on common right and left directions. A protocol is *uniform* if all processors use the same program.

In this paper we answer the following question: *Does a uniform self stabilizing protocol for ring orientation exist?* We begin the presentation by answering this question negatively for deterministic protocols. Then we present a randomized uniform self stabilizing protocol for *ring orientation*. When the protocol stabilizes all processors agree upon a "right" (privileged) direction. The protocol works for a ring of any size and even tolerates dynamic additions and removals of processors as long as the ring topology is preserved. The number of states of each processor is $O(1)$, and its stabilization time is $O(n^2)$, where n is the number of processors in the system.

1 Introduction

A *self-stabilizing* system is a distributed system which can be started in any *possible* global state. Once started the system regains its consistency by itself, without any kind of an outside intervention. Two advantages of self stabilizing systems are:

- Self stabilizing systems need not be initialized globally. Each component can be started separately and in an arbitrary state. The system will self-stabilize into a legitimate configuration.

- The self stabilization property makes the system tolerant to *transient bugs*, bugs in which the state of a component is changed spontaneously while the component is still correct.

A *ring* is a distributed system in which all processors are connected in a circle. A ring is *oriented* if all processors in the ring agree on common right and left directions. A protocol is *uniform* if all processors use the same program. A common problem in distributed protocols is symmetry breaking.

---

[*]Partially supported by Technion VPR Funds - Japan TS Research Fund and B. & G. Greenberg Research Fund (Ottawa).
[‡]Partially supported by a Gutwirth fellowship.

Almost all known self-stabilizing protocols for rings, e.g. [Di-74], [BGW-87], [Bu-87], present self stabilizing protocols for mutual exclusion. In order to avoid dealing with symmetry breaking they make two strong assumptions:

non-uniformity: There is a single "special" processor whose program is different from the program of the rest of the processors.

orientation: The ring is oriented.

The work of of [DIM-89] introduces the use of *registers* for communication between processors. Using registers they design a non uniform self stabilizing protocol for mutual exclusion for general graphs and in particular, for unoriented rings. In [Di-74], Dijkstra has observed that no deterministic uniform self stabilizing protocol for mutual exclusion for rings exists. In [IJ-89] we use randomization to break symmetry and present a randomized uniform self stabilizing mutual exclusion protocol for oriented rings. Thus orientation can be given up in the presence of non-uniformity; while for randomized protocols non-uniformity can be given up in the presence of orientation. It is natural to ask whether both assumptions can be dispensed simultaneously.

In this paper we answer the following question: *Does a uniform self stabilizing protocol for ring orientation exist?* We begin the presentation by answering this question negatively for deterministic protocols. Then we present a randomized self stabilizing uniform protocol for *ring orientation*. When the protocol stabilizes, all processors agree upon a "right" (privileged) direction. The protocol works for a ring of any size and even tolerates dynamic additions and removals of processors as long as the ring topology is preserved. The number of states of each processor is $O(1)$. The expected stabilization time is $O(n^2)$, where n is the number of processors in the system.

The protocol is composed of two "levels", each level consists of a self stabilizing protocol. On the lower level all edges of the ring are directed, not necessarily in a consistent way. Each edge is directed separately by a *randomized* protocol which is run by the processors at its endpoints. The higher level of the protocol is a *deterministic* protocol for orientation of a directed ring (that is a ring whose edges are directed). The final protocol is obtained by using the technique of fair combination of self stabilizing protocols due to [DIM-89].

Recently the works of [BP-88] and [IJ-89] presented uniform self stabilizing protocols for oriented rings. The protocol of [BP-88] works on prime rings in which no symmetric global states exist. The work of [IJ-89] uses *randomization* to break symmetry. The present protocol enables execution of any uniform self stabilizing protocol for oriented rings on any ring. This work together with [IJ-89] are the first to introduce randomization to self stabilizing protocols in which processors communicate using shared memory. A randomized self stabilizing version of the alternating bit protocol appears in [AB-89]. The protocol of [AB-89] uses message passing for communication.

The rest of this paper is organized as follows: in Section 2 the formal model and requirements for self-stabilization and ring orientation are presented. Some impossibility results are brought in Section 3. In Section 4 we present a uniform self-stabilizing protocol for orienting a ring. Concluding remarks appear in Section 5.

2   Model and Requirements

A *uniform ring* consists of n *identical processors*, denoted by $P_0, P_1, ..., P_{n-1}$. Each processor is a *randomized finite state machine*. Processors are *anonymous*, they do not have identities. The subscripts $0, 1, ..., n-1$ are used for ease of notation only. Each processor resides on a node of the system's *communication ring*. Two processors residing on neighboring nodes are called *neighbors*.

Neighbors communicate using *registers*. If e is an edge of the ring between processors P_i and P_j then e is realized by two registers r_{ij} in which P_i writes and from which P_j reads, and r_{ji} in which P_j writes and from which P_i reads. Registers in which a processor P writes are referred to as P's registers. When all edges of the ring are undirected (directed) we call this ring *undirected* (*directed*). The edges of the ring can be either directed or undirected. When an edge of the ring is directed the two processors at its endpoints agree on its direction, one is designated as the edge's "head" while the other is the edge's "tail". An undirected edge is symmetric with respect to its endpoints. In this work the direction of edges is used solely for symmetry breaking; communication is allowed in both directions of an edge regardless of whether it is directed or undirected.

A processor is a finite state machine with *state set M* and transition function δ. The arguments of δ are the state of the processor and the values it reads from the registers of its neighbors. Whenever a processor is activated it executes a single *atomic step* which is determined by the transition function δ. In a single atomic step a processor may read the registers of some of its neighbors, write in some of its registers and then move to its new state. The "size" of each step depends on the type of the adversary and will be discussed below. The transition function δ can be *randomized*; in this case it may enable more than one transition. The transition which is actually executed is chosen with equal probability. A *uniform protocol* is a triplet $<\mathcal{G}, M, \delta>$ where $\mathcal{G}$ is a family of communication graphs, M and δ are the state set and transition function, common to all processors, respectively. If the function δ is randomized the protocol is *randomized*.

The global state of a uniform n processor system is described by its configuration. A configuration of a system is an n-tuple $c = (q_0, q_1, \ldots, q_{n-1})$ where $q_i \in M$ for $0 \leq i \leq n-1$. A processor P is *enabled* in configuration c if $\delta(q, r_1, r_2, \ldots) \neq q$, where q is the state of P in c and $r_1, r_2, \ldots$ are the values read by P from the registers of its neighbors (for randomized protocols we require $\{\delta(q, r_1, r_2, \ldots)\} \neq \{q\}$). If P is not enabled, it is said to be *disabled*. A configuration c is a *deadlock* configuration if no processor is enabled in c.

The behavior of real life distributed systems is modeled by the interleaving model. Processor activity is managed by a *scheduler*. To ensure correctness of the systems, we regard the scheduler as an adversary. We let the scheduler choose its activated processors *on line* using processor states as its input. We do not allow the scheduler to use the results of *future* coin tosses, as this may nullify the extra strength added to the system by randomization. Whenever the adversary activates a processor, the processor executes a single atomic step. The more freedom the adversary has in choosing its activated processors and the smaller the atomic step is, the stronger the adversary is. We hereafter list the most common types of adversaries used in the literature:

(a) The weakest adversary activates processors in sequence, one after the other. Whenever a processor is activated it reads the registers of *all* its neighbors and then it moves to a new state while writing in *all* its registers. This adversary is known as *Central Demon*.

(b) A stronger adversary which is also known as *Distributed Demon* can activate any subset of the system's processors together. Whenever a set of processors is activated by the adversary, all activated processors simultaneously read all the registers of their neighbors. Subsequently all activated processors move to their new state while writing in all their registers.

(c) An even stronger adversary is the *Read/Write Demon*. In a single atomic step it activates a single processor which either reads from a single register, or writes into a single register (but not both).

An adversary is *proper* if it activates only enabled processors as long as the system is not in a deadlock configuration. In this paper we use the **proper distributed demon** as the adversary. An *execution* of a system is a list of configurations $c^1, c^2, \ldots$ where each configuration, c^{i+1}, is obtained from the previous configuration, c^i, by a single activation of some set of processors. The list of subsets of processors activated by the scheduler constitutes a *schedule*.

We proceed by defining the self-stabilization requirements for randomized distributed systems. The set of all possible configurations is denoted by C. Let $L \subset C$ be a set of configurations of a given distributed system. L is called the set of *legitimate* configurations of the system. A system is *self-stabilizing* with respect to L if the following requirements are satisfied:

deadlock - Every deadlock configuration of the system is in L.

closure - For every $c \in L$, and for every $c' \in C$, if c' is reached from c by a single transition then $c' \in L$. (Once the system reaches a legitimate configuration, it will always remain in a legitimate configuration).

randomized no livelock - There is a function f from the natural numbers to the interval $[0,1]$ satisfying $\lim_{k \to \infty} f(k) = 0$ such that for every initial configuration and for every proper scheduler, the probability that the k-th configuration reached is in L is $1 - f(k)$.

Note: In this definition two common requirements are amended. The **no deadlock** requirement is replaced by the **deadlock** requirement which allows legitimate deadlock configurations. The (deterministic) **no livelock** requirement is replaced by the slightly relaxed **randomized no livelock** requirement.

In ring systems each processor has two neighbors. The processor can internally distinguish between a *first* neighbor and a *second* neighbor. Deciding which neighbor is the first and which one is the second is done by the hardware, the correctness and complexity of a protocol should not be affected by any possible assignment of neighbors as first and second. We arbitrarily choose to number processors clockwise (although in unoriented rings processors do not "know" what "clockwise" is). For any processor P_i, P_{i+1} (P_{i-1}) denotes the clockwise successor (predecessor) of P_i, whose subscript is actually $i+1 \pmod n$ ($i-1 \pmod n$).

An *orientation* of a state of a processor in a ring system is a choice of one of the neighbors of that processor. Let R be a ring, a self stabilizing protocol for R in which each processor has an orientation is an *orienting* protocol, if in each legitimate configuration all orientations form a directed cycle and if whenever a legitimate configuration is reached the chosen orientation never changes.

3 Impossibility Results for Ring Orientation

In this section we prove that for some uniform and nonuniform rings, no deterministic orienting protocol exists.

Lemma 1: Let R be a ring of even size in which all processors but one are identical. If each processor communicates with both its neighbors using a single register then R has no self stabilizing orienting protocol under central demon.

Proof: Let $2n$ be the size of the ring. Neighbor ordering is as follows: P_0 is the first neighbor of P_1, P_1 is the first neighbor of P_2,..., and P_{n-2} is the first neighbor of P_{n-1}. P_0 is the first neighbor of P_{2n-1}, P_{2n-1} is the first neighbor of P_{2n-2},..., and P_{n+2} is the first neighbor of P_{n+1} (see Fig 1). There is no requirement on the neighbor ordering of P_0 and P_n. Let c be an initial configuration such that:

$$c_1 = c_{2n-1} \,,\, c_2 = c_{2n-2} \,,\, ..., \, c_{n-1} = c_{n+1}$$

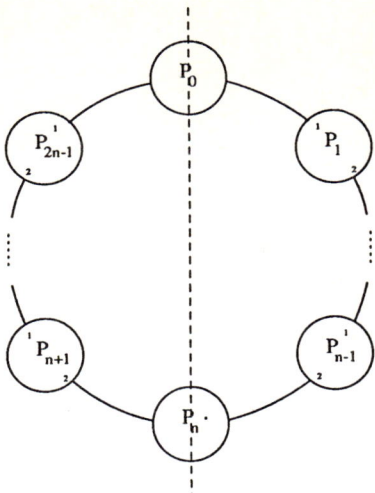

Figure 1: Symmetric neighbor ordering

Thus, we have: $c = c_0 w c_n w^r$, and c has a symmetry axis as shown in Fig 1. It is easy to see that a configuration having such symmetry axis cannot be legal, as edges (P_1, P_2) and (P_{2n-1}, P_{2n-1}) have opposite orientations. Note that for all $i \neq 0$, the state of P_i's first (second) neighbor equals the state of P_{2n-i}'s first (second) neighbor. Thus if we activate P_i and then P_{2n-i}, both will enter the same state, and the configuration will have the same symmetry axis as c. Clearly, activating P_0 or P_n does not break the symmetry. Thus, if the demon activates processors with the following schedule:

$$(P_0 \; P_n \; P_1 \; P_{2n-1} \; P_2 \; P_{2n-2} \; ... \; P_{n-1} \; P_{n+1})^\infty$$

the system will never reach a legal configuration. □

Lemma 2: Let R be a ring of even size. The ring R has no uniform deterministic orienting protocol which is self stabilizing in the presence of distributed demon.

Proof: Let neighbor ordering be as follows: For all $0 < i \leq n$, P_i is the first neighbor of P_{i+1}, and for all $n < i \leq 2n$, P_i is the first neighbor of P_{i-1}. Let c be an initial configuration of the form $c = ww^r$; c has a symmetry axis as shown in Fig 2. The protocol is deterministic and all processors are identical, hence if P_i and P_{2n-i-1} are concurrently activated, then a configuration with the same symmetry will be reached, and therefore the following is a livelock schedule:

$$(\; \{P_0, P_{2n-1}\} \; \{P_1, P_{2n-2}\} \; ... \; \{P_n, P_{n+1}\} \;)^\infty$$

□

4 The Protocol

In this section we present a self stabilizing protocol for orientation of undirected rings, prove its correctness and analyze its complexity. The protocol is composed of two "levels". Each level consists

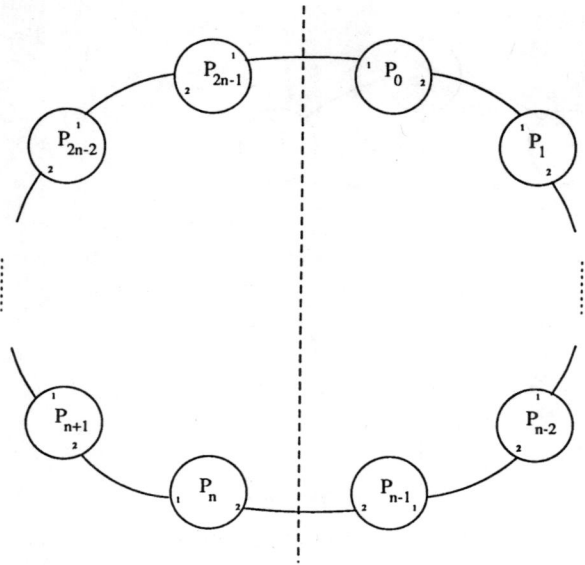

Figure 2: Symmetric neighbor ordering

of a self stabilizing protocol. The lower level directs all edges of the ring, where each edge is directed by a *randomized* protocol which is run by the processors on its endpoints. The higher level of the protocol is a *deterministic* protocol for orientation of a directed ring. The final protocol is obtained as a combination of the two by using the technique of fair combination of self stabilizing protocols presented in [DIM-89].

4.1 Directing a Ring

The n edges of a ring are directed by running n separate copies of a randomized edge directing protocol. The protocol for each edge is executed by its two endpoints. An edge e connecting processors P and Q is implemented by two registers: $P.dir$ which is written by P and read by Q, and $Q.dir$ which is written by Q and read by P. This situation is depicted in Figure 3.

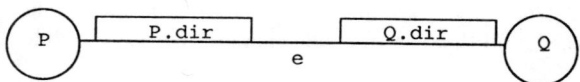

Figure 3: Two processors and edge registers between them

The state set of each processor is $M = Z_3$. The set of system configurations consists of all pairs of numbers from Z_3. We say that the edge e is directed from P to Q if $P.dir + 1 = Q.dir$ (mod 3). The set of legitimate configurations contains all pairs of the form: $\{ (P.dir, Q.dir) \mid P.dir \neq Q.dir \} = \{(0,1), (0,2), (1,0), (1,2), (2,0), (2,1)\}$. Both processors are enabled if the values of the *dir* fields are equal. When a processor is activated, it executes the following statement:

if $P.dir = Q.dir$ **then** $P.dir = \text{random}(Z_3 - \{P.dir\})$

Or in words "if your neighbor's value is the same as yours, then pick something else". Random(A) means choose at random (uniformly) an element from set A. In our protocol, we always have $|A| = 2$.

We now prove that the edge directing protocol is self stabilizing in the presence of a proper distributed demon. The **deadlock** and **closure** requirements are satisfied by the simple observation that all deadlock configurations are legitimate configurations and that all legitimate configurations are deadlock configurations. The **randomized no-livelock** requirement is proved by the following lemma.

Lemma 3:

(a) For any nonlegitimate configuration c, the probability that the system reaches a legitimate configuration within a single transition is $\geq 1/2$.

(b) For any nonlegitimate configuration c, the probability that the system does not reach a legitimate configuration within ℓ transitions is $\leq 2^{-\ell}$.

Proof: In a nonlegitimate configuration, $P.dir = Q.dir$, and both processors are enabled. If the adversary activates a single processor then this processor changes the value of its *dir* field, and a legitimate configuration is reached. If both processors are activated concurrently, they change the value of their *dir* field, and the system reaches either a legitimate configuration or another configuration in which $P.dir = Q.dir$ with equal probability. Thus, the probability that the ℓ-th configuration reached is not legitimate is $\leq 2^{-\ell}$. □

The expected stabilization time of the protocol is $\sum_{l=1}^{\infty} \ell 2^{-\ell} = 2$. To direct a ring of n processors (and n edges) we run n copies of the edge directing protocol, a copy for each edge. Each processor has two *dir* fields, a field for each of its neighbors (there is no requirement on the relative values of these two fields). The total expected number of activations before all edges are directed is $O(n)$.

4.2 The Orienting Protocol

We now introduce a deterministic protocol for orientation of directed rings. The protocol achieves its goal by using tokens. Tokens are not part of the model but an abstract concept which is helpful in protocol design and verification. Each token has a *direction* which is never changed. A token is passed from its holder to the neighbor of the holder which is in the token's direction. This neighbor in its turn passes the token to its other neighbor and so on. Each processor keeps trace of the direction to which the most recent token was passed. This direction constitutes the current orientation of the processor. Whenever two tokens meet one of them is eliminated. Stabilization is achieved when all remaining tokens have the same direction and each processor is visited by at least one of these remaining tokens.

The state of a processor in this protocol consists of two fields: *token* and *orient*. In each state of processor P the values of of these fields are stored in the registers of P. The *token* fields are used to enable token passing, their possible values are I, S or R, which stand for *Idle, Sending token* and *Receiving token*, respectively. The binary *orient* fields in P's registers give the current orientation for P. While the values stored in both *token* fields of P are always equal, the values stored in the two *orient* fields of P are never equal. We will usually represent fields *token* and *orient* by a single letter with an arrow over it, $\overrightarrow{SI}$ representing the situation depicted in Figure 4.

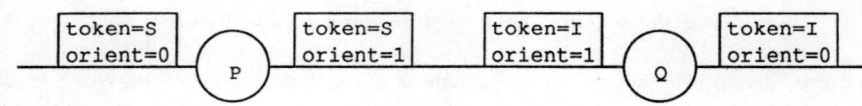

Figure 4: $\vec{S}\overleftarrow{I}$

The orienting protocol is defined by the transition table which appears in Figure 5. When a processor P is activated, it matches its state and the state of its first neighbor Q (which is read from Q's register) against each entry of the table. If a matching entry is found then P executes the transition associated with that entry. The transition specifies the new state of P which is written in its two registers, and may depend on the direction of the edge connecting P and Q. If no match is found, the matching is tried again with Q standing for P's second neighbor. Choosing to start the matching from the first neighbor is an arbitrary decision. The only possible multiple match is in transition 1, preferring the first neighbor over the second may affect the final orientation but does not harm the correctness of the protocol or its stabilization time.

transition #	Q	P	P
1)	$\vec{S}$	I	$\longrightarrow$ $\vec{R}$
2)	$\overleftarrow{R}$	$\vec{S}$	$\longrightarrow$ $\overleftarrow{I}$
3)	$\neg \vec{S}$	$\overleftarrow{R}$	$\longrightarrow$ $\vec{S}$
4)	$\vec{S}$	$\overleftarrow{S}$	$\longrightarrow$ if $e = (P,Q)$ is directed from P to Q then $\vec{R}$
5)	$\vec{I}$	$\overleftarrow{I}$	$\longrightarrow$ if $e = (P,Q)$ is directed from P to Q then $\vec{S}$

Figure 5: Transition table for the orientation protocol

In transition 1, I means "P is either in state $\vec{I}$ or in state $\overleftarrow{I}$". In transition 3, $\neg \vec{S}$ means "Q's state is anything but $\vec{S}$". The direction of the arrow is to be understood with respect to P. For example both situations depicted in Figure 6 match transition 2, but the one in Figure 7 does not. That is, transition 2 matches a situation in which processor P is either in state $\vec{S}$ or in state $\overleftarrow{S}$, Q is the neighbor the arrow points to, and Q is in state R with its arrow pointing away from P.

A processor P_i in state $\overleftarrow{S}$ or $\vec{S}$ *holds a token* whose direction is the arrow direction. If P_{i-1} is not in state $\vec{S}$ and P_i is in state $\vec{R}$ (P_{i+1} is not in state $\overleftarrow{S}$ and P_i is in state $\overleftarrow{R}$) then P_i also *holds a token*, whose direction is the arrow direction. Tokens never change their direction.

A token held by P_i makes a *step* whenever P_i *releases* it by executing transition 2. In this case the token is *passed* to the neighbor of P_i functioning as processor Q for that transition. This could be either P_{i-1} or P_{i+1}. A token is *created* whenever a processor becomes the holder of a token where the token was not passed to it by any of its neighbors. The only transition which involves token creation is transition 5. A token is *eliminated* whenever a processor releases a token which is not passed to any of its neighbors. A processor in state R can only enter state S, so an elimination can take place only if the holder of a token leaves state S (transition 4), or if the holder of a token is in state $\vec{R}$ ($\overleftarrow{R}$) and its left (right) neighbor enters state $\vec{S}$ ($\overleftarrow{S}$) (transition 3).

The set of legitimate configurations contains all configurations in which the ring is oriented, that is all configurations in which all arrows have the same direction:

$$L = (\vec{I} + \vec{S} + \vec{R})^n + (\overleftarrow{I} + \overleftarrow{S} + \overleftarrow{R})^n$$

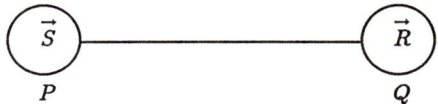

Figure 6: Two situations in which transition 2 is possible

Figure 7: Transition 2 is not possible

We now prove that the orienting protocol is self stabilizing with respect to L and under a distributed demon.

Lemma 4: The **deadlock** requirement is satisfied by the protocol.

Proof: Let c be a configuration in which no processor is enabled. If any processor P is in state R, then either P or one of its neighbors is enabled (transitions 2 and 3). Otherwise, all processors are either in I or in S. Consider a maximal subvector of processors in state S in c. If there is a $\overrightarrow{S}\overleftarrow{S}$ pair in c, then one of these processors in enabled (transition 4). Otherwise the subvector is of the form $(\overleftarrow{S})^i(\overrightarrow{S})^j$. If $0 < i+j < n$, then at one end of the subvector we have a $\overrightarrow{S}\,\overleftarrow{I}$ (or $\overrightarrow{I}\,\overleftarrow{S}$) pair, and then the processor in state I is enabled (transition 1). Hence, either $i+j=0$ or $i+j=n$, that is, either all processors are in state S, or they are all in state I. But neither $\overrightarrow{S}\overleftarrow{S}$ nor $\overrightarrow{I}\overleftarrow{I}$ can appear in c. Therefore, if c is a deadlock configuration, then

$$c \in \{(\overrightarrow{S})^n, (\overleftarrow{S})^n, (\overrightarrow{I})^n, (\overleftarrow{I})^n\} \subseteq L$$

□

Lemma 5: The **closure** requirement is satisfied by the protocol.

Proof: Let c be a legitimate configuration. If c is a deadlock configuration then we are done. Otherwise, the only possible transitions from c are transitions 1, 2 and 3, because transitions 4 and 5 require processors to have arrows in different directions. It is easy to see from the protocol that if transitions 1, 2 or 3 take place when all arrows have the same direction, then after the transition all arrows will still have the same direction. This is true even if many of these transitions take place concurrently. □

We now turn to the **randomized no livelock** proof. Let $E = c^0, c^1, \ldots$ be an execution. Configuration c^0 is called the *initial configuration* of E. A processor P is in its *initial state* in configuration c^i of E if the state of P in c^i is equal to its state in all previous configurations of E. The value of field f of processor P_i in configuration c will be denoted either by $c_i.f$ or by $P_i.f$.

Lemma 6: For any configuration c in execution E, if $(c_i.token, c_{i+1}.token) = (\vec{I}, \overleftarrow{I})$, then both P_i and P_{i+1} are in their initial state in E.

Proof: From the protocol, it is easy to see that P_i cannot enter state $\vec{I}$ if P_{i+1} is in state $\overleftarrow{I}$ and vice versa. Neither can P_i and P_{i+1} enter states $\vec{I}$ and $\overleftarrow{I}$ (respectively) concurrently. □

Lemma 7: In any execution of the protocol E, at most n distinct tokens exist. In other words: The sum of the number of tokens present in the initial configuration of E and the number of tokens created during E does not exceed n.

Proof: A processor can create a token only if it is in its initial state in E and it is *Idle* (because token creation happens only in transition 5). Thus a processor holding a token in the initial configuration of E will not create a token during the execution, and a processor can create at most one token. It follows that in any execution, at most n distinct tokens exist. □

We now introduce some definitions for the following theorem. For any $i \geq 0$ and any configuration c, $Prefix(c, i)$ is the prefix of length i of c, i.e. the states of $P_0, \ldots, P_{i-1}$. $s \sim e$ stands for "s matches the regular expression e".

$$M = \{\vec{I}, \vec{R}, \vec{S}, \overleftarrow{I}, \overleftarrow{R}, \overleftarrow{S}\}$$
$$A_i = \{c \in M^n \,/\, prefix(c, i+1) \sim (\vec{R} + \vec{S} + \vec{I})^{i-1}(\vec{I} + \vec{R})\vec{R}\}$$
$$B_i = \{c \in M^n \,/\, prefix(c, i+1) \sim (\vec{R} + \vec{S} + \vec{I})^i \vec{S}\}$$
$$D = \{c \in M^n \,/\, c_0 = \vec{R} \wedge c_{n-1} \neq \vec{S}\}$$

Theorem 8: (Main theorem) Let $E = c^0, c^1, \cdots, c^k$ be an execution of the system. If a token t which is present in c^0 makes exactly $n-1$ steps, then $c^k \in L$.

Proof: Without loss of generality assume that t is held by P_0 in c^0 and that t moves in E in a clockwise direction (represented by rightward arrows throughout this proof). Configuration c^k is reached when t makes the $(n-1)$-st step. Since we assumed that t is present in c^k we can deduce that t is not eliminated during E.

Claim: Let i_j be the number of steps t made between c^0 and c^j: the holder of t in c^j is P_{i_j}. For all $j \leq k$,

$$\text{if } i_j = 0 \text{ then } c^j \in B_0 \bigcup D$$
$$\text{if } i_j > 0 \text{ then } c^j \in A_{i_j} \bigcup B_{i_j}$$

Proof: By induction on j:
Base: $j = 0$, $i_j = 0$. By our assumption t is held by P_0, so either $P_0.token = \vec{S}$ and $c^0 \in B_0$, or $(P_0.token = \vec{R} \wedge P_{n-1}.token \neq \vec{S})$ and $c^0 \in D$.
Induction Step: Suppose the claim holds for some $j \geq 1$. Consider the following three cases:

(a) $i_j = 0$ and $c^j \in D$: That is $c_{n-1}^j \neq \vec{S}$ and $c_0^j = \vec{R}$. If P_{n-1} enters $\vec{S}$, then t is eliminated. By our assumption this is impossible so we do not consider this case. The only possible transition of P_0 is transition 3 in which it enters $\vec{S}$, thus leading to $i_{j+1} = 0$ and $c^{j+1} \in B_0$. Activation of any other processor results with $c^{j+1} \in D$ with $i_{j+1} = 0$.

(b) $i_j \geq 0$ and $c^j \in B_{i_j}$: That is $c_{i_j}^j = \vec{S}$. Possible transitions for P_{i_j} are transitions 2 and 4. In transition 4, t is eliminated so we do not consider it. If P_{i_j} executes transition 2 then $i_{j+1} = i_j + 1$ and $c^{j+1} \in A_{i_{j+1}}$. Activation of any other processor leads to $i_{j+1} = i_j$ and $c^{j+1} \in B_{i_{j+1}}$.

(c) $i_j > 0$ and $c^j \in A_{i_j}$: That is $c_{i_j}^j = \vec{R}$. The only possible transition for P_{i_j} is transition 2 in which it enters state $\vec{S}$. In this case $i_{j+1} = i_j$ and $c^{j+1} \in B_{i_{j+1}}$. Activation of any other processor in which t is not eliminated leads to $i_{j+1} = i_j$ and $c^{j+1} \in A_{i_{j+1}}$.

The Main Theorem follows from the claim when $j = k$. By definition, $i_k = n - 1$. Thus, $c^k \in A_{n-1} \cup B_{n-1} \subseteq L$. □

We now use Theorem 8 to derive an upper bound on the protocol's stabilization time.

Corollary 9:

(a) The maximum number of token steps made in the system before a legitimate configuration is reached is $n(n-2)$.

(b) The sum of the number of token steps and eliminations happening before the system reaches a legitimate configuration is at most $n(n-1)$.

Proof: By Theorem 8 each token can make at most $n - 2$ steps before a legitimate configuration is reached. In addition each token can be eliminated at most once. □

Lemma 10:

(a) Whenever a processor leaves state S, the token it held either makes a step or it is eliminated.

(b) A processor can make at most 2 transitions without leaving state S.

(c) Let $E = (c^0, \ldots, c^\ell)$ be an execution. If a processor executes j (not necessarily consecutive) transitions in E, then the sum of token steps and token eliminations resulting from these j transitions is at least $\lceil \frac{j-4}{3} \rceil$.

(d) For all positive j, the sum of tokens steps and token eliminations resulting from j activations is at least $\lceil \frac{j-4n}{3} \rceil$.

Proof:

(a) By definition, a processor in state S holds a token. If the processor enters I, it does not hold the token any more. If the processor enters R by transition 5, then it is easy to see from the protocol that it does not hold the token either. Thus, the token the processor held before the transition either has made a step or has been eliminated.

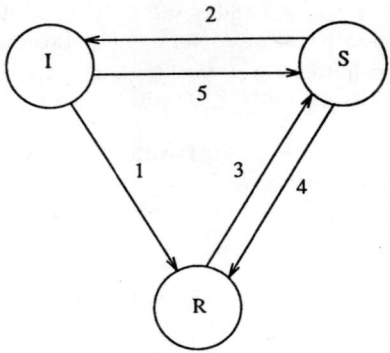

Figure 8: State transitions of a processor

(b) We show all transitions possible in a digraph, in figure 8; arrows over state names are not considered: S means $\vec{S}$ or $\overleftarrow{S}$. Edge labels indicate transition numbers. The longest path not containing an edge going out from node S is $I \xrightarrow{1} R \xrightarrow{3} S$, of length 2.

(c) In the worst case, a processor does not leave state S in its first 2 activations and in its last 2 ones. Among the remaining $j - 4$ activations, at least the first, the fourth, the seventh, etc lead to token steps or eliminations, and the lower bound follows.

(d) Let us denote by j_i the number of activations of P_i; $j = \sum_{i=0}^{n-1} j_i$. We use (c) to find a lower bound on the number of token steps or eliminations that take place:

$$\sum_{i=0}^{n-1} \left\lceil \frac{j_i - 4}{3} \right\rceil \geq \left\lceil \sum_{i=0}^{n-1} \frac{j_i - 4}{3} \right\rceil = \left\lceil \frac{j - 4n}{3} \right\rceil$$

□

Corollary 11: After $3n^2 + n + 1$ transitions take place, the system is in a legitimate configuration.

Proof: By Corollary 9 and Lemma 10(d) with $j = 3n^2 + n + 1$.

Corollary 12: The ring orienting protocol is **Self-Stabilizing**. The protocol's stabilization time is $3n^2 + n + 1 = O(n^2)$.

Proof: By Lemma 4, Lemma 5 and Corollary 11.

4.3 The Combined Protocol

To achieve a self stabilizing ring orienting protocol for undirected rings we use the *fair combination* of self stabilizing protocols due to [DIM-89]. In the combined protocol the ring directing protocol plays the role of the lower level protocol which prepares the system to be in a state in which the upper level protocol, for orientation of directed rings, can be applied. In the combined protocol each register of each processor contains three fields: The *dir* field is written to in the directing protocol and read from in the orienting protocol, and the *token* and *orient* fields which are used solely in the orienting protocol. Whenever a processor is activated in the combined protocol it executes a single step in both lower level and upper level protocols.

The idea here is to use the self-stabilizing properties of each of the protocol to ensure self stabilization of the combined protocol. In our particular case the edge directing protocols do not depend on any external execution and are expected to stabilize after a linear number of activations by any proper adversary. Once the lower level protocol stabilizes the input specifications for the upper level protocol are satisfied. Starting from that configuration the upper level protocol is stabilized within additional $O(n^2)$ activations regardless of the initial configuration.

When a proper scheduler is used it should be ensured that the adversary cannot refrain activating processors which are enabled in the lower level protocol. This can happen if the upper level protocol has infinite runs when the lower level protocol is not stabilized yet. To avoid this we add to the combined protocol a conditional clause which does not allow the orienting protocol to run on any processor whose edges are not directed yet. This will cause tokens to get stuck in front of processors whose edges are not directed until the edge directing protocols are stabilized. The technical details are omitted here.

5 Concluding Remarks

A self stabilizing ring orienting protocol for undirected rings was presented. The correctness of the protocol in the presence of a proper distributed demon was proved and its stabilization time was analyzed. It should be interesting to try to adapt this protocol for read/write atomicity, as well as to try to prove some lower bounds on the stabilization time of ring orienting protocols. We conjecture that an $O(n \log n)$ lower bound for this problem can be proved.

References

[AB-89] Y. Afek and G. M. Brown, "Self Stabilization of the Alternating-Bit Protocol" *Proceedings of the 16th Conference of IEEE in Israel*, March 1989.

[BGW-87] G.M. Brown, M.G. Gouda, and C.L. Wu, "A Self-Stabilizing Token system", *Proc. of the Twentieth Annual Hawaii International Conference on System sciences* (1987), pp. 218-223.

[BP-88] J.E. Burns and J. Pachl, "Uniform Self-Stabilizing Rings", *Aegean Workshop On Computing, 1988, Lecture notes in computer science* 319, pp. 391-400.

[Bu-87] J.E. Burns, "Self-Stabilizing Rings without Demons", Technical Report GIT-ICS-87/36, Georgia Institute Of Technology.

[Di-74] E.W. Dijkstra, "Self-Stabilizing Systems in Spite of Distributed Control", Communications of the ACM 17,11 (1974), pp. 643-644.

[DIM-89] S. Dolev, A. Israeli and S. Moran, "Self Stabilization of Dynamic Systems", to be presented in PODC90.

[IJ-89] A. Israeli and M. Jalfon, "Token Management Schemes and Random walks Yield Self Stabilizing Mutual Exclusion", to be presented in PODC90.

[KP-89] S. Katz and K.J. Perry, "Self-stabilizing Extensions", to be presented in PODC90.

[Kr-79] H.S.M. Kruijer, "Self-stabilization (in spite of distributed control) in tree-structured systems", Information Processing Letters 8,2 (1979), pp. 91-95.

[Tc-81] M.Tchuente, "Sur l'auto-stabilisation dans un réseau d'ordinateurs", RAIRO Inf. Théor. 15 (1981), pp. 47-66.

Memory-Efficient Self Stabilizing Protocols for General Networks

Yehuda Afek
AT&T Bell Laboratories, and
Computer Science Department,
Tel-Aviv University,
afek@taurus

Shay Kutten
IBM T.J. Watson Research Center,
P.O. Box 704, Yorktown Heights,
NY 10598
kutten@ibm.com

Moti Yung
IBM T.J. Watson Research Center,
P.O. Box 704, Yorktown Heights,
NY 10598
moti@ibm.com

EXTENDED ABSTRACT

Abstract

A self stabilizing protocol for constructing a rooted spanning tree in an arbitrary asynchronous network of processors that communicate through shared memory is presented. The processors have unique identifiers but are otherwise identical. The network topology is assumed to be dynamic, that is, edges can join or leave the computation before it eventually stabilizes.

The algorithm is design uses a new paradigm in self stabilization. The idea is to ensure that if the system is not in a legal state (this is a global condition) then a local condition of some node will be violated. Thus the new could restart the algorithm.

The algorithm provides an underlying self-stabilization mechanism and can serve as a basic building block in the construction of self stabilizing protocols for several other applications such as: mutual-exclusion, snapshot, and reset.

The algorithm is memory efficient in that it requires only a linear size memory of words of size $\log n$ (the size of an identity) over the entire network. Each processor needs a constant number of words per incident link, thus the storage requirement is in the same order of magnitude as the size of the traditionally assumed message buffers size. The adversary may be permitted to initiate the values of the variables to any size. Still, in this case the additional memory used by the algorithm is the amount stated above.

Extensions of our algorithm to other models are also discussed.

1 Introduction

In 1974 Dijkstra has introduced a new approach to fault tolerant distributed computing, called *self stabilization* [Dij74]. In this approach protocols are designed with a built-in automatic mechanism that whenever the distributed system is placed in an illegal global state the protocol guarantees that it will converge back to a legal state and will stay legal thereafter.

The first problem for which self-stabilized protocols were designed was the mutual exclusion problem ([Dij74, BGW89, BP89] and others); recently, other tasks have been shown to have this property (e.g., [DIM90, AB89, KP90]). In general one can formulate the question with respect to any distributed computing task and think about a general underlying mechanism which can support several tasks and provide them with a self-stabilization augmentation.

In this work we present a self stabilizing protocol to construct and maintain a rooted spanning tree in a dynamic network. This protocol serves as an underlying primitive for a self-stabilizing compiler; that is, a general mechanism to self stabilize arbitrary distributed algorithms, in the same way as the protocol of [AAG87] is used to run distributed algorithms in a fail-safe fashion (see [KP90] for discussion of such self-stabilizing compilers).

We assume the model of an asynchronous network of processors with unique IDs communicating via shared memory. The IDs are unchangeable throughout the protocol (each is embedded in the machine's hardware).

In [SG89], Spinelli and Gallager present a topology maintenance algorithm for arbitrary dynamic networks. It turns out that their algorithm is easily modified into a self-stabilizing spanning tree algorithm, however it is an expensive algorithm both in its communication complexity (nodes have to periodically read $O(nE)$ memory words, where n and E are the total number of nodes and links in the network, and the $O(nE)$ words are the whole description of the network as viewed by the neighbor), and in its memory requirements (each node has to maintain one copy of the network topology for each of its incident links).

It is well known [Ang80] that no deterministic algorithm can construct a spanning tree in an anonymous (i.e., uniform) symmetric network. Dijkstra has shown that deterministic self-stabilizing protocols for mutual exclusion cannot be constructed in an anonymous ring whose size is a composite number (Burns and Pachl [BP89] gave a uniform protocol for the self-stabilizing mutual exclusion problem on rings whose size is a prime number). Independent of our work, S. Dolev, Israeli and Moran [DIM90] have recently designed a self stabilizing spanning tree protocol. To break the symmetry their protocol assumes the existence of a unique "special" processor in the network (a pre-distinguished leader) and the constructed tree is rooted at that node. Naturally they assume that the adversary cannot change the status of the unique node such that it stops being in the "special" status, neither can it cause any other node to become "special". Note that the network cannot be allowed to be partitioned such that no special node will be included in one of the partitions. In another independent work, Arora and Gouda [AG90] have designed a spanning tree construction self-stabilizing protocol that rather than assuming a predistinguished node assumes that each node has a unique ID (as we do). However they assume a known bound on the network size in order for the algorithm to self stabilize. Moreover, the quiescence time complexity of their algorithm depends on that bound, regardless of the actual size. Note that faults or partitions may decrease the actual size significantly. Our quiescence time complexity is $O(n^2)$ where n is the unknown size of the network (component).

In this paper we use unique IDs at the nodes to break the symmetry but make no farther assumption such as knowledge of the network size, nor that the network is not partitioned. Our approach is that in a fault tolerant network it is more natural not to rely on connectivity to a specific processor (leader), thus we require that connected components of possibly partitioned network will continue the protocol and will independently stabilize as well. This is a crucial difference between a symmetric distributed protocol and an asymmetric protocol that assumes a leader. The price is the need to break symmetry using ID's.

The algorithm design uses a new paradigm in self stabilization. Previous algorithms did not rely on the detection of an illegal state. In the independent work of [KP90] such an illegal state was detected by collecting information about the global state. We detect an illegal global state by ensuring that in such a state a local condition of some node will be violated. Thus the node could restart the algorithm.

The size of the space consumed in a processor by our protocol is exactly what is traditionally the size of a message (and its buffers), i.e., log(Max-ID) times the degree of the processor. Thus, we can view this as an optimal storage requirement, since any link needs a message buffer in any global distributed computing (and if we implement the reading by message passing we assume buffers of this size are available). (The adversary may be permitted to initiate the values of the variables to any size. Still, in this case the additional memory used by the algorithm is the amount stated above.)

Following our deterministic self stabilizing election, a randomized election algorithm for anonymous networks was developed lately by S. Dolev, Israeli and Moran [DIM90a]. Our deterministic protocol has a natural extension (using [AM89]) to run on anonymous networks. In this modular extension we use an additional procedure which carefully draws and continually checks random ID's to verify that they are different. The presentation of the randomized algorithm is postponed to the full paper.

It is also possible to run our algorithm in the message passing model, simulating shared memory with the help of the datalink protocol of [AB89].

2 The Model

We present here the model of a dynamic network of processors communicating via shared memory. The network is represented by a graph (V, E), where V is the set of processors, and E is a set of links. If u and v are processors and (u, v) is in E then we say that there is a link between u and v, and that they are neighbors. Each processor has a unique ID (say, hardwired in its memory), this is the only assumption that makes our model non-uniform. The total number of processors, n, is unknown to the processors and may change dynamically.

Processors communicate only by reading the memory of neighboring processors and by writing each its own local memory. Reads of neighbors memory as well as read and write on local memory are atomic operations. Each processor is a state machine with a bounded number of states (which could be a function of n; the dependency on n can be in the size of its registers). At the beginning of the computation the adversary can put each processor in an arbitrary state. The local computation at each processor consists of a sequence of transitions where each transition consists of an operation that moves the processor from its state before the transition to another (possibly the same) state. Each processor operation is either a local computation step, or an atomic read of a neighbor's memory, or an atomic write of its own memory. The fair scheduler (demon) of the global computation consists of an infinite sequence of processors such that each processor appears in the sequence infinitely often. Whenever a processor appears in the schedule its next transition is performed (every processor always has an operation (e.g. read one of the neighbors memory) that is enabled unless the processor is down). Such an Atomic Read/Atomic Write demon was first suggested by [DIM90].

At any system state, the set of local states of the processors defines a global state. In any self-stabilization problem, a subset of the set of global states is defined as legal global states. In a self-stabilizing protocol the computation moves the system from a legal state only to another (possibly the same) legal state and starting from any state a fair scheduler will eventually bring the system to a legal state.

The network topology is dynamic, that is links and processors can go down and come up an arbitrary number of times. We assume that there is a local self-stabilazing mechanism that eventually updates at each processor the status of its incident links and neighboring processors. When a link is down the processors incident to that link cannot read each other's memory. We further assume that the sequence of topological changes is finite and that eventually topological changes cease [AAG87].

In the spanning tree construction algorithm each node has a Parent variable, that holds an id of a processor. We say that a tree spans the network when the collection of the Parent variables defines such a rooted tree in the natural way. Our main algorithm is a self stabilizing algorithm that computes such a tree. Using this algorithm as a building block it is easy to construct several other self stabilizing algorithms such as: mutual exclusion, snapshot, reset, and leader election.

3   A Self Stabilizing Spanning Tree Algorithm

In a nutshell the algorithm can be described as follows: Every node tries to construct a tree in the network from itself. The process of larger identity nodes overruns the processes of lower identity nodes. Eventually, the tree of the largest Id node overruns all the other trees. A node leaves its tree when it detects a local inconsistency. However, to join another tree whose root Id is larger a node has to propagate a *request* message along the new tree branches to the root and to receive a *grant* message back. This mechanism prevents the necessity to know additional information such as a bound on the network size. Of course these messages propagate through the shared memories of the nodes via a sequence of read and write operations. To maintain self-stabilization we define certain local conditions that if violated in the neighborhood of a node then that node detects it and restarts the algorithm by considering itself the root of a single node tree. (It is guaranteed that if the network is in an illegal state, then some node will notice the violation of these conditions.)

Let us now describe the algorithm in detail. There are three groups of variables at each node:

1. The usual local variables { Id, Edge_list }:
   We denote by v.Id the variable Id in node v. The same convention is used for the other variables.
   v.Id is the read-only identity of node v and
   v.Edge_list is a list of links incident to node v that are operational and such that the processor on the other side is up. This list is maintained by a lower level self-stabilizing protocol which is beyond the scope of this paper. Each change in a link or node status is eventually recorded in Edge_list. If a neighbor u is removed from Edge_list while v is performing a read of the memory of u, then that read may return any value.

2. The variables related to the tree structure { Root, Parent, and Distance} where,
   v.Root is supposed to hold the identity of the root of the tree to which node v belongs;
   (We omit the word "supposed" in the sequel.)
   v.Distance is the distance from node v to its root; and
   v.Parent is the identity of a neighbor of v which is the parent of v in the tree.

3. The variables related to passing the *request* and *grant* messages { Request, To, Direction, and From} where,
   v.Request is either an Id of a node that is currently requesting to join the tree to which v belongs, or equal to v.Id if v itself is trying to join another tree;
   v.From is either the Id of the neighbor from which v copied the value of v.Request, or v.Id if v has initiated a request in an attempt to join a new tree;
   v.To is the name of a neighbor of v through which it is trying to propagate the *request* message;
   v.Direction is either *Ask*, to signify that the node whose Id is in v.Request wishes to join the tree, or *Grant* to signify that this request has been granted.

Definition 3.1 *We say that v is a* child *of w and that w is the* parent *of v if:*
$(w.\text{Id} \in v.\text{Edge\_list}) \wedge (v.\text{Parent} = w.\text{Id}) \wedge (v.\text{Root} = w.\text{Root}) \wedge (v.\text{Distance} = w.\text{Distance} + 1)$.

To achieve self stabilization of the spanning tree construction we define a condition, called $C1$, on the variables at each node such that if (and only if) the condition holds at all the nodes then the network (or component) is in a globally legal state in which a correct spanning tree exists. This condition is checked periodically at all the nodes. The aim of the algorithm is to detect violations of the condition (either because of incorrect initial values of the nodes' variables, or because of topological changes), and to ensure that eventually the network will enter a globally legal state.

Condition C1:
$\{[(\text{Root} = \text{Id}) \wedge (\text{Parent} = \text{Id}) \wedge (\text{Distance} = 0)] \vee$
$[(\text{Root} > \text{Id}) \wedge (\text{Parent} \in \text{Edge\_list}) \wedge (\text{Root} = \text{Parent.Root}) \wedge (\text{Distance} = \text{Parent.Distance} + 1 > 0)]\}$
$\wedge (\text{Root} \geq \max_{x \in \text{Edge\_list}} x.\text{Root})$

A formal description of the algorithm is given in Figure 1. The program consists of a set of actions. Each action is specified as:

$$\langle guard \rangle \rightarrow \langle command \rangle$$

where the *guard* is a Boolean expression and the *command* is a sequence of assignments. An action can only be executed from a state in which its guard is checked and found true. However, since we assume atomicity of a single operation only, some variables already checked can be changed while others are still being read. Thus it it may be the case that the guard is actually false, and still was found true. The execution at each node proceeds by an infinite sequence of operations where in each operation the memory of one neighbor is read atomically, then the guards of the actions are checked. If any guard is true, then the corresponding commands are performed atomically. (It can be shown that the atomicity of the read of one register, is an atomic operation, as well as the write of one register.) Each neighbor in Edge_list after the last topological change, is read infinitely often by the sequence of operations (e.g., the nodes are read in a linear cyclic order).

A node that notices that Condition C1 does not hold, must become a root (Action 1).

Definition 3.2 *Node v is a* **root** *if* $(v.\text{Root} = v.\text{Id}) \wedge (v.\text{Parent} = \text{Id}) \wedge (v.\text{Distance} = 0)$.

By becoming a root condition $C1$ does not necessarily become true, however we define condition $C1'$ which does become true. Moreover, if condition $C1'$ is satisfied at all the nodes then (by transitivity) the Parent variables in the nodes define a spanning forest in the network. If condition $C1'$ holds at a node but $C1$ does not then eventually that node moves on to the process of joining a tree with a larger Root (Action 2).

Condition $C1'$:
[(Root = Id) ∧ (Parent = Id) ∧ (Distance = 0)] ∨
[(Root > Id) ∧ (Parent ∈ Edge_list) ∧ (Root = Parent.Root) ∧ (Distance = Parent.Distance + 1 > 0)]

If node v's Id (which is now also its Root) is not larger than the Roots of its neighbors, it attempts to join another tree. It chooses the neighbor whose Root is the largest among the Roots of its neighbors, and makes a request to join as a child of this neighbor u (Action 2). For that it sets its Request and From to its own Id, its Direction to Ask and its To to u. (Note that some of the operations in the guards are not (and do not need to be) atomic.)

When condition $C1$ holds at node v that node participates in the process of forwarding requests and grants in its tree in order to enable the addition of new nodes to the tree. Its task is to help forwarding requests to join the tree to the root of the tree and grants from the root back to the requesting node (Actions $4-7$). As with the tree related variables, we define Condition $C2$ that is true if the variables related to that process are in a legal state. If $C2$ does not hold then the node must reset those variables (Action 3).

Definition 3.3 *We say that node v is forwarding a request from node w if the following condition holds:*

Condition $C2'$:
$((w.\text{Id} \in v.\text{Edge\_list}) \land$
$(w.\text{Root} = w.\text{Id} = w.\text{Request} = w.\text{From} = v.\text{Request} = v.\text{From} \land$
$(w.\text{To} = v.\text{Id}) \land (w.\text{Direction} = Ask))$
∨
$((w.\text{Id} \in v.\text{Edge\_list}) \land (v \ parent \ of \ w) \land (w.\text{Request} = v.\text{Request} \neq w.\text{Id}) \land$
$(v.\text{From} = w.\text{Id}) \land (v.\text{To} = v.\text{Parent}) \land (w.\text{To} = v.\text{Id}))$

Condition $C2$:

$C2'$ ∨
(v.Request, v.To, v.From, and v.Direction are undefined)

If node v is forwarding a request (i.e., $C2'$ holds at v) and it is a root, it can *grant* the request (Action 6). That is, it changes its Direction to *Grant* (unless its Direction is already *Grant*). A non-root node who is forwarding a request, can forward a grant, provided that its parent is (Action 7):

The program at node v:
do forever: read next neighbor information and

1. $\neg C1 \wedge \neg C1'$ $\rightarrow$ v.Root $:=v$.Id; v.Parent $:=$ Id;
   v.Distance $:= 0$;

2. $C1' \wedge$
   $(u$.Root $= \max_{x \in \text{Edge\_list}} x$.Root$) > v$.Root $\rightarrow$ v.Request $:= v$.From $:= v$.Id;
   v.To $:= u$.Id; v.Direction $:= Ask$;

3. $C1 \wedge \neg C2$ $\rightarrow$ set v.Request, v.From, v.To,
   and v.Direction to undefined

4. $C1 \wedge C2 \wedge \neg C2' \wedge (\exists w \in v$.Edge_list$|$
   $(w$.Direction $= Ask) \wedge (w$.To $= v$.Id$) \wedge$
   $(w$.Request $= w$.Id $= w$.Root $= w$.From$))$ $\rightarrow$ v.Request $:= v$.From $:= w$.Id;
   v.To $:= v$.Parent; v.Direction $:= Ask$

5. $C1 \wedge C2 \wedge \neg C2' \wedge (\exists w \in v$.Edge_list$|$
   $(w\ child\ of\ v) \wedge (w$.To $= $ v.Id$) \wedge$
   $(w$.Direction $= Ask))$ $\rightarrow$ v.Request $:= w$.Request; v.From $:= w$.Id;
   v.To $:= v$.Parent; v.Direction $:= Ask$

6. $C1 \wedge C2' \wedge (v\ is\ a\ root) \wedge (v$.Direction $= Ask)$ $\rightarrow$ v.Direction $:= Grant$

7. $C1 \wedge C2' \wedge (v$.To $= v$.Parent $= u$.Id$) \wedge$
   $(u$.Direction $= Grant) \wedge (v$.Direction $= Ask) \wedge$
   $(u$.Request $= v$.Request$) \wedge (u$.From $= v$.Id$)$ $\rightarrow$ v.Direction $:= Grant$

8. $C1' \wedge \neg C1 \wedge (v$.Direction $= Ask) \wedge$
   $(v$.Request $= u$.Request $= v$.From $= v$.Root $= v$.Id$) \wedge$
   $(u$.From $= v$.Id$) \wedge (u$.Direction $= Grant) \wedge (v$.To $= u$.Id$)$ $\rightarrow$ v.Parent $:= u$.Id;
   v.Distance $:= u$.Distance $+ 1$;
   v.Root $:= u$.Root; reset request variables

Figure 1: The program at node v

1. Forwarding the same request (i.e. the parent and v have equal values in their Request variable); and

2. Has received the request from v (i.e. the parent's From variable is v's Id); and

3. Node v has sent the request to its parent (i.e. v's To is the Id of its Parent); and

4. The parent is forwarding the grant (i.e. the Direction in the Parent is *Grant*).

When a grant has been received by the descendant of a node (i.e., the descendant was requesting from the node, and has now set its direction to grant) that node may reset its requesting related variables (action 3) and service other requests. A node whose request has been granted (Action 8) joins the tree by setting its Root to the tree root, its Parent to its neighbor from which it read the grant and its Distance to be larger by one than that of its parent. In addition its resets its request variables to "undefined".

4 Correctness

In this section we prove the self stabilizing property of the algorithm. Let us first solve one difficulty that will repeatedly appears in the proofs. Intuitively this is the fear that a node "imagines" it read some value, but in fact did not. Note that since a high resulusion atomicity is assumed. Thus this involves also variables not explicityly mentioned in the the code, but also possible temporary variables used by the processor to store intermediate results, e.g. that some term of a computed condition has the value "true".

The formal definition of "imagines" is deferred to the full paper. However, we argue that there exists a time when such "imagination" is no longer possible. This follows from the fact that every node repeatedly reads (and the scheduler is fair). Thus, there exists a time t after the time when the faults cease, that from t on every value that is used in the computation was indeed read, or computed from values that where indeed read. In the following proof we speak only to the time after t. Thus it is not necessary to consider "imagination".

Next let us prove that eventually the node with the highest identity becomes a root, and no higher Root variables ever exist later.

Definition 4.1 *When Condition C1 holds for a node then it and its parent constitute together a* **branch**. *Each connected component of the transitive closure of the branch relation is also called a* **branch**.

A **request interval** *is a maximal path that is included in a branch, such that only the (possibly empty) suffix of the interval (the* Parents' *side) contains a* Grant, *the (possibly empty) prefix contains an* Ask, *and all the* Request *registers in the interval's nodes have the same value.*

Definition 4.2 *Let a* **false root** *be an identity that is a value of a* Root *variable of some node (in a connected component), but is not the identity of any node in the (connected component of the) network.*

Observation 4.3 *No identity that does not exist in the (connected component of the) network at any given time after the changes cease (either as an* Id *or as a* Root*) can become a false root later on.*

Proof Sketch: Notice that a node sets its Root either to its Id or to the Root of its neighbor. ∎

Lemma 4.4 *Consider the time after the changes cease. The number of times a node joins a tree whose* Root *is a false root is bounded.*

Proof Sketch: Let us first discuss only the highest false root (break ties arbitrarily). (If there are false roots then there exists such a Root by Observation 4.3.) The number of request intervals that will ever exist in branches with such a Root is bounded Also, if a request is granted by a grant that existed by mistake than the number of intervals decreases by one.

Consider now the second highest false root. The argument above bounds the number of times a node can join that tree, between times that it joins the tree of the highest false root. This argument can now be applied recursively to all the false roots. ∎

Lemma 4.5 *Eventually in every node* Root *equals some* $w \in V$.

Proof Sketch: By Lemma 4.4 and by Observation 4.3 it suffices to prove that no Root variables can contain a false root forever.

Assume the contrary and consider any time after the changes cease. Let $u \notin V$ be the highest identity such that from some time on there always exists a node $v_0 \in V$ whose Root equals some $u \notin V$.

Consider v_0 at a time after the topological changes. Note that v_0 must have a parent v_1 (i.e. its Parent does not equal its own Id.) whose Root equals u, otherwise, since $v_0 \neq u$, node v_0 would have set its Root to v.

Lev $v_0, v_1, ..., v_q$ be the maximal branch in which Root = u forever. (This is the case at least in v_0. Also, q is finite since v_{i+1}.Distance is smaller than v_i.Distance, and every Distance is not smaller than zero.) However, when v_q will not have a parent whose Root is u, node v_q will have to reset its Root too, contradicting the definition of v_q. This, together with Lemma 4.4 implies the lemma. ∎

Corollary 4.6 *Let u be the node with the highest* Id *in (its connected component of) the network. From some time on u is a root.*

Lemma 4.7 *Eventually every maximal branch contains the node whose* Id *is the value of* Root *in every node on the brach.*

Proof Sketch: Similar to Lemma 4.5. ∎

Let us now proof that a tree rooted at the node with the highest Id eventually spans the network. First let us show that from some time on there are no cycles.

Lemma 4.8 *There exists a time from which on there are no cycles in the parent relation.*

Proof Sketch: Similarly to the proof of Lemma 4.5 every cycle must be broken. New cycles cannot be created (after time t) by the definition of request intervals (that cannot be cyclic because of the Distance register, and the observation that a node joins only when a request interval exists, and it is a root. ∎

Definition 4.9 *Consider the time from which on all* Root *values belong to nodes in the (connected component of the) network. A correct tree is a tree defined by the* Parent *relation, such that its root is the node with the highest identity, and Condition $C1$ holds for all the nodes in the tree.*

Observation 4.10 *There is a time from which on a correct tree exists.*

Proof: Follows from Corollary 4.6 and Lemma 4.5. ∎

Lemma 4.11 *Consider the time from which on a correct tree exists. A node that is not in this correct tree, but is a neighbor of a node in that tree, will eventually try to join the correct tree.*

Proof Sketch: Consider such a node v after the times mentioned in Lemmas reflem:false small root disappear, 4.5 and 4.8. If it is not a root then either its Root is that of the correct tree, or its Root is smaller than the highest. In the second case clearly Condition $C1$ does not hold for it. In the first case if Condition $C1$ did hold for it then it would have belonged to the correct tree. Thus condition $C1$ does not hold for it in both cases, and it must become a root. At that time, the algorithm dictates that it makes a request to join the neighboring tree with the highest Root value, which is the correct tree. ∎

Observation 4.12 *A node that joins the correct tree never leaves it.*

Proof Sketch: The reason is that the correct tree is defined only after there are no false roots, and since a node leaves a tree only when Condition $C1$ does not hold. ∎

Definition 4.13 *An* Id *v in the* Request *variable of a node u in the correct tree is called a* correct request *if*

- *node u is in a request interval that starts in a node w (in the correct tree) that is a neighbor of node v, and*
- *v's request variable is equal to v, and*
- *v's* Direction *variable contains an Ask, and*
- *v is a root.*

Lemma 4.14 *If there are nodes in the correct tree that contain a request that is not a correct request, then this request will eventually disappear from the nodes of the correct tree.*

Proof Sketch: By Observation 4.12 the request interval of this request can change only by growing, and it can grow only to a bounded size. Consider now the prefix of this interval, clearly the node at the beginning (of the prefix) notices that this is an incorrect request, and resets all the variables that are related to the request. Note that the request is copied only from a child to its parent, and not vice versa. Thus a repeated procedure of reset will eventually cause this interval to disappear. ∎

Lemma 4.15 *Consider a time in which the correct tree does not span the entire (connected component of the) network. Eventually another node will join it.*

Proof Sketch: By Lemma 4.11 every neighbor of the correct tree will eventually try to join the correct tree. By Lemma 4.14 incorrect requests will disappear. Note also that the requests are forwarded over a tree. Thus, eventually there must be a branch from some node v not on the correct tree to the root of the correct tree that includes only one request interval – that of the request of v. Thus this request is granted, and v joins the correct tree. ∎

Lemma 4.16 *Eventually a correct tree spans the network.*

Proof Sketch: Follows from Lemma 4.15 and from Observation 4.12. ∎

Theorem 4.17 *Eventually the following holds:*

(convergence) *the (connected component of the) network is spanned by a correct tree, and*

(termination) *no node performs any operation in the tree construction algorithm.*

Proof Sketch: Follows from Lemmas 4.16 and 4.14 and from observation 4.12 ∎

5 Conclusions and Discussion

A motivation for the construction of this algorithm was the possibility to use it as a modular component in other algorithms. One example is the famous token passing problem: Once a spanning-tree protocol is constructed we can achieve mutual exclusion by token passing along a virtual ring embedded in a DFS traversal on the tree. (This idea was independently suggested in [DIM90].) Another example is the randomized self stabilizing symmetry braking in anonymous networks. We can further make the reset procedure presented in [AAG87] self stable. A generalization of the self-stabilized snapshot of Katz and Perry [KP90] can be achieved, which implies in turn that general protocols can be self-stabilized.

As mentioned above our memory is optimal when message buffers are used. It is interesting to find out whether one can show a more memory-efficient protocol. In particular it is interesting to compare memory-efficiency of protocols in the unique ID model to other self-stabilized protocols in other models.

References

[AAG87] Y. Afek, B. Awerbuch, and E. Gafni. Applying static network protocols to dynamic networks. In *Proc. of the 28th IEEE Annual Symp. on Foundation of Computer Science*, pages 358–370, October 1987.

[AB89] Y. Afek and G. M. Brown. Self-stabilization of the alternating-bit protocol. In *Proceedings of the 8th IEEE Symposium on Reliable Distributed Systems*, pages 10–12, October 1989.

[AG90] A. Arora and M. Gouda. Distributed reset. Extended Abstract, 1990.

[Ang80] D. Angluin. Local and global properties in networks of processes. In *Proc. of the 12th Ann. ACM Symp. on Theory of Computing*, pages 82–93, May 1980.

[AM89] Y. Afek, and Y. Matias Simple and Efficient Election Algorithms for Anonymous Networks, *3rd International Workshop on Distributed Algorithms*, Nice, France, September 1989.

[BGW89] G. Brown, M. Gouda, and C.L. Wu. Token systems that self stabilize. *IEEE Transactions on Computers*, 38(6):845–852, 1989.

[BP89] J. E. Burns and J Pachl. Uniform self-stabilizing rings. *ACM Trans. on Programming Languages and Systems*, 11(2):330–344, 1989.

[Dij74] E. W. Dijkstra. Self-stabilizing systems in spite of distributed control. *CACM*, 17:643–644, November 1974.

[DIM90] S. Dolev, A. Israeli, and S. Moran. Self stabilization of dynamic systems assuming read/write atomicity. In *Proc. of the ACM Symp. on Principles of Distributed Computing*, August 1990.

[DIM90a] S. Dolev, A. Israeli, and S. Moran. Private communication

[KP90]   Shmuel Katz and Kenneth J. Perry. Self-stabilizing extensions. In *Proc. of the ACM Symp. on Principles of Distributed Computing*, August 1990.

[SG89]   J. Spinelli and R.G. Gallager. Broadcast topology information in computer networks. *IEEE Transactions on Communication*, 1989.

On the Computational Power Needed to Elect a Leader

(Extended Abstract)

Alon Itai [*]
Department of Computer Science
Technion - Israel Institute of Technology
Haifa 32000, ISRAEL

Abstract

Consider a ring of identical processors that wish to elect a leader. Itai and Rodeh have shown that if n, the size of the ring, is not known then there exists no algorithm in which the processors can sense termination. However, for all $\varepsilon > 0$ there exist leader election algorithms which terminate when all messages arrive. These algorithms elect a leader with error probability $\leq \varepsilon$. They and most subsequent authors have concentrated on the message complexity, and have disregarded the amount of local memory required.

Here we consider a ring in which the amount of local memory in each processor is bounded by a number which is independent of the size of the ring, and depends only on the allowed error rate. We present three algorithms, one in which the probability of error is $O(1/n^\alpha)$, the memory is $O(\log \alpha)$ bits and the communication complexity is $O(n^2 \log n)$ bits, where $\alpha > 1$ is an arbitrary parameter. In the second algorithm, for each $\varepsilon > 0$ the probability of error is $\leq \varepsilon$, $O(\log \log \log(1/\varepsilon))$ bits of memory are required and the communication complexity is $O((n/\varepsilon) \log n \log(1/\varepsilon)(\log \log n + \log(1/\varepsilon)))$ bits. The third algorithm always terminates though its communication complexity may be larger. Since the computation for each message is at most exponential in the number of bits of local memory, the amount of computation is also independent of the size of the ring.

These results are extended to additional topologies.

1 Introduction

We investigate the computational power needed to choose a leader in an anonymous ring and in some other networks.

The *leader election* problem is as follows: Given a network of identical processors, find a distributed algorithm such that on termination one of the processors is designated as the *leader* and the remaining processors know that they are not the leader. A distributed algorithm is a program executed by all the processors (all the processors execute the same program). There are many variants of

---
[*]Supported by Technion V.P.R. – E. & J. Bishop Research Fund.

the problem: If the processors have unique *ids* then the maximum id can serve as the leader – so the problem is reduced to finding the maximum, for which several algorithms exist [4, 5, 8, 11, 14, 16].

If the processors are *anonymous*, (i.e., there are no id's) Angluin [2] showed that there exists no deterministic algorithm to choose a leader. Itai and Rodeh [13] investigated the power of randomized algorithms. They showed algorithms for anonymous rings in which the processors know n – the size of the ring, (actually a bound on n suffices). (Knowing n allows a different algorithm for every n – however, all the proposed algorithms did not depend on n). Furthermore, they defined processor termination and showed that if n is not known then there exists no randomized processor terminating algorithm. An algorithm *processor terminates* if each processor eventually enters a state $\in \{\text{LEADER}, \text{NONLEADER}\}$ from it cannot exit, i.e., once a processor has committed itself as a leader (or nonleader) it cannot change its state. Obviously, the algorithm succeeds if exactly one processor ever enters the LEADER state.

In view of this negative result, it is interesting to consider algorithms in which the processors may leave the LEADER state. Thus the processors don't know whether the algorithm has terminated. To define termination they considered *message driven* algorithms – the processors act only when a message arrives. We say that such an algorithm *message terminates* when there are no more pending messages and all processor have finished processing all their messages. An election algorithm *succeeds* if it always (with probability 1) message terminates and on termination exactly one processor is in the LEADER state.

Since we assume that the memory of each processor is independent of n, we cannot assume that the processors can remember n or have unique id's. (Storing n or an id requires $\Omega(\log n)$ bits.) Thus, in light of Itai and Rodeh's result we must resort to message terminating algorithms.

Itai and Rodeh [13] showed a message terminating algorithm to elect a leader whose communication complexity is $O(n^3 \log(1/\varepsilon))$ bits. Schieber and Snir [17] found a more efficient algorithm that achieves message termination in $O(n \log n + n \log(1/\varepsilon))$ messages of length $O(\log n/\varepsilon)$ bits and with probability $1 - \varepsilon$ elects a unique leader. Matias and Afek [15] gave a simpler algorithm and improved the bit complexity to $O((n/\varepsilon) \log n \log(1/\varepsilon)(\log \log n + \log(1/\varepsilon)))$.

All the above algorithms assumed that all the processors had infinite computational power and unbounded storage. However, a closer look shows that the algorithms guessed (not necessarily unique) id's and the amount of storage is proportional to the length of the maximum such id. Thus Itai and Rodeh need $O(\log n + \log(1/\varepsilon))$ bits, while Matias and Afek need only $O(\log \log n + \log(1/\varepsilon))$ bits.

The question addressed in this paper is whether this is really necessary. How much storage is really needed? We show that the same performance can be achieved even if the storage of each processor is independent of n. Our basic algorithm needs only a constant amount of storage, thus the computational model is equivalent to that of finite automata which have the capability to toss coins.

Subsequent refinements of the basic algorithm allow us to reduce the probability of error from $O(1/n)$ to $O(1/n^\alpha)$ by increasing the number of states by a factor of α. The communication complexity of the basic algorithm is rather poor – $O(n^2 \log n)$ bits. In section 5 we implement Matias and Afek's algorithm, to elect a leader with error probability $\leq \varepsilon$, communication complexity $O((n/\varepsilon) \log n \log(1/\varepsilon)(\log \log n + \log(1/\varepsilon)))$ and $O(\log \log \log(1/\varepsilon))$ bits of memory per processor. Thus the two algorithms exhibit a tradeoff between communication complexity, memory and probability of error. The previous algorithms may, with probability 2^{-n} get into an infinite loop. This situation is ameliorated in Section 6. In Section 7 the results are extended to some additional topologies, and the conclusions follow in Section 8.

2 The model

A PFA – *probabilistic finite automaton* is a sixtuple $A = \langle Q, \Sigma, \delta, \mu, q_0, F \rangle$ where Q is a finite set of states, Σ a finite alphabet, $q \in Q$ the initial state and $F \subset Q$ the set of accepting states. $\delta : Q \times (\Sigma \cup \{\text{WAKEUP}\}) \times \{0,1\} \to Q$ is the transition function. For a given state q, input letter σ and coin toss ρ it determines (deterministically) the next state $\delta(q, \sigma, \rho)$. WAKEUP $\notin \Sigma$ is a special wakeup symbol, enabling some processors to start processing without any input.

$\mu : Q \times \Sigma \cup \{\text{WAKEUP}\} \times \{0,1\} \to \Sigma^*$ is the output function, i.e., a single input bit can produce $m \geq 0$ bits of output.

The processors are arranged in a unidirectional ring: The output of processor v_i is the input of processor v_{i+1}. For $k \geq 1$ processors the input is preceded by the symbol WAKEUP. Our model assumes that the processors act asynchronously. We do not make any explicit assumptions about time, except that every message that was sent eventually arrives and messages travel over the links of the ring in a FIFO manner. Note that since the processors are message driven, this only means that they cannot count time.

We consider rings of n identical PFA's A, each with its own *random tape* – an infinite sequence of $0,1$, from which the processor takes the values of ρ. For a given random tape, the response of each processor to each message is completely determined. Since the edges follow the FIFO discipline and we are dealing with the ring topology, the sequence of random tapes and the distribution of wakeup messages completely determines the output of the algorithm. (For more complicated topologies this is no longer true and we would have to consider different scheduling policies.)

We assume that at least one processor receives the wakeup message WAKEUP. An algorithm A for a ring R in which each processor v_i has a random tape ρ_i *chooses a leader for the random tapes* $(\rho_1, \ldots, \rho_n)$ if for all sets of wakeup messages the algorithm message terminates and on termination exactly one processor is in the LEADER state. We assume that the random tapes have the usual probability distribution – for all $[c_1, \ldots, c_k] \in \{0,1\}^k$ there is probability 2^{-k} that the first k bits of the random tape are equal to $[c_1, \ldots, c_k]$. An algorithm A *chooses a leader for R with probability p* if the probability (over all sequences of random tapes) that A chooses a leader is greater than or equal to p. An algorithm *chooses a leader with probability p* if for all rings R it chooses a leader with that probability. For all $\varepsilon > 0$ there exists an automaton with $\log \log 1/\varepsilon$ states that chooses a leader in a ring with probability $1-\varepsilon$. The process of leader election requires an average of $O(n \log n + n \log 1/\varepsilon)$ bits.

3 The basic algorithm

3.1 An outline

In this section we describe our basic algorithm, which requires $O(1)$ states, has error probability $O(1/n)$ and communication complexity of $O(n^2 \log n)$ bits. We use some ideas of the algorithm of Matias and Afek [15]. We first describe the algorithm assuming each processor has unbounded memory, then we explain how it may be implemented with finite memory.

The processors which receive a wake up message are called *kings*. Each king chooses a random *id*, which is sent around the ring. The length of the id's is also random, there is probability 2^{-i} that the length is i. In [15] it is shown that the expected size of the maximum id is $\log n$, and with probability $> 1 - 1/n$ the largest id is unique. Upon receiving a message carrying an id, a king compares it to its own id. If the id of the message is larger, the message continues around the ring. If the id of the message is smaller, the message is destroyed (not passed further). Finally, if they are equal then the king assumes that its own id has returned, thus it holds the largest id and can nominate the leader. However, this assumption may be wrong, there may be larger messages which have not yet arrived. Thus the leader still participates in the election process and consequently may leave the LEADER state.

3.2 The implementation by automata – an outline

The main difficulty in implementing this algorithm by finite automata is that a king, like all other processors, has only a finite storage space and therefore cannot store its own id. The solution is to use the adjacent processors to store the king's id. Namely, if the id chosen by a king v_k is $[a_1, \ldots, a_\ell]$, v_k stores the id in $\mathcal{K}_k = \{v_k, \ldots, v_{k+\ell}\}$ – its *kingdom*, i.e., a_1 is stored at v_k, a_2 at v_{k+1}, ..., a_ℓ at $v_{k+\ell-1}$. The last processor of the kingdom, $v_{k+\ell}$, holds the symbol '#' to tell us that this processor is the last processor of the kingdom. For simplicity, we first assume that the kingdoms are disjoint. In Section 4 we shall remove this restriction.

The processors of the kingdom send the id to the next kingdom.

When the kingdom $\mathcal{K}_k = \{v_k, \ldots, v_{k+\ell}\}$ gets a message $id' = [b_1, \ldots, b_q]$ it compares it to its own id. If they are equal and no higher id was seen then the last processor of the kingdom, $v_{k+\ell}$, becomes the leader. If $id < id'$ (i.e., $\ell < q$, or $\ell = q$ and $id < id'$ as binary numbers, i.e., $\sum a_k 2^{\ell-k} < \sum b_k 2^{\ell-k}$) then id' continues its traversal of the ring. Thus the maximum id can stop its traversal only when it returns to its originator.

The remainder of this section is rather technical and its purpose is to convince the reader that this protocol can indeed be implemented by PFA's.

3.3 The Sending Protocol

The following protocol and variations of it are used several time in the algorithm. Suppose a message $[m_1, \ldots, m_q]$ is stored at processors $v_1, \ldots, v_q$, i.e., $mem_1 = m_1, \ldots, mem_q = m_q$. To send the message we use the following protocol: v_1 first sends $\langle \text{MESSAGE}, mem_1 \rangle$ and then $\langle \text{MESSAGE}, \text{END} \rangle$. When v_i receives $\langle \text{MESSAGE}, c \rangle$, $(c \neq \text{END})$, it sends $\langle \text{MESSAGE}, mem_i \rangle$, and sets $mem_i := c$. When it receives $\langle \text{MESSAGE}, \text{END} \rangle$ it first sends $\langle \text{MESSAGE}, mem_i \rangle$ and then $\langle \text{MESSAGE}, \text{END} \rangle$.

Theorem 3.1 *If a processor v_i receives the input sequence* $\langle \text{MESSAGE}, m_{i-1} \rangle, \langle \text{MESSAGE}, m_{i-2} \rangle, \ldots,$ $\langle \text{MESSAGE}, m_1 \rangle, \langle \text{MESSAGE}, \text{END} \rangle$ *and $mem_i = m_i$ then it generates the output sequence* $\langle \text{MESSAGE}, m_i \rangle, \langle \text{MESSAGE}, m_{i-1} \rangle, \ldots, \langle \text{MESSAGE}, m_1 \rangle, \langle \text{MESSAGE}, \text{END} \rangle$ *(i.e., the message is prepended by $\langle \text{MESSAGE}, m_i \rangle$).*

Proof (sketch): Show by induction that the t-th message sent by v_{k+i} is $\langle \text{MESSAGE}, m_{i+1-t} \rangle$ and after it is sent $mem_i = m_{i-t}$. □

3.4 The algorithm:

Choosing an id: A king v_k guesses $a_1 \in \{0, 1\}$, stores it in own_k, and sends the message $\langle \text{CHOOSE}, a_1 \rangle$ to v_{k+1}. Each subsequent processor v_{k+i}, when getting the message $\langle \text{CHOOSE}, c \rangle$ first sets $mem_{k+i} := c$ then decides at random whether it is the last processor. If it is, then own_{k+i} is set to '#'. Otherwise, it chooses $own_{k+i} := a_{i+1} \in \{0, 1\}$ and sends $\langle \text{CHOOSE}, a_{i+1} \rangle$.

At the end of this step the id $= [id_1, \ldots, id_\ell]$ is stored both in $own_k, \ldots, own_{k+\ell-1}$ and in $mem_{k+1}, \ldots, mem_\ell$. (We keep it in own_i in order to compare id's and it is kept in mem_i as a preparation to the sending subprotocol which immediately follows.)

Transmitting an id: We use the sending protocol, with the header TRANSMIT. The king v_k initiates the sending protocol with $\langle \text{TRANSMIT}, \text{END} \rangle$. Thus the last processor $v_{k+\ell}$ receives $\langle \text{TRANSMIT}, id_{\ell-1} \rangle, \langle \text{TRANSMIT}, id_{\ell-2} \rangle, \ldots, \langle \text{TRANSMIT}, id_{\ell-1} \rangle, \langle \text{TRANSMIT}, \text{END} \rangle$ and it replaces TRANSMIT by PROBE, i.e., it sends $\langle \text{PROBE}, \# \rangle, \langle \text{PROBE}, id_\ell \rangle, \langle \text{PROBE}, id_{\ell-1} \rangle, \ldots,$
$\langle \text{PROBE}, id_{\ell-1} \rangle,$
$\langle \text{PROBE}, \text{END} \rangle.$

Passing an id: When a processor v_i receives $\langle \text{PROBE}, c \rangle$, it first stores c in mem_i, the next such message $\langle \text{PROBE}, c' \rangle$, initiates the sending protocol.

Compare: When the king receives $\langle \text{PROBE}, \text{END} \rangle$, the id of the message ($id' = [b_1, \ldots, b_q]$) which is stored in $mem_k, \ldots, mem_{k+q-1}$) is compared to $id = [a_1, \ldots, a_\ell]$, which is stored in $own_k, \ldots, own_{k+\ell-1}$. In addition id' is sent (using the sending protocol). The result of the comparison is piggybacked to the sending protocol, i.e., processor v_{k+i} receives $\langle \text{COMPARE}, +, b_i \rangle$ if $i < \ell$ or $[b_1, \ldots, b_i] < [a_1, \ldots, a_i]$, it receives $\langle \text{COMPARE}, -, b_i \rangle$ if $q < i$ or $b_1, \ldots, b_i] < [a_1, \ldots, a_i]$ and finally it receives $\langle \text{COMPARE}, -, b_i \rangle$ if $q < i$ and $[b_1, \ldots, b_i] = [a_1, \ldots, a_i]$.

Ending the comparison: When the last processor of the kingdom, $v_{k+\ell}$, receives message $\langle \text{COMPARE}, e, c \rangle$ ($e \in \{+, -, =\}, c \in \{0, 1\}$), it can conclude the comparison: if $\ell < q$ ($v_{k+\ell}$ knows this since in this case $b_{\ell+1} \neq$ '#' is stored in $mem_{k+\ell}$), the message should continue its probe to the next kingdom, so the message $\langle \text{PROBE}, mem_{k+\ell} \rangle$ is sent and $mem_{k+\ell} := c$ to continue the sending protocol.

A similar action is done when $\ell = q$ and $e = '-'$.

If $\ell > q$ or $\ell = q$ and $e = '+'$ then $id' < id$. In this case the message is destroyed, i.e., $v_{k+\ell}$ does not send any message.

Finally, if $\ell = q$ and $e = '='$ then $id = id'$. In this case, $v_{k+\ell}$ assumes that this is the message generated by v_k which has safely traversed the ring, therefore it was greater than or equal to any other id. The processor assumes that the id of its kingdom has returned, and it is the largest id, so $v_{k+\ell}$ becomes the leader.

Note that even after it is nominated as a leader $v_{k+\ell}$ still compares messages, and if it sees a larger id it resigns.

3.5 Analysis

The internal memory of each processor v_i consists of a single variable mem_i taking values from $\{0, 1, \#\}$ Additional storage is needed to remember the contents of the message. However since the message consists of $O(1)$ bits, the total memory requirements are also $O(1)$.

Suppose κ processors received a wakeup message, [9] proved that the expected maximum length id is $O(\log \kappa)$ and on the average only $O(1)$ processors achieve that length. Since there are $2^{\log \kappa} = \kappa$ different id's of that length, the probability that the maximum id is shared by more than one processor is $1/\kappa$. Thus the greatest error occurs when $\kappa = 2$. (When $\kappa = 1$ then there is no error – only one id is generated, it successfully traverses the ring and when it returns its king nominates the leader).

In a finite automata model each message consists of a single letter, i.e., $O(1)$ bits, thus the message complexity equals the bit complexity. Each of the κ processors with a wakeup message produces an id of expected length $O(\log \kappa)$ bits. Each id travels around the ring, thus the expected communication complexity is $O(\kappa n \log \kappa)$ bits.

4 Overlapping Kingdoms

The previous algorithm assumed that the kingdoms are disjoint, however, this does not necessarily happen and on large rings the probability for this approaches 0. If the kingdoms are not disjoint, a king might have to participate in the generation of an id of another kingdom. It will have to store both its own id and that of the other kingdom. Since there may be up to n kingdoms, we will exhaust any a priori bound on the number of states.

We now outline how the previous algorithm can be modified to handle this eventuality. In the next subsection we assume that the initial distance between kings is greater than or equal to some constant $C \geq 8$. In Section 4.2 we show how this may be achieved. We only outline the protocols, further details and code will appear in the full paper.

4.1 The shifting process

A processor detects that kingdoms are not disjoint if while generating an id of the king v_k the message $\langle \text{CHOOSE}, c \rangle$ reaches the next king v_d requiring it to choose another bit of the id of v_k. The solution is to shift the entire kingdom of v_d one position. (As a result of this shift, the shifted kingdom might bump into the next kingdom, which consequently is also shifted, thus we might get a cascade of shifts).

After v_d's kingdom shifted one position, v_{d+1} becomes the king of v_d's kingdom and the id of that kingdom is stored in $own_{d+1}, \ldots, own_{d+l'}$. Since the processor v_d is no longer a king, it can produce and store the next bit of v_k's id. If it does not choose to be the last processor then another shift is necessary.

Since there is probability 2^{-i} that an id has length i, the expected length of an id is 2 (not including the '#'). Using Chernoff's inequalities we can show that the probability that a shift involves s kingdoms is less than e^{-s}. The expected cost of a shift initiated in a single kingdom is $\sum s \cdot 2 \cdot e^{-s} = O(1)$. Since there are $k = O(n)$ kings, the total cost of all the shifts is $O(n)$ bits.

There is nonzero probability that the total size of all the id's exceeds the size of the ring. In this case the above algorithm fails and its communication complexity is infinite. The following argument shows that the probability of this event is small.

Consider k kingdoms whose total size is m. This is equivalent to Bernoulli trials, in which the k-th success occurs at the m-th experiment. The probability for this is $\binom{m-1}{k-1} 2^{-m}$ [10, p. 226]. Using this, a straightforward calculation shows that if the number of kingdoms is less than $n/4$ then the probability that the total sizes exceeds n is less than $\frac{n}{4} \left(\frac{e}{4} \right)^{n/4}$. Therefore, the probability of this type of error decreases exponentially. In Section 6 we present an algorithm which overcomes this problem.

4.2 Reducing the number of kings

This step is executed by the kings immediately after receiving the wakeup message and before creating the kingdom. This step causes some kings to resign. Only kings which have not resigned create kingdoms.

We use the probabilistic algorithm of the journal version of [13] to reduce the number of kings. In each round, each king chooses a bit and sends it to the next king. The king compares the bit it chose to the bit it received, and if it chose 0 and received 1, it resigns. This protocol is repeated $\log_{4/3} C$ rounds, for some constant $C \geq 8$.

In each round, a king resigns only if the id of its predecessor is larger. Thus not all kings can resign and if there is more than one king, the probability of a king to resign is 1/4. While there are at least two kings, in each round of the protocol the expected number of kings is reduced by a factor of 3/4. If initially there were κ kings, then after $\log_{4/3} C$ rounds the expected number of kings is κ/C.

In the full paper we use Chernoff's inequalities to show that the probability that there remain more than $s/4$ kings in a section of size s decreases exponentially in s. Thus the probability of a cascade of s shifts is also exponentially small, implying as in Section 4.1, that the total cost of the shifts is $O(n)$.

The communication cost of each round is n bits, so the cost of the entire king reduction step is $n \cdot \log_{4/3} C = O(n)$ bits.

4.3 Summary of the basic algorithm

The basic algorithm requires $O(1)$ bits of memory per processor, has communication complexity $O(n^2 \log n)$ and error probability $O(1/\kappa')$, where κ' is the number of kings which survived the reduction process of Section 4.2; the expected number is κ/C, where κ is the original number of kings, (= the number of wakeup messages). Thus the probability of error is $O(1/\kappa)$. This is unacceptable when κ' is small, (e.g. $\kappa' = 2$).

To ensure that the error probability is $O(1/n)$, we can simply wake up all the processors by sending wakeup messages (each processor sends exactly one such message, so the added communication complexity is $O(n)$ bits). The king reduction process is now employed to reduce the expected number of kings to n/C.

Another modification of the basic algorithm decreases the probability of error to $(1/\kappa')^\alpha$. The idea is to increase the length of the id's by a factor of α. This, of course, also increases the communication complexity by a factor of α and each processor needs $\log \alpha$ bits of memory. (The technique of Section 5.2 can be used to reduce the memory requirements to $\log \log \alpha$).

5 Reducing the communication complexity

The communication complexity of the basic algorithm was rather high – $n^2 \log n$ bits. Matias and Afek proposed an algorithm which requires $O(rn \log n(\log \log n + \log r))$ bits of communication, where $r = (40/\varepsilon) \log(1/\varepsilon)$. We first show how Matias and Afek's algorithm can be implemented with $O(\log \log r)$ bits of memory per processor and in Section 5.2 we reduce the memory to $O(\log \log \log r)$.

The main communication cost of the basic algorithm was sending long id's around the ring. Matias and Afek reduced the maximum expected length of the message from $\log n$ to $\log \log n + \log r$. Thus they have $r \log n$ distinct id's, and since each id can travel around the ring at most once, their communication complexity is $O(rn \log n(\log \log n + \log r))$.

5.1 The algorithm

The id's chosen consist of two parts, the first is the length of the id of the basic algorithm and the second is a random integer in the range $0..r$. Thus in order to implement this algorithm, we need to modify the choose part as follows:

Choose 1: Choose an id $[a_1, \ldots, a_\ell]$ as in the basic algorithm.

Choose 2: Count the length of the id's.

Choose 3: Choose another $\log r$ bits.

To implement step **Choose 2**, the king initiates a count by the sending protocol with the caption COUNT. The count is maintained in $v_{k+\ell+1}, \ldots, v_{k+\ell+\log \ell}$, ($v_{k+\ell+1}$ holds the least significant bit). Each COUNT message that arrives increases this number by one (when $v_{k+\ell+i}$ gets such a message, if $own_{k+\ell+i} = 0$, it sets it to 1, and if it was 1 it is set to 0 and a COUNT message is sent).

To implement step **Choose 3**, a counter is added to the CHOOSE message, i.e., the format of the message is $\langle \text{CHOOSE}_3, j, c \rangle$, where j is increased at each processor until it is equal to $\log r$.

Adding the counter to the message increases its length to $\log \log r + O(1)$ bits. If such messages are to be stored by the processors then each of them must have that many bits of memory.

Since each processor can participate in the choose step of at most one kingdom, the entire communication cost of this step is $O(n)$ bits. Using the king reduction technique of Section 4.2 ensures that the expected communication cost of all the shifts is also $O(n)$ bits. To summarize, Matias and Afek's algorithm can be implemented with $O(\log \log r)$ bits of memory, without increasing the expected communication complexity.

5.2 Reducing the number of states

The last algorithm required that the id's have length at least $\log r$. In order to satisfy this requirement, the message $\langle \text{CHOOSE}_3, j, c \rangle$ included a counter j whose maximum value was $\log \log r$.

If we insist that the count be exact then each processor needs $\theta(\log \log r)$ bits. However, we can reduce the number of bits to $\log \log \log r$ by relaxing the condition that id's have length at least $\log r$ and requiring instead that the *expected length* of the id's be $\log r$. The trick is to use the following approximate counting technique due to Flajolet [7].

We now show how to estimate x with only $\log \log x$ bits. On receiving the counter with value j, the counter is increases by one with probability $1/2^j$. Thus, the expected number of attempts until the counter reaches $\log x$ is $\sum_{j=0}^{\log x} 2^j = 2x - 1$. We need only $\log \log x$ bits to hold x.

If the above approximate counting scheme is used instead of exact counting, then the range of the counters in the COUNT$_3$ messages is only $\log \log \log r$ and consequently each processor needs only $\log \log \log r + O(1) = O(\log \log \log(1/\varepsilon))$ bits of memory.

A subtle point of this procedure is that it requires that a coin be chosen with probability $1/2^j$ ($j \leq \log \log r$) – not half, while our model assumes unbiased coins. A biased coin can be simulated by j tosses of unbiased coins (answer 1 only if all the j unbiased tosses return 1). The average number of coin tosses is 2, but the maximum number is $\log \log r$. Thus, in the worst case, the time of each step depends exponentially on the size of the input, (not linearly as in all the previous operations). However since the input consists of $\log \log \log r$ bits, in the worst case the time is $O(\log \log r)$ steps, although (since the average number of coin tosses is 2) the expected time is still $O(1)$.

Also, to count $\log \log r$ successful coin tosses requires $\log \log \log r$ additional bits. Thus the memory requirements are $O(\log \log \log r)$ bits per processor.

In the full paper we show that the expected communication complexity of Matias and Afek's algorithm increases by no more than a constant factor.

6 Finite Communication Complexity

As noted in Section 4 there is a (small) probability that the communication complexity be infinite. The following algorithm overcomes this problem – however another type of error is introduced – there is some probability that no leader is chosen.

To this end, the network is first partitioned into kingdoms then each kingdom chooses an id whose size (in bits) is proportional to the number of processors in the kingdom.

To partition the ring – each processor chooses 0 or 1 with equal probability, and the processors who chose 1 are the last processors of their kingdom. These processors immediately send a message down the ring to notify the next processor that it is the first processor of the next kingdom.

When a processor receives such a message it initiates choosing an id. Each processor of the kingdom chooses $c \geq 3$ bits – thus the size of the id is c times the size of the kingdom.

Each kingdom has size at least 1 and the average size is 2. We claim that with high probability there exists large kingdoms.

Claim 6.1 *With probability greater than $1 - e^{-\sqrt{n}/\log n}$ there exists a kingdom of size $\geq \frac{1}{2}\log n$.*

Proof: If all kingdoms had smaller size there would be at least $2n/\log n$ kingdoms. The probability that a kingdom has size $> i$ is 2^{-i}. Therefore, the probability that a kingdom has size $\leq \frac{1}{2}\log n$ is
$$1 - 2^{-\frac{1}{2}\log n} = 1 - \frac{1}{\sqrt{n}}.$$
The probability that all kingdoms have size $\leq \frac{1}{2}\log n$ is at most
$$\left(1 - \frac{1}{\sqrt{n}}\right)^{2n/\log n} = \left[\left(1 - \frac{1}{\sqrt{n}}\right)^{\sqrt{n}}\right]^{2\sqrt{n}/\log n} > e^{-\sqrt{n}/\log n}.$$
□

Thus, with high probability the maximum id consists of at least $\frac{c}{2}\log n$ bits.

We now use the following claim of Matias and Afek[15]

Claim 6.2 *Let p be the probability that in k drawings from a domain of size d, the largest label (among those that were drawn) was drawn only once. Then, $p \geq 1 - (k/d)$.*

Theorem 6.1 *The probability that the above algorithm choose a leader is at least $1 - n^\alpha$, for some $\alpha > 0$.*

Proof: If not all processors chose one, the maximum size kingdom is $> \frac{1}{2}\log n$ and the maximum is unique, a unique leader is chosen. Use Claim 6.2 with $d \geq 2^{\frac{c}{2}\log n}$, and $k \leq 2n/\log n$ to show that the probability that the maximum is unique is at least $p \geq 1 - (k/d) \geq 1 - \left(\frac{2n}{\log n}n^{-c/2}\right) = 1 - \frac{2}{\log n}n^{-(\frac{c}{2}-1)}$.

Therefore all the above conditions hold with probability at least
$$\left(1 - 2^{-2}\right)\left(1 - e^{-\sqrt{n}/\log n}\right)\left(1 - \frac{2}{\log n}n^{-(\frac{c}{2}-1)}\right),$$
which implies the theorem. □

The computational complexity of the algorithm is $O(n^2)$ since each processor initiates one bit that might traverse the ring. This complexity may be reduced using the techniques of Section 5.

7 Extensions

7.1 Additional Topologies

The above algorithms strongly depend on the topology of the network – the ring. Obviously there is an interest in other topologies. Instead of providing a different algorithm for each topology, we suggest a paradigm for developing algorithms – we use a ring algorithm as a "subroutine" of other algorithms.

As an example consider an $m \times m$ grid:

$$V = \{v_{i,j} : 0 \leq i, j < m\}$$
$$E = \{(v_{i,j}, v_{i,j+1 \bmod m}), (v_{i,j}, v_{i+1 \bmod m, j}) : 0 \leq i, j < m\}.$$

The total number of processors is $n = m^2$ and each processor has two outgoing edges. We assume that a processor $v_{i,j}$ can distinguish between the horizontal and vertical outgoing edge (i.e., between the edge to $v_{i,j+1 \bmod m}$ and that to $v_{i+1 \bmod m, j}$).

For fixed i, the processors $(v_{i,0}, v_{i,1}, \ldots, v_{i,m-1})$ form a ring. Thus any one of the previous algorithms may be applied on each such slice, and then the slices compete among each other. Since each slice has (a collective) memory of $\theta(m)$ bits, the interslice algorithm has no acute memory problems.

In general, consider a strongly connected network which can be (a priori or distributedly) partitioned into subnetworks. The graph obtained by contracting each of the networks is also strongly connected. If a leader can be elected in each network, then a leader of the entire network can be elected by applying any election algorithm on the contracted graph. Moreover, if each subnetwork has size greater than $f(n)$, a leader can be elected in each subnetwork with $O(1)$ bits per processor, and a leader can be elected in the contracted network with $O(f(n))$ bits per processor, then there exists an algorithm to elect a leader in the entire network using $O(1)$ bits per processor.

7.2 Nonoriented rings

In this model the processors are ordered in a ring, but they do not necessarily agree on the notions of right and left. However all processors know from where the message is received. The election algorithm will proceed in two directions. Every processor participates in both algorithms. When it receives a message from its first direction it simulates a unidirectional processor who receives messages from that direction, and when it receives a message from the other direction it simulates a unidirectional processor who receives messages from the other direction.

To choose a leader any processor who wakes up chooses a direction and initiates choosing an id I according to one of the previous algorithms. Since some kings initiate the algorithm in one direction and others in the other direction, we will be comparing id's in both directions and we might run into trouble since the maximum id in one direction is not necessarily maximum in the other direction. We overcome this problem by making sure that all the id's are palindromes – yield the same value when read from left to right and from right to left. To achieve this, the initial id I is concatenated with its reverse thus the new id is II^{rev}. This new id is a palindrome and the maximum in one direction is also the maximum in the other direction. This addition increases the computational complexity by a constant factor.

8   Conclusions

We have investigated the computational power needed to elect a leader in an anonymous ring, and have shown that for a fixed probability of error we need probabilistic finite automata with a finite number of states. We presented two algorithms, which differ in their communication complexity and memory requirements as a function of the probability of error. It remains to show whether the above algorithms are optimal with respect to memory utilization and communication complexity.

Concerning other topologies, we have demonstrated a decomposition method. It is not always clear when this method is applicable, and whether for every family of bounded degree networks there is an $O(1)$ bit election algorithm. Moreover, is there an algorithm which is applicable to all networks. (Or at least, can we construct one for each degree.) Finally, it is interesting to investigate the computational power required by other tasks on distributed networks.

From a more general perspective our algorithms allowed load sharing the memory requirements among several processors. This principle can be applied to other tasks in order to improve memory and computational requirements.

References

[1] K. Abrahamson, A. Adler, L. Higham, and D. Kirkpatrick. Probabilistic solitude verification on a ring. In *5th PODC*, pages 161-173, August 1986.

[2] D. Angluin. Local and global properties in networks of processes. In *12th STOC*, pages 82-93, ACM, April 1980. Los Angeles, California.

[3] J.E. Burns. *A formal model for message passing system.* Technical Report TR-91, Indiana University, September 1980.

[4] E. Chang and R. Roberts. An improved algorithm for decentralized extrema-finding in circular configurations of processes. *Communications of the ACM*, 22:281-283, 1979.

[5] D. Dolev, M. Klawe, and M. Rodeh. An $O(n \log n)$ unidirectional distributed algorithm for extrema finding in a circle. *Journal of Algorithms*, 3:245-260, 1982.

[6] P. Duris and Z. Galil. Two lower bounds in asynchronous distributed computation. In *28 FOCS*, pages 326-330, 1987.

[7] P. Flajolet.

[8] R. Franklin. On an improved algorithm for decentralized extrema-finding in circular configuration of processes. *Communications of the ACM*, 25:336-337, 1982.

[9] A.G. Greenberg and R.E. Ladner. Estimating the multiplicities of conflicts in multiple access channels. In *24th FOCS*, pages 383-392, 1983.

[10] M. A. Golberg. *An Introduction to Probability Theory with Statistical Applications*, Plenum Press, 1984.

[11] D.S. Hirschberg and J.B. Sinclair. Decentralized extrema-finding in circular configurations of processes. *Communications of the ACM*, 23, November 1980.

[12] A. Itai, S. Moran, and N. Navoni. - in preparation.

[13] A. Itai and M. Rodeh. Symmetry breaking in distributed network. In *22nd FOCS*, pages 245-260, 1981. To appear in *Inf. and Comp.*

[14] G. LeLann. Distributed systems - toward a formal approach. *Information Processing Letters*, 77:155-160.

[15] Y. Matias and Y. Afek. Simple and efficient election algorithms for anonymous networks. In *3rd Intl. Workshop on Distributed Algorithms*, pages 183-194, Lecture Notes in Computer Science, Springer-Verlag, Nice, France, September 1989.

[16] G.L. Peterson. An $O(n \log n)$ unidirectional algorithm for the circular extrema problem. *IEEE Transactions on Programming Languages and Systems*, 4:758-762, 1982.

[17] B. Schieber and M. Snir. Calling names on nameless networks. In *ACM Symp. on the Principles of Distributed Computing*, Edmonton, Canada, 1989.

Spanning Tree Construction for Nameless Networks

Ivan LAVALLÉE   Christian LAVAULT

INRIA, Domaine de Voluceau
Rocquencourt, B.P. 105
78153 Le Chesnay Cedex. France
E-mail : lavault@seti.inria.fr*

Abstract

Two types of distributed fully asynchronous probabilistic algorithms are given in the present paper which elect a leader and find a spanning tree in arbitrary anonymous networks of processes. Our algorithms are simpler than in [11] and slightly improve on those in [9,11] with respect to communication complexity. So far, the present algorithms are very likely to be the first fully and precisely specified distributed communication protocols for nameless networks. They are basically patterned upon the spanning tree algorithm designed in [7,8], and motivated by the previous works proposed in [9,11].
For the case where no bound is known on the network size, we give a message terminating algorithm with error probability ϵ which requires $O(m \log \log(nr) + n \log n)$ messages on the average, each of size $O(\log r + \log \log n)$, where n and m are the number of nodes and links in the network, and $r = 1/\epsilon$. In the case where some bounds are known on n ($N < n \leq KN$, with $K \geq 1$), we give a process terminating algorithm, with error probability ϵ, with $O(m + n \log n)$ messages of size $O(\log n)$ in the worst case. In either case, the (virtual) time complexity is $O(D \times \log \log(nr))$. In the particular case where the exact value of n is known, a variant of the preceding algorithm process terminates and always succeeds in $O(m + n \log n)$ messages of size $O(\log n)$.

1 Introduction

In a distributed algorithm, a network of processes collaborate to solve a given problem. In this framework, each site or process acquires, via local interaction with its neighbours, some global information about the system : e.g. the size of the network, its location in a (minimum-weight) spanning tree, the distances to all other processes, etc. Typically, one assumes that the processes have *distinct* identification labels or *identities*, which means that some global coordination between the processes has taken place beforehand. What happens if this assumption is dropped, so that the processes are indistinguishable ? Consider for example regular nameless networks of some fixed degree. The executions of a deterministic algorithm may end with all processes in the same state, irrespectively of the network size or structure ; a deterministic algorithm cannot distinguish between processes of a regular distributed system, nor can it distinguish between distinct regular networks of the same degree (see [4,5]).

The situation is different if processes are assumed to make independent probabilistic choices ; probabilistic choices can be used to break symmetry in anonymous networks of indistinguishable, nameless processes. When designing election algorithms for the leader election problem (*LEP*) and spanning tree construction problem (*STP*) in anonymous networks, one has to consider the following issues : *relative information on the network size* and *termination detection* [1,2,3,5,6,8,9,11].

---
*Supported in part by C^3 (*COPARADIS Group*)

2 Preliminaries and results

2.1 Definitions

We consider here the standard model of *static asynchronous network*. This is a point-to-point communication network, described by an undirected *communication graph* $G = (V, E)$ where the set of nodes V represents processes of the network and the set of edges E represents bidirectional non-interfering communication channels (links) operating between neighbouring nodes ; $|V| = n$ and $|E| = m$. No common memory is shared by the processes. We confine ourselves only to *message-driven algorithms*, which do not have central controller and do not use time-outs, i.e. processes cannot access a global clock in order to decide what to do. In a transition, a process receives a message on one of its links, and changes state ; a transition may be probabilistic. We also assume throughout the processes and the communication subsystem to be error-free, and that the links operate in a FIFO-manner.

The course of any execution of an algorithm is determined by a *scheduler*, that chooses at each step the next message to be received, as a function of the current network state. An algorithm *process terminates* if in every execution all processes reach a special *halting state* ; this corresponds to an algorithm *with termination detection*. An algorithm *message terminates* if in every execution the network reaches a quiescent state where there are no pending messages on the links [11]. In message termination, the processes may ignore that the computation is halted ; this corresponds to an algorithm *without termination detection*. An algorithm has *error probability* ϵ if, for any scheduler and any input, the probability that the algorithm terminates with the right answer is at least $1 - \epsilon$.

2.2 Results

We address the problem of computing a function whose value depends on all the network processes, such as counting the number of nodes in the graph G, or solving *LEP* and *STP* for G. In [6], Itai and Rodeh showed that these problems can be solved on a ring by an algorithm which processor terminates (or *distributively terminates*) and always succeeds *if and only if* the ring size n is known up to a factor of two (see also [1,2,5] for improvements). It was also shown in [6] that it is possible to solve *LEP* in an anonymous network, *with* termination detection and with error probability ϵ *only if* an upper bound on the network size is known.

In [8,9,11] and in the present paper, all of these results were extended to arbitrary networks, while improving some of the bounds. In [11], Schieber and Snir presented efficient schemes of algorithms for *LEP* and *STP* ; in [9], Matias and Afek proposed more detailed schemes of three types of simple algorithms which efficiently solve *LEP*.

Given some $0 < \epsilon < 1$, let $r = 1/\epsilon$. On the assumption that the exact value of n is a priori unknown to any process of the network — and thus without termination detection —, the probabilistic solutions given in [11] require $O(m \log \log(nr) + n \log n)^1$ messages of size $O(\log n + \log r)$, for fixed error probability ϵ. The probabilistic solutions given in [9] require $O(m \log n \times r \log r)$ messages of size $O(\log r + \log \log n)$ on the same assumptions. In the case when n is known up to a factor of $K \geq 1$, that is $N < n \leq KN$, [9] achieves $O(m \times r \log(Kr) \log r)$ messages of size $O(\log n)$, in the worst case, with $N < n \leq 2N$. The time complexity in [9,11] is $O(D)$ and $O(n)$, respectively, where D is the diameter of the network.

Our algorithms are simpler than in [11] and slightly improve on those in [9,11] with respect to communication complexity. Compared to [11], we give overall solutions with improved bit complexity and less bit information per node ; the message complexity in [9] is also higher than ours. Moreover, to the best of the authors' knowledge, the present algorithms are very likely to be the first fully and

---
[1]Throughout the paper, log denotes the base two logarithm and ln the natural logarithm.

precisely specified probabilistic protocols to solve *LEP* and *STP* in nameless networks. On the other hand, our "time" complexity is slightly higher than in [9,11].

For the first algorithm $\mathcal{A}_1$, on the assumption that n is a priori unknown to the processes — and thus without termination detection —, the expected message complexity is $O(m \times \log\log(nr) + n \log n)$, each message of size $O(\log r + \log\log n)$, with probability $> 1 - \epsilon$.

Assuming n is known up to a factor of $K \geq 1$, that is $N < n \leq KN$, the second algorithm $\mathcal{A}_2$ requires — in the worst-case, and with termination detection —, $O(m + n \log n)$ messages, of size $O(\log n)$, with probability $> 1 - \epsilon$.

Whenever the exact value of n at least known to one process — and with no error —, the complexity of $\mathcal{A}_2$ shrinks to $O(m + n \log n)$ messages, with message size $O(\lg n)$

Algorithms $\mathcal{A}_1$ and $\mathcal{A}_2$ run in (virtual) time $O(D \times \log\log(nr))$. $\mathcal{A}_1$ and $\mathcal{A}_2$ also require $O(\log r + \log\log n)$ and $O(\log n)$ bits of state information per node, respectively.

In section 3 we fully describe the message terminating algorithm $\mathcal{A}_1$ which solves *LEP* and *STP* in arbitrary (connected) anonymous networks with fixed error probability ϵ when n is unknown. In section 4 we give the correctness proof of algorithm $\mathcal{A}_1$, followed by a brief analysis of the communication and time complexity in Section 5. Section 6 is devoted to the algorithm $\mathcal{A}_2$. Assuming n is known up to a factor of $K \geq 1$, $\mathcal{A}_2$ is a process terminating protocol which solves *LEP* and *STP* in general (connected) anonymous networks with probability $> 1 - \epsilon$. Description, correctness proof and complexity analysis of algorithm $\mathcal{A}_2$ are given in section 6. The Annex gives a full specification of algorithms $\mathcal{A}_1$ in *CSAP*, an asynchronous variant of *CSP*.

3 The Algorithm $\mathcal{A}_1$

For algorithm $\mathcal{A}_1$, we assume that no bound is known on the size n of the input graph G. In the sequel, we show how the same algorithm $\mathcal{A}_1$ can be modified in $\mathcal{A}_2$ to handle the simpler case when there are bounds on the size n of G.

3.1 High Level Description

At each point in the algorithm a rooted forest of G is maintained which is composed of subtrees building a spanning tree of G. The algorithm start by taking the spanning forest which consists of each one of the nodes as a subtree or *fragment* of size one. Upon termination, there is a single tree spanning the whole network, the root of which is the elected leader, with probability $> 1 - \epsilon$. At each node-process P_i, we maintain the variable id_i. Define id_i, the identity of process P_i, as follows :

Each process tosses a fair coin until a "head" occurs. This experiment is repeated k times. Estimate $\log n$ as j, the longest waiting time for head in any of the k trials. Let t_i be a function of j defined as the number of tosses until P_i tossed a head for the first time in any of the k trials. Finally denote this procedure of choosing t_i as **ESTIMATE** (k,t).

Each process P_i randomly selects a label s_i from some domain I of size d, where $d = O(r \log r)$ (recall $r = 1/\epsilon$; the value of d will be given in the analysis). These n initial random drawings in the range $[1, d]$ take place during the initialization procedure **INIT**.

Finally, let id_i, the identity of process P_i, be the ordered pair $id_i = (s_i, t_i)$ (see procedures **INIT** and **ESTIMATE** (k,t) in 3.3). Note that a lexicographic order on id_i is defined in the natural way.

Initially, each fragment consists of a single node-process which is its own root. Similarly, In the course of the algorithm, the *identification number* of a current fragment F_i is the identity of its root : id_i.

3.1.1 The Merging Process

In the course of the algorithm, every fragment F finds an *outgoing edge* (a, b) (a being an *outgoing son* of F). Eventually, either F gets absorbed or merged into the fragment on the other side of edge (a, b), becoming a subtree within bigger tree, or F captures the latter fragment, this according to Sollin's property [7,8], with probability $> 1 - \epsilon$.

However, in contrast to most distributed spanning tree algorithms which use the technique of *levels* to ensure a balanced growth of the current fragments (thus pseudo-synchronizing the algorithms), our combination of fragments proceeds fully asynchronous as follows :

(i) A given fragment F is candidate for merging *into* some other fragment G : the root of F sends a message of combination request denoted by **comb** to G. The combination process never works in the (usual) reverse direction.

(ii) Any request of F to get merged into G can only be initiated by the one privileged node in F which may send messages **comb**, but it may be accepted (or rejected) by the *first* node of G which receives the message **comb**.

(iii) Consider a chosen outgoing edge (a, b) in the network and suppose it is traversed by a first message **comb** from node a to node b. In our algorithms, rejection of that first message changes link (a, b) into the *virtually directed* link $(b \rightarrow a)$. Such a virtual link reversal works in such a way as to strictly ensure that the next message **comb** sent through this very link shall traverse in the reverse direction from node b to node a. This "flipping-the-edge" mechanism is completed here by updating the boolean array $PORT[\ ]$ at every node (see 3.2.1). Thus the combination process let the fragments grow possibly unbalanced and proceeds fully asynchronous. Yet, it turns out that the fragments do combine and grow properly *altogether* since the combination rules enforce any "worst merging case" to change into a somewhat "best merging case" after one or only very few execution stages.

3.1.2 The Combination Rules

Consider two fragments F and G with identification numbers id_i for the root of F and id_j for the root of G, respectively. Let (x, y) be an outgoing edge between F and G such that $x \in F$ and $y \in G$, and assume that process x owns in fragment F the privilege of emitting the message **comb** (see 3.1.4 for description of this property)

Suppose x sends y a message **comb** asking G to let F get merged into G. Owing to the *partial ordering* in the domain I, two main cases may occur :

Either $id_i \neq id_j$; the merging operation is thus performed (or aborted) as usual, according to the order relation between id_i and id_j : the process y may locally decide to send back either a rejection message **nok** or an acceptance message **ok** to the process x,

or $id_i = id_j$; in this second case, it is impossible to accept *a priori* the combination request of process x because of the possible risk of building a cycle. On the other hand, it is neither possible to simply reject *a priori* this (possibly fair) request.

In such a situation, two conditions may actually arise. Either both processes x and y belong to the same fragment ($F = G$), or not ($F \neq G$). Notice that, owing to their fully local information, no process can distinguish the first condition from the second. Let P_i be the *root* of fragment F. In order to break the symmetry,

- y sends back a message **equal** to x, which passes up the message **equal** to the root P_i.
- Now the behaviour of root P_i depends on the current value of its local variable *credit*, viz. the current number of remaining random changes of identities with which P_i is still credited at this point in the algorithm (see 3.1.3).

    If *credit* > 0, then the procedure **RANDRAW** is called, which makes P_i randomly select a new identity id'_i such that $id'_i > id_i$ (s'_i is randomly chosen in the range $[s_i + 1, s_i + d]$: see 3.1.3 and 3.3). Thereafter, the messages **newroot** update each process' identity within F to the new value id'_i chosen in **RANDRAW** . *credit* is also decremented by 1.

    If *credit* $= 0$, there is no way for P_i to randomly select a new identity since the root is no more credited with any random draw. Thus P_i may regard **equal** as a message received (via x) from some process (y) which actually belongs to its own fragment F.

- P_i sends y a message **cousin** via the same path used by the message **equal** to traverse F. Note that a linking method must be used to keep track of this traversal. The method is fully described in [8] (see also 3.2.1).

- The receipt of a message **cousin** makes it possible for x to clear up the ambiguity of **equal**. So, x locks its port labeled y, and updates its variables $PORT$ and *open*. Then either x keeps on sending messages **comb** to its neighbours as long as there still remains one free neighbour (with *open* $\neq$ false), or x grants its father the privilege of emitting the message **comb** if none of its neighbours is free.

3.1.3 The Notion of Credit

The notion of credit and the integer variable *credit* arise here as a specific *control parameter* which is fixed in advance to *tune the precision* of the algorithm. At each one root and at each point in the algorithm, the local counter variable *credit* is the current number of remaining random drawings the root of the current fragment is granted to randomly choose a new identity if necessary, in using procedure **RANDRAW** . As to the procedure **RANDRAW** , it strictly increases the current random value of the identity id_i of the fragment F_i : viz. a number a being randomly drawn in the range $[1, d]$, where $d = O(r \log r)$, s_i is set to $s_i + a$. We thus ensure that the new identity (s'_i, t_i) keeps the same estimate t_i (see 3.3 for a full specification of **INIT** and **RANDRAW**). Initially, every process is credited with $\lfloor \log r \rfloor$ possible random changes of identity.

Besides, *credit* breaks the tie between the identities of two processes x and y. It makes it possible to determine whether x and y are two cousins within the same fragment F, or actually belong to distinct fragments F and G with the same identifier. In the case when *credit* $= 0$ at the root of the fragment F of x, which is supposed to redraw a new identity, then F and G are regarded as a unique fragment in which x and y would be cousins. Next, the tie is broken by the use of messages **equal** and **ambiguity** (see 3.2.1).

3.1.4 Messages comb and Termination

Within each one fragment, only one single process at a time is allowed to emitt **comb** messages. When two fragments merge, the process which asked for merging, and was allowed to send messages **comb**, looses this privilege upon receipt of message **ok**. The receipt of either **nok** or **cousin**, causes the port used by the message **comb** to get locked with the logical array $PORT[\ ]$ (see 3.2.1). If no more port is available to messages **comb**, the process with the **comb** emission privilege grants its father the privilege, which in turn locks the corresponding port.

Suppose the privilege owner is the root of a fragment. If none of its ports is available any more, then the algorithm message terminates.

3.2 Notations

3.2.1 Messages, Variables and Arrays

Messages are composed of four records and denoted by $< \alpha, \beta, \gamma, \delta >$.

The first record is the identification number of the sending fragment.

The second record is an element of the set {**comb, ok, nok, equal, cousin, merge, newroot, end**}. This record is the message *stricto sensu*.

The third record is a boolean variable which possibly sets the logical state of the consirered port.

The last record is a local logical time (see 3.1.2).
The algorithm uses three local arrays, $PORT$, SON, and $TABLE$, with the following definitions :

1. $PORT[\ ]$ is a logical array with the ports' labels of the current process as indices. Note that each process is assumed to locally distinguish between its different ports. $PORT[\ ]$ makes it possible for the current process to update the labels of the ports through which messages **comb** may still be sent : $PORT[i] = $ true or false, either if the port labeled i is still available, or not, respectively.

2. $SON[\ ]$ is a boolean array with the ports labels of the sons of the current process as indices. $SON[\ ]$ makes it possible for the current process to know its sons : $(SON[i] = 1) \iff$ (The process connected to port i is a son of the current process).

3. $TABLE[\ ]$ is a list data structure with the values of the local logical clock as indices (see 3.1.1, and [8]) : $(TABLE[t] = (j, \tau')) \iff$ (At local time t, the process received a message via its port j, and the process connected to this port j had local time t').

Variables are the following :

- *id* is a local, integer variable the value of which is the identity of the current process.

- *root* is the identity of the root of the fragment to which the current process belongs.

- *father* is a local integer variable with value the label of the port which leads to the current process' father within the subtree.

- τ is a local integer variable with value the last local time computed so far.

- *credit* is a local integer variable with value the number of random drawings with which the current root of a fragment is still credited.

- *req* is a local boolean variable which either allows a current root to send a message **comb** (*req* = 1), or do not (*req* = 0).

- *open* is a local logical variable which determines whether the current process still has free **neighbours** or not (i.e. neighbours which may possibly send back **ok** upon receipt of **comb**).

- *ambiguity* is a local logical variable which gives knowledge that the current process received back **equal** in answer to **comb**, and that some decision of its fragment's root is awaited.

3.2.2 Specification in $CSAP$

$CSAP$ is the acronym for *Communicating Sequential Asynchronous Processes*. The syntax of $CSAP$ is very close to the syntax of CSP. However, the modifications entail important repercussions with respect to the semantics of the language. Thus, the generation of a message is a non-blocking primitive. Denote $P_i!! < >$ the emission of message $< >$ via the port labeled i in a non-blocking way. In such a case, i may be regarded as the identity of the process to which the message is transmitted. This also entails the presence of a buffer at each one process for incoming messages. The corresponding queue must be bounded, which is the case in the present model. Similarly, denote $P_j?? < >$ the receipt of message $< >$ through the port labeled j. As an example of such notations, "$\forall x \in SON, P_x!! < >$" means that message $< >$ is sent from the current process to all its sons via the ports labeled with the indices of the array SON.

3.3 Procedures

The full specification of algorithm $\mathcal{A}_1$ is given in the Annex.

1. **The Procedure $ESTIMATE$**

   Suppose one starts the algorithm by performing the following local probabilistic experiment to estimate $\log n$. The experiment is set so that with high probability at least one process deduces an estimate of a value $\leq \log n$, and no estimate is much larger than $\log n$. Each process P_i uses this estimate to randomly select its label s_i in a domain $\boldsymbol{I}$ whose size d depends on the largest estimate. Thus, in the procedure $ESTIMATE$, each process tosses a fair coin until a "head" occurs. This experiment is repeated k times, and $\log n$ is estimated as j, the longest waiting time for head in any of the k trials.

   Procedure $ESTIMATE\ (k, \boldsymbol{var}\ : max)\ ::$
   $\quad max := 0\ ;$
   $\quad *[k > 0 \rightarrow t := 0\ ;\ x := 0\ ;$
   $\quad\quad *[x = 0 \rightarrow t := t + 1\ ;\ x := \mathbf{random}(\{0, 1\})\ ]\ ;$
   $\quad\quad [t > max \rightarrow max := t]\ ;\ k := k - 1\ ;$
   $\quad].$

2. **The Procedure $INIT$**

   In the case where a process P_i starts executing its algorithm upon receipt of a message sent by process P_j, P_i rightaway considers itself as a node in F_j. Thus, P_i gains its identity and wakes up.
   The variable $TABLE$ is a recursive list type with four records. $id, root, father, credit, \tau$ are integer variables. $req, open, ambiguity$ are boolean variables. SON and $PORT$ are arrays of booleans. $NEIGHB$ is a variable which denotes the set of the neighbours of the current process. id is an ordered pair of integers.

   Procedure $INIT\ (id, \tau, root, father, credit, PORT, req, SON, open, ambiguity)\ ::$
   $\quad$/* performed upon waking up or reception of first message — whichever comes first */
   $\quad[s := \mathbf{random}([1..d])\ ;\ ESTIMATE\ (k, t)\ ;\ id := (s, t)\ ;\ root := id\ ;\ father := id\ ;\ \tau := 0\ ;\ c\tau$
   $\quad SON := \emptyset\ ;\ open := \text{true}\ ;\ \forall x \in NEIGHB, PORT[x] := \text{true}\ ;\ ambiguity := 0].$
   $\quad\quad\quad\quad\quad\quad\quad\quad\quad\quad\quad\quad\quad\quad\quad\quad$/* $d = O(r \log r)$, with $r = 1/\epsilon$ for fixed ϵ */

3. **The procedure $RANDRAW$**

   Procedure $RANDRAW$ $(var : id)$ ::
   $[s := id(1) ; a := \mathbf{random}([1..d]) ; s := s + a ; id(1) := s].$ /* $id(1)$ is the first integer within the ordered pair (s,t) */

4. **The Procedure $SELECT$**

   Procedure $SELECT$ $(var : x, PORT)$ ::
   $x := 0$;
   $\quad [PORT[1] = \text{true} \to x := 1 ;$
   $\quad \blacksquare$
   $\quad PORT[2] = \text{true} \to x := 2 ;$
   $\quad \blacksquare$
   $\quad \ldots\ldots\ldots$
   $\quad \blacksquare$
   $\quad PORT[n] = \text{true} \to x := n ;$
   $\quad].$

5. **The Procedure $UPDT$**

   Procedure $UPDT$ $(y, var : PORT, open)$ ::
   $[PORT[y] := \text{false} ;$ /* $PORT$ is a boolean array which represents
   $open := \vee_k PORT[k] ;$ the state of the port. y is the label of the port
   $].$ through which the current process received a message **nok** or **cousin** */

6. **The procedure $TERM$**

   Procedure $TERM$ (SON) ::
   $[\forall x \in SON, P_x!! < \_, \mathbf{end}, \_, \_ > ; \mathbf{STOP}.].$

4 Correctness

The correctness proof of the algorithm consists of two parts. First, at each point in the algorithm (termination included) a *rooted spanning forest* of G is maintained. Second, the algorithm eventually *message terminates* with a rooted spanning tree with probability $> 1 - \epsilon$.

Owing to the property of strict order relation existing in G, which is a sub-semilattice, we have the following

Lemma 4.1 *In any execution and at each point in the algorithm, the set of current fragments constitute a rooted spanning forest of G, which is an invariant for the algorithm.*

From high level description, and by Lemma 4.1, we can conclude.

Theorem 4.1 *In any execution, the algorithm eventually message terminates. Upon termination, all the fragments constitute a rooted spanning forest of G.*

Let ϵ be the error probability and let $d \geq \lceil 2/\epsilon \rceil$. Recall that the labels are randomly selected from a domain of size d in the procedure $RANDRAW$.

Theorem 4.2 *When the algorithm message terminates, the probability that a spanning tree is found is at least $1 - \epsilon$.*

Proof Suppose that the algorithm terminates with more than one fragment left. By Theorem 4.1, we have that all the roots have the same identity and that $credit = 0$ at each root. The probability that all drawings at $credit = j > 0$ selected the same label (given that more than one drawing occurred) is bounded from above by $1/d^j$. The probability that the algorithm fails at the last value of $credit$ is the probability that all drawings at $credit = 1$ selected the same label (given that more than one drawing occurred). It is whence bounded from above by $\sum_j 1/d^j < 1/(d-1) < \epsilon$. □

5 Analysis

The *worst-case message complexity* of an algorithm (for a given input size) is the maximum over all networks of the given size and over all schedulers, of the largest number of messages sent in any execution of the algorithm. The expected message complexity (for a given input size) is the maximum over all networks of the given size and over all schedulers, of the expected number of messages sent in an execution with this scheduler in the network ; the expectation is over the random choices of the algorithm.

In the following, denote $T = max_i\{t_i\}$, and $M = max_i\{id_i\}$. We first make sure that at least one of the estimates t_i is $\geq \log n$ with probability $> 1 - \epsilon$, and then prove that $T < \log(nr)$ with probability $> 1 - \epsilon$.

Lemma 5.1 *Let t_i be one of the estimates.* $(\exists i \in [1, n])\ Pr\{t_i \geq \log n\} > 1 - \epsilon$.

Proof Assume that for all $i \in [1, n]$, we have $2^{t_i-1} < n \leq 2^{t_i}$. Let $t_i = t$. The probability that the waiting time for head is $\geq t$ is $2^{-t} > 1/2n$. Now the probability that no waiting time of t or more occurred in kn trials at all nodes is $(1 - 2^{-t})^{kn} < (1 - 1/2n)^{kn} \leq e^{-k/2}$. Thus, it is sufficient to take $k = 2\ln(1/\epsilon)$, and for at least one $i \in [1, n]$, $Pr\{t_i \geq \log n\} > 1 - \epsilon$. □

Lemma 5.2 *Let t_i be one of the estimates.* $Pr\{(\exists i \in [1, n])\ t_i < \log(nr)\} > 1 - \epsilon$.

Proof For fixed $i \in [1, n]$, $Pr\{t_i \geq \log(nr)\} = 2^{-\log(nr)} = 1/nr$. Therefore, $Pr\{(\exists i \in [1, n])\ t_i \geq \log(nr)\} \leq 1/r = \epsilon$. □

Corollary 5.1 $(\forall \epsilon > 0)\ Pr\{\log n \leq T < \log(nr)\} > 1 - \epsilon$.

Proposition 5.1 *Let $T = max_i\{t_i\}$, $E[T] = O(\log(nr))$.*

Proof T is the highest estimate computed by a process, viz. T is the maximum of kn waiting times in independent Bernoulli sequences of trials with probability $1/2$ of success. Thus, $E[T] = \sum_j Pr\{T > j\}$, and
$$E[T] < \sum_{j < \log n} Pr\{T > j\} + \sum_{j \geq \log n} Pr\{T > j\}.$$
Since $(\forall \epsilon > 0)\ Pr\{T \geq \log(nr)\} < \epsilon$,
$$(\forall \epsilon > 0)\ E[T] < \sum_{j < \log n} Pr\{T > j\} + \sum_{j = \log n}^{\log(nr)} Pr\{T > j\} + \epsilon,$$
and $E[T] < (\log n)Pr\{T > \log n\} + (\log n + 1)Pr\{T \geq \log n\} + \epsilon$. Hence, $E[T] = O(\log n + \log r)$, and the expected value of T is $O(\log(nr))$. □

The main idea in the procedure **_ESTIMATE_** (see 3.3) is that there is one process with a t_i larger than the others. The following claim and lemma prove the existence, with high probability, of such a *unique* process.

Definition 5.1 *Denote as* **candidate** *a process P_i such that $t_i \geq \log n - t$, for some parameter t to be fixed later. The function $\varphi(t) = \log n - t$ is a* **threshold** *function which identifies processes with large identities, denoted as candidates, from all other processes. The function φ is an integer-valued, nonnegative, monotone nondecreasing function of t.*

Claim 5.1 *Let $M = max_i\{id_i\}$. If the number of candidates is $\leq C$, and s_i was randomly selected from a domain of size Cr, then M is unique, with probability $> 1 - \epsilon$.*

Sketch of Proof Assume that all candidates have the same t_i. If such is not the case, then the definition of a *candidate* should be refined as follows : the candidates are the only processes whose estimate is T. C drawings are completed from a domain I of size $d = Cr$. Now, let p be the probability that in C drawings from I the largest s_i (among those which were drawn) is drawn only once. We know that $p > 1 - C/d$ (see [9, Claim 12] for a detailed proof). Hence, we have that $max_i\{s_i\}$ (over the P_i which are candidates), is unique with probability $> 1 - C/Cr = 1 - \epsilon$. □

Lemma 5.3 *There is a unique process P_i with identity $id_i = M$, with probability $> 1 - \epsilon$.*

Proof The proof of the lemma consists of two parts.

- We first show that (a) there are $\Theta(\log r)$ candidates, with probability $> 1 - \epsilon$. This ensures that, with high enough probability, there is at least one candidate but not too many.

- Second we show that (b) with probability $> 1 - \epsilon$, $max_i\{s_i\}$, where i is over all the candidates, is randomly selected in **_INIT_** by only one among the candidates. This ((a) and (b)) is sufficient to show that there exists a unique process with identity M, with probability $> 1 - \epsilon$.

(a) is proven in using Tchernov's bounds [10]. Let X be the number of candidates. The probability of a process to be a candidate in any of the k trials is $2^{t-\log n} = 2^t/n$ (i.e. the probability to obtain $(\log n - t)$ tails in $(\log n - t)$ tosses. Since, the number of trials is k at each one of the n processes, $c = E[X] = k2^t$.
Now if we let $c = 12\ln(2r)$, by Tchernov's inequalities [10, p. 121],

$$Pr\{c/2 \leq X \leq 3c/2\} > 1 - [exp(-c/12) + exp(-c/8)] > 1 - 2/2r > 1 - \epsilon,$$

and the number of candidates X is $\Theta(\log r)$ with probability $> 1 - \epsilon$.

(b) is a straightforward consequence of Claim 5.1 : Let $C = 18\ln(2r)$, then by Claim 5.1, we have that there exists one unique process P_i with $id_i = M$ (over all the candidates), with probability $> 1 - \epsilon$. This concludes the proof of the lemma □

Remarks
- If we let $C = 18\ln r$, then by the tight inequalities of Tchernov, we know at the same time that
    - the number of candidates is $\Theta(\log r)$ with probability $> 1 - \epsilon$,
    - if the number of candidates is $\leq C$, then M is unique with probability $> 1 - \epsilon$.
- As a direct consequence of Lemma 5.3, we are now able to compute the size d of domain I : since the number of candidates is $\leq C = 18\ln r$, d needs to be $2Cr = 36r\ln r = O(r\log r)$.

- Our average number of candidates is larger than in [9] : $c = k2^t$. This is due to the fact that k initial tosses are experimented at each one process in our procedure ***ESTIMATE*** , which improves the number of processes with estimate closer to $\log n$.

We now turn out to compute the (virtual) running time and expected message complexity of the algorithm, and the size of messages, with probability $> 1 - \epsilon$.

Theorem 5.1 *A rooted spanning tree can be built with probability $> 1 - \epsilon$ in expected message complexity $O(m \log \log(nr) + n \log n)$, with message size $O(\log r + \log \log n)$, and expected (virtual) running time $O(D \times \log \log(nr))$ (where D is the diameter of the network).*

Sketch of Proof Given $E[T]$, we can compute the expected complexity of the algorithm as a function of $E[T]$ and $M = max_i\{id_i\}$. All processes are waken up with $O(m)$ messages in time $O(n)$. The k trials have to be triggered by messages, so that the expected number of messages per node is $O(T)$. Hence, ***ESTIMATE*** requires an expected number of messages of $O(m + n \log(nr))$ in expected running time $O(n + \log(nr))$.

The highest message cost is the number of messages **newroot**, since all identities of processes within *one* of two merging fragments are updated upon receipt of messages **newroot**. Indeed, the total number of other messages is $O(m + n \log(nr))$ on the average. Since the number of candidates is $\Theta(\log r)$ with probability $> 1 - \epsilon$, the upper bound on the number of times a fragment is updated is $\varphi(T \log r)$. Thus, the number of messages **newroot** used in the algorithm is $O(m \times \varphi(T \log r))$. It follows that the total number of messages required by the algorithm is $O(n \log n + m \times \varphi(T \log r))$. Similarly, the (virtual) running time is bounded by $O(D \times \varphi(T \log r))$. Assuming that the threshold function φ is concave and by Proposition 5.1, we have $E[\varphi(T \log r)] \leq \varphi(E[T \log r]) = \varphi(\log r \times E[T]) = O(\varphi(\log r \times \log(nr)))$. Hence, the expected message complexity is $O(n \log n + m \times \varphi(\log r \times \log(nr)))$, and the expected (virtual) running time is $O(D \times \varphi(\log r \times \log(nr)))$.

Using the threshold function $\varphi(x) = \lceil \log x \rceil$ we obtain the results given in Theorem 5.1. The expected number of messages used by the algorithm is $O(n \log n + m \log(\log r \times \log(nr))) = O(n \log n + m \log \log(nr))$, with probability $> 1 - \epsilon$. The expected (virtual) running time is $O(D \times \log \log(nr))$, with error probability ϵ. The maximum number of bits per message is the number of bits in T plus the maximum number of bits in s_i (selected from $[1, d]$, where $d = r \log r$). Therefore, the size of messages is bounded by $O(\log r + \log \log n)$.
Algorithm $\mathcal{A}_1$ requires $O(\log \log n + \log r)$ bits of state information per node, which meets the result in [9]. □

6 The Algorithm $\mathcal{A}_2$

We now address the case when the network is bounded. Assume n is known up to a factor of K : viz. $N < n \leq KN$, with $K \geq 1$. Then Algorithm $\mathcal{A}_2$ is process terminating and solves *LEP* and *STP* with fixed error probability ϵ if $K > 1$, and with no error if n is exactly known.

To handle the case where bounds are known on the size n of the network, algorithm $\mathcal{A}_2$ is designed with slight modifications of $\mathcal{A}_1$, mainly within the procedure ***INIT*** .

The procedure ***ESTIMATE*** is not changed in algorithm $\mathcal{A}_2$. By contrast, the procedure ***INIT*** takes all the additional information about n into account. In words, ***INIT*** uses the new parameters μ and φ (computed in the analysis, in 6.2) to ensure that each process can check if it is a candidate or not. Now, no process but each one *candidate* process can randomly choose its label, and can thereafter trigger the computation by sending a message **comb** to a selected neighbour. Upon receipt of a message **comb**, any noncandidate process gains the identity of the emitting process and is "merged" into the corresponding fragment.

6.1 The Modified Procedure *INIT*

The variable *size* maintained at each process P_i is merely a local integer-valued variable, which counts the number of nodes in the current fragment F_i to which P_i belongs.

Procedure **INIT** $(id, \tau, root, father, credit, PORT, req, SON, open, ambiguity, size) ::$
[**ESTIMATE** (k,t) /* performed upon waking up or upon reception
                                                                of first message — whichever comes first */

$[(d,t) > \mu \rightarrow \forall x \in NEIGHB, P_x!! < \_, \textbf{end}, \_, \_ >$ /* the algorithm fails with probability $< \epsilon$ */
∎

$(d,t) \leq \mu \rightarrow t := t - \varphi$;
[$t > \varphi \rightarrow s :=$ **random**$([1..d])$; $id := (s,t)$; $size := 1$; $root := id$; $father := id$; $\tau := 0$;
$req :=$ false ; $SON := \emptyset$; $open :=$ true ; $\forall x \in NEIGHB, PORT[x] :=$ true ; $credit := \lfloor K \log r \rfloor$;
]
].
/* $K = n/N, d = O(Kr \log r), \mu = (r \log r) \times \log(12Kr \log r)$, and $\varphi = \log N - \log(12 \ln r)$ */

6.2 Analysis

Lemma 6.1 *Let $N < n \leq KN$. The number of candidates is, for some setting of t, $\Theta(K \log r)$ with probability $> 1 - \epsilon$.*

Sketch of Proof Set $t = \log c$, where $c = (n/N)12 \ln r$. Thus the threshold for the candidates is now $\varphi = \log n - \log c = \log N - \log(12 \ln r)$. Similarly to the proof of Lemma 5.3, for $K \geq n/N$, the number of candidates is $c = \Theta(K \log r)$, with probability $> 1 - \epsilon$. □

Let $t'_i = t_i - \varphi$, and $T' = max_i\{t'_i\}$.

Claim 6.1 $T' = O(\log(Kr))$ *with probability* $> 1 - \epsilon$.

Proof $t'_i = t_i - \varphi = t_i - (\log N - \log(12 \ln r))$.
Therefore, $T' = \log(nr) - \log N + \log(12 \ln r) = O(\log(Kr))$. □

Whenever we have this additional information about n, we can use the initial knowledge on N, K, and r to improve the complexity. The threshold $\varphi = \log N - \log(12 \ln r)$ can also be given in advance to all processes. Moreover, the identities will be reduced by φ to yield smaller size messages. By Claim 6.1, for $K = n/N$, we have that $M = max_i\{id_i\} < (r \log r) \times (\log(12Kr \ln r)) = \mu$.

The value of μ can thus be also given to all processes when the algorithm starts, together with N, K, r and φ. All of this information is used to modify the procedures **INIT** in 6.1.

We can now turn out to compute the complexity of algorithm $\mathcal{A}_2$ whenever n is exactly known, and in the cases where $N < n \leq 2N$ and $N < n \leq KN$, with $K > 2$.

Lemma 6.2 *Suppose $N < n \leq K$, with $K \geq 2$. (a) If there are $(\ell - 1)$ current fragments whose size is larger than that of some fragment F, then the latter is bounded by (n/ℓ). (b) The total number of nodes in all fragments such as F is at most $O(n \log n)$.*

We now give the computation of the size d of the domain I whenever the exact value of n is known to the processes or in the case where $N < n \leq KN$, with $K \geq 2$.

In the procedure $INIT$, an initial sequence of random drawings is completed. Let $d = d(r)$ be the size of the domain I from wich the labels are drawn, and where $r = 1/\epsilon$.

1. Assume the exact value of n, the size of the network, is known to at least one process. We can compute the value of d which is sufficient to draw n distinct identities with high probability. Let id_i and id_j be the identities of the two processes P_i and P_j. Let $p = Pr\{id_i \neq id_j\}$. Notice that
$$p = \frac{d(d-1)\cdots(d-n+1)}{d^n}.$$
For all real $x \leq 1/2$, $1 - x > e^{-2x}$. Thus, assuming that $n/d \leq 1/2$, or $d \geq 2n$,
$$p = \prod_{i=0}^{n-1}(1 - i/d) > \prod_{i=0}^{n-1} e^{-2i/d} > e^{-n^2/d}.$$
Since $(\forall\, 0 < \epsilon < 1)\, \ln(1-\epsilon) < -\epsilon$, we have that $p > 1 - \epsilon$ when $d \geq n^2/\epsilon$. Hence, it is sufficient to randomly draw the n initial identities from a domain I of size $d = O(n^2 r)$, for fixed ϵ. In that case, the identities randomly selected in $INIT$ will all be distinct with probability $> 1 - \epsilon$.

2. Assume now n, the size of the network, is only known up to a factor of K, i.e. $N < n \leq KN$. Suppose also that *only the candidates* can randomly select their own labels. Using the fact that $C = \Theta(K\log r)$ (where $K = n/N$, with probability $> 1 - \epsilon$), we can compute the value of d such that two identities are distinct with probability $> 1 - \epsilon$. By Claim 5.1, we know that the size of I can be $d = Cr$. Therefore, it is sufficient to take $d = O(Kr\log r)$ to have all the identities randomly selected in $INIT$ be distinct with probability $> 1 - \epsilon$.

Following the above calculations, we can conclude :

Lemma 6.3 *For fixed error probability ϵ, (i) If n is known in the range $N < n \leq KN$, with $K \geq 2$, then the size of domain I is $d = O(Kr\log r)$. (ii) If the exact value of n is known, then the size of domain I is $d = O(n^2 r)$.*

By Lemmas 6.1 and 6.2, the number of messages is at most $O(Km\log r + n\log n)$. However, this is a very pessimistic upper bound. The main point is that we can use our knowledge on the size of fragments at any point in the algorithm and the upper bound on n to update processes with only a *constant number of messages per link*. This shrinks the message complexity down to $O(m + n\log n)$

The algorithm $\mathcal{A}_2$ solves *LEP* and *STP* with termination detection and with no error provided the variable *credit* is not used as defined for $\mathcal{A}_1$ — or simply removed. Indeed, the (*Las Vegas*) algorithm may be restarted repeatedly till some process eventually learns that its fragment's size is n, and thus spans the whole network. The *process termination* is then completed *with no error*, and with $O(m + n\log n)$ messages each of size $O(\log n)$. This yields the result :

Theorem 6.1 *If the exact value of n is known to at least one process, then $\mathcal{A}_2$ solves LEP and STP with termination detection and with no error. The worst-case message complexity is $O(m + n\log n)$ and each message is of size $O(\log n)$.*

Note that removing *credit* is not really necessary, since one can set the initial value of the variable to a sufficiently large fixed value, $credit = \Omega(n^2)$ for example. The algorithm shall be rerunned till $size = n$, and shall thus eventually succeed.

Theorem 6.2 *If n is known in the range $N < n \leq 2N$, then $\mathcal{A}_2$ solves LEP and STP with probability $> 1 - \epsilon$ and with termination detection. The worst-case message complexity is $O(m + n \log n)$ and each message is of size $O(\log n)$, with fixed error probability ϵ.*

Sketch of Proof If algorithm $\mathcal{A}_2$ is executed with $\epsilon = 1/2$, the labels are randomly selected from a domain of size $d = 2K \log 2 = 2n/N$.

When a fragment F_i merges with another fragment, the new root P_i knows its new fragment's size. If $size \leq N$, then P_i knows for sure that it is not the leader (possibly not yet). If $size > N$, then P_i is still not necessarily the leader, since F_i may be surrounded with other fragments with the same identity. It is also possible that there exists a fragment with larger identity elsewhere in the network. However, F_i can enlarge till either reaching a larger identification number, or till it spans the whole network. In the latter case, F_i wins the game.

With probability $> 1 - \epsilon$, the algorithm eventually process terminates with a spanning tree. Altogether, a *constant* number of messages **newroot** per link is required, and therefore, the worst-case message complexity is $O(m + n \log n)$. The size of messages is $O(\log n)$, for fixed ϵ. □

Theorem 6.3 *If n is known in the range $N < n \leq KN$, with $K > 2$, then $\mathcal{A}_2$ solves LEP and STP with probability $> 1 - \epsilon$ and with termination detection. The worst-case message complexity is $O(m + n \log n)$ and each message is of size $O(\log n)$, with fixed error probability ϵ.*

Sketch of Proof The problem with the case $K > 2$ is that there might exist several fragments with the same identification number, all of size $> N$. However, there are at most K such fragments. Thus, in the same conditions that in the previous case $K = 2$, we are guaranteed with probability $> 1 - \epsilon$ that the labels are all distinct after at most $K/2$ calls of the procedure **RANDRAW**. Hence, a fragment can be surrounded with similar fragments only with probability $< \epsilon$, and if there is a unique leader the algorithm will eventually process terminate.

As to the (worst-case) message complexity, it remains the same as in the case $K = 2$, up to a constant factor, with probability $> 1 - \epsilon$. The size of each message is again $O(\log n)$, for fixed ϵ. Algorithm $\mathcal{A}_2$ requires $O(\log \log n)$ bits of state information per node, which again meets the result in [9]. □

Remark The open problems raised in [11] still remain open. In particular, can one compute spanning trees in anonymous networks of unbounded size with $O(m + n \log n)$ messages, i.e. with a *constant* number of messages per link ? Or, conversely, is it possible to give a lower bound on the number of messages per link ?

References

[1] **K. ABRAHAMSON, A. ADLER, L. HIGHAM and D. KIRKPATRICK**, Probabilistic solitude verification on a ring, *Proc. of the 5th ACM Symp. on Principles of Distributed Computing*, 161-173, Calgary, August 1986.

[2] **K. ABRAHAMSON, A. ADLER, L. HIGHAM and D. KIRKPATRICK**, Randomized function evaluation on a ring, *Distributed Computing* **3**, 107-119, Springer-Verlag, 1989.

[3] **Y. AFEK, M. SAKS**, Detecting global termination conditions in the face of uncertainty, *Proc. of the 6th ACM Symp. on Principles of Distributed Computing*, 109-124, Vancouver, August 1987.

[4] **D. ANGLUIN**, Local and global properties in networks of processors, *Proc. 12th Annual ACM Symposium on Theory of Computing, 82-93, Los Angeles*, May 1980.

[5] **H. ATTIYA, M. SNIR, M. K. WARMUTH**, Computing on an Anonymous Ring, *J. ACM, Vol. 35, No 4, 845-875*, October 1988.

[6] **A. ITAI, M. RODEH**, Symmetry breaking in distributive networks, *Proc. of the 22nd IEEE Symp. on the Foundation of Computer Science, 150-158, Nashville*, October 1981.

[7] **I. LAVALLÉE, C. LAVAULT**, Yet another distributed election and spanning tree algorithm, *R.R. INRIA No 1024*, April 1989.

[8] **I. LAVALLÉE, C. LAVAULT**, Efficient Routing Protocols in Nameless Networks, *R.R. INRIA No 1254*, June 1990.

[9] **Y. MATIAS, Y. AFEK**, Simple and Efficient Election Algorithms for Anonymous Network, *Proc. of the 3rd International Workshop on Distributed Algorithms, 183-194, Nice, LNCS 392, Springer-Verlag*, September 1989.

[10] **A. PAPOULIS**, *Probabilities, Random variables, and Stochastic Processes*, McGraw-Hill, 2nd Edition, 1984.

[11] **B. SCHIEBER, M. SNIR**, Calling Names on Nameless Networks, *Proc. of the 8th ACM Symp. on Principles of Distributed Computing, 319-328, Edmonton*, August 1989.

7 Annex

Specification of Algorithm $\mathcal{A}_1$

Proc id::
INIT $(id, \tau, root, father, credit, PORT, req, SON, open, ambiguity)$;
$*[(root = id \wedge req = false \wedge ambiguity = 0) \rightarrow$
  $[open = false \rightarrow \textbf{TERM}\ (SON)\ ;\ \textbf{SELECT}\ (x, PORT[\ ])\ ;$
    $\tau := \tau + 1\ ;\ req := true\ ;\ P_x!! < id, \textbf{comb}, \_, \tau >$
  $]$
∎
$P_y?? < \alpha, \beta, \gamma, \delta > \rightarrow$
  $[\beta = \textbf{end} \rightarrow \textbf{TERM}\ (SON)$
  ∎
  $\beta = \textbf{comb} \wedge y = father \rightarrow$
    $[open = true \rightarrow \textbf{SELECT}\ (x, PORT[\ ])\ ;\ P_x!! < \alpha, \beta, \gamma, \delta >$
    ∎
    $open = false \rightarrow P_y!! < \_, \textbf{nok}, open, \_ >$
    $]$
  ∎
  $\beta = \textbf{comb} \wedge y \neq father \rightarrow$
    $[\alpha = root \rightarrow P_y!! < \_, \textbf{equal}, \_, \_ >$
    ∎
    $\alpha < root \rightarrow P_y!! < root, \textbf{ok}, \_, \_ >\ ;\ SON := SON \cup \{y\}$
    ∎
    $\alpha > root \rightarrow P_y!! < \_, \textbf{nok}, \_, \_ >$
    $]$

■
$\beta = \mathbf{ok} \rightarrow$
   $[y \in SON \rightarrow SON := \{SON - \{y\}\} \cup \{father\}$;
      $[root = id \rightarrow$
         $root := \alpha$; $\mathbf{UPDT}\ (y, PORT[\ ], open)$; $P_y!! < \_, \mathbf{merge}, open, \_ >$;
         $\forall x \in SON,\ P_x!! < \alpha, \mathbf{newroot}, \_, \_ >$
         ■
         $root \neq id \rightarrow root := \alpha$; $P_{father}!! < \alpha, \beta, \_, \_ >$] ; $father := y$
   ]

■
$\beta = \mathbf{nok} \rightarrow$
   $[y = false \rightarrow \mathbf{UPDT}\ (y, PORT[\ ], open)]$;
   $[root \neq id \rightarrow$
      $[open = true \rightarrow \mathbf{SELECT}\ (x, PORT[\ ])$; $P_x!! < root, \mathbf{comb}, \_, \_ >$
      ■
      $open = false \rightarrow P_{father}!! < \alpha, \beta, \_, \_ >$
   ]
   ■
   $root = id \rightarrow req := false$
   ]

■
$\beta = \mathbf{equal} \rightarrow$
   $[y \in SON \rightarrow [id = root \rightarrow$
         $[credit > 0 \rightarrow \mathbf{RANDRAW}\ (id)$;
            $\forall x \in SON,\ P_x!! < id, \mathbf{newroot}, \_, \_ >$; $root := id$; $credit := credit - 1$
         ■
         $credit = 0 \rightarrow P_y!! < \_, \mathbf{cousin}, \_, \delta >]$
      ■
      $id \neq root \rightarrow \tau := \tau + 1$; $TABLE[\tau] = (y, e)$; $P_{father}!! < \alpha, \beta, \gamma, \delta >$
   ]
   ■
   $y \notin SON \rightarrow ambiguity := 1$; $\tau := \tau + 1$; $TABLE[\tau] := (y, \_)$; $P_{father}!! < \alpha, \beta, \gamma, \delta >$
   ]

■
$\beta = \mathbf{cousin} \rightarrow [ambiguity = 1 \rightarrow$
      $(z, \_) := TABLE[e]$; $\mathbf{UPDT}\ (x, PORT[\ ], open)$;
      $[open = false \rightarrow P_{father}!! < \_, \mathbf{nok}, false, \_ >$
      ■
      $open \neq false \rightarrow \mathbf{SELECT}\ (x, PORT[\ ])$; $P_x!! < root, \mathbf{comb}, \_, \_ >]$; $ambiguity := 0$
   ■
   $ambiguity = 0 \rightarrow (k, l) := TABLE[d]$; $P_k!! < \alpha, \beta, \gamma, \tau >$

■
$\beta = \mathbf{newroot} \rightarrow root := \alpha$; $\forall x \in SON,\ P_x!! < \alpha, \mathbf{newroot}, \_, \_ >$;
   $[\,ambiguity = 1 \rightarrow ambiguity := 0$; $\mathbf{SELECT}\ (x, PORT[\ ])$; $P_x!! < root, \mathbf{comb}, \_, \_ >$

■
$\beta = \mathbf{merge} \rightarrow SON := SON \cup \{y\}$; $PORT[y] := \gamma$; $open := \bigvee_k PORT[k]$;
   $\forall x \in SON,\ P_x!! < \alpha, \mathbf{newroot}, \_, \_ >$; $[\,root \neq id \rightarrow P_{father}!! < \alpha, \beta, open, \_ >]$
]
].

A linear fault-tolerant naming algorithm

Joffroy BEAUQUIER[1]
Paul GASTIN[2]
Vincent VILLAIN[3]

Abstract: We solve the naming problem (how to give a unique identifier to each site of an unknown network), when some sites are supposed to have a faulty behaviour of fail-stop type. The solution uses several tokens, in order to ensure that, despite crash failure of some sites, at least one token will perform a complete traversal of the network. The complexities in time and in number of messages of this algorithm are linear with respect to the size of the network (number of communication lines), which improves the exponential solution already known in the Byzantine case with some special assumptions.

Key words: distributed algorithms, naming problem, fault-tolerance, fail-stop, synchronous message passing.

1. Introduction

Almost all distributed algorithms assume as a precondition that each site has a unique identifier. If we assume no failure, giving to each site an identifier (the naming problem) is straightforward, even if each site does not know about the entire network. All solutions are related to a full traversal of the network.

For instance, a circulating token containing an integer can perform a depth-first traversal of the network, starting with the value 1, and increasing it by 1, each time it leaves a site for the first time. Each site choses as unique identifier the integer carried by the token, at its first arrival. The termination is detected by the initiator of the algorithm when the token comes back while all its neighbors have been visited [Hélary and Raynal 88].

The previous method is sequential, but it is possible to take advantage of the network by using a parallel algorithm and a breadth-first traversal. When a site is identified, it will in turn contact all its neighbors in order to give them an identifier. It is clear that we cannot any more use the increasing integer as identifier since several "tokens" circulate in the network. The solution is then to use as identifier of a site the path from the initiator to this site [Beauquier 89]. The problem of the termination detection is solved by "end" messages that circulate from the leaves of the breadth-first tree back to the root (the initiator). The termination detection is performed by the initiator when it has collected an "end" message from all its neighbors.

---

[1] L.R.I., Université Paris 11, Bât 490, 91405 ORSAY CEDEX, FRANCE.
[2] L.I.T.P., Université Paris 6, 4 place Jussieu, 75252 PARIS CEDEX 05, FRANCE.
[3] L.I.F.A., Université de Picardie, 33 rue Saint Leu, 80039 AMIENS CEDEX, FRANCE.

These methods, like some others ([Cheung 83], [Segall 83]), depend on the fact that each site transmits the tokens correctly. If a given site falsifies the value of a token, two (correct) sites will possibly receive the same identifier. In the case of Byzantine faults [Lamport, Shostak and Pease 80] with some special assumptions, only an exponential algorithm (in the number of exchanged messages with respect to the number of communication lines) is known up to now [Beauquier 89].

In the present paper, as in [Attiya and al.], [Masukawa and al.] or [Bar-Noy and al.] for some other problems, we will consider only fail-stop failures (a site that fails stops forever, but does not send false messages). In this case, the two solutions presented above cannot be trivially generalized.

For the depth-first traversal, there are two main problems. First, if a site crashes when holding the token then the identification will never be achieved. Second, if a site fails during the identification of its sub-tree (in the depth-first covering tree), then again the identification will never be achieved.

With the parallel method, there is no problem with the identification as long as the network is connected. Here the problem comes from the termination detection. More precisely, if a site crashes during the identification of its sub-tree, its "end" message will never be sent. And if the initiator itself fails, which of the sites will detect the termination?

Note that the solution presented in [Beauquier 89] works in the case of fail-stop but is hardly of any use because of its exponential complexity in number of messages.

The solution that we present is based upon a multiple depth-first traversal of the network. Its complexities in time and in number of messages are linear with respect to the size of the network (number of communication lines).

In the following section, we give a general description of the algorithm. In section 3, we present it more formally, then we validate it and study its complexity in section 4.

2. General description of the protocol

Unicity of the initiator.
We assume the existence of a unique initiator (each site knows whether it is the initiator or not). Its role is to start the naming algorithm. Moreover, we assume that the initiator cannot crash during the initialization phase, but it can indeed crash after it.

Connectivity of the network.
As in almost all fault-tolerant algorithms, we assume to be known an upper bound to the number of possible failures : let k be this upper bound. It is clear that the network has to remain connected in order to name all its sites. Therefore, we will assume that the network is k+1 connected, meaning that there exist at least k+1 disjoint paths between each pair of sites.

A multiple depth-first traversal.
In the algorithm presented here, we start from the idea of a token with a value performing a depth-first traversal of the network. As mentioned in the introduction, the death of a site could imply the loss of the token. To solve this problem, we will use several tokens and we will make sure that at least one of them will perform a full traversal of the network. The algorithm is then conceptually simple : the idea is to perform k+1 depth-first traversals in parallel, with

k+1 different tokens, and to manage the token passing in such a way that at most one token could be lost when a site crashes. Unfortunately, this condition implies that a site cannot own several tokens at the same time, which will be ensured by a token passing with acknowledgement.

Synchronous message passing.
If a site tries to send its token to a dead neighbor, it will never receive any acknowledgement. If it waits forever for the acknowledgement, the token will never achieve its traversal and if it is the last token in the network, the identification will never be complete. Hence a site has to be able to detect the death of one of its neighbors within a finite amount of time. It is well known that this detection is not possible if we have an asynchronous message passing. Then in this paper we assume that the message passing is synchronous.

How crashes are dealt with.
The death of a site can involve several problems. Firstly, the tokens owned by this site are lost but this is not a real problem since a site can own at most one token and since we start with more tokens than possible crashes. Secondly, a site can try to send its token to a dead neighbor and will be aware of this death within a fixed amount of time since we have assumed a synchronous message passing. If this happens when the token is going down in the depth first covering tree, there will be in fact no problem and we will go on as if this neighbor does not exist. But if this happens when the token is going up in the depth first covering tree, then we will not be able to complete the depth-first traversal. In this case, we will simply initialize a new traversal for this token.

Deadlocks.
The message passing with acknowledgement can create interferences between the depth-first traversals of the different tokens and this could even lead to a deadlock between several traversals if we did not take some precautions. For instance, let us consider the situation of fig.1 where the sites S1, S2 and S3 want to send their token to the neighbor pointed by the arrow. Since a site cannot own several tokens at the same time, none of them can accept the token of its neighbor. In this case, if each site persists in trying to send its token to the same neighbor, we obtain a deadlock (or more precisely a livelock). This situation can definitely occur if each token tries to perform a depth-first traversal of the network. So at least one token has to give up its depth-first traversal. In the example, we may suppose that S4 has no token and then S1 could send it its token in order to unlock the situation. Note that the depth-first traversal of S1's token has been interrupted. In order to resume it, this token has to go back to S1, which could be very difficult. For this reason, the token will simply initialize a new traversal.

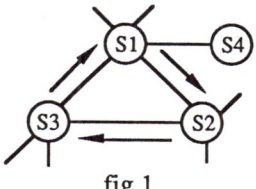

fig.1

Priority on the tokens.

However, the termination detection is obtained when a depth-first traversal is achieved. So we have a problem if each token gives up all its successive traversals. In order to solve it, we introduce a priority on the tokens and manage the token passing in such a way that the highest priority token always follows its way and will never reinitialize its traversal excepted in the case of a death as we have seen above. Moreover, we will use this priority to ensure that the traversal of the highest priority token is not too slowed down. More precisely, we will assure a uniform upper bound to the transmission time of the highest priority token. So, we will easily obtain a time complexity which is linear with respect to the size of the network since the depth-first traversal is of the same complexity.

How to get rid of a token.

A site has to get rid of its token to solve a deadlock situation as mentioned above or, more simply, when a neighbor wants to send it a token with an higher priority. In the example above, if we suppose that S4 does not own any token, it can accept S1's token without any problem. But if we assume that all the neighbors of S1 already own a token, the solution is not so simple. S1 could ask one of its neighbor to get rid of its token but this would delay the solution of the problem and this could even be impossible in some cases. Thus, to ensure the uniform upper bound on the transmission time, we have to solve the problem locally : S1 will either wait with its token or delete it. More precisely, when S1 has to get rid of its token, it enters an alarm state and asks simultaneously to all its neighbors (in fact, it is enough to ask to k+1 neighbors) if they can accept its token. Then three answers are possible : OK: I am free; WAIT: I have a higher priority token or alarm to deal with; NO: I can't accept your token now (I already own one) nor keep you waiting since you have a higher priority (you should perhaps delete your token). If S1 receives at least one OK message, then it will give its token to one of the free neighbors and then it will accept the token from S3. If S1 receives at least one WAIT message then it will keep S3 waiting, because S3's token has not the highest priority in the network. If S1 receives only NO messages, then, in order to solve the problem quickly, it will delete its token and will accept S3's one.

One can possibly think that this destruction can leave fewer tokens than possible future crashes and so make the algorithm fail. But we will prove that this case can only occur if the number of tokens becomes greater or equal to two plus the number of possible future crashes (some sites are dead without loss of token).Therefore one token can be deleted without any problem.

Termination detection.

The termination will be detected by the initiator of a traversal when the token comes back having performed the full traversal. Hence we have to ensure that at least one token does eventually perform a full traversal of the network. The non completion of a traversal is due either to the initialization of a new traversal as we have seen above or to the loss of the token at the time of a site crash. The loss of a token is not a problem since we have more tokens than possible crashes and then at least one token will not be lost. Now, we have seen in previous paragraphs that a site can initialize a new traversal for a token either if the token meets a dead site when it is going up in the depth-first covering tree or if the site has to accept a higher priority token. The situation of the first case can occur only once per death and per token and then cannot be infinitely repeated. Therefore it is not a problem for the termination detection. On the other hand, the second case cannot occur for the highest priority token. Therefore the highest priority token which is not lost will eventually perform a full traversal and the termination will be detected.

3. The protocol

3.1. Hypotheses

- Communication links are bidirectional and no message is lost.
- There is an upper bound k to the number of sites that can possibly crash.
- The network is k+1 connected, meaning that there exist (at least) k+1 disjoint paths between any pair of sites.
- There exists a unique site (the initiator), that initiates the protocol. Moreover, each site knows whether it is the initiator or not.
- The message passing is synchronous. Moreover, in order to simplify the algorithm, we will assume that the network is synchronized and works in phases. Each phase has three steps : reception of the messages of the previous phase, computation, sending of new messages. For example a simple token passing between two sites S1 and S2 is illustrated in figure 2. At the phase P, S1 sends its token to S2, at the phase P+1, S2 receives it and replies "OK, I accept it" to S1 and at the last phase, S1 receives the OK message and S2 tries to send the token to another site.

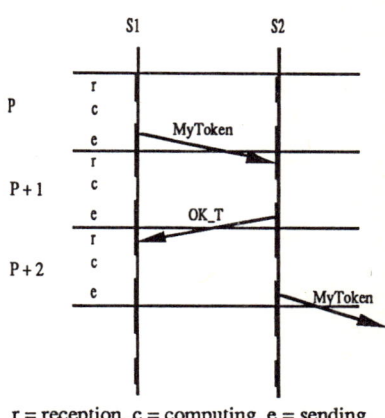

r = reception, c = computing, e = sending

Fig.2

3.2. Some data structures and variables

```
const      k              = the number of possible crashes;
type       T_token        = ( Number, Value, Priority, Trial : integer);
           T_message      = (  Help (priority : integer)1 , Token (T : T_token)1 ,
                              OK_T2 , Wait_T2, OK_H2, Wait_H2, No_H2, End2 );
           T_tokenTable   = array [1..k+1] of
                              (Ident, Trial, CurrentSon, Father : integer;
                               NextSons : set of integer)
var        MyToken        : T_token;
```

---

[1] Help (resp.Token) is the type of the message and priority (resp.T)is its value.

[2] This message is no more than a type. *_T means answer to a token message, *_H means answer to a help message.

```
            NextSite         : integer;   /* number of a communication link */
            AliveNeighbors   : set of integer;
                /* initialized with the numbers of the communication lines of the site */
            TokenTable       : T_tokenTable;
                /* all items are initialized with (?, 0, ?, ?, ?)*/
            HighPriority     : integer;
```

3.3. Primitives and subroutines for identification and graph traversal

primitive send (m : T_message) to (s : integer);
/* sends the message m into the communication link numbered s */

primitive reply (m : T_message) to (m' : T_message);
/* sends the answer m to the sender of message m' */

Function NewTraversal (Token, Neighbors, TokenTable) : integer;
/* initializes a depth-first traversal at the beginning of the algorithm or when a token meets a dead site and returns the first son of the new covering tree */

Procedure Update (Token, Sender, Neighbors, TokenTable)
/* updates the TokenTable entry of this token when it reaches a site for the first time, */

Function Traversal (Token, Sender, TokenTable) : integer;
/* warrants a depth-first traversal of the graph when there is no death or rerouting and returns the next site to be visited */

Procedure Dead (DeadNeighbor, Neighbors, TokenTable)
/* removes the DeadNeighbor from the Neighbors set and from the NextSons sets of the TokenTable entries */

Function Rerouting (Token, PossibleSons, Neighbors, TokenTable) : integer;
/* used in an Alarm state when a token must leave its place for a higher priority token. In this case, a new traversal is initialized for the lower priority token and this function returns the next site to be visited. The single difference with the NewTraversal function is that this next site is chosen in the PossibleSons set instead of in the Neighbors set */

3.4. The protocol

We present the algorithm in three sections : the first one describes the initial behaviours of the initiator and of any other site, the second one describes the general behaviour of any site after its initial behaviour has been achieved and the third one presents a global view of the main part of the algorithm, under the form of a finite state transition system (cf fig.3).

Initial behaviours.
The initial part differs between the initiator and the other sites. The initiator initializes k+1 tokens and sends them to k+1 neighbors, then it waits for the OK_T messages and enters state

Idle1. Note that the initiator should not crash before sending the k+1 tokens and that in this first part we can loose as many tokens as the number of crashed neighbors of the initiator. Thus there remain more tokens than the number of sites which still could crash. Another site is awaked by the first reception of messages (Token or Help) and behaves as in state Idle1 : if the highest priority message is a Token message then it goes in Smooth1 state or else (it is an Help message) it goes in Idle2 state.

General behaviour.

State Idle1   Possible messages: Token, Help, End
    If there is an End message
    then send (End) to every neighbors
        NextState EndIdent
    else **If** the message of highest priority is the Token message T **then**
            Reply (OK_T) to the message T
            Reply (Wait_T) to the other Token messages
            Reply (Wait_H) to the Help messages
            MyToken := T
            Update (MyToken, Sender of T, AliveNeighbors, TokenTable)
            NextSite := Traversal (MyToken, Sender of T, TokenTable)
            NextState Smooth1
        If the message of highest priority is the Help message H **then**
            Reply (OK_H) to the message H
            Reply (Wait_H) to the other Help messages
            Reply (Wait_T) to the Token messages
            NextState Idle2
        If there is no message **then** NextState Idle2

State Idle2   Possible message: End
    If there is an End message
    then send (End) to every neighbors
        NextState EndIdent
    else NextState Idle1

State Smooth1 Possible message: End
    If there is an End message
    then send (End) to every neighbors
        NextState EndIdent
    else **If** NextSite ≠ Me
        then send (MyToken) to NextSite
            NextState Smooth2
        else send (End) to every neighbors
            NextState EndIdent   /* TERMINATION */

State Smooth2   Possible messages: Token, Help, End
    If there is an End message
    then send (End) to every neighbors
        NextState EndIdent
    else   Reply (No_H) to the Help messages with Help.Priority $\geq$ MyToken.Priority
        Reply (Wait_H) to the other Help messages
        Reply (Wait_T) to the Token messages
        Let HighPriority be the highest priority of the Token messages
            /* 0 if there is no Token message */
        If MyToken.Priority > HighPriority
            then NextState Smooth3
            else NextState Alarm1

State Smooth3   Possible message: OK_T xor Wait_T xor none from NextSite, End
    If there is an End message
    then send (End) to every neighbors
        NextState EndIdent
    else   **Case** message **of**
            OK_T   :   NextState Idle1
            Wait_T   :   Send (MyToken) to NextSite; NextState Smooth2
            None   :   /* NextSite has crashed */
                        Dead (NextSite, AliveNeighbors, TokenTable)
                        NextSite := NewTraversal (MyToken, Neighbors, TokenTable)
                        Send (MyToken) to NextSite
                        NextState Smooth2

State Alarm1   Possible message: OK_T xor Wait_T xor None from NextSite, End
    If there is an End message
    then send (End) to every neighbors
        NextState EndIdent
    else   **Case** message **of**
            OK_T   :   NextState Idle1
            None, Wait_T   :   Help.Priority := HighPriority
                              Send (Help) to k+1 neighbors
                              NextState Alarm2

State Alarm2   Possible messages: Token, Help, End
    If there is an End message
    then send (End) to every neighbors
        NextState EndIdent
    else   Reply (No_H) to the Help messages with Help.Priority > HighPriority
        Reply (Wait_H) to the other Help messages
            /* Help.Priority = HighPriority is impossible */
        Reply (Wait_T) to the Token messages

Let HighPriority be the highest priority of the Token messages
/* 0 if there is no Token message */
If HighPriority > MyToken.Priority
then NextState Alarm3
else NextState NoMoreAlarm

State Alarm3 Possible messages: OK_H, Wait_H, No_H, End
  If there is an End message
  then send (End) to every neighbors
    NextState EndIdent
  else **If** we receive only No_H messages **then** NextState Idle1
    If we receive at least an OK_H message **then**
      MyToken.Priority := HighPriority
      NextSite := Rerouting (MyToken, the OK_H senders, Neighbors, TokenTable)
      Send (MyToken) to NextSite
      MyToken.Priority := MyToken.Number
      NextState Smooth2
    Otherwise   /* no OK_H but at least one Wait_H */
      Help.Priority := HighPriority
      Send (Help) to every neighbor
      NextState Alarm2

State NoMoreAlarm Possible messages: OK_H, Wait_H, No_H, End
  If there is an End message
  then send (End) to every neighbors
    NextState EndIdent
  else Send (MyToken) to NextSite
    NextState Smooth2

State EndIdent
    Nothing...

Global view

A global view of the protocol is given by fig.3, it sums up the possible transitions of a site during the protocol. The notation (a, b) means "reception of a and sending of b"; the couple which determines the transition that we will follow is boldfaced.

The cycle "Idle1, Smooth1, Smooth2, Smooth3" (and the cycle "Idle1, Idle2") represents the behaviour of a site when no traffic problem occurs. When a site has to get rid of its token, it enters the cycle with Alarm states (Alarm1, Alarm2, Alarm3, NoMoreAlarm) and according to the solution of this problem, it comes again in the first cycle either by Idle1 state or by Smooth2 state. Note that we distinguish between an arriving token (called Token) and an exiting token (called MyToken).

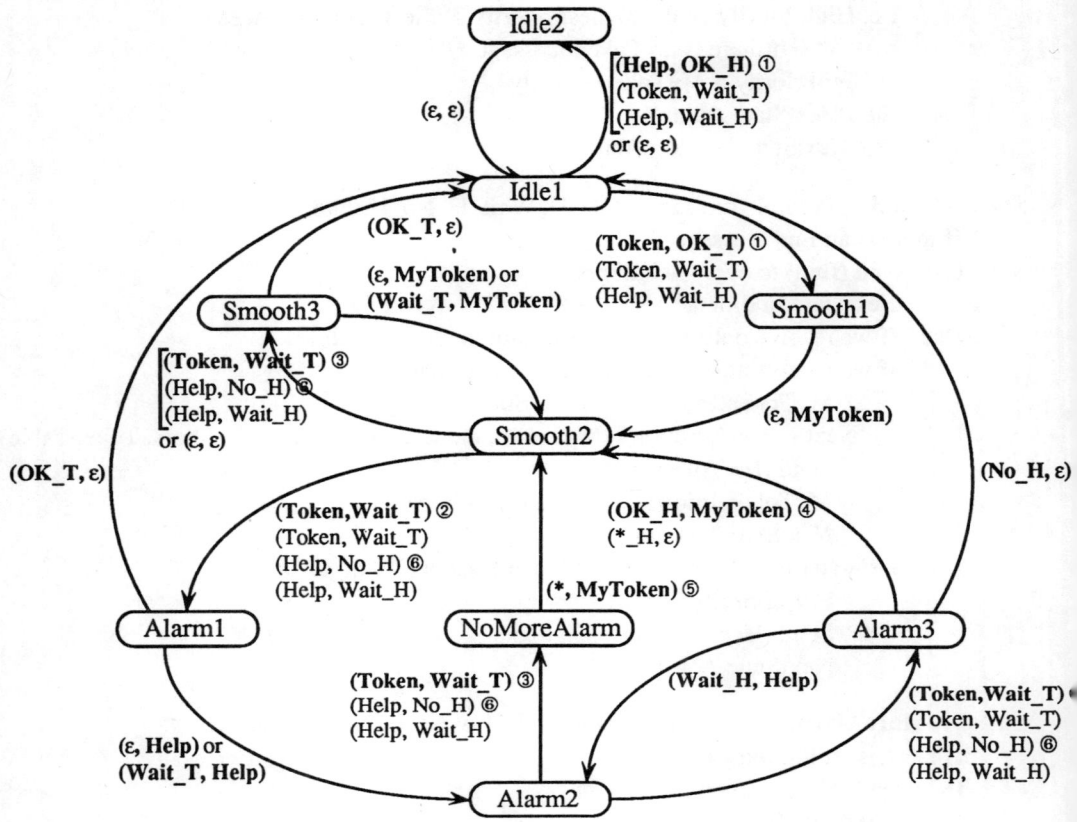

① This is the highest priority message,
② This highest priority token is of higher priority than MyToken,
③ All the Token messages are of smaller priority than MyToken,
④ I send MyToken with HighPriority,
⑤ I don't care for the received messages, I send MyToken to a neighbor,
⑥ I reply No_H only to the Help messages with higher priority than my job.

Fig.3

4. Validation and complexity of the algorithm

Lemma 4.1
When a site deletes a token, all the neighbors which have replied No_H have distinct tokens.

Proof
When a site deletes its token, it has received only No_H messages from its neighbors. Now a neighbor can answer No_H only if it is in state Smooth2 or Alarm2. But two sites in state Smooth2 or Alarm2 cannot own the same token.

Lemma 4.2
If N sites delete simultaneously their tokens, then there is no cycle in the graph induced by the No_H sent by these N sites.

Proof
Suppose there exists a cycle of N' sites ($2 \leq N' \leq N$) in this graph (fig.4):

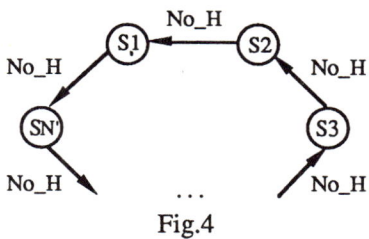

Fig.4

Since S_i deletes its token, S_{i+1} has replied No_H to the Help message of S_i. Thus we have $Help_i.Priority > HighPriority_{i+1}$. Now we have $HighPriority_i = Help_i.Priority$, thus
$$Help_1.Priority > Help_2.Priority > ... > Help_{N-1}.Priority > Help_N.Priority > Help_1.Priority$$
which is impossible.

Lemma 4.3
If N (N>0) sites delete their tokens simultaneously, then there exists a site S which has the following property : $T_S + D_S \geq k+1$, where T_S is the number of distinct tokens owned by its alive neighbors and D_S is the number of its dead neighbors.

Proof
Suppose that $S_1, ..., S_N$ simultaneously delete their tokens during phase p. Let NO_i be the number of neighbors which answer No_H to the site S_i during phase p-1, $Dead_i$ be the number of neighbors which do not answer to S_i during phase p-1, T_i be the number of distinct tokens owned by the alive neighbors of S_i in phase p+1 and D_i be the number of the dead neighbors of S_i in phase p+1. Let C be the number of neighbors of S_i which crash during phases p-1 and p. Clearly, we have $D_i \geq Dead_i + C$ and $NO_i + Dead_i = k+1$.
If N=1, we have $T_1 \geq NO_1 - C$ (cf lemma 4.1) and $T_1 + D_1 \geq NO_1 + Dead_1 = k+1$.
If N>1, let us suppose that $\forall i=1,...,N$ $T_i + D_i < k+1$. So $\forall i$, $T_i < NO_i - C$, therefore a neighbor which has answered No_H to S_i has deleted its token, let us denote S_{i+1} this neighbor. Now we can build a sequence $S_1, S_2, ..., S_i, S_{i+1}, ...$, where S_{i+1} is the neighbor of S_i define above. Since the number of different sites involved is bounded by N, there exist i<j such that $S_j = S_i$. Therefore there is the following cycle
$$S_j \xrightarrow{No\_H} S_{j-1} \xrightarrow{No\_H} ... \xrightarrow{No\_H} S_i = S_j$$
which contradicts lemma 4.2.

Theorem 4.4
At least one token will never disappear.

Proof

First we prove that $\forall t\ T_t+D_t \geq k+1$, where T_t is the number of tokens at the time t and D_t is the number of crashed sites at time t.

At t=0 we have $T_0=k+1$ and $D_0=0$ thus $T_0+D_0 = k+1$. When the initiator sends k+1 tokens to its k+1 neighbors, some of them can crash and cause the loss of the same number of tokens. Then the inequality is still true. Suppose that it is true until t-1 and show that it holds for t. If tokens disappear only because some sites crash, the inequality is obviously true for t (when a site crashes, it can force at most one token to disappear : because in the algorithm, at any time, any site can own at most one token). If tokens are deleted by some sites in Alarm3 state, the lemma 4.3 warrants that there exists a site S such that $T_S+D_S \geq k+1$. Clearly $T_t \geq T_S$ and $D_t \geq D_S$ thus $T_t+D_t \geq k+1$. The result follows from the hypothesis that $D_t \leq k$.

Lemma 4.5

The passing of the highest priority token between two sites takes at most 8+4n phases (where n denotes the number of crashes during this transmission).

Idea of the proof

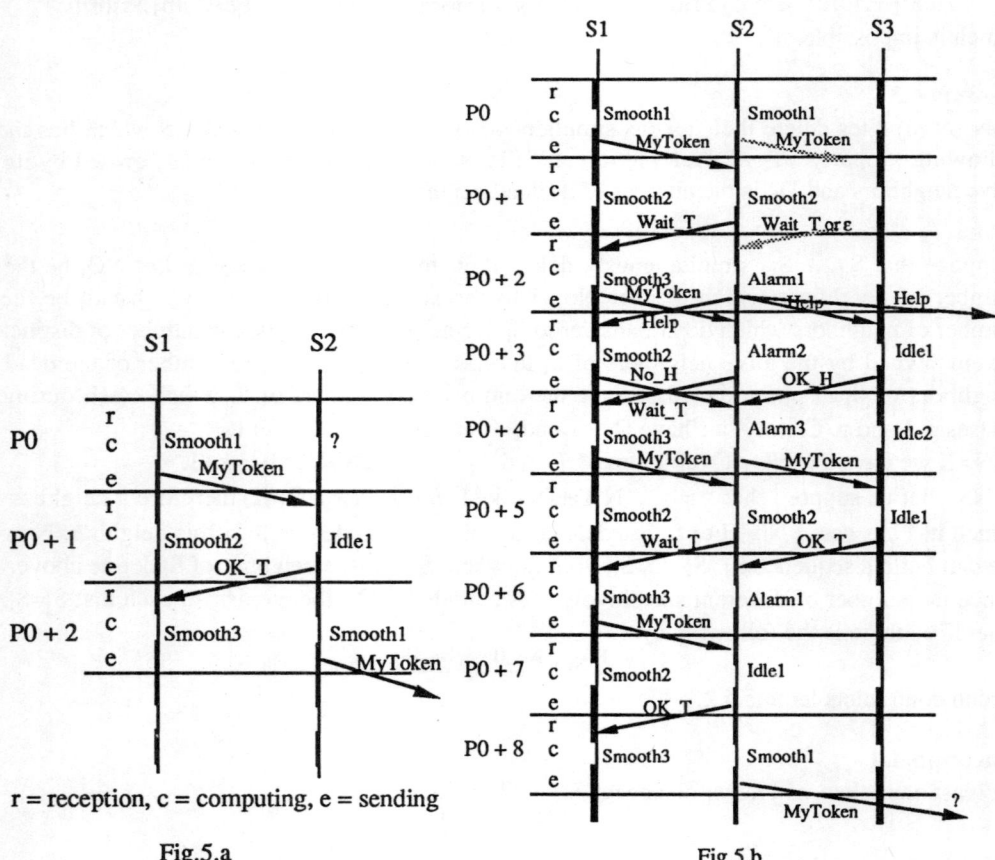

r = reception, c = computing, e = sending

Fig.5.a   Fig.5.b

In Fig. 5.a, we present the best case which takes two phases : S_1 has the highest priority token, and it is immediately accepted by S_2. In fig. 5.b, we have one of the worst cases: S_2 cannot accept immediately the token from S_1, so it alerts k+1 of its neighbors and at least one of them will accept its Help message, and, two phases later, its token. Then S2 can accept the token from S1. In this case, token passing takes eight phases. In fact, it is possible that S3 crashes in phase P0+4, thus it will not accept the token of S2. In this case the four first phases are lost and this explains the 4n additional phases.

Theorem 4.6
At least one token makes a full traversal of the graph.

Proof
First we show that :
1) if the highest priority token does not disappear in a crash of a site, it makes a full traversal of the network :
> the highest priority token cannot be deleted by a site in Alarm3 state. So, according to the above hypothesis, it is straightforward to show that this token performs a full traversal of the network : we have a finite number of sites and the passing of the highest priority token makes a finite time (cf lemma 4.5).

2) when the highest priority token disappears, the disappearance of its messages and the recognition of the next highest priority token take at most two phases :
> every next phase, new messages are sent and the messages concerning previous phases are forgotten. So, when the highest priority token disappears, its last messages do not disturb the new token of highest priority for more than two phases.

So this theorem is a corollary of theorem 4.4, that warrants there always exists at least one token in the network, and of 1) and 2), that warrant that a token performs a full traversal of the network in a finite time.

Theorem 4.7
The identification is full in a finite time and the termination is detected.

Proof
This result is a corollary of theorem 4.6 : at least one token performs a full traversal of the network in a finite time and the last visited site (it is also the initiator of the token traversal) detects the termination.

Complexity
The main advantage of this algorithm is that its complexities in time and in number of messages are linear with respect to the size of the network.

Theorem 4.8
The complexity in time is $O(ae)$ and the complexity in number of messages is $O(k^2 ae)$ where k is the number of possible crashes, a is the number of actual crashes ($a \leq k$) and e is the number of edges in the network.

Proof
Let p be the duration of a phase.

First, we assume that the highest priority token is not lost. The passing of the highest priority token takes at most eight phases when there is no crash (Lemma 4.5). This token makes exactly a depth-first traversal of the graph and so it uses each edge twice. Thus the time needed for the highest priority token to perform a full traversal of the network is bounded by 16ep. Now, if we suppose that the successive highest priority tokens are lost just before achieving their traversal, the total time complexity is bounded by $16(a+1)ep + c$. The complexity in time is lower than $16(a+1)ep+c$ (the constant c stands for the initialization part of the algorithm: 3p; and the extra cost which can be involved by a crash: 4ap (cf lemma 4.5) and 2ap (cf proof of theorem 4.6).

When there is no problem, a site with a token sends just a token message to another site and in the next phase the latter replies while the first one is waiting for the answer. When a site with a token is in Alarm state, it sends at most k+1 help messages and in the next phase its neighbors reply to it. So in the worst case there are k+1 messages per phase and per token. So the complexity in number of messages is lower than $(k+1)^2[16(a+1)e+c] = 16(k+1)^2(a+1)e+c'$.

5. References

Attiya H., Bar-Noy A., Dolev D., Koller D., Peleg D. and Reischuk R., "Achievable Cases in a Asynchronous Environment", Proceedings of the 28th Found. of Comput. Science IEEE, pages 337- 347, 1987.

Bar-Noy A., Dolev D., Koller D. and Peleg D., "Fault-Tolerant Critical Section Management in Asynchronous Networks", in Distributed Algorithms, Lecture Notes in Computer Science 392, pages 13-23, 1989.

Beauquier J., "Distribution d'Identificateurs avec des Processus Byzantins", Actes du séminaire franco-brésilien sur les systèmes répartis, Florianopolis, pages 235-239, 1989.

Cheung T., "Graph Traversal Techniques and the Maximal Flow Problem in Distributed Computation", IEEE Trans. on SE, SE 9 (4), pages 504-512, 1983.

Hélary J.M. and Raynal M., "Assigning Distinct Identities to Sites of an Anonymous Distributed System", Proc. IEEE, Workshop on the Future Trends of Distributed Computing Systems in the 1990s, Hong Kong, pages 82-86, September 1988.

Lamport L., Shostak and Pease, "Reaching Agreement in the Presence of Faults", J.A.C.M., pages 228-234, 1980.

Masukawa T., Nishikawa N., Hagihara K. and Tokura N., "Optimal Fault-Tolerant Distributed Algorithms for Election in Complete Networks with a Global Sense of Direction", in Distributed Algorithms, Lecture Notes in Computer Science 392, pages 171-182, 1989.

Segall A., "Distributed Network Protocols", IEEE Trans. on Inf. Theory, IT 29 (1), pages 23-25, 1983.

Distributed Data Structures: A Complexity-Oriented View

David Peleg *

Abstract

The problem of designing, implementing and operating a data structure in a distributed system is studied from a complexity oriented point of view. Various relevant issues are addressed via the development of an example structure. The structure evolves through a sequence of steps, each oriented towards attacking a different aspect of the problem. The paper concentrates on deterministic structures featuring low memory requirements, memory balance and efficient access protocols. Among the issues treated are centerless organizations of data structures, background maintenance of memory balancing, employing redundancy for increasing search efficiency and concurrent accesses to distributed structures.

1 Introduction

1.1 Background

Efficient data structures are the backbone of any data-manipulating computer system, and effective algorithm design is strongly intertwined with data structure design. This fundamental fact is just as valid for distributed systems. Indeed, distributed data structures are widely used in most existing distributed systems. Some distributed programming languages support distributed data structures explicitly (cf. [CGL86]), and many distributed algorithms make (explicit or implicit) use of such structures for their data-management purposes. But more importantly, such structures are in many cases constructed not as part of any particular algorithm but for "autonomous" use as the building blocks of various permanent storage and retrieval mechanisms, such as distributed dictionaries, name servers in communication networks, bulletin boards, resource allocation managers etc. [FWB85,GS89b,LEH85,MV85,OD81,Ree78,Ter87]. The function common to all of these mechanisms is supplying facilities for storing accumulated information in the system and making it available to potential users throughout the system.

Surprisingly, however, little work was put into furnishing a theoretical framework for the design, analysis and classification of distributed data structures from a complexity

---

*Department of Applied Mathematics, The Weizmann Institute, Rehovot 76100, Israel. peleg@wisdom.bitnet. Supported in part by an Allon Fellowship, by a Walter and Elise Haas Career Development Award and by a Bantrell Fellowship.

oriented point of view. Our goal in this abstract is to initiate the explicit study of these issues.

Distributed data structures can be modeled by the following rather general framework. The distributed system is composed of a collection of sites, interconnected by an (arbitrary) communication network. The processors located at the various sites occasionally generate or accumulate data that might later be needed by processors in other sites. The data is taken from a domain $\mathcal{X}$. Each data item $X \in \mathcal{X}$ consists of two principal fields, $X = (Key_X, Record_X)$, where Key_X is a *key* taken from a totally ordered domain $\mathcal{K}$ and $Record_X$ is a *record* field containing the relevant data. We assume that the size of a data item is at most δ bits of memory. A distributed data structure is composed of a *data organization scheme*, specifying a collection of local data structures storing copies of data items at various sites in the system, coupled with a set of distributed *access protocols* that enable processors to issue modification and query instructions to the system and get appropriate responses. We refer to the processor originating the instruction as the *requesting* processor. In case the request concerns an information item stored at some site, we refer to the processor at that site as the *storing* processor.

In general, the repertoire of instructions may include both *modification* and *query* instructions. A data structure is characterized by the set of primitive operations it supports. For this abstract we concentrate on the *dictionary* structure, which supports the operations *Insert*, *Delete* and *Find*, defined as follows.

- *Find(Key,v)*: Find the data item X such that $Key_X = Key$, and return $Record_X$ to the querying processor v. If no such item exists then return a "not-found" message.

- *Insert(X)*: Insert the data item X into the data structure. This operation is required to be *safe*: if the structure currently stores an item X' with $Key_X = Key_{X'}$ then a "failure" message is returned to the requesting processor.

- *Delete(Key)*: Delete the data item X such that $Key_X = Key$ from the data structure. No reply is sent to the requesting processor.

This set of operations completely specifies the *sequential* behavior of the dictionary structure. However, the *distributed* setting raises several issues that are not encountered in the usual (shared-memory, single-process) setting. We may distinguish between two types of problems: those stemming from the introduction of *concurrency* and those caused by the *distribution* of the data and users. The presence of concurrent accesses creates some serious semantical problems. In fact, the very definition of a data structure, its specification and the characterization of its correct behavior become more complex. A considerable amount of work has been done on these issues in the area of database theory (cf. [BHG87,Her90,LM79,Pap86]). Most approaches are based on some notion of serializability, such as linearizability [HW87] or order preserving serializability [LS88]. (In order to get a feeling for the semantical problems arising in a concurrent environment, consider the following exercise: how should we interpret the word "currently" in the definition of safe insertion, if concurrent operations are allowed? Under an interpretation based on serializability, the way to ensure safe insertions is in essence by preceding the actual insertion with a *Find* operation, which should be an atomic part of the insertion operation. In the full paper we also consider cases in which we may settle for the more basic version of insertion, which simply ignores the issue and leaves the responsibility with the users.)

However, our focus in this paper is mainly on the second type of problems arising in a distributed environment, namely, the implications of distribution on complexity issues. The fact that the data (and the users) are physically distributed in many distant sites raises the issue of the relationships between the topology of the underlying communication network (and its graph-theoretic properties) and efficient schemes for organizing and distributing data in the various sites so as to ensure reasonable complexities. We would like to examine these issues and provide some basic "vocabulary" for classifying such structures from a complexity oriented point of view. The way we chose to present our approach to these tasks in the present abstract is via the development of an example structure, serving as a case study. The structure evolves through a sequence of steps, each oriented towards attacking a different aspect of the problem.

We first describe the access algorithms as if the requests arrive *sequentially*, i.e., each new request arrives only after the processing of the previous one has completed. We begin (in Section 4) with two simple structures: a centralized structure guarantees optimal access complexities but is memory imbalanced, whereas a structure based on emulating a 2-3 tree is memory-balanced but nonoptimal in operation costs. The BIN structure presented next (in Section 5) enjoys optimality in both respects (for a sufficiently large structure) but is still center-based. It is then transformed (in Section 6) into a centerless structure, denoted DIST-BIN. Section 7 introduces the use of data replication in order to improve accessibility, resulting in the CLUSTERED structure. Finally, in Section 8 we discuss removing the assumption of sequential accesses and allowing concurrent instructions.

In order to initiate this investigation of complexity issues, we concentrate in this paper on a somewhat simplified framework. In particular, we make the following assumptions. First, we ignore the important issue of fault-tolerance, and assume that the system operates properly throughout its lifetime. Secondly, in order to sharpen complexity aspects resulting from the distributed nature of the system, we abstract away all site-internal design considerations by making the (standard) assumption that internal processing is negligible compared to communication costs. Studying the more realistic setting in which local costs need to be taken into account is beyond the scope of the current paper. Nevertheless, we argue that divorcing the site-internal and inter-site design issues from each other seems not at all a bad idea, and structures resulting from combining efficient designs on the two levels are likely to have reasonable performances.

1.2 Related work

Our problem intersects several wide research areas, such as ordinary data structures and data types, distributed database and concurrency control theory. We shall make no attempt to view the relevant literature here. Let us mention some pointers to work on problems that are more directly related to ours. A distributed implementation of extendible hashing is described in [Ell85]. Distributed directories based on local caches are analyzed in [GK89]. Implementations of distributed dictionaries are described (implicitly or explicitly) in [DMadH90,KU86,Ran87,UW87], in the context of simulating PRAMs in actual parallel computers. We discuss some of the differences between these works and ours in Section 8. Distributed data structures based on *distance-dependent* strategies, attempting to relate the communication cost of searches to the distance between the searching processor and the origin of the searched data item, are discussed in [Pel89].

Two related research areas have attracted considerable attention in the last decade, namely, concurrent data structures and VLSI data machines. The term "concurrent data structures" [BS79,Ell80a,Ell80b,Her90,HW87,KL80,LS88,Man86,ML84,RK88,Sag85] refers to data structures stored in common (shared) memory but accessible by many processes concurrently. The fact that the structure is maintained in one memory module means that our special topology-originated problems are irrelevant in this context.

The design of special purpose VLSI machines for implementing data structures, notably dictionaries, is studied in [AK85,CCIR86,DS85,DS87,Lei79,ORS82,SL87]. The VLSI model usually assumes a tightly-coupled machine of fixed topology (e.g., a complete binary tree, a mesh or a cube), dedicated to the activity of maintaining the data structure. It has a single I/O port, through which it receives a stream of instructions and returns the replies, processing the operations in a pipelined fashion.

2 Preliminaries

Let us first define some basic graph concepts. For two vertices u, w in G, let $dist_G(u, w)$ denote the length of a shortest path in G between those vertices. The degree (number of neighbors) of each vertex $v \in V$ is denoted by $deg_G(v)$. Let $D(G)$ denote the *diameter* of the network, i.e., $\max_{u,v \in V}(dist_G(u,v))$. (In all of the above notations, we sometimes omit the reference to G where no confusion arises.)

A distributed system is composed of a collection of sites connected through an (arbitrary) asynchronous point-to-point message-passing communication network (following the standard model, cf. [GHS83]). The network is described by a connected undirected graph $G = (V, E)$, $|V| = n$. A processor may communicate directly only with its neighbors, and messages to nonneighboring processors u and v are sent along some path connecting them in the graph. All the processors have distinct identities. There is no common memory, and algorithms are event-driven (i.e., processors cannot access a global clock in order to decide on their action). Messages sent from a processor to its neighbor arrive within some finite but unpredictable time.

Let us now define our complexity measures. For the data organization scheme, our primary concern is to maintain the data structure using reasonable overall space requirements. The memory requirement of a data structure, denoted M, is the total amount of memory bits it uses in the processors of the network. This measure is generally dependent on the number of information items stored in the structure, denoted by m. It is commonly assumed that $m \geq n$. Actually, in typical situations it may safely be assumed that $m >> n$.

No less important is the need to *balance* the memory loads over the sites of the system; future systems are expected to carry enormous loads of data, and a single site can hardly be expected to function as the sole storing site for a large data structure. The balance level of the structure can be quantified by comparing the maximal amount of memory required at a site, denoted $\hat{M}$, with the optimal ratio $Load = \lfloor \frac{m}{n} \rfloor$.

The memory requirements of the organization scheme can also be evaluated from a complementing angle, namely, the *redundancy* level of the structure. Each data item can be replicated in multiple copies at different sites. The *redundancy* of a data item X in the structure is the number of sites storing a copy of X. The maximal (respectively, minimal) redundancy of a data item in the data structure is denoted $\rho_{\max}$ (resp., $\rho_{\min}$).

Next consider the access protocols associated with the structure. In a distributed system, access and retrieval are costly due to the inherent communication involved. This problem can be handled to some extent by introducing a certain degree of replication in the structure (thus raising interesting tradeoffs with the memory requirements of the structure). We consider the measures of *time complexity* and *communication complexity*, defined as follows. The *communication complexity* of an algorithm A, C_A, is the worst-case total number of bits (i.e., the sum of the sizes of all messages) sent during the run of the algorithm. For simplicity we assume that the allowable message size is larger than δ, so a data item can be transmitted in a single message. The *time complexity* of an algorithm A, T_A, is the worst-case number of time units from the start of the run to its completion, assuming that each message incurs a delay of at most one time unit over any edge it traverses. This assumption is used only for performance evaluation. For any operation Op of the data structure, denote the communication complexity of Op by C_{Op} and its time complexity by T_{Op}. Note that this worst-case complexity is defined over all possible configurations of the data structure and its contents and all possible choices of the requesting processor and the involved item. We may sometimes consider the worst-case *amortized* complexity of an operation, namely, we measure the average cost over a sequence of operations performed consecutively on the data structure and take the worst of all possible sequences.

We assume the existence of full communication facilities in the system, specifically including the services of *broadcast*, enabling a processor to broadcast a message throughout the network, *convergecast* (or PIF), enabling a processor to query all sites of the network and collect responses (see [Seg83,Awe85]), and *routing*, enabling a processor to send a message to some other processor for whom it has sufficient routing information (e.g., its routing label). We further assume that these facilities are as efficient as possible. Specifically, it is assumed that broadcast and convergecast both require $O(n)$ messages and $O(D)$ time for an n-vertex network of diameter D, and routing requires $O(d)$ messages and time whenever the sender and the receiver are at distance d of one another. In general, these assumptions are reasonable, since in most cases such facilities exist in most systems, and serve a host of functions that is much wider than the support of any particular data structure. In the full paper we shall discuss cases in which the above assumptions cannot be made (for example, when designing a data structure as part of the routing mechanism itself, e.g., for storing information required for various routing tables and name-address dictionaries). In particular, we shall analyze the types of services needed and the extent to which they are needed, and describe the appropriate implementations (based in part on techniques developed in [ABLP89,PU89,SK85]).

3 Some basic bounds

Let us first furnish some terminology and establish basic bounds on memory and access costs. Call a distributed data structure *compact* if it stores exactly one copy of each data item (i.e., $\rho_{\max} = 1$). Otherwise we say that the data structure is *redundant*. A data structure is *memory-balanced* if the amount of storage required at the various sites in the system is roughly the same (i.e., $\hat{M} = O(Load)$).

Memory balancing depends on the *insertion function* f_I used to specify the storing location(s) of each data item. This function is implicitly defined by the protocol implementing the insertion operation of the data structure. The next easy lemma says that an "oblivious" insertion function cannot guarantee balanced distribution. (All proofs are omitted from the abstract.)

Lemma 3.1 Consider a distributed data structure supporting the $Insert(X)$ operation. If the insertion function f_I defined by the insertion protocol depends only on the inserting processor v and the key of X, then $\hat{M} = \Omega(m\delta)$, assuming sufficiently large $\mathcal{K}$. ∎

Thus, in order to reach balanced memory distribution it is necessary to use less oblivious insertion protocols, relying for instance on the number of items currently stored in the structure, the current contents of the structure or possibly its history.

Let us now consider the cost of access instructions. To begin with, the following lemma establishes that safe insertions require $\Omega(D)$ communication and time.

Lemma 3.2 For every graph G and for every data structure supporting safe $Insert$ operations on G, $T_{Ins} = \Omega(D(G))$ and $C_{Ins} = \Omega(D(G)\delta)$. ∎

As for the $Find$ instruction, our expectations should depend on the level of replication allowed in the structure. For instance, in a compact structure (where $\rho_{\max} = 1$) the best bound we may hope to have for the complexity of the $Find$ operation is proportional to the diameter of the network. This is implied by the following lemma.

Lemma 3.3 For every graph G and for every data structure supporting the $Find$ operation on G,
$T_{Find} \cdot \rho_{\min} = \Omega(D(G))$ and $C_{Find} \cdot \rho_{\min} = \Omega(D(G)\delta)$. ∎

The above lemmas therefore direct us towards setting the goal of designing a compact, memory-balanced structure with complexity $O(D(G))$ for all three instruction types.

Discussion: Our definitions are oriented towards the more commonly used types of structures, which explicitly refer to "copies" of the data items stored at the various sites.

Likewise, the entire concept of redundancy is defined based on the same restrictive assumption. In a broader setting, one may consider more general types of structures, in which data may be coded in sophisticated ways or stored implicitly. For such structures, the concept of redundancy is ill-defined, and consequently Lemma 3.3 is rendered meaningless. Furthermore, our lower bound proofs need to take this type of schemes into account. Nevertheless, Lemma 3.2 can be rigorously proved in the strongest model. As for Lemma 3.3, one can reformulate an analogous statement in which space is substituted for redundancy.

Lemma 3.4 For every graph G and for every data structure supporting the $Find$ operation on G, $C_{Find} \cdot \frac{M}{m\delta} = \Omega(D(G))$. (Similarly for T_{Find}). ∎

4 Two departure points

A natural starting point in the evolution of distributed data structures is to consider the centralized organization. The CENTRAL structure is based on storing a single copy of all data items at one central processor r. This structure can be set up using minimal space. However, it is clearly not memory balanced (in accordance with Lemma 3.1). The access costs of this structure depend on the distance of the requesting processor from the center r, which is bounded by $D(G)$. To summarize, recalling our assumption regarding the existence of efficient routing services in the network, we get

Proposition 4.1 It is possible to implement a compact dictionary based on the CENTRAL structure with $T_{Op} = O(D)$ and $C_{Op} = O(D\delta)$ for any operation Op. The memory requirements of the resulting structure satisfy $M = \hat{M} = O(m\delta)$. ∎

Our second point of departure involves design strategies based on trees. Tree structures are traditionally used for storing ordered data, and specifically for implementing a dictionary (cf. [AHU83]). For the sake of illustration we describe tree structures that attempt to maintain perfect balancing, meaning that each of the n sites stores either $Load$ or $Load + 1$ items. In our later constructions we relax this requirement.

Broadly speaking, tree-based strategies fall into two main categories. The first type is what we call *static* trees. In such strategies, the tree structure itself is explicit and fixed, i.e., it corresponds to some specific spanning tree of the network. It has n nodes, each corresponding to some vertex of the network and possibly storing many items. However, the data allocation is dynamic, and data items keep flowing along the edges of the tree in order to guarantee memory balancing.

Discussion: Static trees are well suited for VLSI dictionary machines, discussed in Section 1.2. In the distributed setting, a static tree can be used to implement a priority queue with balanced memory and $O(D)$ time and communication per instruction. The construction resembles the priority queue machine of [Lei79], and will be described in the full paper. As for the use of static trees for implementing a dictionary, insertions create a problem due to the need to shift long data chains from time to time, in order to maintain memory balance. This results in $O(n)$ communication complexity for insertions. There are various possible measures that can be taken in order to reduce the impact of this effect, and they will be discussed in the full paper. ∎

The other type, on which we shall concentrate for the rest of this abstract, is referred to as *dynamic* trees. Here we are essentially concerned with simulating traditional tree-based data structures in the network. For concreteness let us consider the 2-3-TREE structure. The topology of such a tree is dynamic in nature. Nodes may join the tree or be deleted, split or be merged. The tree is "virtual", and does not correspond to the topology of any particular subnetwork. It has m nodes, each corresponding to a single item. Thus several nodes may reside in the same site in the network. The embedding dilation of this virtual tree in the network may be greater than 1, that is, some of the edges of the tree may connect nonneighboring vertices of the network. Basically, the structure is obtained by viewing the sites of the network as "memory blocks" and trying to implement the traditional 2-3 tree.

All queries and modification instructions are cleared through a central coordinator, r. This coordinator holds a pointer to the root of the tree, and performs the modifications by simply simulating the standard algorithm. That is, it traces pointers in search of the right insertion location, splits, adds and releases nodes as necessary, and so on.

There is an additional component to this structure (and later constructions), namely, the list of "free" vertices available for storing data. This component consists of a substructure called LIST. It has the form of a doubly-linked list of vertices, connected by pointers (which are simply the addresses of the next vertex on the list). The LIST structure is managed by the central coordinator.

We omit the operation details of the 2-3-TREE structure from the abstract. The only complications involve the need to maintain memory balance in the vertices, which requires some additional activities.

Since the 2-3-TREE has m nodes that are stored at arbitrary locations in the network, every pointer traversal may require $O(D(G))$ messages. Each instruction requires $O(\log m)$ basic manipulation operations on the tree or the LIST structure. The complexities of the structure are therefore as follows.

Proposition 4.2 It is possible to implement a compact dictionary based on the 2-3-TREE structure with $T_{Op} = O(D \log m)$ and $C_{Op} = O(D \log m \delta)$ for every operation Op. The memory requirements of the resulting structure satisfy $M = O(m\delta + n \log n)$, $\hat{M} = O(m\delta/n + \log n)$. Thus the dictionary is memory-balanced as long as $m = \Omega(n \log n)$. ∎

5 The BIN dictionary

In this section we describe a compact dictionary structure, called BIN, that guarantees both memory-balancing and optimal complexities ($O(D(G)\delta)$) for all three operations simultaneously, provided m is sufficiently large (specifically, $m = \Omega(n^2)$). The basic idea behind this structure is implementing a variation of the well-known multiway tree or B-tree (cf. [Knu73]).

The solution is based on a "flat" tree consisting of two levels: a central vertex r serving as a "directory", and a collection of bins $B_1, \ldots, B_p$ (each maintained in some vertex) storing the data in ordered fashion. Vertices that currently do not store a bin are kept in a linked list of "free bins", using the LIST structure described earlier.

It is convenient to describe the solution first assuming that the *Load* parameter does not change throughout the execution. Later we describe how to modify the solution to remove this restriction.

Each of the bins B_i stores a consecutive segment of k_i data items, $1 \leq k_i \leq 2(Load + 1)$. That is, bin B_1 stores the first k_1 items, bin B_2 stores the next k_2 items, and so on. A bin B_i is said to be *full* if it stores $k_i = 2(Load + 1)$ items. The central manager r stores, for every bin B_i, $i > 1$, the extreme key $Low(B_i)$. This is the lowest key among all items stored at B_i. The center also stores the identity of the site where the bin B_i is stored.

It is clear that locating the appropriate placement of a key (in either of the three operations) can be done in $O(D\delta)$ messages, since the central manager r is able to determine that location locally using the *Low* pointers. An item with key Key is inserted to the bin B_i such that $Low(B_i) \leq Key < Low(B_{i+1})$ or to B_1 if there is no such i. (We assume the convention that $Low(B_{p+1}) = \infty$.)

In order to guarantee balanced distribution we require the dictionary to maintain the following *density invariant*:

(DI1)   At any given time, for any $1 \leq i < p$, either B_i or B_{i+1} are full.

Maintaining the density invariant guarantees that the number of occupied bins, p, is at most n. It is also clear that each vertex but the center r stores no more than $O(Load)$ items. The center may store up to $O(n)$ pointers, hence the structure is memory balanced when $m \geq 2n^2$. (We comment that it is possible to devise a variant k-BIN of the BIN data structure corresponding to a k-level multiway tree, that guarantees balanced memory when $m = O(n^{1+1/k})$. However, we shall not describe this variant in the abstract.)

It remains to describe how to perform update operations efficiently (i.e., with $O(D\delta)$ messages) while maintaining this invariant. Insertion is performed as follows. The item to be inserted is sent to r, which forwards it to the appropriate bin B_i. This bin stores the item locally in the right place. If B_i was not full, then the operation is completed. Now suppose that B_i was full before the new item was inserted. Then B_i needs to move one of its extreme items to an adjacent bin. If B_{i-1} or B_{i+1} are not full, the appropriate item can be moved there. Otherwise, a new bin is created (using a "free" vertex that was not used so far, taken from the LIST structure storing the free bins). This bin can be created either between B_{i-1} and B_i or between B_i and B_{i+1}. The insertion and consequent moves may also change the relevant *Low* pointers at the center r. Deletion is performed analogously.

The entire operation is coordinated by the center r, including determining which item to move and to which neighbor. The center therefore needs to keep count of the current load in each bin. It also manages the operations of the LIST structure. Note that all the additional bookkeeping operations required for the update operations can be carried out using $O(D\delta)$ additional messages. Thus as long as *Load* does not change, all update operations can be performed correctly with worst-case communication complexity $O(D\delta)$ per operation (i.e., a non-amortized bound).

Handling load changes: Special care has to be taken when *Load* increases due to an increase in the number of items stored in the dictionary. When this happens, all full bins suddenly cease to be full. (Analogously, when *Load* decreases, all full bins overflow.) Insisting on fixing the situation whenever that happens might prove extremely expensive if *Load* "thrashes" up and down when m is nearly a multiple of n and the stream of instructions contains alternating insertions and deletions.

The way to overcome this problem is by introducing an extra degree of freedom into our density invariant. Let us allow bins to contain up to $2Load + 6$ items each, and call a bin B_i *near-full* if it stores $2Load + 2 \leq k_i \leq 2Load + 6$ items. The density invariant is modified to state that:

(DI2)   At any given time, for any $1 \leq i < p$, either B_i or B_{i+1} are near-full.

Insertions and deletions are performed as before (replacing "full" with "near-full" as required), except that whenever a bin B_i that is currently not near-full needs to be made near-full, its load is set to the middle point of $2Load + 4$.

We now argue that repeated dangling of *Load* between two values ℓ and $\ell + 1$ cannot affect bins that do not otherwise modify their data. In other words, each balancing operation can be "charged" to some specific operation carried out in the bin in question. Details of the argument are omitted. The conclusion is that update costs due to *Load* changes can be amortized over previous operations, resulting in an additional amortized cost of $O(D\delta)$ per operation.

Proposition 5.1 It is possible to implement a compact dictionary based on the BIN structure with amortized worst-case complexities $T_{Op} = O(D)$ and $C_{Op} = O(D\delta)$ for every operation Op. The memory requirements of the resulting structure satisfy $M = O(m\delta)$, $\hat{M} = O(m\delta/n + n\delta)$. Thus the dictionary is memory-balanced as long as $m = \Omega(n^2)$. Furthermore, for $m = \Omega(n^{1+1/k})$ (for fixed $k \geq 1$) it is possible to implement a compact, memory-balanced variant k-BIN with kD replacing D in the complexity bounds. ∎

In subsequent work [PZ90] we handle the issue of devising a data structure that enjoys similar properties while eliminating the constraint $m = \Omega(n^2)$.

Background balancing and actual amortization: The above solution has the problem that a change in *Load* requires a burst of activity. This activity is composed of two parts: moving items between neighboring bins in order to regain the density invariant, and informing the center r of the new *Low* pointers. The first type of activity is less of a problem, as every vertex is required to participate in sending / receiving only $O(1)$ messages. However, the center may need to process up to $O(n)$ messages. It is preferable to have "smoother" behavior of the structure, in which the worst-case complexity of each individual operation is bounded individually. This can be achieved upon noting that the type of balancing activity required after a change of *Load* can actually be postponed, or deferred, over a longer period of time, and performed in the background. This demonstrates what one may term "actual amortization." The idea is that instead of just amortizing operations "virtually" in the usual accounting sense, in a distributed environment it is possible to defer maintenance operations (such as balancing) till later or spread them over a longer period of time during which they are to be performed "in the background". We omit a more detailed description of how this can be done in our specific case without violating the correct behavior of the structure.

Proposition 5.2 It is possible to implement the BIN dictionary with the same bounds on the individual (rather than amortized) worst-case complexities of the operations. ∎

6 Distributing the center

A problematic point with all of the solutions discussed so far is that they place an extreme work load on some central manager. We now consider a variation of the BIN structure that does away with the need to use a central manager. The basic idea is to nominate *several* centers for the structure, let all of them store identical information, and split the "costumers" evenly among the centers. Pushing this idea to its natural extreme, we may cancel the center altogether and let each processor serve as its own manager (storing the pointer $Low(B_i)$ and the identity of the storing site of B_i for each bin B_i). This change alone reduces the load as far as $Find$ operations are concerned, since each processor is involved (in its capacity as a "center") in processing only its own instructions (naturally, it is otherwise involved in other's requests if it resides on the route between a requesting processor and the appropriate storing bin or if it stores this bin itself). However, insertions and deletions may require updating all processors, since the Low keys of some bins may change. This means that the work load on processors due to modification operations remains as bad as the load on the single center of the original scheme, and worse, the total cost of a modification operation increases to $O(nD\delta)$.

It is thus necessary to change the scheme further. The idea is to enable "most" modification operations to be performed without any "spillovers" between bins, and hence without having to update all processors. A bin is now allowed to store up to 5γ items, for $\gamma = 2(Load + 1)$. We classify bins according to their load as follows. The bin B_i, storing k_i items, is said to be *near-full* if $\gamma + 1 \leq k_i \leq 5\gamma$, and *near-empty* if $1 \leq k_i \leq \gamma$. The density invariant is modified once again, to state that:

(DI3)   At any given time, for any $1 \leq i \leq p$, bin B_i is near-full.

The rules governing changes in the structure of bins are:

1. Whenever a bin overflows, it splits into two near-equal bins storing 2.5γ or $2.5\gamma + 1$ items each.
2. Whenever a bin B_i becomes near-empty, it is balanced with a neighboring bin B (either B_{i-1} or B_{i+1}) as follows. If B has fewer than 3γ items, then B_i is cancelled and its items are moved to B. Otherwise, B moves γ of its items to B_i.

These changes are now initiated and managed by the involved bins themselves, since there is no responsible center. The LIST structure of free bins is also managed by all vertices.

Note that the pointers $Low(B_i)$ remain fixed between any two consecutive structural changes. In particular, it is possible that the item whose key is $Low(B_i)$ is deleted from B_i at some stage, but the pointer does not change until the next split or merge involving this bin.

Lemma 6.1 After each bin split or merge, every involved bin stores $2\gamma \leq k \leq 4\gamma$ bins. ∎

It follows that a bin has to be split or merged at most once per γ modification operations accessing it. Whenever that happens, it is indeed necessary to update all processors regarding the new boundary pointers of the bins involved in the change and the status of the LIST structure. This is done by broadcasting this information throughout the network, which costs $O(n\delta)$ messages and $O(D)$ time. Amortized over the *Load* modification operations preceding it, this amounts to another $O(1)$ messages per operation (recalling $Load \geq n$). Again, "virtual" amortization can be replaced by an "actual" one, performing balancing operations in the background.

Proposition 6.2 It is possible to implement the (centerless) DIST-BIN dictionary with the same complexity bounds as the BIN dictionary. ∎

7 Introducing redundancy

The concept of *redundancy* or *replication* plays a vital role in the design of distributed data structures. A major motivation for using redundancy is fault-tolerance (which, as mentioned earlier, is mostly ignored here), namely, overcoming crashes of processors and storage devices and enhancing data availability in the face of communication failures, including possible network partitions. Two other purposes (which are addressed here) are reducing access costs by placing the data closer to its potential users, and work load balancing. In order to use redundancy, it is necessary to develop techniques for efficient decomposition of the network into logical regions, both for data placement and for distributing the tasks and responsibilities involved in managing the data. In this section we illustrate these potential uses of redundancy.

The most extensive form of redundancy, forming an opposite extreme to the class of compact structures, is presented by what we call the TOTAL structure, which is based on storing a copy of the entire data bank in each site of the system. This structure enjoys high availability, but suffers extremely high memory requirements. A reasonable approach is to try to strike a balance between memory requirements and efficiency of searches by adopting an intermediate level of redundancy. This is the direction taken in the sequel.

The first thing we need for that purpose is a partitioning of the network into "zones", or clusters. Formally, a *cluster* is a subset of vertices $S \subseteq V$ such that the subgraph

$G(S)$ induced by S in G is connected. A *partition* of V is a collection of clusters $\mathcal{S} = \{S_1, \ldots, S_\ell\}$ such that $\bigcup_i S_i = V$ and $S_i \cap S_j = \emptyset$ for every $1 \leq i < j \leq \ell$. Let $D(S)$ denote the diameter of $G(S)$. Given a partition $\mathcal{S}$, let $D(\mathcal{S}) = \max_i D(S_i)$. The *minimal size* of a partition $\mathcal{S}$ is $size(\mathcal{S}) = \min\{|S| \mid S \in \mathcal{S}\}$.

Now, it is possible to view each cluster as an autonomous subnetwork that has to maintain its own copy of the data structure locally. Let us now describe the structure CLUSTERED, based on using the DIST-BIN directory inside clusters. Let $t = |\mathcal{S}|$ and $n_c = size(\mathcal{S})$. Define $Load_c = \lfloor m/n_c \rfloor$. In each cluster $S_i \in \mathcal{S}$, $1 \leq i \leq t$, we store an identical copy of the entire collection of p bins, denoted $B_1^i, \ldots, B_p^i$. The definitions of near-full and near-empty bins (and the implied density invariant) are now based on $Load_c$. The modified density invariant guarantees that $p \leq n_c$. A *Find* instruction is carried out by querying the local copy of the appropriate bin. *Insert* and *Delete* instructions are sent to all copies of the relevant bin, in all clusters. In case there is a need to merge or split the bin, this is done separately in each of the clusters. Operational details and analysis are omitted. We get

Proposition 7.1 Using a partition $\mathcal{S}$ for the graph G, and assuming $m = \Omega(n^2)$, it is possible to implement a dictionary based on the CLUSTERED structure with $T_{Ins} = T_{Del} = O(D(G))$, $T_{Find} = O(D(\mathcal{S}))$, $C_{Ins} = C_{Del} = O(\min\{n\delta, |\mathcal{S}|D(G)\delta\})$ and $C_{Find} = O(D(\mathcal{S})\delta)$. The memory requirements of the resulting structure satisfy $\rho_{\max} = |\mathcal{S}|$, $M = O(|\mathcal{S}|m\delta)$ and $\hat{M} = O(m\delta/size(\mathcal{S}))$. ∎

We would therefore like to base our structure on a partition $\mathcal{S}$ with relatively small parameters $|\mathcal{S}|$ and $D(\mathcal{S})$ and a sufficiently large parameter $size(\mathcal{S})$. We construct the desired partition via a two-step process. First, we pick a set C of vertices that are well-spread in the network. For such a set C of centers, let $D(C)$ denote $\max_{v \in V} \{\min_{c \in C} \{dist(v, c)\}\}$. For our purposes we need to get simultaneously small $|C|$ and $D(C)$. The associated optimization problem is hard (in the NP sense). We may rely on various approximation algorithms for fixing one of these parameters and seeking to minimize the other ([Lov75,HS84], see also [ABLP89,BP90] for a more detailed discussion).

In the second step we construct a cluster around each of the centers of C. This requires us to take care of the third parameter, namely, make sure that $size(\mathcal{S})$, the size of the minimal "constituency" in $\mathcal{S}$, is also not too small. Clearly, if one insists on guaranteeing that the partition has diameter $\leq \ell$, then the constituency cannot in general be made larger than $O(\ell)$ (as can be verified by considering graphs consisting of a single path). The following lemma establishes that this bound can be met.

Lemma 7.2 For every graph G and every $C \subseteq V$, there exists an (efficiently constructible) partition $\mathcal{S}$ satisfying $|\mathcal{S}| \leq |C|$, $D(\mathcal{S}) \leq 3D(C)$ and $size(\mathcal{S}) \geq D(C)$. ∎

Corollary 7.3 Using a partition constructed as in Lemma 7.2 based on any set of centers C, it is possible to implement the CLUSTERED dictionary structure (assuming $m = \Omega(n^2)$), with $T_{Ins} = T_{Del} = O(D(G))$, $T_{Find} = O(D(C))$, $C_{Ins} = C_{Del} = O(n\delta)$ and $C_{Find} = O(D(C)\delta)$. The memory requirements of the resulting structure satisfy $\rho_{\max} = |C|$, $M = O(|C|m\delta)$ and $\hat{M} = O(m\delta/D(C))$. ∎

Discussion: In the area of replicated data it is common to study such schemes under the assumption that only part of the copies are updated, and look for ways to maintain consistency under such an assumption, usually based on quorum voting (cf. [BHG87,Gif79]). Basically, there is a tradeoff between the costs of queries and updates; the more copies are updated, the less need to be read. Thus, at the other extreme one may consider a scheme in which *all* copies are read when executing a *Find* operation, and only one needs to be touched when updating the record (time-stamps may be used to distinguish between incompatible copies). In this abstract we chose to focus on efficient *Find* operations, and ignore the rest of the spectrum. ∎

8 Incorporating concurrent accesses

In all the discussion so far, we assumed that the data structure at hand is to be accessed *sequentially*, and instructions are issued and processed one by one. However, an essential requirement from a practical distributed data structure is that it support multiple simultaneous accesses (of both queries and modifications). In this section we briefly discuss some of the issues involved in allowing concurrent executions of instructions.

When trying to accommodate for concurrent accesses, our main concern involves the correctness aspect of the problem, namely, how to ensure and maintain the consistency of the data. This aspect was treated extensively in database theory, using techniques known collectively as concurrency control (cf. [BHG87,Her84,Her87,LM79,Pap86,Wei84]). The typical approach is to try to guarantee the *serializability* of executions for a given sequence of operations. For general data structures we may actually require a stronger consistency property, called "order preserving serializability" in [LS88] (and reminiscent of the definitions of Lamport, cf. [Lam86]). Essentially, this property requires that if, in the actual computation, an instruction I is completed before another instruction I' starts, then this order is preserved in the serialization.

In structures that are coordinated by a central processor, serializability can be enforced by the coordinator simply by handling the requests in order of arrival, starting to process the next instruction only after completing the previous one. Things become more interesting when we consider structures that are not centrally coordinated. The types of problems that may arise in this case revolve around inconsistent views of the current contents and internal organization of the structure resulting from different arrival orders of nearly simultaneously issued modification instructions or messages. In the full paper we describe a "truly concurrent" version of the DIST-BIN structure and the CLUSTERED structure based on it. The transformation is based on standard methods, essentially enforcing a limited subset of possible orderings of the sub-operations, using lock-protected bin reconfiguration procedures and restricted communication mechanisms such as the multicast primitive (cf. [BJ87,CM84,GS89a]). The access costs remain bounded as before. Details are omitted. To summarize,

Proposition 8.1 It is possible to implement the concurrent version of the CLUSTERED dictionary with the same bounds on communication complexity and memory as the sequential version. ∎

It should be stressed, however, that when it comes to *time* complexity, the concurrent setting raises some interesting (and difficult) complexity questions. One may divide these issues into two main categories. The first involves the need to minimize the extent of using mutual exclusion and the duration of locking periods. For this purpose it is possible to use techniques developed for concurrent data structures in shared memory. For example, instead of locking the entire 2-3-TREE throughout the duration of an operation, it is possible to lock only the necessary sections of the tree, releasing sections that are no longer necessary, and thus enabling other transactions to proceed independently.

The second category is unique to the distributed setting, and involves the avoidance or sufficient control of traffic bottlenecks, the control of "hot-spots" [PN85] and balancing processor work loads. In order to gain some intuition for this problem, let us suppose that n instructions are issued in the system simultaneously, one at each site, and define the *concurrency load* CL as the maximum number of different instructions whose processing requires access to a particular processor. (Note that we tacitly assume that accesses, i.e., messages and internal searches, cannot be "combined".) We now observe that deterministic strategies cannot guarantee small concurrency load. First, assume that the insertion function f_I and the search function f_F depend only on the processor issuing the instruction and the key involved in it (i.e., $f_I, f_F : V \times \mathcal{K} \mapsto 2^V$), and that the search function is *rigid*, meaning that executing $Find(v, Key_X)$ always requires accessing *every* vertex in $f_F(v, Key_X)$. Under these severe assumptions, it is claimed that

Lemma 8.2 In a dictionary structure based on an insertion function f_I and a search function f_F such that $f_I, f_F : V \times \mathcal{K} \mapsto 2^V$ and f_F is rigid, $CL = \Omega(\sqrt{n})$. ∎

The upper bound may be achieved by using insertion and search protocols based on functions derived from two constructions of [MV85] for distributed match making, namely, the symmetric construction and the projective plane based construction. Proof of the lower bound is omitted.

More generally, the following lower bound can be stated.

Lemma 8.3 In any dictionary structure, $CL \cdot \rho_{\max} = \Omega(n)$ (assuming sufficiently large m). ∎

Discussion: Note that these bounds hold even in a complete network, where communication bottlenecks are not a problem. These results suggest that in order to get reasonably bounded time complexities for the concurrent case, one is forced to resort to randomized constructions, for which *probabilistic* bounds can be proved. This approach was taken in the context of parallel computing (specifically, in simulating PRAM algorithms on realistic parallel computers), cf. [DMadH90,KU86,Ran87,UW87]. That context is different from ours in that the underlying network is assumed to have some specific topology (e.g., AKS, hypercube or complete network), and the machine operates synchronously. Rather tight complexity bounds are given for both time and memory balance, although both are of a probabilistic nature. It is plausible that analogous results (with appropriate slowdown) can also be achieved in the distributed (asynchronous, arbitrary topology) framework. This line of attack is left (together with the issue of fault tolerance) as a major direction for future research. ∎

9 Conclusion

The paper presents a detailed example of a distributed data structure. The structure is developed via a process consisting of several stages. Each of these stages is geared towards handling a particular aspect of the problem, such as low memory requirements, memory balance and efficient access protocols, centerless organization, employing redundancy for increased search efficiency and concurrent accesses.

Although the paper concentrates on a single example for illustration purposes, it should be clear that the general approach and most of the high-level considerations apply to a wide range of data structure types. In the full paper we discuss methods for implementing other basic types of structures besides dictionaries (e.g., stacks, queues and priority queues), as well as several other design strategies that do not fall under the categories discussed earlier. These include local organizations, in which every inserted data item is kept at (or near) the inserting site, "hybrid" structures composed of local storage plus some global "directory" mechanism supporting efficient searches, and round-robin insertion policies.

The general approach proposed in this paper is well suited for discussing permanent, autonomous data structures. Similar complexity considerations apply also to "short-lived" data structures that are internal to some specific distributed algorithm. However, for this type of structures, other considerations might be present and possibly dominant. Consequently, it may be necessary to develop an entirely different class of strategies for such structures. This direction, too, is left for future research.

Acknowledgments

I would like to thank Baruch Awerbuch, Judit Bar-Ilan, Israel Cidon, Oded Goldreich, Shay Kutten, Nancy Lynch and Yoram Moses for helpful discussions.

References

[ABLP89] B. Awerbuch, A. Bar-Noy, N. Linial, and D. Peleg. Compact distributed data structures for adaptive routing. In *Proc. 21st ACM Symp. on Theory of Computing*, pages 230–240, Seattle, Washington, May 1989.

[AHU83] A.V. Aho, J.E. Hopcroft, and J.D. Ullman. *Data Structures and Algorithms*. Addison-Wesley Publishing Co., Reading, MA, 1983.

[AK85] M.J. Atallah and S.R. Kosaraju. A generalized dictionary machine for VLSI. *IEEE Trans. on Computers*, C-34:151–155, 1985.

[Awe85] B. Awerbuch. Complexity of network synchronization. *J. of the ACM*, 32:804–823, 1985.

[BHG87]   P.A. Bernstein, V. Hadzilacos, and N. Goodman. *Concurrency Control and Recovery in Database Systems*. Addison-Wesley Publishing Co., Reading, MA, 1987.

[BJ87]   K.P. Birman and T.A. Joseph. Reliable communication in the presence of failures. *ACM Trans. on Comput. Syst.*, 5:47–76, 1987.

[BP90]   J. Bar-Ilan and D. Peleg. *How to Assign Network Centers*. Technical Report CS90-20, The Weizmann Institute, August 1990.

[BS79]   R. Bayer and M. Schkolnick. Concurrency of operations on B-trees. *Acta Informatica*, 9:1–21, 1979.

[CCIR86] J.H. Chang, M.J. Chung, O.H. Ibarra, and K.K. Rao. Systolic tree implementation of data structures. In *Proc. IEEE Int. Conf. on Parallel Process.*, pages 669–671, IEEE, 1986.

[CGL86]   N. Carriero, D. Gelernter, and J. Leichter. Distributed data structures in Linda. In *Proc. 13th ACM Symp. on Principles of Prog. Lang*, pages 236–242, ACM, 1986.

[CM84]   J. Chang and N.F. Maxemchuk. Reliable broadcast protocols. *ACM Trans. on Comput. Syst.*, 3:251–273, 1984.

[DMadH90] M. Dietzfelbinger and F. Meyer auf der Heide. How to distribute a dictionary in a complete network. In *Proc. 22nd ACM Symp. on Theory of Computing*, 1990. to appear.

[DS85]   W.J. Dally and C.L. Seitz. *The Balanced Cube: A Concurrent Data Structure*. Technical Report 5174:TR:85, California Institute of Technology, 1985.

[DS87]   F. Dehne and N. Santoro. Optimal VLSI dictionary machines on meshes. In *Proc. IEEE Int. Conf. on Parallel Process.*, pages 832–840, IEEE, 1987.

[Ell80a] C.S. Ellis. Concurrent search and insertion in 2-3 trees. *Acta Informatica*, 14:63–86, 1980.

[Ell80b] C.S. Ellis. Concurrent search and insertion in AVL trees. *IEEE Trans. on Computers*, C-29:811–817, 1980.

[Ell85]   C.S. Ellis. Distributed data structures, a case study. *IEEE Trans. on Computers*, C-34:1178–1185, 1985.

[FWB85]   A.J. Frank, L.D. Wittie, and A.J. Bernstein. Maintaining weakly-consistent replicated data on dynamic groups of computers. In *Proc. IEEE Int. Conf. on Parallel Process.*, pages 155–162, IEEE, 1985.

[GHS83]   R.G. Gallager, P.A. Humblet, and P.M. Spira. A distributed algorithm for minimum weight spanning trees. *ACM Trans. on Programming Lang. and Syst.*, 5:66–77, 1983.

[Gif79]   D.K. Gifford. Weighted voting for replicated data. In *Proc. 7th ACM Symp. on Operating Systems Principles*, pages 150–162, ACM, Pacific Grove, CA, December 1979.

[GK89]   P.M. Gopal and B.K. Kadaba. *Analysis of a Class of Distributed Directory Algorithms*. Research Report RC-14488, IBM Yorktown, March 1989.

[GS89a]   H. Garcia-Molina and A. Spauster. Message ordering in a multicast environment. In *Proc. 9th IEEE Conf. on Distributed Computing Systems*, Newport Beach, CA, 1989.

[GS89b]   D. Ginat and A.U. Shankar. Decentralized ordering of contending nodes in a distributed system. 1989. unpublished manuscript.

[Her84]   M. Herlihy. *Replication Methods for Abstract Data Types*. Technical Report TR-319, MIT, Lab. for Computer Science, May 1984.

[Her87]   M. Herlihy. Concurrency versus availability: atomicity mechanisms for replicated data. *ACM Trans. on Comput. Syst.*, 5:249–274, 1987.

[Her90]   M. Herlihy. A methodology for implementing highly concurrent data structures. In *Proc. ACM PPoPP*, ACM, 1990.

[HS84]   D.S. Hochbaum and D. Shmoys. Powers of graphs: a powerful technique for bottleneck problems. In *Proc. 16th ACM Symp. on Theory of Computing*, pages 324–333, ACM, April 1984.

[HW87]   M. Herlihy and J. Wing. Axioms for concurrent objects. In *Proc. 14th ACM Symp. on Principles of Prog. Lang*, pages 13–16, ACM, January 1987.

[KL80]   H.T. Kung and P.L. Lehman. Concurrent manipulation of binary search trees. *ACM Trans. on Programming Lang. and Syst.*, 5:339–353, 1980.

[Knu73]   D.E. Knuth. *The Art of Computer Programming, Vol. 3*. Addison-Wesley Publishing Co., Reading, MA, 1973.

[KU86]   A. Karlin and E. Upfal. Parallel hashing - an efficient implementation of shared memory. In *Proc. 18th ACM Symp. on Theory of Computing*, May 1986.

[Lam86]   L. Lamport. On interprocess communication, parts I and II. *Distributed Computing*, 1:77–101, 1986.

[LEH85]   K.A. Lantz, J.L. Edighoffer, and B.L. Histon. Towards a universal directory service. In *Proc. 4th ACM Symp. on Principles of Distributed Computing*, pages 261–271, August 1985.

[Lei79]   C.E. Leiserson. *Systolic Priority Queues*. Technical Report CMU-CS-79-115, Carnegie-Mellon University, 1979.

[LM79]   N. Lynch and M. Merritt. Introduction to the theory of nested transactions. In *Int. Conf. on Database Theory*, pages 278–305, Rome, Italy, 1979.

[Lov75]   L. Lovász. On the ratio of optimal integral and fractional covers. *Discrete Mathematics*, 13:383–390, 1975.

[LS88]   V. Lanin and D. Shasha. Concurrent set manipulation without locking. In *Proc. 7th ACM Symp. on Principles of Database Systems*, pages 211–220, ACM, 1988.

[Man86]   U. Manber. On maintaining dynamic information in a concurrent environment. *SIAM J. on Comput.*, 15:1130–1142, 1986.

[ML84]   U. Manber and R.E. Ladner. Concurrency control in a dynamic search structure. *ACM Trans. on Database Syst.*, 9:439–455, 1984.

[MV85]   S.J. Mullender and P.M.B. Vitányi. Distributed match-making. *Algorithmica*, 3:367–391, 1985.

[OD81]   D. Oppen and Y.K. Dalal. *The Clearinghouse: A Decentralized Agent for Locating Named Objects in a Distributed Environment*. Technical Report OPD-T8103, Xerox Corp., October 1981.

[ORS82]   T.A. Ottman, A.L. Rosenberg, and L.J. Stockmeyer. A dictionary machine for VLSI. *IEEE Trans. on Computers*, C-31:892–897, 1982.

[Pap86]   C.H. Papadimitriou. *The Theory of Concurrency Control*. Computer Science Press, Rockville, MD, 1986.

[Pel89]   D. Peleg. *Distance-Dependent Distributed Directories*. Technical Report CS89-10, The Weizmann Institute, May 1989.

[PN85]   G.F. Pfister and V.A. Norton. "hot spot" contention and combining in multistage interconnection networks. *IEEE Trans. on Computers*, C-34:943–948, 1985.

[PU89]   D. Peleg and E. Upfal. A tradeoff between size and efficiency for routing tables. *J. of the ACM*, 36:510–530, 1989.

[PZ90]   D. Peleg and K. Ziegler. A compact, memory-balanced, deterministic distributed dictionary. 1990. In preparation.

[Ran87]   A. Ranade. How to emulate shared memory. In *Proc. 28th IEEE Symp. on Foundations of Computer Science*, pages 185–194, October 1987.

[Ree78]   D.P. Reed. *Naming and Synchronization in a Decentralized Computer System*. PhD thesis, MIT, Dept. of Electrical Engineering, 1978.

[RK88]   V.N. Rao and V. Kumar. Concurrent insertions and deletions in a priority queue. In *Proc. IEEE Int. Conf. on Parallel Process.*, pages 207–211, IEEE, 1988.

[Sag85]   Y. Sagiv. Concurrent operations on B-trees with overtaking. In *Proc. 4th ACM Symp. on Principles of Database Systems*, pages 28–37, ACM, January 1985.

[Seg83]   A. Segall. Distributed network protocols. *IEEE Trans. on Info. Theory*, IT-29:23–25, 1983.

[SK85]   N. Santoro and R. Khatib. Labelling and implicit routing in networks. *The Computer Journal*, 28:5–8, 1985.

[SL87]   A.M. Schwartz and M. Loui. Dictionary machines on cube-class networks. *IEEE Trans. on Computers*, C-36:100–105, 1987.

[Ter87]   D.B. Terry. Cashing hints in distributed systems. *IEEE Trans. on Software Eng.*, SE-13:48–54, 1987.

[UW87]   E. Upfal and A. Wigderson. How to share memory in a distributed system. *J. of the ACM*, 34:116–127, 1987.

[Wei84]   W.E. Weihl. *Specification and Implementation of Atomic Data Types*. Technical Memo MIT/LCS/TM-314, MIT, Lab. for Computer Science, March 1984.

An Improved Algorithm to Detect Communication Deadlocks in Distributed Systems

B. Kröger[*], R. Lüling, B. Monien, O. Vornberger[*]

Department of Mathematics and Computer Science
University of Paderborn, West Germany
e-mail : berti@dosuni.uucp, rl@uni-paderborn.de

Abstract

This paper presents a new algorithm for the detection and resolution of communication deadlocks in distributed systems. The algorithm is based on some well known concepts for distributed deadlock detection and adds some new features to reduce message- and space complexity. It was implemented on a transputer network and shown to be more efficient than previously published algorithms.

1 Introduction

In message-based distributed systems the occurrence of communication deadlocks is an inherent problem. A synchronizing exchange of messages between two processes forces the process which arrives at its communication statement first to wait as long as the other processing unit is ready to perform the corresponding communication statement. A number of processing units which are directly or indirectly waiting for each other and whose waiting is endless regardless of any communication which might subsequently be initiated are said to be in a deadlock.

In general we distinguish deadlocks in the AND/OR - model [5]. In the AND - model it is necessary to allocate all the required recources, whereas in the OR - model it is sufficient to allocate one of these resources to finish the blocking of a process. The communication deadlock which is dealt with in this paper corresponds to a deadlock in the OR - model. At least one of the required resources (i.e. here: the readiness to perform a corresponding communication) has to become available for a waiting process. If a processing agent can request exactly one resource at a time the deadlock detection is reduced to a circle detection amongst the processing units [8]. In [2] a deadlock detection algorithm for the general *n-out-of-m model* is presented to which all other models are adaptable. In this paper the term *deadlock* will always be used according to the general OR - model and refers to a communication deadlock like above.

[*]Now with the Department of Mathematics and Computer Science at the University of Osnabrück, West-Germany

Deadlocks can be handled in different ways [4] : A deadlock *prevention* as well as a deadlock *avoidance* strategy restricts the communication behaviour of a distributed calculation to ensure its deadlock-freeness. A deadlock *detection* strategy however, determines only those communications which actually cause a deadlock. These deadlocks might be removed by a following *resolution* phase. Since the occurance of deadlocks is undecidable by any a-priori analysis [6], deadlock detection of this kind can only be performed while running the distributed algorithm to be observed.

As a consequence, this paper deals with a distributed algorithm to detect and remove each deadlock which appears at runtime of a distributed computation. Some pioneering work referring to this topic has be done by Chandy/Misra/Haas [3] and Natarajan [7]. They introduced distributed protocols which run in parallel to the real work of a processing agent and determine at least one agent from a set of deadlocked units to report this fact.

The algorithm in [7] is based on a periodic protocol which elects exactly one process from a deadlocked set to resolve it. In the algorithm of [3] each blocked process builds its own distributed spanning tree on the set of blocked processes to verify an assumed deadlock. That's why this algorithm needs O(n) space complexity for each processing unit in a network of n processes. Another disadvantage of this algorithm is the missing support of deadlock resolution, since several processes might detect the same deadlock and no processing unit is elected to start a resolution phase. The proposed algorithm overcomes all these disadvantages. Its event driven protocol selects exactly one process from a deadlocked set to reveal a deadlock. In a subsequent step this process either resolves the deadlock directly or it starts a more sophisticated method to determine one process to resolve it [1]. All this is done with space complexity O(d), where d is the number of neighboring processes. Furthermore an additional lookahead technique ensures an extended detraction of control messages.

The algorithm was integrated in a simulation environment for distributed computations which runs on a network of transputers. The algorithms of Chandy/Misra/Haas [3] and Natarajan [7] were also included to qualify the efficiency of our algorithm. Experimental results show that the proposed algorithm behaves superior to the ones in [3, 7].

The paper is organized as follows: The next section introduces the communication model taken as a basis for our algorithm and formally defines a communication deadlock. Our distributed deadlock detection and resolution algorithm is presented in section three. In the fourth part it is compared to different algorithms which were also implemented in OCCAM on a transputer network. Concluding remarks will finish this paper.

2 Fundamentals

A distributed computation consists of a set of autonomous, communicating processes whose communication is exclusively done by the exchange of messages (i.e. no global memory is available). All communications take place on a communication network which connects pairs of processing units.

The communication network can be viewed as an undirected graph $G = (V,E)$ with the set of nodes $V = \{v_1, ..., v_n\}$ representing the processing units. An edge $\{v_i,v_j\} \in E$ connects two different nodes illustrating a communication link between them. Each node v_i is assigned a constant number $deg(v_i) \in I\!N^+$ of *ports* by which the corresponding channels can be linked to v_i. Then, an edge $\{v_i,v_j\} \in E$ is described more precisely as $\{(v_i,p),(v_j,q)\}$ with $p \leq deg(v_i)$ and $q \leq deg(v_j)$.

This sight of channels enables the communication network to form a multigraph, where two nodes are connected by several edges and each edge indicates a bidirectional point-to-point connection.

The communication mechanism has to satisfy the following attributes:

i) None of the messages gets lost, i.e. communication is safe.

ii) All channels obey the FIFO-rule, i.e. messages being sent by an agent v_i to v_j via channel c reach v_j in the same order they were sent by v_i.

iii) Communication synchronizes the processes being involved.

iv) Each communication either allows to send a message to exactly one neighboring unit or to receive a message alternatively from one out of several neighbors. This implies that waiting on the input of a message is finished as soon as one of the candidates transmits a message.

The properties iii) and iv) suggest the definition of another graph, named *TWG (Transaction−Wait−For−Graph)* in the literature. At each time t the TWG $G_t = (V, E_t, f_t)$ describes a directed graph whose underlying undirected graph forms a subgraph of the communication network $G = (V,E)$ and which models the waiting relation at time t. By means of the function f_t each node is appointed one of the states {send, receive, terminated, computing} depending on its current state at time t. The set E_t of edges has to correspond to the communication properties which were explained above. This is ensured by defining $((v_i,p),(v_j,q)) \in E_t$ iff either $f_t(v_i) =$ send and sending is done via port p or $f_t(v_i) =$ receive and receipt *can* be done via port p. In addition, E_t has to exclude each pair of processing units which have initiated matching transactions, i.e. E_t does not contain two different edges $((v_i,p),(v_j,q))$, $((v_j,q),(v_i,p))$ with $f_t(v_i) \neq f_t(v_j)$ and $f_t(v_i), f_t(v_j) \in$ {send, receive}. This demand does not restrict a processors possibilities to communicate, it merely frees the TWG from all communications which can take place and thus are negligible to a deadlock detection.

Obviously, a TWG G_t depends on the dynamical behavior of the distributed system and therefore is changed when going from time unit t to time unit t+1. A TWG G_t, for example, has to be enlarged by some additional edges to a TWG G_{t+1} if some processing units leave their state computing in G_t and initiate a communication transaction. Let $G_t = (V, E_t, f_t)$ denote a TWG and let $v_i \in V$ be an arbitrary process with $f_t(v_i) \in$ {send, receive}. Then the following definitions hold:

- a processing unit $v_j \in V$ is called a *candidate* of v_i at t, iff there exists $((v_i,p),(v_j,q)) \in E_t$.

- $cand_t(v_i) := \{v_j \mid v_j$ is a candidate of $v_i\}$ denotes the set of candidates of v_i.

- $candports_t(v_i)$ denotes the set of all ports p of v_i from which an edge $((v_i,p),(v_j,q)) \in E_t$ leads to a $v_j \in cand_t(v_i)$.

- a processing unit $v_i \in V$ is called to be in state *communicating* at t, iff $f_t(v_i) \in$ {send, receive}; it is called *terminated*, iff $f_t(v_i) =$ terminated.

Let $G_t = (V, E_t, f_t)$ be a TWG, $v_i \in V$ a processing unit. Then v_i is defnded to be in a deadlock iff all paths in the TWG which start from v_i are components of paths which either lead back to v_i or end with a terminated process.

This definition is an extension to the ones given in [3, 7]: Natarajan [7] defines a process v_i to be in a deadlock iff all paths in the TWG starting at v_i lead back to v_i. Chandy/Misra/Haas [3] already consider a process v_i as a member of a deadlocked set iff all processes that v_i can reach in the TWG are in state communicating. Thus, working on the same TWGs the set of processes which are deadlocked by the definition of [3] forms a superset of those which are deadlocked by Natarajan's and the above definition.

The definition of [3] complicates an efficient resolution phase because a deadlocked process is not necessarily able to resolve this deadlock by aborting its communication, whereas the definition of Natarajan enables such simple deadlock resolution.

3 The Distributed Deadlock Detection Algorithm

This section presents an improved, distributed, message-based algorithm to detect communication deadlocks. Its main characteristics are the following:

- Avoidance of periodically sending control messages (used in [7]) to reduce the amount of messages.

- Space complexity $O(deg(v_i))$ for each processing unit v_i (different from [3] where space complexity $O(|V|)$ is needed).

- Detection of deadlocks by exactly one process of each set of deadlocked processes to support an efficient resolution.

- Support of a repeated deadlock detection to ensure the practicability of the algorithm.

- Integration of the "basic" communications of a distributed computation in the protocol.

Since the deadlock detection has to be done at runtime there exist two levels of distributed computation: The first one is the computational level which has to be controlled by the deadlock detection. The corresponding algorithm is called "basic algorithm" and its communication ("basic communication") is observed by a supervisor process performing the deadlock detection. The communication which is required for the deadlock detection algorithm is called "control communication" in the following. One supervisor process is attached to each component process of the basic algorithm. Each time the basic algorithm arrives at a communication statement, the communication is delayed, since the corresponding supervisor process has to be informed first about the mode (sending or receiving) and the candidates of the communication. By performing the deadlock detection algorithm the supervisor subsequently verifies whether the current basic communication leads to a deadlock. If a deadlock has occured and the related process is elected to resolve it, the basic algorithm is reactivated and continues without performing its communication. When no deadlock has occured the supervisor releases the desired communication.

3.1 Main Ideas of the Algorithm

The distributed deadlock detection algorithm consists of two phases: By an initial election-phase a set T of processes satisfying a necessary condition for the existence of a deadlock is determined. Exactly one of the processes from T succeeds during the second phase of the algorithm in building a distributed spanning tree on all processes from a deadlocked set in order to verify the existence of a deadlock.

The essential idea of our protocol is, that each processing unit only initiates or propagates one of these phases, if it fails to exclude the existance of a deadlock by its local "knowledge". As an example, a processing unit does not propagate one of the two phases, if for one of its candidates v_j holds: $f_t(v_j)$ = computing.

Consequently, determining the states of its candidates has to be done by each communicating unit during the first phase of the algorithm. A communicating agent v_i obtains its local knowledge by analyzing those processes which v_i can reach in the TWG, i.e. by looking ahead along the edges of the TWG. Considering only paths of length $\leq k$, $k \in I\!N$, to examine the states of the corresponding processes is termed a $k-edge-lookahead$. For simplicity reasons this paper only deals with the 1-edge-lookahead.

Then each communicating unit v_i assigns one of the states from {safe, endangered, terminated} to each port $p \in$ candports$_t(v_i)$. Let $((v_i,p),(v_j,q)) \in E_t$ for a TWG G_t. Then we define :

port p is safe $\quad\Leftrightarrow v_i$ "knows": $f_t(v_j)$ = computing

port p is endangered $\Leftrightarrow v_i$ "knows": v_j is communicating and the transactions of v_i and v_j do not match

port p is terminated $\Leftrightarrow v_i$ "knows": v_j is terminated

Each time a basic communication of the distributed computation is initiated by v_i, an unique identification number *cid* (**current identity**) is assigned to it. Neighboring communicating processes exchange their cids and each agent keeps the maximum of all cids, it has received, in a local variable *did* (**deadlock identity**). The same maximum principle is used in [7].

To ensure the correct detection of each repeated deadlock, a cid being generated by an unit v_i at time t has to exceed the values of all cids or dids which v_i has received or generated before.

3.1.1 Election Phase

For requesting the states of its candports and for announcing itself to be a predecessor in the TWG, each processing unit v_i delivers (INGOING, cid)-messages to its candidates v_j after each initiation of a basic communication. According to the state of v_j, v_i receives a SAFE, TERMINATION.REPLY or (ENDANGERED, did) reply and modifies the states of the corresponding ports in accordance.

When none of v_i's candports (1-edge-lookahead) remains in the state safe, but at least one of these ports is endangered (i.e. not all candports are terminated), the occurrence of a deadlock cannot be excluded by v_i's local knowledge. As a consequence, v_i participates in the election phase and delivers (MAXNR, did)-messages to all of its predecessors in the TWG. These messages are propagated by all those processing units which themselve cannot disprove the existence of a deadlock. Another essential condition for propagating an ingoing (MAXNR, dref)-message, (where dref is the did of the sending process) demands that the value of dref exceeds the did which was forwarded by MAXNR-signals before. Thus, in a deadlocked set of processing units the value of the local dids will converge towards c_{max} which is the maximal cid of all processes from the corresponding set of connected deadlocked processes. The process v_i whose cid equals c_{max}, particularly will receive c_{max} by ENDANGERED- or MAXNR-messages from all of its candidates. This event not necessarily means that v_i is already deadlocked, it is only a necessary condition for the existence of a deadlock. That's why there is the need for the second phase, called verify-phase.

The below example presents a deadlockfree distributed computation, since process e is in the state computing. Here process a ownes the maximum of all dids as its own cid and will receive it via the path (a, f, g, c, b, a) from its only candidate b. Because of the 1-edge-lookahead no process on this way can exclude the existence of an deadlock. When using a 2-edge-lookahead process b would not propagate a's did to its predecessor in the TWG and process a consequently would not start trying to verify a deadlock.

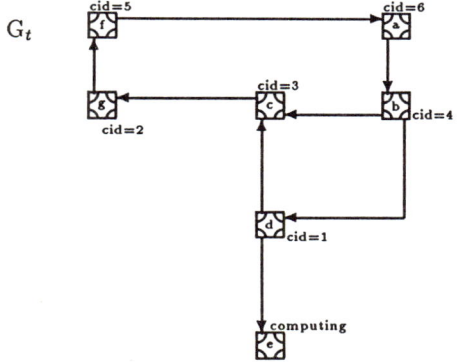

3.1.2 Verify Phase

During the verify-phase a distributed spanning tree is built, including for each deadlocked set at most all processors which cannot exclude the existence of a deadlock. This phase is similar to [3]. In difference to [3], not each processing unit tries to create a spanning tree, but only those units v_i, having previously received their local cid from all of their candidates, initiate the construction of these trees. The algorithm does not guarantee that only one of the processes from a deadlocked set of units initiates the construction of a spanning tree (with the initiating process as root of the tree), but in no case all members of this set will start building their trees. To start the verify phase, (VERIFY, cid) messages are forwarded to all candidates.

On receipt of a (VERIFY, dref) signal a processing unit propagates it to all of its candidates, if dref equals its local did and exceeds the value which was propagated last by a VERIFY-signal. Otherwise the received message is skipped. Thus, even when several different spanning trees are in construction, only the one whose value of cid is largest will be able to reach all units of a deadlocked set of processes. Building the spanning tree stops if a (VERIFY, dref) message reaches a terminated process or a process which has initiated or propagated this message before. On these events REVERIFY-signals are generated which traverse the distributed spanning tree in the opposite direction as it was built.

The propagation rule used here works as follows: After having received a (REVERIFY, dref)-message from each candidate and dref being equal to the did which was propagated by the VERIFY-signals last, the processing unit transmits a (REVERIFY, dref)-signal to its predecessor in the spanning tree.

Detection of a deadlock is completed when one of the initiators of the VERIFY-messages (which is the root of the spanning tree) receives its currend cid by (REVERIFY, cid) signals from all of its candidates. One way to resolve this deadlock is to cancel the communication of the detecting process.

3.2 Algorithm

For the distributed deadlock detection algorithm each processing unit v_i maintains several variables, which are at a time t defined as follows:

- *maxnum* keeps the maximum of all communication identities v_i has received or initiated up to time t. Initial: maxnum = 0.
- *did.of.port[k]* stores the cid which v_i has last recently received via ENDANGERED- or MAXNR-messages over port k. Initial: did.of.port[k] = 0.
- *last.propagated* saves the value of the did which was propagated last by MAXNR-signals to v_i's predecessors in the TWG. Initial: last.propagated = 0.
- *latest.verify* specifies the value of the did which was forwarded last in along with a VERIFY-signal to v_i's candidates. Initial: latest.verify = 0.

In addition each processing unit organizes three boolean variables *prop.cond*, *ver.cond* and *term.cond* which specify certain conditions on which the algorithm has to react. These variables are newly evaluated after each receipt of the corresponding messages:

prop.cond = true $\Leftrightarrow \forall\, p \in$ candports$_t(v_i)$: stateofport[p] $\in$ {endangered, terminated}

ver.cond = true $\Leftrightarrow \forall\, p \in$ candports$_t(v_i)$:
 did.of.port[p] = cid or stateofport[p] = terminated

term.cond = *true* $\Leftrightarrow \forall\, p \in$ candports$_t(v_i)$: stateofport[p] = terminated

The algorithm listed below runs in parallel to the basic computation on each processing unit, it ensures that all basic communications which are carried out remain deadlock-free. On each initialization of a basic communication as well as on each termination of one of the processing units, the corresponding "active" part of the algorithm is performed. The "reactive", passive part of the algorithm is permanently able to receive messages of the deadlock detection protocol.

On initiation of a basic communication by process v_i :
    communicating:=true; $\forall$ p $\in$ {1,...,deg(v_i)} : stateofport[p]:= safe; did.of.port[p]:= 0
    if communication is matching with one of the candidates v_j
        inform candidate v_j; communicating:= false; perform desired basic communication;
        stateofport[$p_{to\_v_j}$]:= safe
    if communication is not matching with one of the candidates v_j
        calculate new cid; maxnum:= did:= cid; latest.verify:=0;
        $\forall$ p$\in$candports(P_i): p ! (INGOING, cid)

On termination of a process v_i :
    terminated:= true; communicating:=false
    $\forall$ predecessors p in the TWG and stateofport[p]$\neq$ terminated do p ! (TERMINATION)

On receipt of a message via port p, v_i reacts as follows :
    if message = $(INGOING$, dref)
        maxnum:= max{maxnum, dref}; note p as a predecessor in the TWG
        if communicating
            p ! (ENDANGERED, cid)
        if ($\neg$ communicating) and ($\neg$ terminated)
            p ! (SAFE)
        if terminated
            p ! (TERMINATION.REPLY)

    if message = (ENDANGERED, dref) or message = (TERMINATION.REPLY)
        if message = (ENDANGERED, dref)
            stateofport[p]:= endangered; did.of.port[p]:= dref; did:= max{did, dref};
            maxnum:= max{maxnum, dref}
        if message = (TERMINATION.REPLY)
            stateofport[p]:= terminated; did.of.port[p]:= 0
        if prop.cond and communicating and ($\neg$ ver.cond) and ($\neg$ term.cond)
            last.propagated:= did; $\forall$ predecessors q in the TWG: q ! (MAXNR, did)
        if ver.cond and communicating and ($\neg$ term.cond)
            if last.propagated < did
                last.propagated:= did; $\forall$ predecessors q in the TWG: q ! (MAXNR, did), if did.of.port[q] $\neq$ did
            latest.verify:= did; num:= |cand(P_i)|; $\forall$ q$\in$candports(P_i): q ! (VERIFY, did)
        if term.cond and communicating
            {deadlock detected}
            communicating:= false; $\forall$ predecessors q in the TWG: q ! (SAFE)

    if message = (SAFE)
        stateofport[p]:= safe

    if message = (TERMINATION)
        stateofport[p]:= terminated
        if term.cond and communicating
            {deadlock detected}; communicating:= false; $\forall$ predecessors q in the TWG: q ! (SAFE)
        if communicating and ($\neg$ term.cond)
            if ver.cond
                if last.propagated < did
                    last.propagated:= did; $\forall$ predecessors q in the TWG: q ! (MAXNR, did), if did.of.port[q] $\neq$ did
                latest.verify:= did num:= |cand(v_i)|; $\forall$ q$\in$candports(v_i): q ! (VERIFY, did)
            if $\neg$ ver.cond
                calculate new cid; maxnum:= did:= cid; latest.verify:=0
                if prop.cond
                    $\forall$ predecessors q in the TWG: q ! (MAXNR, did)
                if $\neg$ prop.cond
                    $\forall$ q$\in$candports(v_i) und stateofport[q]=safe: q ! (INGOING, did)

    if message = (KILL.INGOING)
        remove p as a predecessor in the TWG

```
if message = (MAXNR, dref)
   stateofport[p]:= endangered; did.of.port[p]:= dref; maxnum:= max{maxnum, dref}
   if (dref = latest.verify) and (dref = did)
      p ! (VERIFY, did)
   if (dref ≠ latest.verify) or (dref ≠ did)
      did:= max{did, dref}
      if prop.cond and communicating and (last.propagated < did)
         last.propagated:= did; ∀ predecessors q in the TWG: q ! (MAXNR, did)
      if (dref = did) and (dref = cid) and communicating and ver.cond
         latest.verify:= did num:= |cand(v_i)|; ∀ q∈candports(v_i): q ! (VERIFY, did)

if message = (VERIFY, dref)
   maxnum:= max{maxnum, dref}
   if terminated
      p ! (REVERIFY, dref)
   if communicating and (dref = did) and (latest.verify = dref)
      p ! (REVERIFY, dref)
   if communicating and (dref = did) and (latest.verify < dref)
      latest.verify:= dref; note p as predecessor in the verify-tree;
      num:= |cand(v_i)|; ∀ q∈candports(v_i): q ! (VERIFY, did)

if message = (REVERIFY, dref)
   maxnum:= max{maxnum, dref}
   if communicating and (dref = did = cid) and (latest.verify = dref)
      num:= num - 1
      if num = 0              {Deadlock detected}
         communicating:= false; ∀ q∈candports(v_i): q ! (KILL.INGOING);
         ∀ predecessors q in the TWG: q ! (SAFE)
   if communicating and (dref = did) and (dref ≠ cid) and (latest.verify = dref)
      num:= num - 1
      if num = 0
         send (REVERIFY, dref) to the predecessors in the verify-tree
```

3.3 Analysis

The correctness proof of the proposed algorithm, which is out of scope for this paper, is given in [6]. Principles of this proof are based on the corresponding proofs in [3] and [7].

To analyze the worst case complexity of the proposed algorithm we assume $G_t = (V, E_t, f_t)$ to be a static TWG (i.e. all processing units v_i enter their state $f_t(v_i)$ at time t-1 and none of the the processing units changes its state). Let $\{d_1, ..., d_k\}$ denote the cids of all processes which are communicating at time t (w.l.o.g. $d_i > d_j$ $\forall i<j$). In the worst case, the did d_i is propagated over $|E_t|-i+1$ edges, which results in a message complexity of $O(|V| * |E_t|)$ for the election phase. The message complexity of the verify-phase is $O(|V| * |E_t|)$, too, since with $O(|V| * |E_t|)$ messages each processing unit could span "its" tree over $|V|$ processes. The worst case time complexity of the election phase is determined by the time which passes by until all local dids equal d_{max} (the maximum cid in the deadlock) and the process v_j whose cid equals d_{max} has received d_{max} from all its not terminated candidates. In the worst case it takes $O(|cand_t(v_l)|)$ time until a further process changes his did to d_{max}. Thus, the election phase needs time complexity $O(|E_t|)$. The time complexity of the verify phase is dominated by the time which is spent by the process whose cid is the largest in the set of deadlocked processes for building the spanning tree. This requires time $O(|E_t|)$.

4 Experimental Results

For an empirical analysis of the algorithm, the distributed deadlock detection was integrated in a simulation environment for general distributed computations, which was implemented on a transputer network. These simulations aimed at:

- comparing our algorithm to other existing algorithms
- measuring the runtime overhead being imposed by a deadlock detection tool on deadlock-free distributed computations

We compared our algorithm to those of [3, 7]. For this, the algorithms of Natarajan and Chandy/Misra/Haas had to be adjusted to our deadlock model. An exact explanation of these modifications can be found in [6].

To simulate a distributed computation eight communication processes are generated at random. For each of these processing units a sequence of actions from :

[send, p_i] : send a message via port p_i

[receive, $p_1,\ldots,p_i$] : receive a message alternatively via one of the ports $p_1,\ldots, p_i$

[compute, t] : simulate a non-communicating computation for t seconds.

is chosen randomly. Their ratio and the duration of the computing times can be influenced by some parameters. The resulting sequence of actions are given to the corresponding processes, which terminate after having sequentially worked out their stimuli.

Three parameters determine the characteristic of a testpattern class $T_{i.j.k}$. Parameter i determines wether deadlocks may occur during the simulation of a pattern from this class, or not. If i equals 0, it is guaranteed that the stimuli to be worked out remain deadlock-free. If i equals 1 deadlocks are very likely to occur since all operations are randomly chosen. Parameter j∈ {0, ..., 100} fixes the ratio between communication operations and internal (non-communicating) computations, i. e. it provides that j percent of the operations each process has to work out are internal computations. The duration of these internal computations is specified by a parameter k=(u, o), which indicates that for all computations times t, u≤t≤o must hold.

For our simulations we used the parameters i∈{0,1}, j∈ {1, 5, 10, 20, ..., 90} and k∈(1,1), (2,4), (1,10)}. Each class $T_{i.j.k}$ contains 50 randomly generated testpatterns, each assigning 10 operations to every process.

Let $T_{i.j.k}$ denote a class of testpatterns, $|T_{i.j.k}|$ the size of this class and $\gamma \in T_{i.j.k}$ a single testpattern. To analyze the alogrithms, let MRun(γ) be the maximum runtime needed by one of the processes to work out testpattern γ (in 1/1000 seconds), and Mess(γ) the sum of control messages which all processes have sent while simulating γ. Let further GRun($T_{i.j.k}$) and GMess($T_{i.j.k}$) be the average value of all MRun(γ) resp. GMess(γ), with $\gamma \in T_{i.j.k}$

Table 1 shows some experimental results for the algorithms of Natarajan, Chandy/Misra/Haas

k			10	20	30	40	50	60	70	80	90	avg
(1,1)	Natarajan	GRun	11152	16343	22176	26146	29309	33225	32938	30886	25117	25255
	CMH	GRun	6478	12137	17431	22065	25679	28721	29931	28818	23071	21592
	Alg	GRun	6443	12281	17623	22040	25622	28729	29584	28728	22933	21553
	Natarajan	GMess	8376	10445	12995	14490	15254	16603	14717	11342	5912	12237
	CMH	GMess	1194	970	798	785	587	450	321	201	98	600
	Alg	GMess	878	742	607	557	433	337	226	133	53	440
(2,4)	Natarajan	GRun	27867	44446	57187	72470	87732	92763	95552	93117	72918	71561
	CMH	GRun	21008	37960	50304	66926	79980	87252	90079	86312	68041	65318
	Alg	GRun	20896	36740	50292	66225	79411	87418	88539	87441	68019	64997
	Natarajan	GMess	18249	26071	31694	38478	43760	41144	40556	35091	15463	32278
	CMH	GMess	1285	1080	897	707	603	402	333	203	96	622
	Alg	GMess	988	786	703	506	440	305	231	140	51	461
(1,10)	Natarajan	GRun	40949	79121	102050	135742	154232	181784	170992	171810	135179	130206
	CMH	GRun	32933	70623	92626	124968	135999	163833	159991	161548	129630	119127
	Alg	GRun	33133	70305	91816	123807	136387	164409	161131	162589	129627	119245
	Natarajan	GMess	26094	43539	59854	67906	76335	90775	76477	63764	31286	59558
	CMH	GMess	1154	990	981	677	544	472	302	212	105	604
	Alg	GMess	866	776	709	511	409	348	213	143	56	447

Table 1: Results for $T_{1,j,k}$, average runtime and messages

k		10	20	30	40	50	60	70	80	90	avg
(1,1)	Natarajan	16.0	11.7	9.5	10.8	9.7	7.5	7.5	6.1	6.0	9.42
	CMH	5.2	3.0	1.5	3.2	2.6	1.1	1.3	0.1	0.1	2.01
	Alg	5.4	3.1	1.6	3.2	2.6	1.1	1.3	0.1	0.1	2.05
(2,4)	Natarajan	13.8	15.9	12.2	9.1	9.2	7.7	7.3	6.4	6.0	9.73
	CMH	5.4	7.8	4.9	2.1	2.3	1.2	0.8	0.3	0.2	2.77
	Alg	5.5	7.9	4.9	2.2	2.3	1.2	0.8	0.3	0.2	2.81
(1,10)	Natarajan	11.9	11.7	10.1	9.0	8.3	6.6	9.8	6.9	5.9	8.91
	CMH	3.8	4.2	2.9	2.1	1.6	0.2	3.3	0.9	0.1	2.12
	Alg	3.8	4.2	2.9	2.1	1.6	0.2	3.3	0.9	0.1	2.12

Table 2: Results for $T_{0,j,k}$, runtime overhead

and the one proposed here, named *Alg* in the following. These results indicate that the periodical sending of messages in the algorithm of Natarajan induces great overheads in runtime. The overhead increases if the internal computation time is high. The number of messages needed for the algorithms of Chandy/Misra/Haas exceeds the corresponding number of the proposed algorithm for 35 percent, but this has no effect on the runtime due to the fast communication capabilities of transputers.

Table 2 shows the percentual overhead in runtime which the deadlock detection algorithm imposes on deadlockfree computations. For this, all stimuli were additionally worked out without any deadlock detection. For realistic distributed computations the overhead which is caused by a deadlock detection and resolution is obviously negligible since it imposes only an overhead of 2 percent.

5 Concluding Remarks

In this paper a distributed, message-based algorithm to detect and resolve communication deadlocks in distributed systems is presented. The implementation of the proposed algorithm is compared with other well known protocols.

For a practical use of a deadlock detection tool, it is essential to influence the communication behaviour of the distributed computation as less as possible. Only the algorithm being presented here and the one of Chandy/Misra/Haas ([3]) are shown to fulfill this demand. Natarajan's deadlock detection is not able to accomplish this requirement because of its periodical transmission of QUERY- and DETECT-signals. This approach might be helpful in networks whose communication is not guaranteed to be error free.

The most serious disadvantages of Chandy/Misra/Haas' deadlock detection algorithm are the space complexity $O(n)$ for each of the n processes of a distributed computation and the missing support of an efficient deadlock resolution. The space requirement is reduced by our algorithm to $O(d)$ for each process where d denotes the number of neighbors in the communication network. To resolve the deadlock, our algorithm elects exactly one process from a deadlocked set which either resolves the deadlock directly by cancelling its basic communication or which could start a more sophisticated deadlock resolution.

References

[1] B. Awerbuch, S. Micali, *Dynamic Deadlock Resolution Protocols*, ACM Symposium on Foundations of Computer Science 1986, pp. 196 - 207

[2] G. Bracha, S. Toueg, *Distributed deadlock detection*, Distributed Computing (1987) 2, pp. 127 - 138

[3] K. M. Chandy, J. Misra, L.M. Haas, *Distributed Deadlock Detection*, ACM Transactions on Computer Systems, vol. 1, no.2, May 1983, pp. 144 - 156

[4] A. K. Elmagarmid, *A survey of distributed deadlock detection algorithms*, SIGMOD Records, vol. 15, no. 3, Sept. 1986, pp. 37 - 45

[5] E. Knapp, *Deadlock Detecton in Distributed Databases*, Technical Report, Department of Computer Science, University of Texas, Jan. 1988

[6] B. Kröger, R. Lüling, *Communication Deadlocks inDistributed Systems*, Master Thesis, Department of Mathematics and Computer Science, University of Paderborn, Aug. 1988

[7] N. Natarajan, *A Distributed Scheme for Detecting Communication Deadlocks*, IEEE Transactions on Software Engineering, vol. SE-12, no. 4, April 1986, pp. 531 - 537.

[8] B. A. Sanders, P. A. Heuberger, *Distributed Deadlock Detection and Resolution with Probes*, 3rd Intern. Conference on Distributed Algorithms, Lect. Notes in Computer Science 392, Sept. 1989, pp. 207 - 218

On the Average Performance of Synchronized Programs in Distributed Networks
(Preliminary Version)

Sergio Rajsbaum*   *Moshe Sidi***

Technion- Israel Institute of Technology
Haifa, Israel 32000

ABSTRACT

A synchronizer is a compiler to transform a program for a synchronous network into a program for an asynchronous network. A simple synchronizer, which also represents a basic mechanism for distributed computing, was studied by Even and Rajsbaum in [ER1] and [ER2]. They studied the behavior of the synchronizer in networks with constant message transmission delays and processing times. Here we study the case in which the transmission delays and the processing times are random.

1. INTRODUCTION

Consider a network of processors which communicate by sending messages along communication links. The network is *synchronous* if there is a global clock whose beats are heard by all the processors simultaneously, and the time interval between clock beats is long enough for all messages to reach their destinations and for local computational steps to be completed before the clock beats again. The network is *asynchronous* if there is no global clock, and the transmission times of messages are unpredictable.

In general, a program designed for a synchronous network, will not run correctly in an asynchronous network. Instead of designing a new program for the asynchronous network, it is possible to use a *synchronizer*, [A1], i.e., a compiler that converts a program designed for a synchronous network, to run correctly in an asynchronous network.

Synchronizers provide a useful tool because programs for synchronous networks are easier to design, debug and test than programs for asynchronous networks. Awerbuch later [A2] demonstrated another important use of synchronizers, namely, the design of asynchronous algorithms, that were more efficient than any previously known.

---
* Department of Computer Science
Electronic address:
BITNET: RAJSBAUM@TECHUNIX
  ** Department of Electrical Engineering
Electronic address:
BITNET: SIDI@TECHUNIX
Csnet: sidi@sel.technion.ac.il

The (worst case) time complexity of a distributed algorithm is usually computed assuming that message transmission delays are 1 unit. The aim of this paper is to study the effect of random processing times and transmission delays on the performance of synchronous programs running in an asynchronous network under the control of a simple synchronizer. We compare the results with the deterministic case considered by Even and Rajsbaum [ER1], [ER2], in which message delays, as well as processing times are constant (or bounded).

The problem of designing efficient synchronizers has been studied in the past (e.g. [A1], [AP90], [PU89]). Here we investigate the average performance of networks whose operation is controlled by a simple synchronizer. The operation of the synchronizer is as follows: Each processor waits for a message to arrive on each of its in-comming links before it performs the next computational step, and then sends one message on each of its out-going links; it is assumed, for example, that every message is followed by an end-of-message marker, even if the message is empty.

We believe that this synchronizer is the simplest one to capture the essence of the synchronizer methodology; it ensures that a processor does not initiate a new phase of computation before it knows that all the messages sent to it in the previous phase have arrived. The end-of-message markers model the flow of information that must exists between every pair of processors p and q, such that $p \to q$ is a link [A2].

The synchronizers suggested by Awerbuch [A1] require that all links are bidirectional. Our synchronizer is similar to α in [A1], but can be be used also in directed networks. Many distributed protocols are based on the same mechanism; for example, the snapshot algorithm of Chandy and Lamport [CL85] and clock synchronization (e.g. [BS88], [OG87]); in [RM], Rajsbaum and Malka show how this synchronizer can model the behavior of any marked graph (e.g. Commoner, et. al. [CHEP]), of the synchronizers of [A1], and of distributed schedulers in [BG89], [MMZ88]. Thus, our work should be relevant to problems in stochastic petri nets, where, due to the huge size of the state space, the solution techniques often rely on simulation (e.g. Molloy [M1], [M2], Marsan [Ma89]).

A model similar to ours is considered in [BT89], where it is claimed that the rate is $\theta(1 / \log \delta_{out})$, for regular networks with out-degree equal to δ_{out}, identically exponentially distributed transmission delays with mean 1, and negligible processing times. In [BS88] it is claimed only the lower bound of $1 / \delta_{in}$ on the rate, for regular networks with in-degree equal to δ_{in}, with negligible transmission delays, and identically exponentially distributed processing times.

Main Results

This paper is devoted to the performance analysis of directed networks controlled by the synchronizer, in which transmission delays, as well as the time it takes a processor to complete a computational step are random variables. We concentrate on the *rate* of a network R_v, i.e. the average number of computational steps executed by a processor in the network, per unit time. To facilitate the presentation, we first assume that the transmission delays are negligible, and only at the end of the paper describe how to extend the results for networks with non-negligible delays.

In Section 3 we study the case in which the random variables have general probability distributions. First (Section 3.1) we analyze the effect of the topology on the rate. We present a technique to compare the rate of networks with different topologies. We give examples of networks with different topologies, but with the same rate. Then (Section 3.2) we analyze networks with the same topology but different processing times. By defining a partial order on the set of distributions, we show that deterministic (i.e. constant) processing times maximize the rate of computation. For this case, Even and Rajsbaum show in [ER1] that if the processing times are equal to λ^{-1}, the rate of the network is λ, regardless of the number of processors in the network or its topology. In this paper we show that in case the processing times are random and unbounded, the rate may be degraded by a logarithmic factor in the number of processors. This occurs in the case of exponentially distributed processing times. However, we show that the exponential is the worst, among a large and natural class of distributions (is the maximum w.r.t. the partial order).

In Section 4 we concentrate on the case of exponentially distributed random variables with mean λ^{-1}. We prove that the rate is between $\lambda / 4log (\Delta + 1)$ and $\lambda / \log (\delta + 1)$, where Δ (δ) is the maximum (minimum) vertex in-degree or out-degree. Hence, for regular-degree (either in or out-degree) networks, the rate is $\Theta(\lambda / \log (\delta + 1))$. We compute the exact rate and the stationary probabilities for the extreme cases of a directed cycle and a complete graph. Finally, we study the effect of having one processor that runs slower than the rest of the processors. The result is that in some sense, the directed cycle network is more sensitive to such a processor than a complete network.

In the last section we show that it is easy to extend the results to networks with non-negligible transmission delays. Consider the exponential case. Adding transmission delays to a regular degree network may reduce its rate by at most a constant factor, provided that they are not larger (w.r.t. the partial order) than the processing times. In networks with processing times exponentially distributed with mean 1, and larger delays with mean λ^{-1}, we compare the results with those of [ER2], where it was shown that for the corresponding deterministic case the rate is λ. In the probabilistic case of a regular-degree network, the rate is at least $\lambda / (\log \delta)$. Thus, in both cases (small and large delays), the rate of a bounded degree network is reduced only by a constant factor.

For lack of space, some of the proofs are omitted in this Preliminary Version.

2. THE MODEL

The *network* is modeled by a (finite) directed, strongly connected graph $G(V, E)$, where $V = \{1,2,...,n\}$ is the set of vertices of the graph and $E \subset V \times V$ is the set of directed edges. A vertex of the graph corresponds to a processor that is running its own program, and a directed edge $u \to v$ corresponds to a communication link from processor u to processor v. In this case, we shall say that u is an *in-neighbor* of v, and v is an *out-neighbor* of u in the network. The processors communicate by sending messages along the communication links. To facilitate the presentation, we assume that the message transmission delays are negligible. At the end we briefly discuss the case of non-negligible transmission delays.

Initially, all processors are in a *quiescent state,* in which they send no messages and perform no computations. Once a processor leaves the quiescent state, it never reenters it and is considered *awake*. When awakened, each processor operates in phases as described in the sequel. Assume that at an arbitrary time, $t(v)$, processor v leaves the quiescent state and enters its first *processing state*, PS_0 (this may be caused by a message from another processor, or a signal from the outside world, not considered in our model). Then, processor v remains in PS_0 for $\tau_0(v)$ units of time and then transits to its first *waiting state*, WS_0. From this time on, let PS_k and

WS_k, $k \geq 0$, denote the processing state and the waiting state, respectively, for the kth phase. Observe that we are concerned with the rate of computation of the network; the nature of the computation is of no concern to us here. Thus we take the liberty of denoting with the same symbol the k-th processing state of all the processors.

The transition rules between states are as follows: If a processor v transits from state PS_k to WS_k, it sends one message on each of its outgoing edges. These messages are denoted by M_k. Note that this labeling is not needed for the implementation of the protocol; it is used only for its analysis. When v sends the M_k messages, we say that v has completed its k-th *processing step*.

If a processor v is in state WS_k, and has received a message (M_k) on each of its incoming edges, it removes one message from each of its incomming edges, transits to state PS_{k+1}, remains there for $\tau_{k+1}(v)$ units of time and then transits to state WS_{k+1}. Otherwise, (at least on one incoming edge, M_k has not yet arrived) processor v remains in state WS_k until it receives a message from each of its in-neighbors, and then operates as described above.

The states PS_k, $k \geq 0$, correspond to the computation steps during the k-th phase. The *processing times*, $\tau_k(v)$, correspond to the time it takes to processor v to complete the k-th computation step. The processing times $\tau_k(v)$, $k \geq 0$, $v \in V$, are positive, real-valued random variables defined over some probability space. The states WS_k, correspond to the waiting of v until the completion of the k-th phase by all of its in-neighbors.

For $k \geq 0$, let $t_k^G(v)$ (or $t_k(v)$, whenever G is understood) be the k-th completion time, i.e., the time at which processor v sends messages M_k in network G. Let the *in-set* of a vertex v in G, $IN^G(v)$ (or simply $IN(v)$), be the set of vertices in G that have an edge to v, including v itself, that is, $IN(v) \stackrel{\Delta}{=} \{ u : u \to v \in E \} \cup \{v\}$. With this notation, the operation of processor $v \in V$ is as follows. Once v has sent a message M_k at time $t_k(v)$, it waits until all processors with an edge to it send message M_k, and then starts its $(k+1)$-st computation step; that is, after the maximum of $t_k(u)$, $u \in IN(v)$, it starts the $(k+1)$-st computation step, which takes $\tau_{k+1}(v)$ units of time, and then sends out M_{k+1}. Hence, the evolution of the network can be described by the following recursions:

$$t_0(v) = t(v) + \tau_0(v)$$

$$t_{k+1}(v) = \max_{u \in IN(v)} \{ t_k(u) \} + \tau_{k+1}(v), \qquad k \geq 0.$$

(1)

It is interesting to note that the completion times $t_k(v)$ have a simple graph theoretic interpretation. For a vertex v, let $S_k(v)$ be the set of all directed paths of length k ending in v, assuming each vertex has a self-loop (an edge $v \to v$). For $k = 0$, the only path of length 0 ending in v consists of v itself. For a path $P_k = v_0 \to v_1 \to \cdots \to v_k(=v)$, let $T(P_k) \stackrel{\Delta}{=} t(v_0) + \sum_{i=0}^{k} \tau_i(v_i)$, and $T(S_k(v)) \stackrel{\Delta}{=} \{ T(P) : P \in S_k(v) \}$. Thus, $T(S_k(v))$ is a set of random variables; each one is the sum of $k+1$ random variables. Note that these random variables are not independent, even if the $\tau_i(v)$'s are independent. The explicit computation of $t_k(v)$ is as follows.

Theorem 2.1: For every $v \in V, k \geq 0$, $\quad t_k(v) = \max T(S_k(v))$.

The Performance Measures

The most important performance measures investigated in this paper are the completion times $t_k(v)$, $k \geq 0$, $v \in V$. A related performance measure of interest is the counting process $N_t^G(v)$ (or simply $N_t(v)$), associated with processor v defined by

$$N_t(v) \stackrel{\Delta}{=} \sup \{ k : t_k(v) \leq t \},$$

that is, $N_t(v)$ is the number of computation steps (minus 1) completed by v up to time t, or the highest index of an M_k message that has been sent by v up to time t. Similarly, $N_t \stackrel{\Delta}{=} \sum_{v=1}^{n} N_t(v)$ denotes the total number of processing steps (minus n) executed in the network up to time t. The following claim indicates that no processor can advance (in terms of executed processing steps) too far ahead of any other processor.

Claim : Let d be the diameter of a directed, strongly connected graph G. Then for all $u, v \in V$, and $t \geq 0$, $|N_t(u) - N_t(v)| \leq d$.

Another important performance measure is the *computation rate*, $R^G(v)$, (or simply $R(v)$) of processor v in network G, defined by

$$R(v) \stackrel{\Delta}{=} \lim_{t \to \infty} \frac{N_t(v)}{t},$$

whenever the limit exists. Note that $R(v)$ is a number, not a random variable. Similarly, the computation rate of the network is defined by

$$R \stackrel{\Delta}{=} \lim_{t \to \infty} \frac{N_t}{t}.$$

Claim 1 implies that for every $u, v \in V$, $R(u) = R(v)$, and therefore $R = n \cdot R(v)$.

3. GENERAL PROBABILITY DISTRIBUTIONS

In this section we develop several techniques to compare the performance of different networks, with general distributions of the processing times $\tau_k(v)$. We first show that adding edges to a network with an arbitrary topology slows down the operation of each of the processors in the network. We show how the theory of of graph embeddings can be used to compare the rates of different networks. As an example we present graphs, which have the same rate (up to a constant factor) for general distributions, although they have different topologies. Finally, we compare networks with the same (arbitrary) topology but different distributions of the processing times. Specifically, we show that determinism maximizes the rate, and exponential distributions minimize the rate, among a large class of distributions.

3.1 Topology of the Network

Monotonicity

Here we show that adding edges to a network with an arbitrary topology slows down the operation of each of the processors in the network. The basic methodology used is the *sample path* comparison; that is, we compare the evolution of message transmissions in different networks for every instance, or realization, of the random variables $\tau_k(v)$. This yields a stochastic ordering between various networks [Ro83], [S84].

Theorem 3.1: Let $G(V, E)$ be a graph, and $E' \subset V \times V$ be a set of directed edges. Let $H(V, E \cup E')$ be the graph obtained from G by adding edges E'. Assume that processor v, $1 \leq v \leq n$ awakens in both G and H at the same time $t(v)$. For every realization of the random variables $\tau_k(v)$, $k \geq 0$, $1 \leq v \leq n$, the following inequalities hold

$$t_k^G(v) \leq t_k^H(v),$$

for all $k \geq 0$, $1 \leq v \leq n$.

The above theorem can be proved in a way similar to Theorem 3.3. It implies immediately

Corollary 3.1: Under the conditions of Theorem 3.1 we have that $N_t^G(v) \geq N_t^H(v)$ and $R^G(v) \geq R^H(v)$ (when the limits exist) for all $v \in V$. Also $N_t^G \geq N_t^H$.

Remark 3.1: Notice that no assumption was made about the random variables $\tau_k(v)$. In particular, they need not be independent.

Remark 3.2: The sample path proof above implies that the random variables N_t^G is stochastically larger than the random variable N_t^H, denoted $N_t^G \geq_d N_t^H$, i.e. $Prob\{N_t^G \geq \alpha\} \geq Prob\{N_t^H \geq \alpha\}$ for all α.

Remark 3.3: The above implies that if one starts with a simple, directed cycle (a strongly connected graph with the least number of edges) and successively adds edges, a complete graph is obtained, without ever increasing the rate.

Embeddings

The theory of graph embeddings has been used to model the notion of one network simulating another on a general computational task (see for example [R88]). Here we show how the notion of graph embeddings can be helpful in comparing the behavior and the rates of different networks controlled by the synchronizer.

An *embedding* of graph G in graph H is specified by a one-to-one *assignment* α of the nodes of G to the nodes of H: $\alpha : V_G \to V_H$, and a *routing* ρ of each edge of G along a distinct path in H: $\rho : E_G \to \text{Paths}(H)$. The *dilation* of the embedding is the maximum amount that the routing ρ "stretches" any edge of G:

$$dilation(\alpha, \rho) = \max_{u \to v \in E_G} \text{length}(\rho(u \to v)).$$

The dilation is a measure of the delay incurred by the simulation according to the embedding. The following theorem is a generalization of Theorem 3.1.

Theorem 3.2: Let (α, ρ) be an embedding with dilation D, of a graph $G(V_G, E_G)$ in a graph $H(V_H, E_H)$. Assume $t(v) = t(\alpha(v))$, for all $v \in V_G$, and that $\tau_k(v)$ and $\tau_{kD}(\alpha(v))$ for all $k \geq 0$, $v \in V_G$, have the same distribution. For every realization of the random variables $\tau_k^G(v) = \tau_{kD}^H(\alpha(v))$, $k \geq 0$, $v \in V_G$, the following inequalities hold

$$t_k^G(v) \leq t_{kD}^H(\alpha(v)), \qquad k \geq 0, v \in V_G.$$

Corollary 3.2: Under the conditions of Theorem 3.2 we have that $D \cdot N_t^G(v) \geq N_t^H(\alpha(v))$ and $D \cdot R^G(v) \geq R^H(\alpha(v))$ (when the limits exist) for all $v \in V_G$.

Remark 3.4: Notice that no assumption was made about the processing times of the processors of G (and H). In particular, they need not be independent.

A simple corollary of Theorem 3.2 is that if G is a subgraph of H, $N_t^G(v) \geq_d N_t^H(\alpha(v))$. This is because if G is a subgraph of H, then there is an embedding from G in H with dilation 1. In addition, if the number of vertices in G and H are equal, and the dilation of the embedding is D, then G is a D–*spanner* of H (e.g. [PS89], [PU89]), and we have the following.

Corollary 3.3: If H has a D-spanner G, then $R^G / D \leq R^H \leq R^G$.

A motivation for the the theory of embeddings is simulation. Namely, one expects that if there is an embedding (α, ρ) from G in H with dilation D, then the architecture H can simulate T steps of the architecture

G on a general computation in order of $D \cdot T$ steps, by routing messages according to ρ. In our approach, we compare the performance of G and of H under the synchronizer, without using ρ; the embedding is used only for the purpose of proving statements about the performance of the networks. Consider for example the following two results of the theory of embeddings [R88].

Proposition 1: For all $n \geq 1$: One can embed the order n Shuffle-Exchange graph in the order n deBruijn graph with dilation 2. One can embed the order n deBruijn graph in the order n Shuffle-Exchange graph with dilation 2.

Proposition 2: For all $n \geq 1$: One can embed the order n Cube-Connected-Cycles graph in the order n Butterfly graph with dilation 2. One can embed the order n Butterfly graph in the order n Cube-Connected-Cycles graph with dilation 2.

By Theorem 3.2, the average rate of the graphs of Proposition 1 (2) are equal up to a constant factor of 2, provided that the processing times of corresponding processors have the same distributions (regardless of what are these distributions).

3.2 Probability Distributions

Deterministic Processing Times

Now we compare networks, say $G(V, E)$ and $H(V, E)$, having the same (arbitrary) topology, but operate with different distributions of the random variables $\tau_k(v)$. To that end, we assume that the processing times $\tau_k^G(v)$, $k \geq 0$, $v \in V$ are independent and have finite mean $E[\tau_k^G(v)] = \lambda_v^{-1}$.

We say that λ_v is the *potential rate* of v, as this would be the rate of v if it would not have to wait for messages from its in-neighbors. The processing times in H are distributed as in G except for a subset $V' \subset V$ of processors, for which the processing times are assumed to be deterministic, i.e. $\tau_k^H(v) = \lambda_v^{-1}$, $v \in V'$, for $k \geq 0$. We let $\tau_k^H(v) = \tau_k^G(v) = \tau_k(v)$, $k \geq 0$, $v \notin V'$, be any specific realization of the random variables in G. Again, it is assumed that the processors are awakened at the same time in both networks.

Theorem 3.3: Under the above conditions we have that

$$t_k^H(v) \leq E[t_k^G(v)],$$

for all processors v, and $k \geq 0$. The expectation is taken over the respective distributions of processing times of processors in of G in V'.

Proof: The proof is by induction on k. For the basis, $k = 0$ observe that

$$E[t_0^G(v)] = t_0^G(v) = t(v) + \tau_0(v) = t_0^H(v),$$

for $v \notin V'$, and

$$E[t_0^G(v)] = t(v) + \lambda_v^{-1} = t_0^H(v),$$

for $v \in V'$.

The induction hypothesis is $t_k^H(v) \leq E[\, t_k^G(v)\,]$, and we need to show that $t_{k+1}^H(v) \leq E[\, t_{k+1}^G(v)\,]$, for all $v \in V$.

From (1) we have that

$$t_{k+1}^G(v) = \max_{u \in IN(v)} \{t_k^G(u)\} + \tau_{k+1}^G(v),$$

for $v \in V$. Jensen's inequality implies

$$E[\, t_{k+1}^G(v)\,] \geq \max_{u \in IN(v)} \{E[\, t_k^G(u)\,]\} + E[\, \tau_{k+1}^G(v)\,].$$

By induction hypothesis,

$$E[\, t_{k+1}^G(v)\,] \geq \max_{u \in IN(v)} \{t_k^H(u)\} + E[\, \tau_{k+1}^G(v)\,] = t_{k+1}^H(v),$$

since $E[\, \tau_{k+1}^G(v)\,] = \tau_{k+1}(v)$ for $v \notin V'$, and $E[\, \tau_{k+1}^G(v)\,] = \lambda_v^{-1} = \tau_{k+1}^H(v)$, for $v \in V'$. ∎

Remark 3.5: Theorem 3.3 holds also if the processing times $\tau_k^H(v)$ of processors v of H, in V', are deterministic, but not necessarily the same for every k.

When all processing times in the network H are deterministic, the computation of the network rate is no longer a stochastic problem, but a combinatorial one. Thus, a conclusion of Theorem 3.3 is that in this case, the computation rate of H, obtained via combinatorial techniques ([ER1] and [ER2]), yields an upper bound on the average rate of G. Furthermore, if the times $t_k^H(v)$ are computed, they give a lower bound on $E[\, t_k^G(v)\,]$, for every $k \geq 0$.

More Variable Processing Times

More general, we study the effect of substituting a random variable in the network (e.g. the processing time of a given processor, for a given computational step) with a given distribution, for a random variable with another distribution on the rate of the network, and define an ordering among probability distributions.

Recall that a function h is *convex* if for all $0 < t < 1, x_1, x_2, h(tx_1 + (1-t)x_2) \leq th(x_1) + (1-t)h(x_2)$. A random variable X with distribution F_X is said to be *more variable* than a random variable Y with distribution F_Y, denoted $X \geq_c Y$ or $F \geq_c G$, if $E[\, h(X)\,] \geq E[\, h(Y)\,]$ for all increasing convex functions h. The partial order $\leq_c$ is called *convex* order (e.g. [Ro83], [S84]). Intuitively X will be more variable than Y if F_X gives more weight to the extreme values than F_Y; for instance, if $E[X] = E[Y]$, then $Var(X) \geq Var(Y)$, since $h(x) = x^2$ is a convex function.

Here we compare networks, say $G(V, E)$ and $H(V, E)$ having the same arbitrary topology, but some of the processing times in G are more variable than the corresponding processing times in H, i.e., for some k's and some v's, $\tau_k^G(v) \geq_c \tau_k^H(v)$, while all other processing times have the same distributions in both graphs. When $t^G(v) = t^H(v)$ and all processing times in G (H) are independent of each other, the following holds.

Theorem 3.4: Under the above conditions the following holds for all processors v, and $k \geq 0$

$$t_k^H(v) \leq_c t_k^G(v).$$

Proof: From Theorem 2.1 we have

$$t_k(v) = \max\{ T(P_k) : P_k \in S_k(v) \},$$

where $P_k = v_0 \to v_1 \to \cdots \to v_k (= v)$ is a directed path of length k ending in v, and $T(P_k) = t(v_0) + \sum_{i=0}^{k} \tau_i(v_i)$. From the fact that the τ's are positive and max and $\sum$ are convex increasing functions, it follows that $t_k(v)$ is a convex increasing function of its arguments $\{\tau_i(u), 0 \leq i \leq k, u \in P_k, P_k \in S_k(v)\}$. Now we can use Proposition 8.5.4 in [Ro83]:

Proposition 8.5.4: If $X_1, X_2, \ldots, X_n$ are independent r.v., and $Y_1, Y_2, \ldots, Y_n$ are independent r.v., and $X_i \geq_c Y_i$, $i = 1, 2, \ldots, n$, then $g(X_1, X_2, \ldots, X_n) \geq_c g(Y_1, Y_2, \ldots, Y_n)$ for all increasing convex function g which are convex in each of its arguments.

The proof of the theorem now follows since by assumption the τ's in G are independent, the τ's in H are independent, and $\tau_k^H(v) \leq_c \tau_k^G(v)$, $k \geq 0$, $v \in V$. Note that the random variables $T(P_k)$ are not independent. ∎

Corollary 3.4: Under the above conditions $N_t^G(v) \leq N_t^H(v)$, $R^G(v) \leq R^H(v)$ and $R^G \leq R^H$.

In Section 4 we show that if the processing times are independent and have the same exponential distribution with mean λ^{-1}, then the rate of *any* network is at least $\lambda |V| / \log |V|$. We conclude this subsection by characterizing a set of distributions for which the same lower bound holds.

Assume that the expected time until a processor finishes a processing step given that it has already been working on that step for α time units is less or equal to the original expected processing time for that step. Namely, we assume that the distributions of the processing times $\tau_k(v)$, for all $v \in V$, $k \geq 0$, are *new better than used in expectation* (NBUE) (e.g. [Ro83], [S84]), so that if τ is a processing time, then

$$E[\tau - a \mid \tau > a] \leq E[\tau] \quad \forall\, a \geq 0.$$

Let $G_d(V, E)$ be a network with deterministic processing times, let $G_e(V, E)$ be a network with corresponding processing times with the same mean, but independent, exponentially distributed, and let $G(V, E)$ be a network with corresponding processing times with the same mean and independent, but with any NBUE distribution. The following theorem follows from the fact that the deterministic distribution is the minimum, while the exponential distribution is the maximum with respect to the ordering $\leq_c$, among all NBUE distributions [Ro83], [S84].

Theorem 3.5: For every $v \in V$, $k \geq 0$, $t_k^{G_d}(v) \leq_c t_k^G(v) \leq_c t_k^{G_e}(v)$.

Some examples of distributions which are less variable than the exponential (with appropriate parameters) are the Gamma, Weibull, Uniform and Normal.

4. EXPONENTIAL DISTRIBUTIONS

In this section we assume that the processing times $\tau_k(v)$, $k \geq 0$, $v \in V$ are independent and exponentially distributed with mean λ^{-1}. We first consider general topologies and derive upper and lower bounds on the expected values of $t_k(v)$, and thus obtain upper and lower bounds on the rate of the network. These bounds depend on the in-degrees and out-degrees of processors in the network, but not on the number of processors itself. Then we derive the exact rates of two extreme topologies- the directed ring and the fully connected (complete) network. For these two topologies we study also the effect of having a single slower processor within the network.

4.1 Upper and Lower Bounds

Denote by $d_{out}(v)$ ($d_{in}(v)$) the number of edges going out of (into) v in G plus 1, and let

$$\Delta_{out} = \max_{v \in V} d_{out}(v), \quad \Delta_{in} = \max_{v \in V} d_{in}(v);$$

$$\delta_{out} = \min_{v \in V} d_{out}(v), \quad \delta_{in} = \min_{v \in V} d_{in}(v).$$

Lemma L (Lower Bound):
(i) For every $k \geq 0$ there exists a processor $v \in V$ for which

$$E[\,t_k(v)\,] \geq \max_{u \in V} t(u) + \lambda^{-1}[1 + k \cdot \log \delta_{out}].$$

(ii) For every $k \geq 0$, and every $v \in V$, the following holds

$$E[\,t_k(v)\,] \geq \min_{u \in V} t(u) + \lambda^{-1}[1 + k \cdot \log \delta_{in}].$$

Proof: We present a detailed proof for part (i) only; the proof of part (ii) is discussed at the end. We start by proving that for every $k \geq 0$, there exists a (not necessarily simple) path $v_0 \to v_1 \to \cdots \to v_k$, such that

$$E[\,t_{i+1}(v_{i+1})\,] - E[\,t_i(v_i)\,] \geq \lambda^{-1} \log \delta_{out}, \quad 0 \leq i < k.$$

We assume the statement holds for $k \geq 0$, and prove it for $k + 1$. The proof of the basis is identical. Let v_{k+1} be the processor for which the processing time during the $(k+1)$th computational step is maximum, among the out-neighbors of v_k, including v_k itself, i.e.,

$$\tau_{k+1}(v_{k+1}) = \max_{v_k \to v} \tau_{k+1}(v),$$

where we assume that for all $u \in V$, $u \to u \in E$, for convenience of notation. Since v_{k+1} will not start the $(k+1)$st computational step before v_k finishes the kth computational step, we have that $t_{k+1}(v_{k+1}) - t_k(v_k) \geq \tau_{k+1}(v_{k+1})$. The quantity $\tau_{k+1}(v_{k+1})$ is equal to the maximum of at least δ_{out} independent and identically distributed exponential random variables with mean λ^{-1}. It is well known (e.g. [BT89], [D70]) that the mean of the maximum of c such random variables is at least $\lambda^{-1} \log c$. It follows that

$$E[\,t_{k+1}(v_{k+1})\,] - E[\,t_k(v_k)\,] \geq \lambda^{-1} \log \delta_{out}.$$

We can chose v_0 to be the one with latest waking time $t(v_0)$, and thus $E[t_0(v_0)] = t(v) + \lambda^{-1}$. Therefore, for every $k \geq 0$, there exists a processor v such that

$$E[\,t_k(v)\,] \geq \max_{u \in V} t(v) + \lambda^{-1}[1 + k \cdot \log \delta_{out}].$$

completing the proof of (i). The proof of part (ii) evolves along the same lines, except that we start from v_k and move backwards along a path. ■

Remark 4.1: From its proof, one can see that Lemma L holds for any distribution F of the processing times, for which the expected value m_c of the maximum of c independent r.v. with distribution F exists. In this case it implies that $R_v \leq 1 / m_c$, with $c = \delta_{out}$ or $c = \delta \in$.

Remark 4.2: Lemma L implies that for the exponential case, the slowdown of the rate is at least logarithmic in the maximum degree of G. By Remark 4.1, there are distributions for which the slowdown is larger; an example is $F(x) = 1 - 1/x^2$, $x \geq 1$, for which the slowdown is at least the square root of the maximum degree of G ([D70] pp.58).

Lemma U

(i): For every $k \geq 1$, for every processor v,

$$E[\,t_{k-1}(v)\,] \leq \max_{u \in V} t(u) + \frac{4}{\lambda}(1 + k \cdot \log \Delta_{in}).$$

(ii): For every $k \geq 1$, for every processor v,

$$E[\,t_{k-1}(v)\,] \leq \max_{u \in V} t(u) + \log |V| + \frac{4}{\lambda}(1 + k \cdot \log \Delta_{out}).$$

Proof Sketch: Again we restrict ourselves to the proof of part (i). Recall that Theorem 2.1 states that for every $v \in V$, $k \geq 0$, $t_k(v) = \max T(S_k(v))$. Also, for a path $P_k = v_0 \to v_1 \to \cdots \to v_k$, $T(P_k) = t(v_0) + \sum_{i=0}^{k} \tau_i(v_i)$, but for the moment let $t(v) = 0$, for every v.
By Proposition D.2 of the Appendix,

$$Pr\left[T(P_{k-1}) \geq \frac{ck}{\lambda} \log \Delta_{in}\right] \leq e^{-\frac{ck}{4} \log \Delta_{in}},$$

for every $c > 4$, since $\log 2 / \log \Delta_{in} \leq 1$. It follows that

$$Pr\left[t_{k-1}(v) \geq \frac{ck}{\lambda} \cdot \log \Delta_{in} \right] \leq \Delta_{in}^k \cdot e^{-\frac{ck}{4} \log \Delta_{in}} = e^{-k(\frac{c}{4} - 1)\log \Delta_{in}},$$

for every $c > 4$, and

$$E[t_{k-1}(v)] \leq \int_0^{\frac{4k}{\lambda} \log \Delta_{in}} 1 \, dt + \int_{\frac{4k}{\lambda} \log \Delta_{in}}^{\infty} e^{-k(\frac{c}{4} - 1)\log \Delta_{in}} \frac{4}{\lambda} \log \Delta_{in} dc = \frac{4k}{\lambda} \log \Delta_{in} + \frac{4}{\lambda}.$$

∎

Theorem 4.1:

$$\frac{\lambda}{4\log \min(\Delta_{out}, \Delta_{in})} \leq R_v \leq \frac{\lambda}{\log \max(\delta_{out}, \delta_{in})}.$$

4.2 Exact Computations

Theorem 4.1 implies the following bounds for the rate of a directed cycle C_n, and of a complete graph K_n, where n is the number of processors:

$$0.36\lambda \leq R^{C_n}(v) \leq \lambda,$$

$$\frac{\lambda}{4 \log n} \leq R^{K_n}(v) \leq \frac{\lambda}{\log n}.$$

In this section we shall compute the exact values for the rates of C_n and K_n. To that end we consider the Markov chain associated with the network This Markov chain is denoted by $X(t) = (X_1(t), X_2(t), ..., X_m(t))$, where $X_i(t)$, is the number of messages stored in the buffer of edge i at time t, and m is the number of edges in the network. Note that a processor with positive number of messages on each of its in-comming edges is in a processing state. When such a processor completes its processing (after an exponential time), one message is deleted from each of its in-comming edges and one message is put on each of its out-going edges. We denote by s_0 the state in which $X_i(0) = 1$, $1 \leq i \leq m$. Thus, the network can be represented as a Marked Graph (e.g. [CHEP]).

The number of states in the Markov chain is finite, say N, because a transition of the chain does not change the total number of messages in a circuit in the network. Moreover, if the network is strongly connected, then the Markov chain is irreducible. Therefore, the limiting probabilities P_i, $1 \leq i \leq N$, of the states s_i of the chain exist, they are all positive and their sum is equal to 1 (e.g. [C67],[Ro83]).

Let G_X denote the transition diagram (directed graph) of the Markov chain X. Consider a *BFS* (breadth first search) tree of G_X, rooted at s_0. The level $L(v)$ of a vertex v will be equal to the distance from s_0 to v. Thus, $L(s_0) = 0$. Denote by L_i, $i \geq 0$, the set of vertices at level i, and by L the number of levels of G_X.

A Simple Directed Cycle

We study the performance of a simple directed cycle of n processors $C_n = p_1 \to p_2 \to \cdots \to p_n \to p_1$. It is not difficult to observe that the Markov chain associated with C_n, corresponds to that of a closed queuing network; we return to this approach later. Here we choose to use a combinatorial approach.

Theorem 4.2

(i) All the states associated with C_n, have the same limiting probability.

(ii) For any graph G which is not a simple directed cycle (i) does not hold.

The next theorem states that each processor of C_n works at least at half of its potential rate λ, regardless of the value of n.

Theorem 4.3: The rate $R(v)$ of a processor in C_n is

$$R(v) = \lambda \cdot \left[1 - \frac{n-1}{2n-1} \right] - \frac{\lambda}{2},$$

$$N = \frac{(2n-1)!}{n!(n-1)!}$$

and the limiting probability of each state is $1/N$, where N is the number of states in the associated chain.

A Complete Graph

Let K_n be a complete graph with n processors. Recall that N is the number of states in the associated Markov chain, and let s_0 be the state in which each edge has one token. A state is at level l, $0 \leq l \leq n-1$, if it can be reached from s_0 by the firing of l processors. The limiting probability of a state at level l is denoted by $P(l)$.

Theorem 4.4: The rate of a processor in K_n is

$$R(v) = \lambda / \sum_{i=1}^{n} \frac{1}{i} - \frac{\lambda}{\log n}.$$

Proof Sketch: A simpler proof can be derived as in the proof of Theorem 4.6; here we give a combinatorial proof which also yields the number and the limiting probabilities of the states of the associated Markov chain.

We consider a Markov chain T, similar to the Markov chain associated with network K_n. The root of T, s_0, is the state with a message in each edge. A state s will have one son for each one of the enabled processors at state s; a son of s corresponds to the state arrived from s by the firing (completion of a processing step) of one of the enabled processors in state s. Note that in chain T there are several vertices corresponding to the same state of the chain associated with K_n.

In T, the number of states in level l is $n!/(n-l)!$, because each time a processor fires it can not fire again until the rest of the processors have fired. Thus, the number N^T, of states in T is

$$N^T = \sum_{i=1}^{n} \frac{n!}{i!}$$

The number of states in which a given processor is enabled at level l, $en(l)$ (edges from level l to level $l+1$), is

$$en(l) = \frac{1}{n}\frac{n!}{(n-l-1)!},$$

because at level l there are $n!/(n-l-1)!$ enabled processors, and by symmetry, each processor is enabled the same number of times at each level.

Let us denote by P_l^T, the limiting probability of a state of T in level l. One can show that $P_l^T = (n-l-1)!/K$, where

$$K = \sum_{l=0}^{n-1} \frac{n!}{(n-l)!} P_l^T = n! \sum_{i=1}^{n} \frac{1}{i}.$$

It follows that the percent of time that a processor is enabled is

$$ut = \sum_{l=0}^{n-1} ut(l) = \frac{1}{\sum_{i=1}^{n} \frac{1}{i}},$$

where $ut(l) = en(l)P_l^T$, and its rate is $\lambda \cdot ut$.

∎

Corollary 4.1: For a network K_n,

$$N = 2^n - 1,$$

$$P_l = \frac{l!(n-l-1)!}{n! \sum_{i=1}^{n} \frac{1}{i}}.$$

Proof: As noted before, it may be that two states of T correspond to the same state, say s, of K_n. In fact, if a state of T is reached from s_0 by firing a sequence of processors of length k, then all $k!$ permutations of the processors in this sequence constitute a valid firing sequence, which leads to the same state s. Thus, the limiting probability of a state s at level l is

$$P_l = l! P_l^T = \frac{l!(n-l-1)!}{n! \sum_{i=1}^{n} \frac{1}{i}},$$

The number of different states at level l is $n!/l!(n-l)!$, and the total number of different states is

$$\sum_{l=0}^{n-1} n!/l!(n-l)! = 2^n - 1.$$

■

Corollary 4.2: Asymptotically, the rate of any network of n processors is between $\lambda n/2$ and $\lambda n/\log n$.

Observe that the best possible rate of a processor is $2/3$ of the potential rate, in the case of a cycle of two processors; adding more processors can only lower this rate, but not below $1/2$. Yet, the rate of the network grows linearly with n. In the case of a complete graph, the rate of a processor reduces as n grows, but also here the total number of computational steps executed per unit time ($n/\log n$) grows with n.

4.3 Bottlenecks

Suppose that the potential rate of all processors of a graph is λ, except for one, which has a lower rate μ. We shall now show that such a bottleneck has a stronger effect in a network which is a directed cycle, than in one which is a complete graph.

Consider the case of a simple directed cycle with n vertices CB_n, where $n-1$ processors have rate λ, and one processor has rate μ. Using standard techniques of Queuing Theory, we prove the following.

Theorem 4.5: The rate of a processor in CB_n is

$$\mu\left[1 - \frac{\binom{2n-2}{n}\rho^n}{\sum_{i=0}^{n}\binom{n+i-2}{i}\rho^i}\right]; \quad \rho = \frac{\lambda}{\mu}.$$

Several conclusions can be derived from Theorem 4.5. First, observe that the rate of the cycle cannot exceed μn, thus the slow processor bounds the rate of the network. Moreover, for a fixed n and a very slow processor ($\mu \to 0$ or $\rho \to \infty$), the rate of the network is $\mu n[1 - \binom{2n-2}{n}\rho^{-n}]$, namely, as ρ increases, the rate approaches its upper bound μn.

Next, we consider the case where the graph is a complete graph KB_n. We continue to assume that the rate of $n-1$ processors is λ and the n-th processor is slower, operating at rate μ. We shall show that, for fixed μ and λ, as the number of processors n grows to infinity, the influence of the slow processor diminishes, and in the limit, the rate of the network is the same as that of a network with all processors running with the same rate λ.

Theorem 4.6: The rate of a processor in KB_n is at least $\dfrac{\lambda}{\rho + \log n}$, $\rho = \lambda/\mu$.

Proof: Suppose that the network is in state s_0 at a given time, and after some time T_1 it returns to that state; then after some time T_2 it returns again, and so on. Then, $\{T_i, i = 1, 2, \cdots\}$ is a sequence of nonnegative independent random variables with a common distribution F, and expected value $E[T_i]$.

Denote by $N(t)$ the number of events (returning to s_0) by time t. The counting process $\{N(t), t \geq 0\}$ is a renewal process. Therefore, with probability 1,

$$\frac{N(t)}{t} \to \frac{1}{E[T_i]} \quad \text{as } t \to \infty.$$

(See, for example [Ro83]). Moreover, since each time the process returns to s_0, each processor of the network has completed exactly one computational step, it follows that the rate of the network is $1/E[T_i]$. We proceed to bound $E[T_i]$.

The expected time of T_i, that takes to return to s_0 is of the form:

$$\alpha_j = \frac{1}{(n-1)\lambda + \mu} + \frac{1}{(n-2)\lambda + \mu} + \cdots + \frac{1}{(n-j)\lambda + \mu} + \frac{1}{(n-j)\lambda} + \frac{1}{(n-j-1)\lambda} + \cdots + \frac{1}{\lambda},$$

for some $1 \leq j \leq n$, depending on when the slow processor completes a computational step. If the system leaves s_0 because the slow processor completed a computational step, $E[T_i]$ is α_1. In general, if the j-th ($1 \leq j \leq n$) processor to complete a computational step, after leaving s_0, is the slow processor, then $E[T_i]$ is α_j.

The probability of $E[T_j]$ being equal to α_j, is not necessarily the same for every j, but for the case $\lambda > \mu$, it holds that $\alpha_j < \alpha_{j+1}$. Thus,

$$\alpha_n = \sum_{k=0}^{n-1} \frac{1}{\lambda \cdot k + \mu}$$

gives an upper bound on the time $E[T]$ that takes to return to s_0, and $1/\alpha_n$ is a lower bound on the rate of a processor in the network.

We have

$$\alpha_n = \sum_{i=1}^{n-1} 1/(\lambda i + \mu) \leq \frac{1}{\mu} + \frac{1}{\lambda}\sum_{i=1}^{n-1}\frac{1}{i} \leq \frac{1}{\mu} + \frac{1}{\lambda}\log n.$$

We see that $E[T_i] = O(\dfrac{1}{\mu} + \dfrac{1}{\lambda}\log n)$ and thus $R(KB_n)$ is at least $\dfrac{1}{\dfrac{1}{\mu} + \dfrac{1}{\lambda}\log n}$. ∎

For fixed μ, R_v is $\Theta(\lambda/\log n)$, but observe that the rate of the network cannot exceed μ. However, when the number of processors n increases to infinity, the rate of the network decreases in proportion to $1/\log n$, as if the slower processor is not in the network.

5. NON-NEGLIGIBLE TRANSMISSION DELAYS

We briefly discuss the case of non-negligible transmission delays. In this model the processing times are random, as before, but the transmission delays are also random. Denote the transmission delay of message M_k, $k \geq 0$, along edge $u \to v$, by $\tau_k(u, v)$, and let $\tau_k(u, u) = 0$, for all $u \in V$. It follows that the behavior of the system is described by the recursions

$$t_0(v) = t(v) + \tau_0(v)$$

$$t_{k+1}(v) = \max_{u \in IN(v)} \left\{ t_k(u) + \tau_k(u, v) \right\} + \tau_{k+1}(v) \quad k \geq 0.$$

Note that this system is not equal to the one of [BT89], in which the processing times are negligible, and the delays non-negligible, with a self-loop in each processor (to model its processing delay).

Let $P_k = v_0 \to v_1 \to \cdots \to v_k \,(=v)$ be a path of length k. It is easy to see how to modify the definition of $T(P_k)$:

$$T(P_k) \stackrel{\Delta}{=} t(v_0) + \sum_{i=0}^{k-1} [\tau_i(v_i) + \tau_i(v_i, v_{i+1})] + \tau_k(v_k).$$

Thus a theorem similar to Theorem 2.1 holds, and the corresponding results for general distributions follow.

Consider the case in which the processing times, as well as the transmission delays are exponentially distributed, with the same mean, say 1. It is easy to see that Lemma L still holds, and that Lemma U holds up to a factor of 2. Namely, by Theorem 3.4, a regular network with non-negligible delays runs at the same rate that the same network, up to a constant factor, provided that the delays are less or equal (in the convex order) than the processing times. In [ER1] we show that for a network with negligible delays and deterministic processing times equal to 1, the rate of any network is equal to 1. Thus, in this case, random processing times degrade the rate by at most a logarithmic factor in the maximum degree of a processor.

Now, consider the case in which all processing times have mean 1, but the delays have mean λ^{-1} greater than 1, both exponentially distributed. The rate in the deterministic case is equal to λ [ER2], and thus, by Theorem 3.2, in our case the rate is at most λ. One can prove (using Proposition D.2), that also in the case of non-negligible delays, the rate is degraded by at most a logarithmic factor in the maximum degree of a processor, with respect to the (optimal) deterministic case, for any NBUE distribution.

As for the exact computations for networks with average processing times and delays exponentially distributed with mean 1, the rate of a simple cycle can be computed using the same tools of Queuing Theory that we used in the case of negligible delays. To compute the rate of a complete network K_n things are not as straightforward; the structure of the Markov process is more complicated, but by the arguments above, we have that the rate is between $1/8\log n$ and $1/\log n$. However, using the ideas of embeddings, let us show that the rate of K_n, is at least $1/4\log n$. Let K'_n be a complete network with negligible delays. Construct G, from K'_n by inserting one vertex in each of its edges. By Theorem 3.1 (or also 3.4), the rate of any processor in G is at least

$$\frac{1}{\log(n + n(n-1))} = \frac{1}{2\log n}.$$

One can show that the rate of any processor v in K_n is greater or equal than half the rate of the corresponding processor in G, using using the fact that there is an embedding of K_n into G of dilation 2. Therefore, the rate of v in K_n is at least $1 / 4\log n$.

APPENDIX

The following proposition (similar to pp. 672 in [BT89]) is used to prove the lower bounds on the rate of a network.

Proposition D.2: Let $\{X_i\}$ be a sequence of independent exponential random variables with mean λ^{-1}. For every positive integer k and any $c > 4\log 2$,

$$Pr\left[\sum_{i=1}^{k} X_i \geq \frac{ck}{\lambda}\right] \leq e^{-\frac{ck}{4}}.$$

Acknowledgments

We would like to thank Gurdip Singh and Gil Sideman for helpful comments.

References

[A1]   B. Awerbuch, "Complexity of Network Synchronization ", JACM, Vol. 32, No. 4, Oct. 1985, pp. 804-823.

[A2]   B. Awerbuch, "Reducing Complexities of Distributed Max-Flow and Breadth-First-Search Algorithms by Means of Network Synchronization ", Networks 15, 1985, pp. 425-437.

[AP90]   B. Awerbuch, D. Peleg, "Network Synchronization with Polylogarithmic Overhead", Proc. IEEE FOCS, 1990.

[BG89]   V.C. Barbosa, E. Gafni, "Concurrency in Heavily Loaded Neighborhood- Constrained Systems", ACM Trans. on Programming Languages and Systems, Vol. 11, No. 4, Oct. 1989, pp. 562-584.

[BS88]   P. Berman, J. Simon, "Investigations of Fault-Tolerant Networks of Computers", Proc. of the 20th ACM STOC, 1988.

[BT89]   D. P. Bertsekas, J. N. Tsitsiklis, *Parallel and Distributed Computation*, Prentice-Hall, N.J. 1989.

[C67]   K. L. Chung, *Markov Chains With Stationary Transition Probabilities*, Springer-Verlag, 2nd edition, 1967.

[CL85]   K. M. Chandy and L. Lamport, "Distributed Snapshots: Determining Global States of Distributed Systems", ACM Trans. on Computer Systems, Vol. 3, No 1, Feb. 1985.

[CHEP]   F. Commoner, A.W. Holt, S. Even, A. Pnueli, "Marked Directed Graphs", J. of Computer and System Sciences, Vol, 5, No 5, Oct. 1971.

[D70]   H.A. David, *Order Statistics*, John Wiley & Sons, 1970.

[ER1]   S. Even, S. Rajsbaum, "Lack of Global Clock Does Not Slow Down the Computation in Distributed Networks", TR #522, Department of Computer Science, Technion, Haifa, Israel, Oct. 1988. The first part of this paper appears with the title "Unison in Distributed Networks" in *Sequences: Combinatorica, Compression, Security, and Transmission*, R.M. Capocelli (ed.), Springer-Verlag, 1990.

[ER2]   S. Even, S. Rajsbaum, "The Use of a Synchronizer Yields Maximum Rate in Distributed Networks", Proc. of the 22nd ACM STOC, 1990.

[Ma89] M.A. Marsan, "Stochastic Petri Nets: An Elementary Introduction", in Advances in Petri Nets 1989, Lecture Notes in CS 424, Springer-Verlag, 1989, pp. 1-29.

[MMZ88] J. Malka, S. Moran, S. Zaks, "Analysis of a Distributed Scheduler for Communication Networks", TR-495, Department of Computer Science, Technion, Haifa, Israel, Feb. 1988. Also in Lecture Notes on CS, Vol. 319, pp. 351-360, Springer Verlag, 1988.

[M1]   M. K. Molloy, "Performance Analysis Using Stochastic Petri Nets", IEEE Trans. on Computers, Vol. c-31, No. 9, Sep. 1982, pp. 913-917.

[M2]   M. K. Molloy, "Fast Bounds for Stochastic Petri Nets", International Workshop on Timed Petri Nets, Torino, Italy, July 1985, pp. 244-249.

[OG87] Y. Ofek, I. Gopal, "Generating a Global Clock in a Distributed System", IBM Research Report, 1987.

[PS89] D. Peleg and A. A. Schaffer, "Graph Spanners", J. of Graph Theory, Vol. 13, No. 1, 1989, pp. 99-116.

[PU89] D. Peleg and J. D. Ullman, "An Optimal Synchronizer for the Hypercube", SIAM J. Computing, Vol. 18, No. 4, August 1989, pp. 740-747.

[R88]   A. L. Rosenberg, "Shuffle-Oriented Interconnection Networks", COINS Technical Report 88-84, Univ. of Massachusetts, 1988.

[RM]   S. Rajsbaum, Y. Malka, "Synchronizers, Schedulers and Marked Graphs", in preparation.

[Ro83] S. M. Ross, *Stochastic Processes*, J. Wiley, 1983.

[S84]   D. Stoyan, *Comparison Methods for Queues and Other Stochastic Models*, English Translation (D.J. Daley, Ed.), J. Wiley & Sons, New York, 1984.

Distributed Algorithms for Reconstructing MST after Topology Change*

Jungho Park, Toshimitsu Masuzawa, Ken'ich Hagihara and Nobuki Tokura

Faculty of Engineering Science, Osaka University
1-1 Machikaneyama, Toyonaka, 560, JAPAN

ABSTRACT

This paper considers the *Updating Minimum-weight Spanning Tree Problem* (UMP), that is, the problem to update the *Minimum-weight Spanning Tree* (MST) in response to topology change of the network. This paper proposes the algorithm which reconstructs the MST after several links are deleted and added. Its message complexity and its ideal-time complexity are $O(m + n \log(t + f))$ and $O(n + n \log(t + f))$ respectively, where n is the number of processors in the network, t (resp. f) is the number of added links (resp. the number of deleted links of the old MST), and $m = t + n$ if $f = 0$, $m = e$ (i.e. the number of links in the network after the topology change) otherwise. The last part of this paper touches on the algorithm which deals with deletion and addition of processors as well as links.

1. Introduction

Consider an asynchronous network where a cost or weight (representing, for example, usage fee or delay) is associated with every link. For the purpose of disseminating information in the network, it is advantageous to broadcast it over the *Minimum-weight Spanning Tree* (MST), since information will be delivered to every processor with small communication cost. Thus, constructing the MST of a network is one of basic network problems, and many distributed algorithms have been proposed for the problem. Most of them consider the *initial* problem to construct the MST, that is, construction of the MST starts *from scratch*. In a real network, topology of a network often changes because of processor (or link) deletion (e.g. failure) and addition (e.g. recovery). When a network N changes to N' by topology change, the MST T of N may not be the MST of N'. For example, when a link of T is deleted by the topology change, T is divided into two trees, and the new MST must be reconstructed to broadcast or collect messages efficiently. This paper considers the *Updating MST Problem* (UMP), that is, the problem to update the MST in response to topology change. In UMP, it is assumed that topology change does not occur during the execution of an algorithm. In the following, N (resp. N') represents a network before topology change (resp. a network after topology change), and T (resp. T') represents the MST of N (resp. N').

---
* This work was supported in part by the Scientific Research Grant-in-Aid from Ministry of Education, Science and Culture of Japan, the Mazda Foundation's Research Grant and the Inamori Foundation's Research Grant.

It is obvious that UMP can be solved by the known algorithms which construct the MST from scratch. For the problem to construct the MST from scratch, Gallager et al. presented a distributed algorithm and showed that its message complexity is optimal within a constant factor[3]. Moreover, Awerbuch improved the algorithm and obtained an optimal algorithm in both the message complexity and the ideal-time complexity[1]. By applying Awerbuch's algorithm to the network after topology change, UMP can be solved with $O(e + n \log n)$ messages and $O(n)$ (ideal-) time, where n (resp. e) is the number of processors (resp. links) in the network N'.

However, most part of the new MST T' may coincide with the old MST T, and it is natural to assume that each processor knows the *old solution*, that is, each processor knows which incident links are those of the old MST T. This raises a question : *How efficiently UMP can be solved by utilizing the old solution?* This is an interesting subject of study.

This paper proposes the algorithm which reconstructs MST T' after several links are deleted and added. Its message complexity is $O(m + n \log(t+f))$ and its ideal-time complexity is $O(n + n \log(t+f))$, where t (resp. f) is the number of added links (resp. the number of deleted links of the old MST T), and where $m = n + t$ if $f = 0$, $m = e$ otherwise. Thus, our algorithm is superior to Awerbuch's algorithm in the message complexity. In the last part of this paper, we will touch on the algorithm which deals with deletion and addition of processors as well as links.

Up to date, a number of algorithms have been proposed for UMP. However, they consider *either* link deletion *or* link addition, and no algorithm considers the situation where *both* link deletion and link addition occur. Hence, our algorithm is the first algorithm that considers both link deletion and link addition.

To show the efficiency of our algorithm, the rest of this section compares our algorithm with the previous results in the following two cases (see Table 1):
(a) the case that only link addition occurs, and
(b) the case that only link deletion occurs.

Table 1. Complexities of algorithms for UMP.

	j Links Deletion		t Links Addition	
	Paper [2]	Our Result	Paper [5]	Our Result
Message Complexity	$O(j^2 p)$	$O(e + n \log f)$	$O(nt + t^2)$	$O(n + t + n \log t)$
Bit Complexity	$O(j^2 pe \log n)$	$O(e \log n + n \log f \log n)$	$O(nt \log n + t^2 \log n)$	$O(n \log t \log n + (n+t) \log n)$
Ideal-time Complexity	$O(j^2 p)$	$O(n + n \log f)$	$O(nt + t^2)$	$O(n + n \log t)$
Space Complexity	$O(ne \log n)$	$O(e)$	$O(e)$	$O(e)$

n : The number of processors.
e : The number of links in the network after topology change.
f : The number of deleted links of the old MST T.
p : The length of the longest cycle in the network before topology change.

(a) The case that only link addition occurs.

For UMP after t links addition, Tsin presented an algorithm[5]. Its message complexity and its ideal-time complexity are both $O(nt+t^2)$. In this case, the message complexity and the ideal-time complexity of our algorithm are $O(n+t+n\log t)$ and $O(n+n\log t)$, respectively. Therefore, our algorithm is superior to Tsin's algorithm in both the message complexity and the ideal-time complexity.

(b) The case that only link deletion occurs.

For the case of j links deletion, Cimet et al. presented an algorithm (hereinafter denoted as CK)[2]. Its message complexity and its ideal-time complexity are both $O(j^2 p)$, where p is the length of the longest cycle in a network N. For this case, the message complexity and the ideal-time complexity of our algorithm are respectively $O(e+n\log f)$ and $O(n+n\log f)$. (Recall that f represents the number of deleted links of the old MST T, while j represents the number of deleted links including the non-tree links. This implies that $f \leq j$ holds.) It depends on the parameters n, e, f, p and j whether or not our algorithm is better than CK in the message complexity and the ideal-time complexity. The bit complexity of CK is $O(j^2 pe \log n)$ since the length of each message used in CK is $O(e \log n)$ bits. On the other hand, the bit complexity of our algorithm is $O(e \log n + n \log f \log n)$ since the length of each message used in our algorithm is $O(\log n)$ bits. Therefore, our algorithm is better, in the bit complexity, than CK. Furthermore, CK utilizes the auxiliary information, called a *replacement set*, in addition to the old solution. To store the replacement set, CK needs the storage of $O(ne \log n)$ bits in the whole network. On the other hand, the space complexity (the total storage in the whole network) of our algorithm is $O(e)$ since our algorithm needs no auxiliary information except for the old solution. Therefore, our algorithm is superior to CK in the space complexity.

2. Model

Our model is standard one, that is, **(A1)** through **(A5)** are assumed.

(A1) The processors are connected by *bidirectional* communication links and the processors communicate only by passing messages along the links. A network N is denoted as N=(P,L) where P is the set of processors and L is the set of links. In the followings, n (resp. e) stands for the number of processors (resp. links), that is, $|P|=n$ and $|L|=e$ hold.

(A2) The network is *asynchronous*, that is, the time to transmit a message along a link is finite but unpredictable.

(A3) Each processor u has a unique *identity number* (i.e. processor number) $ID(u)$, and every identity number is represented in $O(\log n)$ bits. Each link (u, v) has a unique *weight* $W(u, v)$, and every weight is represented in $O(\log n)$ bits.

(A4) The processors all perform the same program. The program executed in each processor includes (a) its internal operations, (b) *send* operations to send messages via its ports, and (c) *receive* operations to receive messages from its ports. (Each processor can distinguish its ports each other.)

(A5) Any non-empty subset of processors may start the algorithm spontaneously, and each of other processors starts the algorithm when it receives a message.

MST Updating Problem (MUP)

Let N be an arbitrary network and assume that the MST T of N is already constructed, that is, each processor knows which incident links are those of T. The *MST Updating Problem* (MUP) is the problem to reconstruct the MST T' after N changes to N' by topology change. In order to solve UMP efficiently, some auxiliary information (e.g. the replacement set [2]) may be utilized. In this case, the auxiliary information needs to be updated so as to correspond to N'. Formally, MUP is the problem to obtain the following final configuration from the following initial configuration :

Initial configuration : Each processor knows
(a) the *old solution*, that is, each processor knows which incident links are those of the old MST T,
(b) the deleted links and added links incident to itself, and
(c) some *auxiliary information* (such as the replacement set [2]), if necessary.

Final configuration : Each processor knows
(a) the *updated solution*, that is, each processor knows which incident links are those of the new MST T', and
(b) *updated auxiliary information*, if some auxiliary information is utilized.

In this paper, MUP is considered under the following assumption.
(A6) The topology change does not occur during the execution of an algorithm.

This paper mainly considers only deletion and addition of links, and touches on deletion and addition of processors in the last part.

Throughout this paper, N=(P,L) (resp. N'=(P,L')) represents a network before topology change (resp. a network after topology change), and T=(P,L_T) (resp. T'=(P,L_T')) represents the MST of N (resp. N'). Moreover, L_a (resp. L_d) represents the set of added links (resp. the set of deleted links of T), and let $t=|L_a|$ and $f=|L_d|$. Note that L_d does not contain the deleted links which are not the links of T. We will pay only a little attention to the deleted links not included in L_T.

Measures of algorithms' efficiency

In this paper, to measure the efficiency of an algorithm, we use the following measures.

(A) *Message complexity* : The (worst case) message complexity is the maximum total number of messages transmitted during any execution of the algorithm.

(B) *Bit complexity* : The (worst case) bit complexity is the maximum total number of bits transmitted during any execution of the algorithm.

(C) *Ideal-time complexity* : The (worst case) ideal time complexity is the maximum number of time units from start to the completion of the algorithm, assuming that the propagation delay of every link is one time unit of some global clock. This assumption is used only for the purpose of evaluating the ideal-time complexity, since the network is asynchronous.

(D) *Space complexity* : The (worst case) space complexity is the total amount of storage of all processors in the whole network.

3. The idea of our algorithm

In this section, we describe the idea of our algorithm to update MST after several links are deleted and added.

3.1 Application of Gallager's method

Our algorithm is partly based on the technique Gallager *et al.* proposed[3]. At first, we shortly illustrate the Gallager's method.

Gallager's method

Any connected subgraph of the MST is referred to as a *fragment*. Define a link as an *outgoing link* of a fragment if one adjacent processor is in the fragment and the other is not. Gallager's algorithm, denoted as GHS hereinafter, starts with each individual processor as a fragment, that is, there exist n fragments each of which consists of one processor in the beginning of GHS. GHS enlarges each fragment by finding its minimum-weight outgoing link and combining the fragment with the fragment at the other end of the link, then enlarges each of the new fragments again in the same way, and so forth. GHS repeats the above procedure until there exists only one fragment, that is the MST. In order to reduce the message complexity, GHS enforces a balanced growth of fragments by utilizing the *level* of each fragment, which reflects the size (i.e. the number of processors) of the fragment.

Applying Gallager's method to UMP

When topology change occurs, most part of the new MST T' may coincide with the old MST T. For example, consider the topology change where *several links of T are deleted and no link is added*. By the topology change, the old MST T is divided into several connected components, and then each connected component is a fragment of the new MST T'. This fact implies that an algorithm for UMP can start with enlarged fragments, that is, each of the fragments may contain more than one processor. By the algorithm exactly similar to GHS, we can efficiently construct MST from the initial configuration. We denote the algorithm as PMHT.

[Property 1][4] If the algorithm PMHT starts in the situation that there are r fragments, then it constructs the MST with the message complexity $O(n \log r + e)$ and the ideal-time complexity $O(n \log r + n)$. □

When f links of the old MST T are deleted, T is divided into $f+1$ connected components. Therefore from Property 1, the algorithm PMHT constructs the new MST with the message complexity $O(n \log f + e)$ and the ideal-time complexity $O(n \log f + n)$, after f links of the old MST T are deleted (and no link is added).

Consider the case where *both link addition and link deletion occur*. In this case, the old MST T is divided into several connected components by the deleted links. But all connected components of T-L_d(=(P,L_T-L_d)) may not be fragments of the new MST T' since there exist added links. Therefore, the algorithm PMHT can not be simply applied to this case.

How can we solve UMP after several links are added and deleted? We can pick up an idea in the algorithm Tsin proposed[5].

3.2 Tsin's method

Tsin [5] proposed an algorithm to reconstruct MST after *several links are added and no link is deleted*. We briefly describe the algorithm.

Removal of the maximum-weight link in a cycle

When a link (u, v) is added, (u, v) and the u-v path in T form a cycle. In this case, the new MST T' is obtained by removing the maximum-weight link in the cycle from $T + \{(u, v)\}$ $(= (P, L_T \cup \{(u, v)\}))$. When several links are added, $T + L_a$ $(=(P, L_T \cup L_a))$ has several cycles. Tsin's algorithm reconstructs the new MST T' as follows.
(1) Let T_{work} be $T+L_a$.
(2) Repeat the following (i) and (ii) until there exists no cycle in T_{work}.
   (i) Find a cycle of T_{work}, and find the maximum-weight link l in the cycle.
   (ii) Let T_{work} be $T_{work} - \{l\}$.
(3) T_{work} is the new MST T'. (Note that T_{work} has no cycle in (3).)

Tsin's algorithm cannot *efficiently* reconstruct the new MST T', because it *sequentially* processes the cycles in $T+L_a$. Therefore the authors of this paper tried to process the cycles *concurrently*, but it seems to be difficult to do so in the reasonable cost (the number of messages). In order to solve this difficulty, we introduce new idea.

3.3 Partition links (Our new idea)

The algorithm PMHT is an efficient algorithm to reconstruct T' from the initial configuration where there exist several enlarged fragments. In order to apply PMHT to the case where *both link addition and link deletion occur*, we newly introduce the *partition links* (the set of partition links is denoted by L_p throughout this paper). The definition of the partition links is described later. The key point is that the partition links satisfy the following properties.

[Property 2] The partition links are links of the old MST T, that is, $L_T \supseteq L_p$ holds. □

[Property 3] Let l be any link of the old MST T. If l is deleted by the topology change, l is a partition link, that is, $L_p \supseteq L_d$ holds. □

[Property 4] Each connected component of $T-L_p$ is a fragment of the new MST T', that is, L_p contains the maximum weight link in each cycle of $T+L_a$. □

[Property 5] The number of connected components in $T-L_p$ is $O(t+f)$. □

[Property 6] The partition links can be found efficiently. □

Our algorithm first finds the partition links and changes them to non-tree links, then all tree links (except for partition links) of the old MST T form several connected components of the old MST T. Property 4 implies that each connected component is a fragment of the new MST T'. Therefore in the latter half of our new algorithm, PMHT is applied to construct the new MST T'. Property 5 is needed to bound the message complexity of the latter half of our algorithm, because the message complexity of PMHT depends on the number of fragments in the initial configuration. In order to bound the complexity of the former half of our algorithm, Property 6 is needed.

The partition links are defined as follows. For simplicity, the MST is regarded as a rooted tree in what follows.

[Definition] An *adjacent processor* is a processor incident to an added link or a deleted link. Define a processor u as a *branch processor*, if there exist two processors in u's sons such that each of them has an adjacent processor in its descendants. In other words, u is a branch processor if u is the nearest common ancestor of some two adjacent processors. A *marked processor* is a processor that is an adjacent processor or a branch processor. (We use the term a marked processor when there is no need for distinguishing a branch processor from an adjacent processor.) For any marked processors u and v, the u - v path in T is called a *primary path* if no other marked processor exists on the u - v path. Define a link as a *partition link*, if it is the maximum-weight link in a primary path (Figure 1). □

It is clear that the partition links defined above satisfy Property 2. Since a processor incident to a deleted link is an adjacent processor, a deleted link of T itself forms a primary path. This implies that the partition links defined above satisfy Property 3. The following two lemmas show that Property 4 and Property 5 hold. Property 6 is examined later.

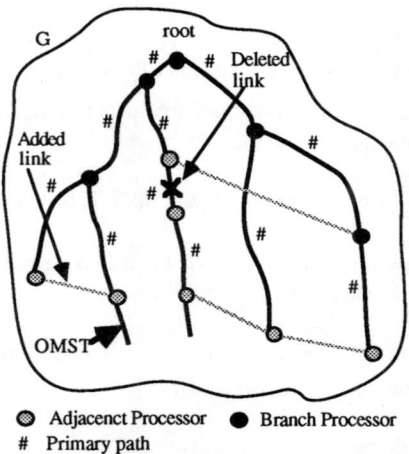

Figure 1. Adjacent processors, branch processors and primary paths

[Lemma 1] Each connected component of $T-L_p$ is a fragment of the new MST T'.

(Proof) Assume that there is a connected component in $T-L_p$ which is not a fragment of the new MST T'. Then, it contains a link (u, v) such that it is not a link of T', that is, $(u, v) \notin L_{T'}$. Let v be a son of u in the old MST T. By removing (u, v) from T, T is divided into two trees, the subtree with the root v (denoted by T_v) and the other part $T-T_v$ (denoted by G').

(Case 1) There exists an added link between T_v and G', that is, an added link which connects a processor in T_v and a processor in G' (Figure 2).

Since there is an adjacent processor in each of T_v and G', there is a marked processor which is an ancestor of u. Let t be the nearest marked processor to u among ancestors of u. (t may be u itself.) Similarly, let s be the nearest marked processor to v among descendants of v. (s may be v itself.) From the definition of the partition link, the maximum-weight link in the t-s path is the partition link. Let (g, g') be the maximum weight link and g be a son of g'. Since (u, v) is a link in $T-L_p$, (u, v) is not a partition link, that is, $(u, v) \neq (g, g')$ holds. Without loss of generality, assume that (g, g') is a link on the u-t path. When (u, v) and (g, g') are removed from T, T is divided into three connected components; the connected component containing u and g (denoted by G1), the connected component containing g' and the subtree with a root v. From the definition of t, there exists no marked processor on u-t path other than t. It follows that there is no marked processor in G1, that is, there is no added link connecting to a processor in G1. This fact implies that $W(z, z') > W(u,v)$ or $W(z,z') > W(g, g')$ holds for any link (z, z') between G1 and $T-G1$. Since (g, g') is the maximum-weight link on the t-s path, $W(g, g') > W(u, v)$ holds. Therefore, $W(z, z') > W(u,v)$ holds (**Claim 1**). When (u, v) is added to the new MST T', (u, v) and the u-v path in T' form a cycle. On the cycle, there exists a link (p, q) between G1 and $T-G1$ except for (u, v). Since $W(p, q) > W(u, v)$ holds from Claim 1, (T-$\{(p, q)\}) + \{(u, v)\}$ is a spanning tree and its total sum of weight is smaller than that of T'. This contradicts the fact that T' is the new MST of N'.

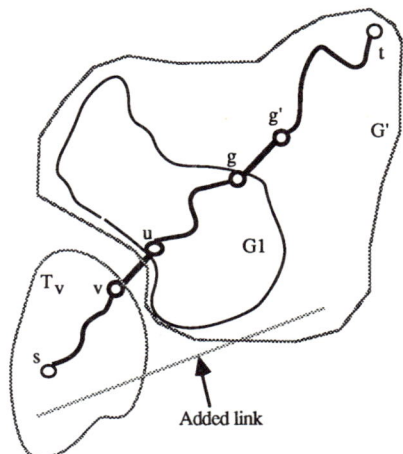

Figure 2. Network in the proof of Lemma 1

(case 2) There exists no added link between T_v and G'.

Since (u, v) is a link of the old MST T, $W(p, q) > W(u, v)$ holds for any link (p, q) between T_v and G' **(Claim 2)**. When (u, v) is added to T', (u, v) and the u-v path in T' form a cycle. On the cycle, there exists a link (p, q) between T_v and G' except for (u, v). Since $W(p, q) > W(u, v)$ holds from Claim 2, $(T' - \{(p, q)\}) + \{(u, v)\}$ is a spanning tree and its total sum of weight is smaller than that of T'. This contradicts the fact that T' is the new MST of N'. □

[Lemma 2] The number of connected components in $T-L_p$ is $O(t+f)$.

(Proof) Consider the graph G'=(V', E') where

V'={u | u is a connected component in $T-L_p$}, and

E'={(u, v) | There exists a partition link (s, t) such that s (resp. t) belongs to the connected component u (resp. v) in $T-L_p$}.

Clearly, G' is a tree. From the definition of the partition link, **(a)** each connected component corresponding to a leaf of G' contains one adjacent processor, and **(b)** each connected component corresponding to an internal node of G' has at least two sons **(Claim 3)**. From Claim 3, we can show that the number of the connected components in $T-L_p$ (i.e. the number of vertices in G') is $O(t+f)$. □

Consider the case that no link of the old MST T is deleted, that is, $L_d = \emptyset$ holds. Then, it is obvious that the following Lemma 3 holds. This observation saves messages in this case. It follows from Lemma 3 that we can ignore all links which belong to N but not belong to T. In other words, no message is sent along those links in our algorithm when no link of T is deleted.

[Lemma 3] The new MST T' of N' coincides with the MST of $T+L_a$, if there exists no deleted link of T. □

4. The outline of our algorithm

In this section, we describe the outline of our algorithm. Our algorithm consists of four phases.

(Phase 1) Check whether there exists a deleted link of T or not. If there exists no deleted link of T (i.e. $L_d = \emptyset$), ignore all links in $L-L_T$ (links which belong to N but not belong to T) in the following three phases.

(Phase 2) Elect a leader in each connected component of $T-L_d$. In the next phase, the connected component of $T-L_d$ is regarded as the rooted tree whose root is the leader elected in this phase.

(Phase 3) Find the partition links and change them into non-tree links.

In this phase, the partition links are found in each connected component of $T-L_d$ as follows.

(3.1) Find the marked processors (the adjacent processors and the branch processors) in the bottom-up fashion from leaves of the connected component of $T-L_d$. This step proceeds as follows.

(i) Each leaf of the connected component of $T-L_d$ decides whether it is an adjacent processor or not, and then sends a message to its parent to inform whether the leaf is an adjacent processor or not.

(ii) When each internal-processor receives messages from all sons, it decides whether it is a marked processor or not, that is, it decides to be a marked processor if and only if (a) it is an adjacent processor or (b) there exist two sons which inform that there exist adjacent processors in their descendants. Then, it sends a message to its parent to inform whether there exists an adjacent processor in its descendants or not.

(3.2) Every marked internal-processor is the upper end of a primary path, if it receives the message telling that its son has an adjacent processor in its descendants. In order to find the partition link in the primary path, the marked internal-processor sends a message to every son that informs there exists an adjacent processor in its descendants. The message is transfered to a marked processor, which is the other end of the primary path, and the message carries the maximum weight of the link which it has ever traversed. When the message reached the other end of the primary path, the processor finds the weight of the partition link of the primary path, and the message is forwarded upward to the processors incident to the partition link. The processor incident to the partition link changes the link to a *non-tree* link.

(Phase 4) Apply the algorithm PMHT to the network N' with the initial configuration where each connected component of $T-L_p$ is a fragment of the new MST T'.

5. Correctness and complexities of our algorithm

Correctness of our algorithm

It is obvious that Phases 1, 2 and 3 terminate within a finite time, and the partition links are correctly found in Phase 3. From Lemma 1, each connected component of $T-L_p$ is a fragment of the new MST T'. Thus, by applying PMHT, the new MST T' of N' can be reconstructed within a finite time in Phase 4.

Complexities of our algorithm

[Theorem 1] The message complexity of our algorithm is $O(n \log(t+f) + m)$ and the bit complexity is $O(n \log(t+f) \log n + m \log n)$, where $m = n + t$ if $f = 0$, $m = e$ otherwise.

(Proof) The message complexity of Phase 1 is $O(n)$, and the message complexity of Phase 2 is $O(n)$ if $f = 0$, $O(e)$ otherwise. In Phase 3, a constant number of messages are sent through each remaining link of T. Thus, the message complexity of Phase 3 is $O(n)$. From Lemma 2, there exist $O(t+f)$ fragments in the beginning of Phase 4. Therefore, it follows from Property 1 that the message complexity of Phase 4 is $O(n \log(t+f) + e)$, if $f \neq 0$. If $f = 0$, from Lemma 3 we ignore all links in $L-L_T$, and no message is sent along these links. Thus the message complexity is $O(n \log t + (n+t))$ if $f = 0$.

Each message of our algorithm is $O(\log n)$ bits long, hence, the bit complexity of our algorithm is $O(n \log(t+f) \log n + m \log n)$, where $m = n + t$ if $f = 0$, $m = e$ otherwise. □

[Theorem 2] The ideal-time complexity of our algorithm is $O(n \log(t+f) + n)$.

(Proof) The ideal-time complexity of Phases 1, 2 and 3 is $O(n)$. From Property 1, the ideal-time complexity of Phase 4 is $O(n \log(t+f) + n)$. □

6. Addition and deletion of processors and links

We can easily modify our algorithm so that it can reconstruct the new MST after addition and deletion of processors as well as links occur. We have only to modify the definition of an *adjacent processor*. In the modified algorithm, a processor u is defined as an *adjacent processor*, if u is incident to an added or deleted link, or if u is adjacent to an added or deleted processor. When a processor v is added or deleted, it causes at most d *adjacent processors* where d is the degree of v (i.e. the number of links incident to v). Therefore there exist $O(g+t+f)$ *adjacent processors*, where g is the sum of degree over all added or deleted processors. From this observation, the following theorem can be proved.

[Theorem 3] After processors and links are deleted and added, the new MST T can be reconstructed with the message complexity $O(n'\log(g+t+f)+r)$ and the ideal-time complexity $O(n'\log(g+t+f)+n')$. Here, n' is the number of processors in the network after topology change, g is the sum of degree over all added or deleted processors, and $r=n+g+t$ if there exists no deleted processor nor deleted link, $r=e'$ (i.e. the number of links in the network after topology change) otherwise. □

REFERENCES

[1] B. Awerbuch : "Optimal Distributed Algorithms for Minimum Weight Spanning Tree, Counting, Leader Election and related problems", Proceedings 19th Annual ACM Symposium on Theory of Computing, pp.230-240(1987).

[2] I.Cimet and S.P.Kumar : "A Resilient Distributed Algorithms for Minimum Weight Spanning Trees", Proceedings of the 1987 International Conference on Parallel Processing., pp.196-203(1987).

[3] R. Gallager, P. Humblet and P. Spira : "A Distributed Algorithm for Minimum Weight Spanning Trees", ACM TOPLAS, 5, 1, pp.66-77(1983).

[4] J.Park, T.Masuzawa, K.Hagihara and N.Tokura : "Distributed Algorithms for Reconstructing Minimum Spanning Tree - The Case of Link Deletions -", Tech. Rep. IECEJ, COMP-89-25(in Japanese) (1989).

[5] Y.H.Tsin : "An Asynchronous Distributed MST Updating Algorithm for Handling Vertex Insertions in Networks", Proc. of the International Conference on Parallel Processing and Applications, pp.221-226(1987).

EFFICIENT DISTRIBUTED ALGORITHMS FOR SINGLE–SOURCE
SHORTEST PATHS AND RELATED PROBLEMS ON PLANE NETWORKS [1]

Ravi Janardan

Siu Wing Cheng

Department of Computer Science, University of Minnesota

Minneapolis, MN 55455, U.S.A.

Abstract. An efficient distributed algorithm is given for computing a single-source shortest path tree in an asynchronous planar network. The algorithm has message and time complexity $O(pn)$ on an n-node network, where p is the smallest number of faces needed to cover all the nodes, taken over all possible plane embeddings of the network. The complexity of the algorithm ranges from $O(n)$ to $O(n^2)$ as p ranges from 1 to $\Theta(n)$. The algorithm incorporates optimal distributed solutions to a number of interesting subproblems including: (i) decomposing the plane embedding into $\Theta(p)$ outerplane graphs with favorable properties; (ii) a single-source algorithm for outerplane graphs; and (iii) identifying any edge in an outerplane graph whose cost exceeds the distance between its endpoints. As an application, an efficient message routing scheme is presented which adapts to changing link conditions and routes along near-shortest paths.

1 Introduction

Given a distributed network with positive edge costs, the *single-source shortest path* problem is to determine a shortest path from a designated source node to each of the other nodes. Distributed algorithms for the single-source problem are used in many well-known networks to compute shortest paths over which messages are routed to and from the source [11]. Moreover, since this computation is done very frequently (sometimes as often as once every few seconds) in response to changes in edge costs, efficiency is a very important con-

---

[1] Research supported in part by a grant-in-aid of research from the Graduate School of the University of Minnesota. The first author was also supported in part by NSF grant CCR–8808574. Authors' e-mail addresses: janardan@umn-cs.cs.umn.edu; scheng@umn-cs.cs.umn.edu.

sideration. We refer the reader to [11] for a comprehensive discussion of the use of shortest path algorithms in message routing.

In a distributed algorithm for the single-source problem, each node has only local information about the network (i.e., knowledge of the incident edges) and has a copy of the algorithm. The algorithm is initiated by the source and the nodes compute the shortest paths cooperatively by executing their local algorithms and exchanging messages with their neighbors. At termination, the shortest paths define a spanning tree of the network rooted at the source, and each node knows its parent and its children in the tree. The computation at each node proceeds asynchronously. Thus, there is no global clock in the network and, furthermore, the delay experienced by a message from the time it is transmitted over an edge to the time it is processed at the other end is unpredictable (but bounded). Computation time at a node is assumed to be negligible compared to message delay and all message delays are normalized to 1. The efficiency of the algorithm is measured by the total number of messages exchanged and the total execution time. A message has length $O(\log(\max(n, W)))$ bits, where n is the number of nodes and W is the largest edge cost.

Efficient distributed algorithms for the single-source problem have been proposed recently in [1, 3, 7]. The algorithms presented in [1, 3] are applicable to networks of general topology. The algorithm in [3] has message and time complexity $O(e2^{O(1)(\log n)^{3/4}} \log W)$, where e is the number of edges. The algorithm in [1] has message complexity $O(n^2)$ and time complexity $O(n \log n)$. The algorithm in [7] is designed specifically for planar networks and has a message and time complexity of $O(n^{5/3})$.

The main result of this paper is a distributed single-source algorithm of message and time complexity $O(pn)$ for a planar network. Here p is the smallest number of faces needed to cover all the nodes, taken over all possible plane embeddings of the network. We call the embedding that realizes p a *p-plane embedding*. Figure 1 illustrates a 4-plane embedding of a planar network, where the covering faces are shown bold. Our result is especially significant for classes of planar networks for which p is a constant since our algorithm then achieves a message and time complexity of $O(n)$. In general, the value of p can range from 1 to $\Theta(n)$ for an n-node network, so that the complexity of our algorithm ranges from $O(n)$ to $O(n^2)$. Our algorithm is more efficient than the one in [7] when p is $o(n^{2/3})$ and more efficient than the one in [3], when applied to planar networks, when $p = o(2^{O(1)(\log n)^{3/4}})$. However, our algorithm does require some additional information at the nodes. Specifically, the edges incident at any node v are given in the clockwise order in which they appear at v in the p-plane embedding. Furthermore, v is also given the name of the covering face that

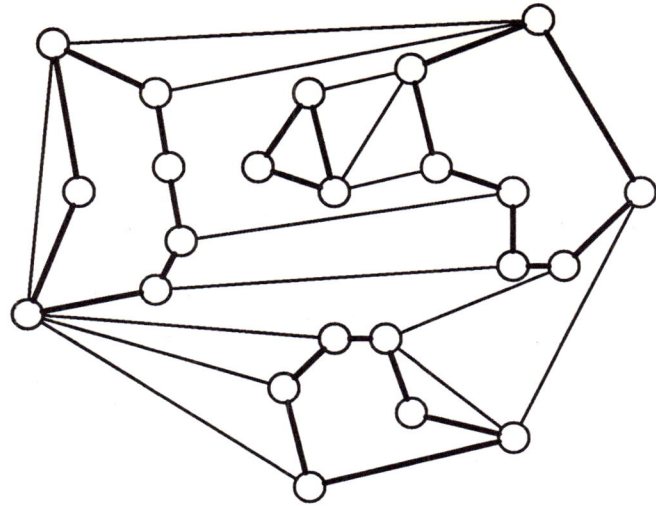

Figure 1: Example of a 4-plane network, with the covering faces shown bold

it belongs to, an integer between 1 and p. (We assume that v lies on only one covering face. This assumption is for simplicity only and can be removed without much difficulty.) Having the clockwise ordering at each node is equivalent to having the p-plane embedding, even though the physical layout of the network may not correspond to this embedding or, indeed, even be plane.

The additional information at the nodes can be computed centrally and then distributed. This is done when the network is set up and subsequently if the topology changes due to the introduction or removal of nodes and edges. The cost of computing and distributing the additional information can be significant. However, the topological changes that trigger this computation occur relatively infrequently in comparison to the frequency with which edge costs change and necessitate running the shortest path algorithm. Thus it is well worth computing the additional information because of the resulting improvement in the efficiency of the shortest path algorithm. We note that computing the optimal set of covering faces is NP-hard in the sequential model [6]. However, for our purposes it is sufficient to have a set of covering faces that is at most a constant times larger than optimal. To this end, the centralized computation is done using the linear-time algorithm given in [8], which produces an embedding and a corresponding set of covering faces that is at most four times as large as optimum.

In closing this section we remark that p-plane embeddings have been the subject of much recent study [6, 8, 9] because of their interesting properties. In particular, in [9] they have been shown to be an attractive choice of network topology because they allow all-pairs shortest path routing information to be encoded succinctly as intervals, which results in a substantial savings in space over complete routing tables.

2 An overview of the single-source algorithm and associated results

Throughout the paper, we will consider a p-plane embedding of the given planar network and model this by a connected, n-node plane graph, $G = (V, E)$, which we call a *p-plane graph*. Our single-source algorithm requires efficient distributed solutions to a number of interesting subproblems. For each we present a message- and time-optimal distributed algorithm.

We first give a distributed algorithm of message and time complexity $O(n)$ to compute an outerplane decomposition of G, defined as follows: An *outerplane graph* is a plane graph in which a single face covers all the nodes. An *outerplane decomposition* of G is a grouping of its nodes and edges into graphs $G_i = (V_i, E_i)$ with the following properties:

1. Each G_i is a biconnected outerplane graph consistent with the embedding G.

2. Every node and every edge of G is contained in at least one of the G_i. Furthermore, $\sum_i |V_i|$ and $\sum_i |E_i|$ are $\Theta(n)$.

3. There are $\Theta(p)$ graphs G_i.

4. There are $\Theta(p)$ nodes in G, called *gateway nodes*, such that any path in G that is not wholly contained in some G_i must contain at least one gateway node.

Figure 2 shows an outerplane decomposition of the 4-plane graph in Figure 1. A similar decomposition was first used in [8] for the planar all-pairs shortest paths problem in the sequential model. However, the decomposition algorithm in [8] is quite different from ours and cannot be implemented efficiently in the distributed setting. Our algorithm is simple, involving essentially a distributed depth-first search, during which certain pairs of covering faces are processed carefully to identify the outerplane graphs.

We next devise a single-source algorithm for biconnected outerplane graphs which uses $O(m)$ messages and time for an m-node graph. Our algorithm exploits the tree-structured adjacency of the faces in an outerplane graph to completely eliminate all synchronization with the source, which is the computational bottleneck in most single-source algorithms.

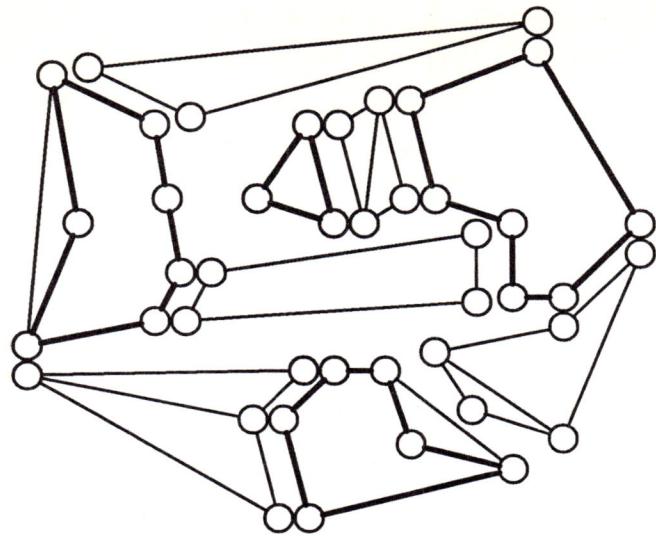

Figure 2: Outerplane decomposition of the graph of Figure 1

Instead, our algorithm searches for shortest paths by carefully processing the faces of the outerplane graph in the tree order, which makes it enough to synchronize only locally on each face.

The single-source algorithm for G is now as follows. We first compute distributively an outerplane decomposition of G. Then we apply the single-source algorithm for outerplane graphs iteratively. At each iteration, we select a gateway node to which the distance from the source has already been found and then run the outerplane single-source algorithm from this gateway node in all the outerplane graphs that it belongs to. Each iteration yields improved distance estimates in G. Our single-source algorithm for G requires only $\Theta(p)$ synchronizations with the source, each taking $O(n)$ messages and time, but pays for this by having to invoke the outerplane shortest path algorithm $\Theta(p)$ times. However, each invocation costs only $O(n)$ and so the overall complexity is $O(pn)$.

We also present a distributed algorithm of message and time complexity $O(m)$ to identify edges in an m-node outerplane graph that do not satisfy the generalized triangle inequality. An edge satisfies the *generalized triangle inequality* (*gti*, for short) if its cost is at most the distance between its endpoints. Edges that violate the gti need to be identified before our outerplane single-source algorithm can be applied, because the violating edges could shield faces that contain shortest paths. These edges are identified in a distributed preprocessing phase and our outerplane single-source algorithm then simply bypasses them. Our algorithm to enforce the gti is a distributed implementation of a recursive algorithm for the same problem in the sequential model [9]. An interesting feature of our algorithm is that it it illustrates how to simulate recursion distributively on an outerplane graph.

Our single-source algorithm also yields a distributed breadth-first search algorithm of message and time complexity $O(pn)$. Previous breadth-first search algorithms when specialized to planar networks have complexity $O(n^{3/2})$ [7], $O(n2^{(\log n \log \log n)^{1/2}})$ [4], $O(n^{8/5})$ [5], and $O(n2^{O(1)(\log n)^{3/4}})$ messages and $O(D2^{O(1)\log D/(\log n)^{1/4}})$ time [3], for a network of diameter D.

As an application of our single-source algorithm, we show how to design an all-pairs routing scheme that adapts efficiently to changing link conditions and routes along near-shortest paths. The link conditions that we consider are links becoming faulty and faulty links becoming operational at their original cost. Our scheme uses $O(tpn)$ messages and time, where t is the current number of link faults, and generates routings that are at most three times longer than optimal. The complexity of our scheme ranges from $O(pn)$ to $O(pn^2)$ as t ranges from 1 to $\Theta(n)$. By contrast, computing all-pairs shortest path routing information by running our single-source algorithm from each node takes $O(pn^2)$ messages and time, independent of t. Our scheme is also space-efficient and extends the utility of the space-efficient interval routing scheme of [9] which could not handle even simple changes in link condition.

In response to changes in link status, the nodes distributively compute some *recovery information*, which they then use in conjunction with interval routing information for the fault-free network to do the routings. The dominant costs in computing the recover information are finding certain shortest path trees and propagating the recovery information through the network. We show that the latter cost can be reduced to $O(tpn)$ by taking advantage of an interval property of p-plane networks to encode the recovery information compactly as it is propagated. By using our single-source algorithm, the former cost can also be reduced to $O(tpn)$, thus giving an overall bound of $O(tpn)$.

The rest of this paper is organized as follows. Section 3 proves the existence of the decomposition and shows how to compute it distributively. Section 4 describes the shortest path algorithms for outerplane graphs and for p-plane graphs and also contains the gti algorithm. The adaptive routing scheme is discussed in Section 5. For lack of space, most details and proofs are omitted here, but can be found in [10].

3 Outerplane decomposition of a p-plane graph

Wlog, assume that the infinite face of G is not a covering face. The edges of G are classified as either intraface edges or as interface edges. An *intraface edge* has both endpoints on the same covering face, whereas an *interface edge* has its endpoints on different covering faces. To simplify our discussion here, we will assume that for all interface edges incident with any node of G, their other endpoints lie on a common covering face. This assumption can be removed easily, as shown in [10]. A node is an *intraface node* if all its incident edges are intraface edges. Otherwise, it is an *interface node*. Two covering faces are *adjacent* if there is an interface edge with one endpoint on each face. For an outerplane graph, the face that covers all the nodes is called the *exterior face*; any other face is an *interior face*. Any edge that is on the exterior face is an *exterior edge*; any other edge is an *interior edge*. Any n-node plane graph has $O(n)$ edges.

3.1 Existence of the decomposition

Theorem 1 *Let G be an n-node, p-plane graph. There exists an outerplane decomposition of G.*

Proof (sketch) For $1 \leq i \leq p$, let G_i consist of the nodes and the intraface edges of F_i. Each G_i, $i \leq p$, is an *intraface graph*. To include the interface edges, define graphs G_i, $i > p$, as follows. Let F and F' be adjacent covering faces of G. Let $v_1, v_2, \ldots, v_r$, $r \geq 1$, and $u_1, u_2, \ldots, u_s$, $s \geq 1$, be clockwise sequences of consecutive nodes on F and F', respectively, that are maximal with respect to the following:

(i) v_1 is adjacent to u_s and v_r is adjacent to u_1.

(ii) For $1 \leq i \leq r$ (resp. $1 \leq j \leq s$), v_i (resp. u_j) is either an intraface node, or has interface edges leading to nodes in $u_1, u_2, \ldots, u_s$ (resp. $v_1, v_2, \ldots, v_r$) only.

For some index $i > p$, let G_i be the subgraph of G that is induced by $v_1, v_2, \ldots, v_r$ and $u_1, u_2, \ldots, u_s$. Each G_i, $i > p$, is an *interface graph*.

Properties 1 and 2 of an outerplane decomposition can be verified easily. We prove Property 3 as follows: Let $\hat{G} = (\hat{V}, \hat{E})$ be the plane, p-node multigraph in which each node v_F corresponds to a covering face F and each edge corresponds to an interface graph in the obvious way. For any unordered pair, $v_F, v_{F'}$, of adjacent nodes, let $m(v_F, v_{F'})$ be one less than the number of multiedges joining v_F and $v_{F'}$ and let m be the sum of $m(v_F, v_{F'})$, taken over all such pairs. The number of graphs in the decomposition is $p+|\hat{E}|$, and $|\hat{E}|$ is m plus $O(p)$, since $\hat{G}$ is plane. Thus it suffices to prove that m is $O(p)$.

Consider any nodes v_F and $v_{F'}$ with at least two multiedges joining them. Let e_i and e_j be two of these multiedges such that they enclose a finite region of the plane and this region contains no other multiedge joining v_F and $v_{F'}$. Let e_i precede e_j clockwise around v_F. The region enclosed by e_i and e_j must contain at least one edge with one endpoint being either v_F or $v_{F'}$, wlog v_F, and the other being a node different from v_F and $v_{F'}$. Otherwise, the graphs G_i and G_j in G could have been merged into a single graph. Let $e_k = (v_F, v_{F''})$ be such an edge, where e_k is chosen to be the successor of e_i in a clockwise ordering of the edges incident with v_F. See Figure 3a.

We claim that there cannot exist another pair of nodes v_L and $v_{L'}$ such that $v_{F''}$ is also in the region enclosed by consecutive multiedges joining v_L and $v_{L'}$. Suppose that such nodes v_L and $v_{L'}$ exist. Figures 3b and 3c illustrate the possibilities. In Figure 3b, neither v_L nor $v_{L'}$ is equal to v_F; in Figure 3c $v_{L'} = v_F$ and $v_L \neq v_F$ (or vice versa). Note that in Figure 3c, because of our choice of e_k, neither of the two edges joining v_L and $v_{L'} = v_F$ can meet v_F at any point that is in between the meeting points of e_i and e_k with v_F. Thus, in both cases, e_k must cross an edge joining v_L and $v_{L'}$, which contradicts the fact that $\hat{G}$ is plane.

We charge e_j to $v_{F''}$. We repeat this for each pair of consecutive multiedges between v_F and $v_{F'}$ and then do this for all unordered pairs, $v_F, v_{F'}$, of adjacent nodes. The total charge to the nodes is m, and by the above claim, any node of $\hat{G}$ is charged at most once. Thus $m \leq p$.

Property 4 follows once we pick v_1, v_r, u_1, and u_s from each interface graph as the gateway nodes. □

3.2 An efficient distributed algorithm to compute the decomposition

Assume that each node v has a list of its incident edges in clockwise order around v. This information enables v to send a message around the boundary of any incident face. For each G_i that v is in, the algorithm labels with i the edges of G_i incident with v, identifies the interior faces of G_i incident with v, and marks v as a gateway node if it is one.

The algorithm has two phases. In the first phase, v classifies each edge as interface or intraface, and labels the intraface edges with the index of its covering face. This can be done by a distributed depth-first search [2]. In the second phase, the interface graphs are identified. Let G_{i-1}, $i > p$, be the highest-numbered graph identified so far. Let F and F'

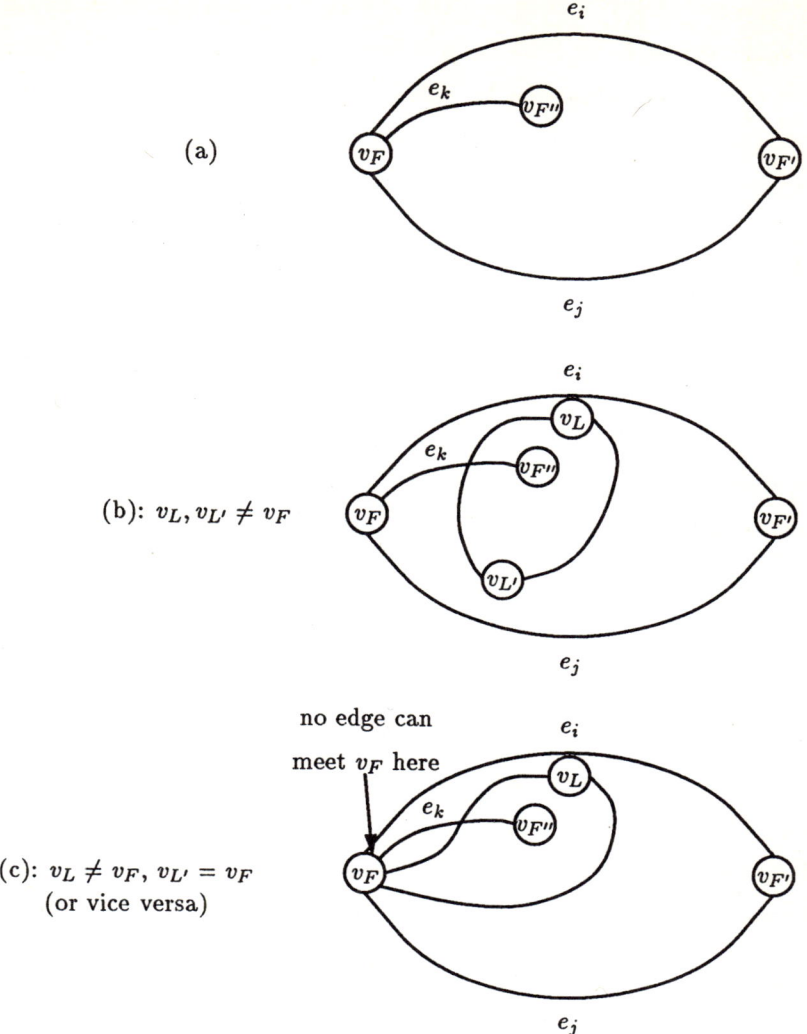

Figure 3: Illustration for proof of Property 3, Theorem 1

be adjacent covering faces. The nodes on F and F' find maximal clockwise sequences of nodes, as in Theorem 1, and each node in the sequences labels with i each incident intraface and interface edge whose other endpoint is also in the sequences.

In more detail, the second phase involves a distributed depth-first search [2] of G, modified to traverse the boundary of a covering face completely before transferring to a new covering face. Let F be the current face. By sending a suitable message around F, a maximal clockwise sequence, $x, \ldots, y$, of nodes on F is identified, such that x and y are interface nodes and all interface edges from $x, \ldots, y$ lead to the same covering face F'. Let x' and y', respectively, be the most clockwise and most counterclockwise neighbors of y and x on F'. If all interface edges from $x', \ldots, y'$ lead to F, then $x, \ldots, y$ and $x', \ldots, y'$ are the desired maximal sequences, and the graph they induce is an interface graph. Otherwise, by suitable message passing from x' to y' on F', the sequence $x', \ldots, y'$ is split into maximal subsequences whose endpoints are interface nodes and all of whose interface edges lead to F. Each subsequence is matched with a subsequence of $x, \ldots, y$, and all intraface and interface edges with endpoints in the matched subsequences are labelled with the integer i. (i is carried along "globally" in the messages and incremented after each outerplane graph has been found.) The above is repeated for all maximal sequences, $x, \ldots, y$, on F.

Let v be any node, on covering face F. After the second phase terminates, for the intraface graph G_i that it belongs to, v marks the two incident edges of F as exterior edges and all other incident edges belonging to G_i as interior edges. For each interface graph G_i that it belongs to, v marks as exterior the first and last edges in the sequence of incident edges that are labelled i, and marks all others interior. In any G_i, the interior faces incident to v are those that are bounded by consecutive incident edges that are labelled i and are not both exterior edges. The face bounded by the exterior edges is the exterior face. Finally, v marks itself as a gateway node, if it is an interface node and has at most one edge of F incident with it.

Theorem 2 *Let G be an n-node, p-plane graph. An outerplane decomposition of G can be identified in $O(n)$ messages and time.* □

4 Distributed single-source shortest path algorithms

4.1 A single-source algorithm for biconnected outerplane graphs

Let H be an m-node biconnected outerplane graph with positive edge costs, and let r be any node of H. We give an $O(m)$-message and time distributed algorithm, $Op\_Shortest(r, H)$,

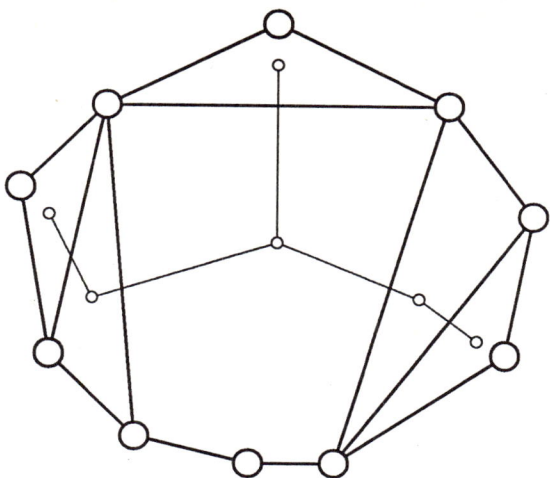

Figure 4: An outerplane graph and its tree-structured dual (The dual vertex corresponding to the exterior face is omitted.)

to find a shortest path tree, T, in H, rooted at r. For now, assume that H satisfies the generalized triangle inequality (gti). For any graph, I, let $c(I)$ denote the sum of the costs of the edges of I. Thus, in particular, $c((u,v))$ denotes the cost of edge (u,v).

Each node maintains the following information: $d(v)$, the current best estimate of the distance from r to v in H; and $p(v)$, the tentative parent of v in T. Initially, $d(r) = 0$, $d(v) = \infty$ for all $v \neq r$, and $p(v)$ is undefined for all v. At termination, $d(v)$ is the distance from r to v in H and, for all $v \neq r$, $p(v)$ is the parent of v in T.

Op_Shortest finds distances by processing interior faces in an order that is governed by the tree-structured adjacency of interior faces in an outerplane graph. This tree structure is apparent if the dual of H, excluding the vertex corresponding to the exterior face, is considered (Figure 4). All nodes on a processed face will have $d(\cdot)$ and $p(\cdot)$ set correctly. An interior face f is processed by a pair of adjacent nodes, x, y, on it, as follows. Nodes x and y send a message *process_face(D)* over f, in opposite directions. Here D is a real-valued field, initialized to $d(x)$ and $d(y)$, respectively. The messages accumulate distances along f in D. Let v be any node on f, different from x and y, which receives a *process_face(D)* message from a neighbor, u, on f. If $D + c((u,v)) < d(v)$, then v does the following: It resets $d(v)$ to $D + c((u,v))$; if $p(v)$ is not undefined, then v sends a message, *not_child*, to $p(v)$; it then resets $p(v)$ to u and sends a message, *child*, to u. Irrespective of whether or not $D + c((u,v)) < d(v)$ was true above, v sets D to $D + c((u,v))$ and sends the message, *process_face(D)*, to its other neighbor on f. When a neighbor of v receives a *child* (resp.

not_child) message from v, it marks v as a child (resp. as not a child) in T. The processing of f terminates when x and y each receive a *process_face* message over f.

Op_Shortest begins when r, playing the roles of both x and y, processes an incident interior face f_0. Suppose that an interior face, h, has been processed by nodes a and b. Next, for each interior edge (x, y) on h, where $(x, y) \neq (a, b)$, x and y process the interior face, f, which shares (x, y) with h. Thus the dual vertex corresponding to f_0 becomes the root of the tree-structured dual and control passes down this tree in the natural root-to-leaf order. Termination is accomplished easily in a leaf-to-root pass in the shortest path tree.

The generalized triangle inequality guarantees that the shortest path from x and y (hence from r) to a node v on f lies on f. To ensure that the gti holds, r preprocesses H using the distributed algorithm *Enforce_gti(H)* given in Section 4.2, which runs in $O(m)$ messages and time. Then *Op_Shortest* is run as before, except that the violating edges are ignored.

Theorem 3 *Let H be a biconnected, m-node outerplane graph with positive edge costs. Let r be any node of H. Op_Shortest(r, H) computes a shortest path tree in H, rooted at r, using $O(m)$ messages and time.* □

We illustrate *Op_Shortest* on the outerplane graph in Figure 5, with node 1 as source. First, node 1 processes the face $\langle 1, 10, 9, 1 \rangle$, and shortest paths to nodes 9 and 10 are found. Then 1 and 9 process $\langle 9, 6, 5, 4, 1, 9 \rangle$, to find shortest paths to 6, 5, and 4. Then 9 and 6 process $\langle 9, 6, 7, 9 \rangle$ and 1 and 4 process $\langle 1, 2, 4, 1 \rangle$ to find shortest paths to 7 and 2, respectively. Finally, 9 and 7 process $\langle 9, 7, 8, 9 \rangle$ and 2 and 4 process $\langle 2, 4, 3, 2 \rangle$ to discover shortest paths to 8 and 3, respectively.

4.2 A distributed algorithm to enforce the generalized triangle inequality

Enforce_gti(H) is based on the following recursive, sequential algorithm [9]. Call an interior face of H a *leaf face* if it has exactly one interior edge. For each interior face of H, precompute its cost and identify the interior faces that share an edge with it. If H has no leaf face, then it is a cycle. Delete the maximum cost edge on the cycle if its cost exceeds that of the remainder of the cycle, and return. Otherwise, pick a leaf face, f, with interior edge e. If $c(e) > c(f - e)$, then delete e, merge the interior faces that shared e into a single face and recurse on the resulting graph. Otherwise, delete $f - e$, recurse on the resulting graph, and add back $f - e$ into the graph H' returned. Let f' be the interior face created

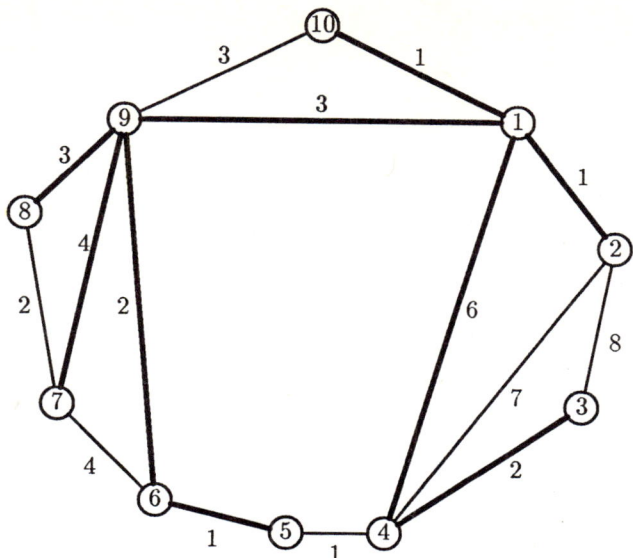

Figure 5: Example outerplane graph for illustrating $Op\_Shortest$. A shortest path tree rooted at node 1 is shown bold.

by this addition, and let e' be a maximum cost edge of $f - e$. If $c(e') > c(f' - e')$, then delete e' and return.

$Enforce\_gti$ requires additional ideas, however, primarily because the recursion in the algorithm of [9] cannot be implemented directly in the distributed model. $Enforce\_gti$ achieves the same effect by a different approach. Moreover, operations such as merging two interior faces along an edge and keeping track of the cost of the resulting face need to be handled with care in the distributed setting, in order to obtain the $O(m)$ message bound.

$Enforce\_gti(H)$, as invoked from r, is as follows. In a preliminary step of message passing, the nodes compute the cost of their incident interior faces and mark their incident interior edges as *new*. The algorithm proper has two phases. In the first phase, the exterior face of H is traversed clockwise from r, leaf faces f are successively identified, and either e is found to violate the gti and marked as *violating* or $f - e$ is marked as *reprocess* for later processing. In the second phase, the exterior face of H is traversed counterclockwise from r, and any segments $f - e$ that are marked as *reprocess* are processed in the reverse order of their marking, by checking a maximum cost edge for violation of the gti and marking it as *violating* if so. The two phases effectively simulate the desired recursion, since the second phase unravels in reverse order certain actions taken by the first phase. Edges marked *violating* are ignored subsequently, except for the purpose of traversing the exterior face of

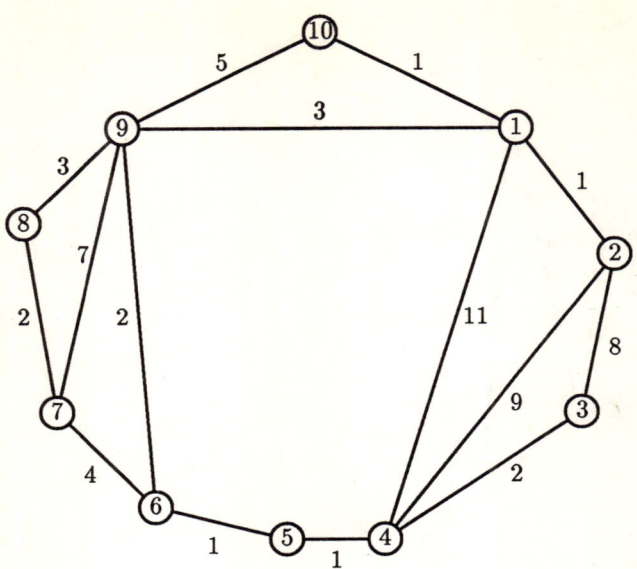

Figure 6: Example outerplane graph for illustrating *Enforce_gti*

H. Edges marked *reprocess* are also ignored until such time as when they are reprocessed.

When an edge e is marked *violating*, the faces g and h that share e are effectively merged into one face k. The cost, $c(k)$, of k can be computed at an endpoint of e, as $c(g) + c(h) - 2c(e)$. Now $c(k)$ must be made available to other nodes on k. The obvious approach of sending $c(k)$ all the way around k can be expensive since k can in turn get merged, and so on. The overall message complexity can be $O(m^2)$ in worst case. Instead, $c(k)$ is sent around k during the traversal of the exterior face only as long as the traversal stays on k. If the traversal leaves k at some node v, then it does so to traverse some interior face which shares an edge (v, u) with k. Then v sends $c(k)$ ahead to u on (v, u). Thus, when the traversal eventually reaches u, $c(k)$ will be available.

Theorem 4 *Let H be an m-node biconnected outerplane graph. Algorithm Enforce_gti(H) identifies the edges of H that violate the generalized triangle inequality in $O(m)$ messages and time.* □

We illustrate *Enforce_gti* on the outerplane graph in Figure 6, with node 1 as r. In the first phase, at node 4, edges (4,2) and (4,1) are considered, in that order, and segment $4 - 3 - 2$ is marked *reprocess* (by marking (4,3)), and (4,1) is marked *violating*. Faces $\langle 4, 2, 1, 4 \rangle$ and $\langle 4, 1, 9, 6, 5, 4 \rangle$ are merged into face $\langle 4, 2, 1, 9, 6, 5, 4 \rangle$, of cost 17. This cost is then carried along in the clockwise traversal to nodes 5 and 6. At 6, the cost is sent ahead to

9 over (6,9), since the traversal leaves ⟨4, 2, 1, 9, 6, 5, 4⟩. At 9, (9,7) is marked *violating* and faces ⟨9, 8, 7, 9⟩ and ⟨9, 7, 6, 9⟩ are merged into ⟨9, 8, 7, 6, 9⟩. Then segments $9-8-7-6$ and $9-6-5-4-2-1$ are marked as *reprocess*, in that order. When the clockwise traversal reaches 1, the current graph is face $l = \langle 1, 10, 9, 1 \rangle$, and (10,9) is marked *violating*. The first phase then terminates. In the second phase, at 9, segment $9-6-5-4-2-1$ is reprocessed and (4,2) is marked *violating*, and then $9-8-7-6$ is reprocessed uneventfully. At 4, $4-3-2$ is reprocessed uneventfully. The counterclockwise traversal then reaches 1 and the phase terminates.

4.3 A single-source algorithm for p-plane graphs

We give a distributed algorithm, $Gen\_Shortest(s, G)$, to find a shortest path tree, T, in G, rooted at any node s. Assume that an outerplane decomposition of G has already been computed. Let $\rho(\cdot, \cdot)$ be the distance function for G.

The idea behind $Gen\_Shortest$ is as follows. Let v be any node of G, P a shortest (s, v)-path in G, and u the last gateway node on P before v. The (u, v)-segment of P is wholly contained in some outerplane graph, H, to which u belongs, and is a shortest (u, v)-path in H. If u invokes $Op\_Shortest(u, H)$, with $d(u) = \rho(s, u)$, then the distance found to v is $\rho(s, u) + \rho(u, v) = \rho(s, v)$. Thus we need to know $\rho(s, u)$ for all gateway nodes u. For the gateway node closest to s this can be found by running $Op\_Shortest(s, H)$, with $d(s) = 0$, in every outerplane graph, H, that s belongs to. In general, the distance to a gateway node will be known once $Op\_Shortest$ has been run from all closer gateway nodes.

In more detail, each gateway node is in one of two states: *open* or *closed*. Initially, all gateway nodes are *open*. Each node v maintains two fields: $d(v)$, the current estimate of $\rho(s, v)$; and $p(v)$, the tentative parent of v in T. These fields are modified by the various invocations of $Op\_Shortest$. Initially, $d(s) = 0$, $d(v) = \infty$ for all $v \neq s$, and $p(v)$ is undefined for all v. When $Gen\_Shortest(s, G)$ terminates, $d(v)$ and $p(v)$ are $\rho(s, v)$ and the parent of v in T, respectively, for all v.

Wlog, let s be a gateway node. It repeatedly identifies an *open* gateway node, u, with $d(u)$ minimum, and sends out a message, $close(u)$, down the current tree. When u receives $close(u)$, it marks itself *closed* and runs $Op\_Shortest(u, H)$, with $d(u) = \rho(s, u)$, in all outerplane graphs, H, that it belongs to. After all invocations of $Op\_Shortest$ have terminated, u sends a message, *over*, to s, up the current tree. When s receives *over*, it searches for the *open* gateway node to close next. Briefly, a leaf-to-root pass is done in the current shortest path tree. Each leaf v that is an *open* gateway node sends the pair $(v, d(v))$

to $p(v)$. Any other leaf sends the pair (v, ∞) to $p(v)$. At an internal node $v \neq s$, if v is an *open* gateway node, then let y be the smaller of $d(v)$ and the smallest distance received by v from its subtree. Otherwise, let y be the smallest distance received by v. Let x be the node such that $d(x) = y$. Then v sends the pair (x, y) to $p(v)$. If the smallest of the distances received in the *echo* messages at s is not ∞, then s selects the node corresponding to this distance as the next node, u, to close. Otherwise, $Gen\_Shortest(s, G)$ terminates, as there are no more *open* gateway nodes.

Theorem 5 *Let G be an n-node, p-plane graph with positive edge costs. Let s be any node of G. When algorithm $Gen\_Shortest(s, G)$ terminates, for all v in G, $d(v)$ is the distance from s to v in G, and $p(v)$ is the parent of v in a shortest path tree, T, rooted at s. The algorithm uses $O(pn)$ messages and $O(pn)$ time in worst case.* □

5 Adaptive routings in dynamic p-plane networks

5.1 An adaptive scheme for edge faults

In [9], a space-efficient all-pairs shortest path routing scheme, called an *interval routing scheme*, is given for a p-plane graph G. Given a naming of the nodes from 1 to n, clockwise around each covering face, each edge incident with any node, v, can be labelled with at most $\lfloor 3p/2 \rfloor$ intervals, such that a shortest routing from v to any node u uses the edge whose label contains u. However, this scheme cannot handle even simple network changes such as edges failing or failed edges recovering at their original cost. We present a scheme in which the network adapts to such changes by computing some recovery information, which is used in conjunction with interval routing to restore near-shortest routings.

When an edge (or edges) changes status, its endpoints either mark the edge as faulty and set its cost to ∞, or mark it as operational and assign it its original cost. Let $e_i = (u_i, v_i), i = 1, 2, \ldots, t$ be the faults currently in the network. Each node $w \in \{u_1, \ldots, u_t, v_1, \ldots, v_t\}$ computes distributively a shortest path tree, T_w, in G, and sets up recovery information for routing down T_w. There are two computational bottlenecks: finding T_w and setting up the recovery information. We use $Gen\_Shortest(w, G)$ to find T_w, at a cost of $O(pn)$. To route down T_w, each node needs to know which descendant is in which subtree. Collecting this information by propagating each descendant's name up T_w can take $\Omega(n^2)$ messages. We reduce this to $O(pn)$ by using the following lemma to encode the names as intervals, as they are sent up T_w. (Note that the lemma does not follow from the results in [9].)

Lemma 1 *Let G be a p-plane graph with a clockwise naming of its nodes. Let T_w be a shortest path tree in G, rooted at some node w. For any node u, the names of the nodes contained in the subtree of T_w rooted at u can be encoded into at most $\lfloor 3p/2 \rfloor$ intervals.* □

In any routing, each participating node v uses interval routing to find the incident edge to route over. If the edge is not faulty, then v uses it. Otherwise, v routes to d over T_v.

Theorem 6 *Let G be an n-node, p-plane graph, currently containing t edge faults. The above scheme recovers from these faults using $O(tpn)$ messages and time, and $O(tpn)$ recovery information. Any routing is at most 3 times longer than a shortest routing.* □

5.2 Handling a single fault more efficiently

For $t = 1$, the above scheme can be improved as follows. Let $e = (a, b)$ be the faulty edge and let P be a shortest (a, b)-path in $G - e$. As before, interval routing is attempted. If the routing fails, wlog at a, then e is bypassed by routing from a to b over P. From b interval routing is done to d. The key to routing over P is the following lemma.

Lemma 2 *Let $e = (a, b)$ be any edge of G and let P be a shortest (a, b)-path in $G - e$. There exist consecutive nodes, x and y, on P, with x preceding y on P from a to b, such that the interval routings from a to x and from y to b do not use e.* □

So, the routing over P is done in three stages: interval routing from a to x; from x to y over edge (x, y); and, interval routing from y to b. The names x and y for e are either precomputed at a or computed distributively, as follows. When e fails, a and b set its cost to ∞. Then a determines P using *Gen_Shortest* and sends the $\lfloor 3p/2 \rfloor$ or fewer intervals in its interval routing information for e over P. Node y is the first node on P whose name is in one of the intervals, and x is y's predecessor on P.

Theorem 7 *Let G be an n-node, p-plane graph with positive edge costs. The above scheme recovers from any single edge fault by distributively computing $O(1)$ recovery information, using $O(pn)$ messages and time. Alternatively, $\Theta(n)$ information for all edges can be precomputed at network setup time, with no communication overhead. In either case, any routing is at most 3 times longer than a shortest routing.* □

References

[1] B. Awerbuch. Complexity of network synchronization. *Journal of the ACM*, 32(4):804–823, 1985.

[2] B. Awerbuch. A new distributed depth-first search algorithm. *Information Processing Letters*, 20:147–150, 1985.

[3] B. Awerbuch. Distributed shortest paths algorithms. In *Proc. 21st ACM Symposium on Theory of Computing*, pages 490–500, Seattle, May 1989.

[4] B. Awerbuch and R. Gallager. Distributed BFS algorithms. In *Proc. 26th IEEE Symposium on Foundations of Computer Science*, pages 250–256, Portland, OR, October 1985.

[5] B. Awerbuch and R.G. Gallager. A new distributed algorithm to find breadth-first search trees. *IEEE Transactions on Information Theory*, 33(3):315–322, 1987.

[6] D. Bienstock and C.L. Monma. On the complexity of covering vertices by faces in a planar graph. *SIAM Journal on Computing*, 17:53–76, 1988.

[7] G.N. Frederickson. A single source shortest path algorithm for a planar distributed network. In *Proc. 2nd Symposium on Theoretical Aspects of Computer Science*, pages 143–150, 1985.

[8] G.N. Frederickson. A new approach to all pairs shortest paths in planar graphs. In *Proc. 19th ACM Symposium on Theory of Computing*, pages 19–28, New York, May 1987. Revised version available as: "Planar graph decomposition and all pairs shortest paths", TR–89–015, ICSI, Berkeley, March 1989.

[9] G.N. Frederickson and R. Janardan. Designing networks with compact routing tables. *Algorithmica*, 3:171–190, 1988.

[10] R. Janardan and S.W. Cheng. Efficient distributed algorithms for single-source shortest paths and related problems on plane networks. Technical Report TR–89–47 (1st revision), University of Minnesota, July 1989. Submitted.

[11] M. Schwartz and T.E. Stern. Routing techniques used in computer communication networks. *IEEE Transactions on Communications*, 28(4):539–552, 1980.

Stepwise Development of a Distributed Load Balancing Algorithm

Peter Grønning   *Thomas Qvist Nielsen*
Hans Henrik Løvengreen

Department of Computer Science
Building 344, Technical University of Denmark
DK-2800 Lyngby, Denmark

Abstract

This paper describes an algorithm which has been used to implement load balancing for the problem-heap paradigm on a distributed system of transputers. The algorithm is relatively simple due to the fact that only local balance is sought. Emphasis is put on the stepwise development of the algorithm. First an abstract algorithm is presented and its safety and liveness properties are proved. Then a concrete, distributed algorithm is shown to correctly implement the abstract one. Experimental results of the transputer implementation are also reported.

Keywords: Load balancing, distributed data structures, formal methods, transputer.

1 Introduction

The *problem-heap paradigm* [11] is a technique to exploit the potential parallelism in divide-and-conquer algorithms[1]. In such algorithms a problem is divided into a number of smaller, independent problems until the problems can be readily solved. A *problem-heap* is a collection of such sub-

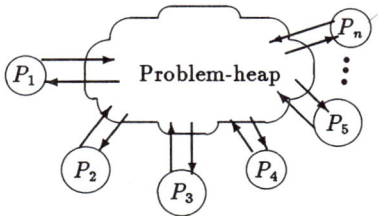

Figure 1
General problem-heap

problems which may be accessed by a number of *solver processes* corresponding to the number of available processors. Each process repeatedly obtains a problem and either solves it (if it is simple) or divides it into sub-problems and returns these to the heap. The processes run until the heap is exhausted indicating that the initial problem has been solved (see figure 1).

This approach is well-suited for a multi-processor architecture with shared memory, but for larger numbers of processors, the shared data-structure (the single heap) could become a bottle-neck.

---

[1]This paradigm is also known under many other headings, such as *problem farming*, *task queues* or *tuple spaces*.

To overcome this problem, the heap may be divided into a number of sub-heaps, each residing in the local memory of one of the processors. This solution fits a distributed architecture.

By dividing the data-structure a new problem has been introduced, viz. the distribution af problems between the heaps. Since problems are consumed and created dynamically, the distribution must be dynamic, always trying to keep the heaps non-empty. We therefore need a mechanism to perform this load balancing as indicated in figure 2.

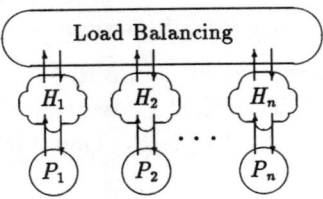

Figure 2

Distributed problem-heap

The subject of this paper is to devise such a mechanism. In section 2, we give an informal presentation of the distribution problem. Some basic notions are introduced in section 3 and used in section 4 to formalise the problem. We then present a solution in two steps. First, in section 5, we propose a simple abstract solution based on non-local information. Secondly, in section 6, we present a truly distributed algorithm. The formal correctness of the algorithm is elaborated upon in section 7. In section 8, we briefly describe the actual implementation of the algorithm on a transputer system and state some experimental results. Finally, in section 9 our approach is discussed.

2 Problem Statement

We shall make the following abstractions of the above problem:

- The distributed architecture will be represented by an undirected graph whose nodes represent processors and whose edges represent communication links.

- Since the actual nature of the problems is of no concern for the distribution, each heap will be represented by a *count* of problems.

- The consumption and creation of problems on each node will be represented by operations increasing and decreasing the problem count by choice.

- The basic operation for distributing problems will be the transfer of a single problem along a communication link.

The state of the system may be formalised as follows:

Definition 1 Let the system be represented by a connected graph G with nodes $V = \{v_1, \ldots, v_n\}$. A *load distribution* is a mapping $l : V \to N_0$. Usually, we let l_i denote $l(v_i)$, i.e. the number of problems at node v_i. ∎

We stipulate that an ideal distribution is one where the loads at the various nodes are as close as possible. Since problems are discrete, we cannot demand them all to be equal. Instead, we define:

Definition 2 A load distribution l is *globally k-balanced* iff for any two nodes $(v_i, v_j) : |l_i - l_j| \leq k$. ∎

Thus, a globally 1-balanced distribution is the best we can hope for.

Globally balanced distributions, however, may be expensive to achieve in distributed systems since they require global information. In order to arrive at a simple solution, we relax our distribution requirement and base it on the following notion:

Definition 3 A load distribution l is *locally k-balanced* iff for any two neighbour nodes (v_i, v_j): $|l_i - l_j| \leq k$. ∎

Note that local balance does not imply global balance. It is the kind of distribution obtained by piling up sand, soil, or other non-floating materials. As more sand is added to a pile, it will grow higher until a certain critical slope (corresponding to k) is reached. Then, the pile will give way and start to spread.

Although a locally balanced distribution is not necessarily globally balanced, we obviously have that for any two nodes $(v_i, v_j) : |l_i - l_j| \leq d \cdot k$, where d is the diameter[2] of the graph.

Our distribution mechanism should strive for locally k-balanced distributions for some k. Since problems are consumed and created dynamically, however, we may never be able to reach such a distribution. As an approximation, we shall instead require the following:

- Each solver-process is assumed to be in one of two states. In the *passive* state, it cannot consume or create problems. In the *active* state, it may consume and create problems by its own choice.

- If at any time all the solver-processes remain passive forever, the distribution of the remaining problems will eventually become locally balanced and then remain stable (i.e. no further problems will be transferred).

Since we do not want the distribution mechanism to depend on the internal state of the solver-processes it must continuously seek to achieve a locally balanced distribution from the current one. In [5] this property is treated under the heading *self-stabilisation*.

3 Basic Notions

Transition Systems

The behaviour of a system will be represented by the standard interleaving model of a labelled transition system [13]. Thus, the system is given by a four-tuple (S, A, T, s^I), where S is the set of *states* of the system, A is the set of *actions* which may be performed in the system, $T \subseteq S \times A \times S$ is the *transition relation* which determines the possible changes of states due to performance of actions, and $s^I \in S$ is the *initial state*.

As usual we write $s \xrightarrow{a} s'$ instead of $(s, a, s') \in T$. An action a is said to be *enabled in state* s iff there exists an s' such that $s \xrightarrow{a} s'$. A *terminal* state is one in which no action is enabled.

We now define an *execution* of the system to be an infinite sequence σ of transitions:

$$\sigma = s_0 \xrightarrow{a_0} s_1 \xrightarrow{a_1} s_2 \xrightarrow{a_2} \ldots$$

where $s_0 = s^I$ and each transition $s_i \xrightarrow{a_i} s_{i+1}$ either belongs to T or is a *silent transition* of the form $s \xrightarrow{\epsilon} s$. Furthermore, there is an (implicit) *progress requirement* stating that if the state is not terminal, some action will eventually be performed. Note that we allow silent transitions to be arbitrarily interspersed in the computations ("stuttering") to ease the refinement of systems [8, 12]. The progress requirement guarantees that the stuttering will be finite unless the system is in a terminal state.

---

[2]The *diameter* of a graph is the maximal length of the shortest path between any two nodes.

Like in UNITY [4], our algorithms will be described by declaring a number of *state variables* [3] determining the state space of the system and listing a set of *conditional assignment statements* of the form:

$$a : \langle\, b \rightarrow \overline{x} := \overline{e}\,\rangle$$

where b is a state predicate known as the *precondition* and $\overline{x} := \overline{e}$ is a (multiple) assignment. Each such statement can be seen as a *transition scheme* since it determines the set of a-transitions as those of the form $s \stackrel{a}{\rightarrow} s'$ where b is true on s and s' is the result of executing $\overline{x} := \overline{e}$ on s. As indicated by the angle brackets, the conditional assignment is thereby assumed to be atomic.

Temporal Logic

For expressing properties of the system behaviour we use standard linear time temporal logic (TL) [13]. Assertions are constructed using the following temporal operators:

$\quad\quad\quad\quad \Box P \quad$ P is true from now on.
$\quad\quad\quad\quad \Diamond P \quad$ P will eventually become true (or is true now).

Liveness properties may be specified using the usual *leads-to* operator defined by:

$$P \rightsquigarrow Q \stackrel{\text{def}}{=} \Box(P \Rightarrow \Diamond Q)$$

The atomic formulas may be either *state predicates* over the program state and global variables or *action predicates* of the form a with the interpretation that the action a is (about to be) performed. For a given execution such as σ above, the truth-value of a formula φ (denoted by $\sigma \models \varphi$), is defined over the corresponding sequence of *observations*:

$$\rho = (s_0, a_0), (s_1, a_1), (s_2, a_2), \ldots$$

The semantics of the boolean connectives and the temporal operators is the standard one, e.g.:

$\quad\quad\quad\quad \rho \models P \quad$ iff $\quad \rho_0 \models P \quad$ if P is atomic
$\quad\quad\quad\quad \rho \models \Box P \quad$ iff $\quad (\forall i)\rho^{+i} \models P$
$\quad\quad\quad\quad \rho \models \Diamond P \quad$ iff $\quad (\exists i)\rho^{+i} \models P$

where ρ^{+i} is ρ taken from the i'th element. The truth-value of atomic formulas is defined over a single observation (s_i, a_i):

$\quad\quad\quad\quad (s_i, a_i) \models \mathcal{P} \quad$ iff $\quad [\mathcal{P}](s_i)$
$\quad\quad\quad\quad (s_i, a_i) \models a \quad$ iff $\quad a = a_i$

where $\mathcal{P}$ is a state predicate and $[\mathcal{P}]$ its semantics.

These definitions imply that we talk about the system when it is just about to perform its next action.

As usual in linear time logic, a formula φ is *valid* for a system if $\sigma \models \varphi$ holds for all executions σ of the system.

We use the state predicate $Enabled(a)$ to denote that the action a is enabled in a given state s. Thus, we may express the progress property of the system by:

$$(\exists a \in A)(Enabled(a)) \rightsquigarrow (\exists a' \in A)\, a' \quad\quad\quad\quad (P)$$

---

[3]State variables will be written as capitalised *italic* words, e.g. L_i, NC.

4 Properties of the Distribution

The load distribution of the system will be represented by a state variable L_i for each node v_i. Together they determine a distribution denoted by L.[4] Under normal *active* operation the loads may be changed by the solver processes by *up* and *down* actions on each node i:

$$up_i : \quad \langle\, L_i := L_i + 1 \,\rangle$$
$$down_i : \quad \langle\, L_i > 0 \rightarrow L_i := L_i - 1 \,\rangle$$

For a load distribution l, we let Σl denote the summation over all components:

$$\Sigma l \stackrel{\mathrm{def}}{=} \sum_{i=1}^{n} l_i$$

We also define the predicate *Balanced* over l to denote that l is in local balance for some given k. Letting $nb(i,j)$ denote that the nodes v_i and v_j are neighbours, we get:

$$Balanced(l) \stackrel{\mathrm{def}}{=} (\forall i, j)(nb(i,j) \Rightarrow |l_i - l_j| \leq k)$$

The properties of the distribution rely on the assumption that from some moment of time, no change of the loads are made by the solver processes. Formally, this is expressed by:

$$\Box([\Box(\forall i)(\neg(up_i \lor down_i))] \Rightarrow Distr) \qquad (*)$$

where *Distr* is the conjunction of the following properties:

$$\Sigma L = m \Rightarrow \Box\, \Sigma L = m \qquad \text{(A1)}$$
$$\Diamond\, Balanced(L) \qquad \text{(A2)}$$
$$\Box((Balanced(L) \land L = l) \Rightarrow \Box\, L = l) \qquad \text{(A3)}$$

(A1) states that the total load remains constant. (A2) expresses that a locally balanced state will eventually be reached and hence by (A1) having the same total load as the initial one. The last property (A3) says that once balanced, the load will remain stable. (A1) and (A3) are safety properties, whereas (A2) is a liveness property.

To prove $(*)$ for some program system consisting of a solver part including the operations *up* and *down* plus a distribution part, it is sufficient to prove the three distribution properties for the distribution part alone, but starting in *any* state which is reachable in the original program.

5 The Abstract Algorithm

The idea of the algorithm is simply to move a problem from one node to a neighbour node when the load of the first one exceeds the load of the second one by more than k. This is expressed by a distribution part accessing the state variables L_i. It consists of the following conditional assignment for each pair of neighbour nodes v_i, v_j:

$$ex_{ij}: \langle\, L_i - L_j > k \rightarrow L_i, L_j := L_i - 1, L_j + 1 \,\rangle$$

Note that the algorithm is topologically distributed, but requires atomic access to the state of two neighbour nodes. The algorithm might be called *semi-distributed*.

---
[4] L is not a state variable, but a derived state function.

Correctness

We now show that the algorithm satisfies (A1–A3) for any initial load distribution. Since ex_{ij} atomically increments L_j and decrements L_i by one, the total load will remain constant and thus (A1) is trivially satisfied.

To show the liveness property (A2), we propose the following *metric function* on a distribution l into N_0:

$$\mathcal{M}(l) \stackrel{\text{def}}{=} \sum_{i=1}^{n} l_i^2$$

Let l and l' be the values of L before and after the execution of an action ex_{ij}. Since only L_i and L_j are changed, we have:

$$\begin{aligned}\mathcal{M}(l') &= (l_i')^2 + (l_j')^2 + \sum_{r \neq i,j} (l_r')^2 \\ &= (l_i - 1)^2 + (l_j + 1)^2 + \sum_{r \neq i,j} l_r^2 \\ &= l_i^2 + l_j^2 + 2(l_j - l_i + 1) + \sum_{r \neq i,j} l_r^2 \\ &< l_i^2 + l_j^2 + \sum_{r \neq i,j} l_r^2 \\ &= \mathcal{M}(l)\end{aligned}$$

where $<$ holds iff $(l_j - l_i + 1) < 0$, or equivalently:

$$l_i - l_j > 1$$

Since the precondition for ex_{ij} requires $l_i - l_j > k$ this is obviously satisfied for $k \geq 1$.

Due to the progress property, some action will be executed in a nonterminal state and since any action will decrease the value of $\mathcal{M}$ then, starting in an arbitrary state, a terminal state will eventually be reached. In the terminal state, all preconditions of all actions will be false which is equivalent to the load being locally k-balanced. Thus, (A2) follows.

Also (A3) is satisfied since in a locally k-balanced distribution, no action will be enabled.

6 Distributing the Algorithm

Since the abstract algorithm accesses the state of two nodes atomically, it cannot be directly implemented on a distributed system. In this section, we propose a truly distributed algorithm where balancing is performed by a number of distribution processes D_i interacting by message passing.

Informal description

The idea of the algorithm is still to transfer problems when the problem count difference exceeds k, but based on estimates of the counts of the neighbour nodes. A node with an estimate of one of its neighbour nodes telling that the neighbour has at least k problems more, must take the initiative to transfer a problem by sending a request to the neighbour. To make sure that the transfer takes place only when the difference in counts exceeds k, the requesting node needs to inform the other node about its count, which is done by including the count in the request. By restricting a node to request only one neighbour node at a time, the recipient of the request is able to decide whether the transfer should take place. If this decision turns out to be negative, the returned rejection will include the actual count on the neighbour node as an argument for the rejection and to prevent further requests.

Beside the information exchanged during requests and rejections, the estimates are based on direct information exchange over a special information channel. Information is, however, only sent to a neighbour when the count of the node exceeds the last value sent out to that neighbour. This means that decrease of the count is not discovered by the neighbour until a request for a problem is rejected.

Formal specification

Obviously the node D_i needs to remember the estimates of the neighbour counts and the last value sent to each neighbour. This is recorded in the variables NC_i and LS_i mapping neighbours to counts respectively last value sent. Thus, D_i's estimate of the neighbour D_j's count is denoted by $NC_i(j)$ and the last value sent from D_i to D_j by $LS_i(j)$. D_i's problem count is held in the variable C_i.

To carry the information between every two neighbour processes D_i and D_j, we have four channels in each direction. The channels are indexed by i,j to denote that messages are sent from D_i to D_j. $Inform_{ij}$ is used to disseminate count information. Req_{ij} is used by D_i to request a problem from D_j. D_j answers a request by sending a signal along a channel $Transfer_{ji}$ denoting the transfer of a problem, or by sending its actual count along channel Rej_{ji} indicating a rejection.

Furthermore, there is a channel $Solver_i$ connecting D_i with the local solver-process.

The algorithm is specified in figure 3 using a CSP-like notation [6]. The communication, however, is assumed to be asynchronous (buffered) in order to avoid deadlock problems at this stage. (Implementing the algorithm using synchronous communication is discussed in section 8.)

$$
\begin{aligned}
&\textbf{process } D_i \triangleq \\
&\quad \textbf{declare} \\
&\qquad NC_i := [\, j \mapsto 0 \mid j \in nb_i \,] \;\textbf{type} : \mathrm{N}_0 \rightarrowtail \mathrm{N}_0 \\
&\qquad LS_i := [\, j \mapsto 0 \mid j \in nb_i \,] \;\textbf{type} : \mathrm{N}_0 \rightarrowtail \mathrm{N}_0 \\
&\qquad C_i := 0 \;\textbf{type} : \mathrm{N}_0 \\
&\qquad Busy_i := \textbf{false type} : \mathrm{BOOL} \\
&\quad *[\; \neg Busy_i \wedge Solver_i?Up() \quad &&\to\; C_i := C_i + 1 \\
&\quad \|\; C_i > 0 \wedge Solver_i?Down() \quad &&\to\; C_i := C_i - 1 \\
&\quad \|_{j \in nb_i} C_i > LS_i(j) \quad &&\to\; Inform_{ij}!C_i;\; LS_i(j) := C_i \\
&\quad \|_{j \in nb_i} Req_{ji}?n \wedge C_i - n > k \quad &&\to\; C_i := C_i - 1;\; Transfer_{ij}!Problem() \\
&\quad \|_{j \in nb_i} Req_{ji}?n \wedge C_i - n \le k \wedge C_i \ge LS_i(j) \quad &&\to\; Rej_{ij}!C_i \\
&\quad \|_{j \in nb_i} Req_{ji}?n \wedge C_i - n \le k \wedge C_i < LS_i(j) \quad &&\to\; LS_i(j) := C_i;\; Rej_{ij}!C_i \\
&\quad \|_{j \in nb_i} \neg Busy_i \wedge Inform_{ji}?n \quad &&\to\; NC_i(j) := n \\
&\quad \|_{j \in nb_i} Transfer_{ji}?Problem() \quad &&\to\; C_i := C_i + 1;\; Busy_i := \textbf{false} \\
&\quad \|_{j \in nb_i} Rej_{ji}?v \quad &&\to\; NC_i(j) := v;\; Busy_i := \textbf{false} \\
&\quad \|_{j \in nb_i} \neg Busy_i \wedge NC_i(j) - C_i > k \quad &&\to\; Req_{ij}!C_i;\; Busy_i := \textbf{true} \;]
\end{aligned}
$$

Figure 3

The Concrete Algorithm

$Busy$ is a flag indicating that D_i has made a request and is expecting an answer. The constant nb_i denotes the set of neighbour nodes for D_i.

Observe that $LS_i(j)$ may be increased when j is explicitly informed about the count, but not when implicitly informed through a rejection. This detail will prove essential in the correctness proof.

7 Correctness

In this section we will investigate what it means for the CSP-specification to correctly implement the abstract algorithm.

We show how the CSP-specification can be seen as a transition system. This enables us to talk about a correspondence between the two specifications and thereby to define a notion of correctness.

The correspondence between the two systems will be given as a functional relationship. Given that two conditions called *Safe* and *Loyal* are satisfied by the two systems under this correspondence, we show that any property φ enjoyed by the abstract system can be *inherited* as a similar property $\widehat{\varphi}$ enjoyed by the concrete system. In this sense the concrete system will correctly implement the abstract one.

7.1 Transition System Model

The CSP-like specification language was chosen for readability. By considering the channels as unbounded FIFO-queues, we are, however, able to translate the specification into a transition system.

We use the following notation for queues: Let Q be a queue (channel). Then $|Q|$ denotes its length and $\textbf{hd}\,Q$, $\textbf{tl}\,Q$ and $\textbf{lst}\,Q$ denote the head, tail and last element of Q. Finally, $Q \frown R$ denotes concatenation of Q and R.

Naturally, CSP-input corresponds to removing the head of a non-empty queue and CSP-output to appending an element to the queue.

The CSP-like program for D_i is a repetition of a set of guarded commands each containing only one input operation. Since the variables are private for D_i and since output operations can never be delayed, each guarded command can be transformed into an atomic conditional assigment statement cf. [9]. E.g. the guarded command

$$Req_{ji}?n \wedge C_i - n > k \rightarrow C_i := C_i - 1;\ Transfer_{ij}!Problem()$$

is transformed into

$$\langle |Req_{ji}| \neq 0 \wedge (C_i - \textbf{hd}\,Req_{ji}) > k \rightarrow$$
$$Req_{ij}, C_i, Transfer_{ij} := \textbf{tl}\,Req_{ji}, C_i - 1, Transfer_{ij} \frown <Problem()>\rangle$$

In order to refer to these actions below, we often use the communication events they perform. E.g. the action above will be referred to as $!Transfer_{ij}$, i.e. the action where output to $Transfer_{ij}$ occurs.

7.2 Correspondence

Let $\mathcal{A} = (S_\mathcal{A}, A_\mathcal{A}, T_\mathcal{A}, s_\mathcal{A}^I)$ be the abstract system and $\mathcal{C} = (S_\mathcal{C}, A_\mathcal{C}, T_\mathcal{C}, s_\mathcal{C}^I)$ be a concrete implementation thereof. We use the indices $\mathcal{A}, \mathcal{C}$ to indicate which system an element belongs to.

The correspondence may in general be given by a relation between states of the two systems and between actions of the two systems. For our purposes, it suffices to consider only a functional relationship. Thus, we assume the existence of two *functions* to *retrieve* the abstract image of a concrete execution.

$\mathcal{R}_s : S_\mathcal{C} \rightarrow S_\mathcal{A}$ The correspondence between the states.
$\mathcal{R}_a : A_\mathcal{C} \rightarrow A_\mathcal{A} \cup \{\epsilon\}$ The correspondence between the actions.

Figure 4 shows how the retrieve functions map a concrete transition into an abstract one; its *abstract image*.

Theorem 4 For any concrete execution $\sigma_\mathcal{C}$ with the abstract image $\sigma_\mathcal{A}$:

$$\sigma_\mathcal{A} \models \varphi \quad \textbf{iff} \quad \sigma_\mathcal{C} \models \widehat{\varphi}$$

where $\widehat{\varphi}$ is constructed by substituting for each abstract state variable $X_\mathcal{A}$ its definition in $\mathcal{R}_s$ and for each abstract action predicate $a_\mathcal{A}$ the disjunction of the concrete action predicates $a_\mathcal{C}$ that are mapped into $a_\mathcal{A}$.

Proof: By strutural induction on φ. ∎

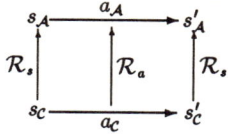

Figure 4
Correspondence between transitions

7.3 Correctness conditions

According to the definition of transition systems there are intuitively two requirements which must be satisfied to inherit any property:

- Every performed action in the concrete system must have a corresponding abstract action.

- If something will eventually happen in the abstract system, something will eventually happen in the concrete system.

These conditions will be formalised as the properties *Safe* and *Loyal* below:

The *Safe* property states that the initial states are related and that any concrete transition is mapped into an abstract or silent transition:

$$Safe \stackrel{def}{=} \mathcal{R}_s(s_\mathcal{C}^I) = s_\mathcal{A}^I \ \land$$
$$[s_\mathcal{C} \stackrel{a_\mathcal{C}}{\to} s'_\mathcal{C} \in T_\mathcal{C} \land a_\mathcal{A} = \mathcal{R}_a(a_\mathcal{C}) \Rightarrow$$
$$\mathcal{R}_s(s_\mathcal{C}) \stackrel{a_\mathcal{A}}{\to} \mathcal{R}_s(s'_\mathcal{C}) \in T_\mathcal{A} \cup \{(s, \epsilon, s) | s \in S_\mathcal{A}\}]$$

We may easily show that the *Safe* property alone is sufficient to ensure that properties of the form $\varphi = \Box I$, i.e. *invariants*, can be inherited.

The *Safe* property guarantees that the abstract image $\sigma_\mathcal{A}$ of any concrete execution $\sigma_\mathcal{C}$ consists of ϵ-transitions or transitions from $T_\mathcal{A}$. If furthermore we can prove that the progress property (P) is satisfied by $\sigma_\mathcal{A}$, then $\sigma_\mathcal{A}$ will be an execution of $\mathcal{A}$. This is what will be expressed by the *Loyal* property defined below.

Let $\Sigma(s)_\mathcal{C}$ denote the set of all execution sequences that can be generated by $\mathcal{C}$ with s as initial state. Furthermore, let $Enabled(a, s)$ be the truth-value of the state predicate $Enabled(a)$ in state s and let $Act(\sigma(i))$ be the action of the i'th transition in σ. We may now state *Loyal* as:

$$Loyal \stackrel{def}{=} (\exists a_\mathcal{A} \in A_\mathcal{A})(Enabled(a_\mathcal{A}, \mathcal{R}_s(s_\mathcal{C}))) \Rightarrow$$
$$(\forall \sigma \in \Sigma(s_\mathcal{C})_\mathcal{C})$$
$$(\exists a_\mathcal{C})((\mathcal{R}_a(a_\mathcal{C}) \neq \epsilon) \land (\exists i)(Act(\sigma(i)) = a_\mathcal{C}))$$

Informally, *Loyal* guarantees that from any concrete state $s_\mathcal{C}$ where some abstract action is enabled in the corresponding state $s_\mathcal{A}$, a concrete action $a_\mathcal{C}$ whose corresponding abstract action is not silent will eventually be executed. Thereby, the abstract image will satisfy the progress property.

Now, assume *Safe* and *Loyal* to be satisfied and let $\sigma_\mathcal{A}$ be the abstract image of any concrete execution $\sigma_\mathcal{C}$. Assume also φ to be satisfied by any abstract execution, then φ will be satisfied by $\sigma_\mathcal{A}$, and therefore, by theorem 4, $\widehat{\varphi}$ will be satisfied by $\sigma_\mathcal{C}$. Thus, we have the following theorem:

Theorem 5 *If Safe and Loyal are satisfied by $\mathcal{A}$ and $\mathcal{C}$ with respect to $\mathcal{R}_s$ and $\mathcal{R}_a$, then for any property φ that holds for $\mathcal{A}$, the inherited property $\widehat{\varphi}$ holds for $\mathcal{C}$.* ∎

The above justifies the following correctness criteria:

Definition 6 *If there exist retrieve functions $\mathcal{R}_s$ and $\mathcal{R}_a$ from a system $\mathcal{C}$ to a system $\mathcal{A}$ such that the Safe and Loyal properties are satisfied, then $\mathcal{C}$ correctly implements $\mathcal{A}$ with respect to $\mathcal{R}_s$ and $\mathcal{R}_a$.* ∎

7.4 Proving Correctness

In the following we will sketch the proof that the CSP-specification correctly implements the abstract algorithm. The details can be found in the appendix.

Retrieve functions

As described in section 5, the abstract state is given by L. The concrete state consists of the state variables declared in each D_i and the contents of each channel.

The retrieve function $\mathcal{R}_s$ is defined by:

$$L_i \stackrel{\text{def}}{=} C_i + \sum_{j \in nb_i} |Transfer_{ji}|$$

I.e L_i equals C_i plus the number of problems on their way towards $node_i$.

Let ex_{ij} denote the abstract action:

$$\langle L_i - L_j > k \rightarrow L_i, L_j := L_i - 1, L_j + 1 \rangle$$

and $!Transfer_{ij}$ denote the concrete action:

$$Req_{ji}?n \wedge C_i - n > k \rightarrow C_i := C_i - 1; Transfer_{ij}!Problem()$$

We now define $\mathcal{R}_a$ by:

$$\mathcal{R}_a(a_C) \stackrel{\text{def}}{=} \begin{cases} ex_{ij} & \text{if } a_C = !Transfer_{ij} \text{ for some } i, j \\ \epsilon & \text{Otherwise} \end{cases}$$

Proving the Safe property

To prove the *Safe* property, it suffices to show for every concrete statement with an empty abstract image that the assignment does not change the abstract state, and likewise for every concrete statement with an non-empty abstract image, that the precondition of the concrete statement is true only when the abstract precondition is and that the two assignments have the same effect on the abstract state.

Since we at this point assume the solver-processes to remain passive, the only actions with empty images that might possibly change the abstract state (i.e. change C_i or $Transfer_{ji}$) are receipts of problems $?Transfer_{ji}$. That $?Transfer_{ji}$ does not change L_i follows from the definition of $\mathcal{R}_s$, because incrementing C_i and decrementing the length of $Transfer_{ji}$ atomically leave L_i unchanged.

For the only statement with a non-empty image, $!Transfer_{ij}$, we have to check:

Preconditions: If the action $!Transfer_{ij}$ takes place, its precondition $|Req_{ji}| = 1 \wedge C_i - \mathbf{hd}Req_{ji} > k$ must be satisfied. By the invariants $L_i \geq C_i$ and

$$|Req_{ij}| = 1 \Rightarrow \mathbf{hd}Req_{ij} \geq L_i$$

(follows from (I3) and (I4) in the appendix) we obtain

$$C_i - \mathbf{hd}Req_{ji} > k \wedge \mathbf{hd}Req_{ji} \geq L_j \wedge L_i \geq C_i$$

which implies the precondition of the corresponding abstract action ex_{ij}:

$$L_i - L_j > k$$

Assignments: From the fact that $!Transfer_{ij}$ atomically decrements C_i and increments the length of $Transfer_{ij}$, it follows directly from the definition of $\mathcal{R}_s$ that L_i is decremented because all $Transfer_{j'i}$ are left untouched. Also, L_j is incremented because $|Transfer_{ij}|$ is incremented and C_j is left untouched.

Proving the Loyal property

In the abstract algorithm an action is enabled only if the state is not balanced. Since the only concrete action that has a non-empty abstract image is $!Transfer_{ij}$ we need to prove:

$$\neg Balanced(L) \rightsquigarrow \exists i,j : !Transfer_{ij}$$

The proof is by contradiction, assuming that no problem will ever be transferred. This assumption will eventually lead to a quiescent state where no communication will ever take place but where each process D_i believe its neighbours to have at least as many problem as they actually have. With this belief, D_i would request for a problem if the difference really were greater that k, which contradicts the assumption.

8 The Implementation

The distribution algorithm has been built into a general problem-heap system implemented in occam on a hyper-cube[5] of 16 transputers.

The implementation has been programmed almost directly from the concrete algorithm, however certain aspects had to be considered:

Finite buffer capacity

In an occam implementation channels must be modelled as buffer processes of finite capacity. Fortunately, as a consequence of the invariants proved in the appendix, the channels need not to be infinite:

- *Req*, *Rej* and *Transfer*: Due to the fact that each node only issues one request at a time, there can at most be 1 element buffered on each channel conveying either requests or answers.

- *Inform*: The channels transferring counts also only need to contain one element, viz. the latest one since all the others are outdated.

Resolution of non-determinism

In the CSP-specification a node may have a choice of neighbours to request for a problem. We have used the simple heuristic of sending the request to the neighbour which is believed to have the highest load, to resolve this non-deterministic choice.

The problem-heap

In addition to the distribution process, processes for problem division/solving and processes administering the combination of sub-problem solutions were needed.

In order to avoid filling up the internal buffers with problems to be solved, we have imposed a depth-first strategy on the system, in the way that the most difficult problems (those which will probably generate most sub-problems) will be sent to the neighbours and the easier problems will be solved at the current node.

The level at which a problem is considered so simple that it should be solved directly without further division must be carefully chosen. If problems are divided too far, the overhead of dividing a problem, distributing its sub-problems and combining their solutions becomes significant compared to the time taken to solve the problem directly. On the other hand a certain number of problems must be generated in order to utilise the distributed architecture.

---

[5]Notice that the hyper-cube reduces the diameter of the network to 4 – which gives a better theoretical load distribution (see section 2).

8.1 Empirical results

In spite of the many processes running on each transputer the implemented system has proved to be efficient.

Benchmarks were carried out for $k = 1$ using the the well-known *fibonacci*-function. The speedup given by using 16 transputers instead of 1 was measured by comparing the execution time on the transputer network T_{16} by the execution time T_1 of computing the fibonacci-function by a standard recursive algorithm on a single transputer. We use *prim* to indicate the level where problems are not

prim	T_{16} sec.	P_{trans}	P_{gen}
19	200.0	5222	92735
20	198.4	4340	57313
21	197.6	1778	35421
22	197.9	1372	21891
23	197.9	798	13529
24	199.5	658	8361
25	201.8	308	5167

Table 1

Variation of *prim* for calculating $Fib(41)$.

divided further, i.e. $Fib(prim)$ is solved directly. Table 1 shows how the choice of *prim* influences the execution time on 16 transputers, the total number of problems transferred P_{trans} and the total number of problems generated P_{gen}. Notice that although the number of problems varies drastically, the impact on the execution time is not significant. Table 2 shows that the network becomes more

n	T_1 sec.	T_{16} sec.	P_{trans}	P_{gen}	Speedup
35	171.2	11.9	112	1973	14.38
36	277.0	18.9	294	3193	14.66
37	448.2	29.9	308	5167	14.99
38	725.2	47.6	602	8361	15.24
39	1173.4	76.2	882	13529	15.40
40	1898.7	122.6	1120	21891	15.49
41	3072.1	197.6	1778	35421	15.55
42	4970.8	319.4	3080	57313	15.56
43	8042.9	515.7	6636	92735	15.60
44	13013.8	833.8	9128	150049	15.61

Table 2

Variation of n with $prim = 21$ for calculating $Fib(n)$.

efficient with harder problems reaching a speedup on 15.61 for calculating $Fib(44)$. Beside this, table 2 as well as table 1 show that the fraction of problems transferred is fairly constant, varying from 5% to 9% of the total number of generated problems.

This is an encouraging result especially when taking into account that the times on the network include overhead of running more than a dozen processes on each transputer, and overhead of combining solutions of sub-problems.

9 Conclusion

We have presented a stepwise development of an algorithm for dynamic load distribution. By not insisting on global balancing, we were able to start from a very simple abstract version of the algorithm whose correctness was easily proven. Although the abstract version was distributed to some extent, it had to be further refined in order not to refer to the state of different nodes atomically. By generally proving the implementation to be *safe* and *loyal* wrt. the abstract algorithm, the abstract properties could be inherited. Finally, the transformation to the actual transputer implementation was sketched.

It should be noted that even though the algorithm has been cast in the problem heap terminology, it is applicable for any kind of load balancing where the load can be transferred in units of one.

One quality of the algorithm is that it is independent of the network topology. It is also interesting that although the abstract algorithm is very simple, the empirical results show that the implementation works efficiently without having to resort to complex heuristics.

The ideas of stepwise development used here are similar to the ones presented in eg. [3, 4, 7, 8, 10, 14]. For a collection of recent papers, see [2]. In fact, we see our work at a case-study in application of these ideas.

The work shows that it is indeed possible to use a stepwise and formal approach resulting in an implemented distributed algorithm whose correctness is ensured and which is sufficiently effective. There are, of course, still a lot of methodological problems. Of these, the most interesting and difficult one is probably how to get from an abstract, non-distributed algorithm to a concrete distributed version. In the present work, this step is solely based on intuition and good ideas, but in general it would be nice to have a systematic, correctness preserving way to transform the abstract algorithm into a distributed one. Suggestions for such transformations have been presented in e.g. [1] and it could be interesting to see if such an approach can result in reasonable efficient implementations. For the moment, however, we see a large gap between the transformational approach and our, more ad hoc, technique.

Acknowledgements

We would like to thank Nils Klarlund and Anders Gammelgaard for many helpful comments on the draft paper and Johnny Jensen for providing the experimental results.

References

[1] R.J.R. Back & R. Kurki-Suonio: *Decentralization of process nets with centralized control.* Distributed Computing, Vol. 3, No. 2, 1989, Pages 73–87.

[2] J.W. de Bakker, W.-P. de Roever & G. Rozenberg (Eds.): *Stepwise Refinement of Distributed Systems.* Lecture Notes in Computer Science, Vol. 430, Springer 1990.

[3] Mani Chandy & Jayadev Misra: *An Example of Stepwise Refinement of Distributed Programs: Quiescence Detection.* ACM Transactions on Programming Languages and Systems, Vol. 8, No. 3, July 1986, Pages 326–343.

[4] Mani Chandy & Jayadev Misra: *Parallel Program Design – A Foundation.* Addison-Wesley Publishing Company Inc. 1988.

[5] Edsger W. Dijkstra: *Self-stabilizing Systems in Spite of Distributed Control.* Communications of the ACM, Vol. 17, No. 11, 1974, Pages 643–644.

[6] C.A.R. Hoare: *Communicating Sequential Processes.* Communication of ACM, Vol. 21, No. 8, August 1978, Pages 666–677.

[7] Bengt Jonsson: *Compositional Verification of Distributed Systems.* Ph.D. Thesis, Uppsala DoCS 87/09 1987.

[8] Leslie Lamport: *An Assertional Correctness Proof of a Distributed Algorithm.* Science of Computer Programming, Vol. 2, 1982, Pages 175–206.

[9] Leslie Lamport: *A Theorem on Atomicity in Distributed Algorithms.* Digital Systems Research Center, Report No. 28, 1988.

[10] Leslie Lamport: *A simple approach to specifying concurrent systems.* Communications of the ACM, Vol. 32, No. 1, 1989, Pages 32–45.

[11] Peter Møller-Nielsen & Jørgen Staunstrup: *Problem-heap: A paradigm for multiprocessor algorithms.* Parallel Computing, Vol. 4, 1987, Pages 63–74.

[12] Van Nguyen et. al.: *A model and temporal proof system for networks of processes.* Distributed Computing, Vol. 1, 1986, Pages 7–25

[13] A. Pnueli: *Applications of Temporal Logic to the Specification and Verification of Reactive Systems: A Survey of Current Trends.* Lecture Notes in Computer Science, Vol. 224, Springer 1986, Pages 510–584.

[14] Eugene W. Stark: *Foundations of a Theory of Specification for Distributed Systems.* Ph.D. Thesis, Report No. MIT/LCS/TR-342, Massachusetts Institute of Technology 1984.

A Correctness Proof

A.1 Invariants

To prove the correctness of the concrete algorithm, we shall need a number of invariants for the algorithm:

Invariant (I1)

$$|Busy_i| = \sum_{j \in nb_i} |Transfer_{ji}| + \sum_{j \in nb_i} |Rej_{ji}| + \sum_{j \in nb_i} |Req_{ij}|$$

<u>Initial condition</u>: Initially all queues are empty, and $Busy_i$ is false i.e. (I1) is satisfied.

<u>Preservation</u>: Given a process D_i and one of its neighbours D_j, both must preserve the invariant. Let *rhs* be an abbreviation for the right hand side of the invariant.

- D_i can affect (I1) only by changing $Busy_i$, writing on Req_{ij}, reading from $Transfer_{ji}$ or reading from Rej_{ji}, i.e. executing one of the last three actions in the algorithm :

  - ? $Transfer_{ji}$: By atomic reading from $Transfer_{ji}$ and setting $Busy_i$ false, the invariant is preserved.
  - ? Rej_{ji}: By atomic reading from Rej_{ji} and setting $Busy_i$ false, the invariant is preserved.
  - ! Req_{ij}: The precondition secures that *rhs* is zero when executing ! Req_{ij}. By atomic writing on Req_{ij} and setting $Busy_i$ true, the invariant is preserved.

- D_j can affect (I1) only by reading from Req_{ij}, writing on $Transfer_{ji}$ or writing on Rej_{ji}, i.e. executing one of the actions with $Req_{ji}?n$ in the guard. All these actions read from Req_{ji} and write on either $Transfer_{ij}$ or Rej_{ji} so that the invariant is preserved.

Invariant (I2)

For each pair of neighbour nodes i, j:

$$|Inform_{ij}| + |Rej_{ij}| = 0 \Rightarrow NC_j(i) \geq LS_i(j)$$

Thus, if no counts are on their way from i to j, $node_j$ is more optimistic about $node_i$'s count than $node_i$ thinks it is.

Proof: We start by defining the following notational abbreviations : Let $\overline{I}$ and $\overline{R}$ denote the predicates $|Inform_{ij}| = 0$ and $|Rej_{ij}| = 0$, and I and R denote the negated properties (i.e. that message queues are non-empty).

We may now formulate (I2) by stating that for any pair of neighbour nodes i,j, the formula $\mathcal{I}_{ij}$ is an invariant where:
$$\mathcal{I}_{ij} \stackrel{def}{=} \overline{I}\,\overline{R} \Rightarrow NC_j(i) \geq LS_i(j)$$
where $\overline{I}\,\overline{R}$ denotes the conjunction of $\overline{I}$ and $\overline{R}$.

It is obvious that the queues $Inform_{ij}$ and Rej_{ij} can be in 4 different states, viz. $\overline{I}\,\overline{R}$, $I\overline{R}$, $\overline{I}R$ and IR.

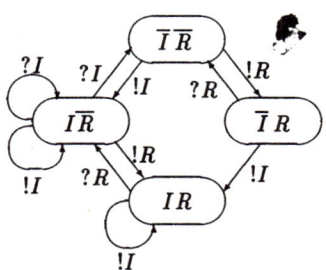

Figure 5
State-diagram

Figure 5 defines the possible transitions between the four states. I and R abbreviates $Inform_{ij}$ and Rej_{ij} respectively. Note that when a rejection is on its way, no further rejections are possible due to (I1).

To show that $\mathcal{I}_{ij}$ is an invariant, it is necessary to prove a stronger invariant which implies $\mathcal{I}_{ij}$. The stronger invariant $\mathcal{I}'_{ij}$ is given as the conjunction of the invariants (i1–i4):

$\overline{I}\,\overline{R} \Rightarrow NC_j(i) \geq LS_i(j)$ (i1)
$I\overline{R} \Rightarrow \underline{\text{lst}}\,Inform_{ij} \geq LS_i(j)$ (i2)
$\overline{I}\,R \Rightarrow \underline{\text{lst}}\,Rej_{ij} \geq LS_i(j)$ (i3)
$I\,R \Rightarrow \underline{\text{lst}}\,Inform_{ij} \geq LS_i(j)$ (i4)

Initial condition: In the initial state, all queues are empty, and all tables are initialised to 0. Thereby $\mathcal{I}'_{ij}$ holds initially.

Preservation of $\mathcal{I}'_{ij}$: It has to be proved that each of the 10 possible transitions in the state-diagram preserves the invariant.

Each transition is easily proved: Assume that $\mathcal{I}'_{ij}$ holds before the transition and let v' denote the value of a variable v after the transition. We have for instance:

- $I\overline{R} \xrightarrow{?Inform} \overline{I}\,\overline{R}$:
  The change of state implies that $Inform_{ij}$ only contained one element. The value sent over the channel is placed in $NC_j(i)$ on node$_j$. This implies $NC_j(i)' = \underline{\text{lst}}\,Inform_{ij}$. Furthermore the value of $LS_i(j)$ is left unchanged. From this and (i2) it follows that the rhs. of (i1) and thereby $\mathcal{I}_{ij}$ are satisfied after the transition.

- $I\overline{R} \xrightarrow{!Rej} IR$:
  The concrete action in node$_i$ that sends out rejections never increases the value of $LS_i(j)$ nor does it change the value of $\underline{\text{lst}}\,Inform$. Therefore the rhs. of (i4) is easily seen to be satisfied after the transition, if (i2) was satisfied before.

The other eight transitions can be proved similarly. This completes the proof of $\mathcal{I}_{ij}$ and thus (I2).

The following invariants express properties between the concrete state variables and the corresponding abstract ones.

Invariant (I3)

$$L_i \geq C_i$$

Proof: Follows directly from the definition of $\mathcal{R}_s$.

Invariant (I4)

$$|Req_{ij}| = 1 \Rightarrow \mathbf{hd}\, Req_{ij} \geq L_i$$

i.e. if one request is on its way from i to j the count included in the request is at least the actual count of node i.

Proof: From (I1) we get the following lemma:

$$|Req_{ij}| = 1 \Rightarrow \sum_{z \in nb_i} |Transfer_{zi}| + \sum_{z \in nb_i} |Rej_{zi}| + \sum_{z \in nb_i \setminus \{j\}} |Req_{iz}| = 0 \wedge Busy_i$$

Now assume $|Req_{ij}| = 1$ to be satisfied. The only actions that can change the truth of $\mathbf{hd}\, Req_{ij} \geq L_i$ are $!Transfer_{ij}$, $?Transfer_{ji}$, $!Req_{ij}$, and Up_i. But the only action enabled is $!Transfer_{ij}$, and this preserves (I4) by only decrementing L_i.

Now assume $|Req_{ij}| = 1$ to be false. The only action that can change the truth of this is $!Req_{ij}$, which is enabled only when $Busy_i$ is false, i.e. when no problem can be on its way to D_i (follows from (I1)). If $!Req_{ij}$ is executed, we therefore have that $C_i = L_i$, and since this value is written on the queue Req_{ij}, we have $\mathbf{hd}\, Req_{ij} = L_i$ after the action so that (I4) is preserved.

Remark

The above invariants hold from the very beginning, i.e. also while the solver-processes are active. The *Safe* and *Loyal* properties hold between the concrete and abstract systems from the moment that the *Solver* queues remain empty.

A.2 Proof of the Loyal Property

In this section we use the following abbreviations where (i,j) ranges over neighbour pairs and ch ranges over one of the channels *Transfer*, *Inform*, *Req*, *Rej*:

$$
\begin{aligned}
!ch &\stackrel{\text{def}}{=} (\exists i,j)\, !ch_{ij} \\
Comm &\stackrel{\text{def}}{=} (\exists ch)\, !ch \\
Empty &\stackrel{\text{def}}{=} (\forall ch)\, |ch| = 0 \\
Quiescence &\stackrel{\text{def}}{=} \Box(\neg Comm \wedge Empty)
\end{aligned}
$$

For example, $!Transfer$ will denote the transfer of a problem somewhere in the system.

We first introduce some useful lemmas.

Lemma 1
$$\Box \neg ! \mathit{Transfer} \Rightarrow (\Box \Diamond ! \mathit{Req}_{ij} \Rightarrow \Box \Diamond ? \mathit{Inform}_{ji})$$

This lemma states that if no problems are transferred, the dispatch of an infinite number of requests implies that an infinite number of counts will be read.

Proof: We assume that no problems are sent and show that once a request has been sent, a count must be read from Inform_{ji} before another request can be sent.

When node_i sends a request, no problem can be on its way towards node_i (I1). Since no problems are ever sent, the value of C_i will therefore remain constant.

If a request with the value C_i sent from node_i to node_j is not answered, no new request can be sent. If it is answered, it will be rejected, otherwise a problem would be sent. This means that the current value v of C_j which is sent together with the rejection satisfies: $v - C_i \leq k$ (C_i has not changed). Since $NC_i(j)$ is set to v upon receiving the rejection, $NC_i(j) - C_i \leq k$. This means that no request can be sent before either $NC_i(j)$ or C_i changes value. Since C_i remains constant, $NC_i(j)$ must be changed but this can only occur when a count is read from Inform_{ji}.

Thus, in order to send more that one request from i to j, a count will have to be read from Inform_{ji}. From this, lemma 1 follows.

Lemma 2
$$\Box \neg ! \mathit{Transfer} \leadsto \Box \neg \mathit{Comm}$$

I.e. if no problems are transferred, then after a while, no messages are sent at all. We prove this by proving that only a finite number of messages will be sent on each group of queues:

- On *Transfer*: This is our assumption.

- On *Inform*: Information about a count is sent only after the count has been increased. A count is increased only by receiving a problem. Since no new problems are sent, only a finite number of count-messages will be sent on the *Inform*-channels.

- On *Req*: Since the number of count-messages is finite, it follows from Lemma 1 that only a finite number of requests can be sent on each *Request*-channel.

- On *Rej*: As the sending of a rejection requires a simultanous consumption of a request, also the number of rejections is finite.

Lemma 3
$$\Box \neg \mathit{Comm} \leadsto \mathit{Quiescence}$$

When all communication ceases, the queues will eventually be emptied and (since no more messages are sent) will remain empty (*Quiescence*).

Proof: As long as there are messages on any *Transfer*, *Rej*, or *Req* channel, input from these channels will be enabled and therefore eventually they are emptied. When this has happened, (I1) guarantees that all *Busy* variables are and remain false, and therefore the *Inform* channels will be emptied too.

Lemma 4
$$\mathit{Quiescence} \Rightarrow \Box (\forall i, j) C_i \leq LS_i(j)$$

Proof: By contradiction: Assume that the system is in a quiescent state and for some i, j, $C_i > LS_i(j)$. Since nothing is sent, the values of C_i and $LS_i(j)$ never change. This implies that the guard for sending count-information in D_i is enabled. Then the progress property ensures that this (or some other enabled statement) is performed violating our quiescence assumption. Therefore, the rhs. is true and will remain true due to the quiescence.

Proof of the loyal property

According to section 7.4, the *Loyal* property holds if the following can be proved:

$$\neg Balanced(L) \rightsquigarrow \; !Transfer$$

Proof: The proof is by contradiction: Assume that we have reached a state which is not balanced and where no more problems are ever transferred:

$$\neg Balanced(L) \wedge \Box \neg ! Transfer$$

Since no problems are transferred, we can assume $\neg Balanced(L)$ for the rest of this proof.

From Lemma 2 and 3 we conclude that the system will eventually become quiescent and when this has happened, Lemma 4 gives:
$$\Box C_i \leq LS_i(j)$$

The invariant (I2) together with *Quiescence* implies

$$\Box LS_i(j) \leq NC_j(i)$$

Combining the two inequations above gives us:

$$\Box C_i \leq NC_j(i)$$

As we assume $\neg Balanced(L)$ then for some i, j, $L_i - L_j > k$ and since $C_i = L_i$ under quiescence ($\mathcal{R}_*$), we have

$$\Box NC_j(i) - C_j > k$$

Since this enables the sending of a request to node$_i$ from node$_j$, quiescence is a contradiction. This concludes the proof of the *Loyal* property.

A.3 Concrete properties

It has been proved that balance will be reached with respect to the abstract model. This means that the retrieved counts L_i will be balanced and remain unchanged, but it does not imply that the concrete counts C_i will be balanced and remain unchanged, as expressed by:

$$(\exists c)(\Diamond(C = c \wedge Balanced(c) \wedge \Box(C = c))) \tag{C1}$$

To prove (C1) it is sufficient to show:

$$Balanced(L) \rightsquigarrow \Box(C = L) \tag{LC}$$

(LC) implies (C1) because the *Safe* property secures that after balance is reached, L remains unchanged and thereby also C will eventually be balanced and then remain unchanged.

Now assume that we have reached an abstractly balanced distribution. This distribution remains unchanged i.e. $\Box \neg ! Transfer$. This property together with lemma 2 and 3 gives:

$$Balanced(L) \rightsquigarrow Quiescence$$

The definition of $\mathcal{R}_*$ enables us to conclude:

$$Quiescence \Rightarrow \Box C = L$$

From this (LC) follows and we may conclude:

The concrete system will reach a balanced distribution and will then remain unchanged.

Greedy Packet Scheduling

Israel Cidon*   Shay Kutten[†]   Yishay Mansour[‡]   David Peleg[§]

Abstract

Scheduling packets to be forwarded over a link is an important subtask of the routing process both in parallel computing and in communication networks. This paper investigates the simple class of *greedy* scheduling algorithms, namely, algorithms that always forward a packet if they can. It is first proved that for various "natural" classes of routes, the time required to complete the transmission of a set of packets is bounded by the sum of the number of packets and the maximal route length, for any greedy algorithm (including the arbitrary scheduling policy). Next, tight time bounds of $\Theta(n)$ are proved for a specific greedy algorithm on the class of shortest paths in n-vertex networks. Finally it is shown that when the routes are arbitrary, the time achieved by various "natural" greedy algorithms can be as bad as $\Omega(n^{1.5})$, when $O(n)$ packets have to be forwarded on an n-vertex network.

---

*IBM T.J. Watson Research Center, P.O. Box 704, Yorktown Heights, NY 10598, and Faculty of Electrical Engineering, The Technion, Haifa 32000, Israel. cidon@techsel.bitnet

[†]IBM T.J. Watson Research Center P.O. Box 704, Yorktown Heights, NY 10598, kutten@ibm.com

[‡]Laboratory for Computer Science, MIT, Cambridge, MA 02139. Partially supported by IBM graduate fellowship. mansour@theory.lcs.mit.edu

[§]Department of Applied Mathematics, The Weizmann Institute, Rehovot 76100, Israel. peleg@wisdom.bitnet. Supported in part by an Allon Fellowship, by a Walter and Elise Haas Career Development Award and by a Bantrell Fellowship. Part of the work was done while visiting IBM T.J. Watson Research Center.

1 Introduction

The task of managing the delivery of packets in a distributed communication network is intricate and complex. Consequently, many routing strategies incorporate design choices directed at simplifying the process. One prime example for this type of choice is the decision to create a clear distinction between two subtasks, namely, *route selection* and *packet scheduling*. The first subtask involves selecting for each packet the route it should use from its source to its destination. This selection is done in advance, before the packet actually leaves its source. The second subtask concerns the transmission stage itself, and involves deciding on the schedule by which the different packets are to be forwarded over each edge along their routes. At this stage, the packets are restricted to their predetermined routes, and cannot deviate from them. This paper concentrates on routing strategies adopting this separation, henceforth referred to as *fixed-route strategies*, and in particular on the scheduling subtask.

A second type of design choice, aimed at simplifying the scheduling process considerably, is to make scheduling decisions *locally* and per packet, rather than globally. The scheduling policy is thus restricted to the selection of local rules for managing the queues on outgoing links, namely, resolving the conflicts between the different packets that need to be advanced on the same outgoing edge. Intuitively, a *local* algorithm has the property that the rules used by a node in order to schedule awaiting packets rely only on information concerning these packets (typically contained in the packets' headers), such as the identity of the source and destination, the distance traversed by the packet so far, the arrival time at the current node etc. In contrast, a global algorithm can base its decisions on additional global information on the status of the network, such as the current distribution of packets in the network and the routes of these packets.

Although the two design decisions discussed above may not generally lead to a globally optimal algorithm, they are both widely used. In fact, one of the main distributed network strategies for packet routing is *virtual circuit switching* [CGK88, Mar82, BG87], which is based on fixing a single predetermined *logical circuit* from the given source to the given destination, and transmitting *all* packets between them on this circuit. While setting up the circuit, the sender and the destination do not know which other logical circuits will overlap their own circuit in the duration of its existence. Nontheless, similar considerations apply also for the second common routing strategy, known as *packet switching* [MRR80]. In a fixed-route packet switching strategy, different packets going from the same source to the same destination may traverse different paths. Thus in such strategy, congestion over a link does influence the selection of routes for later packets, although it cannot change the routes of packets that are already in transit.

Fixed-route strategies are employed in communication networks such as SNA [Mar82], APPN [BGGJP85] and TYMNET [BG87]. As for the scheduling policy, most networks use a combination of FIFO, certain priority parameters, and flow control information, to determine the next packet to be forwarded. All of these mechanisms are "approximately" local greedy (although flow control adds some global flavor). The main reasons for these choices are based on their advantages from an engineering point of view, namely, their simplicity and low complexity (compared to the global approach). The costs of the alternative, global approach are high due to the need to synchronize the network (so that schedules for entire routes be meaningful), the need to accumulate information about packets that are to be sent (and the associated delay until the information is accumulated), and related factors.

Packet scheduling algorithms for fixed-route strategies were studied by Leighton, Maggs and Rao [LMR88]. Although motivated by routing problems in specific networks realizing parallel machines, the paper studies the problem on networks of arbitrary topology. The results of [LMR88] demonstrate the fact that fixed-route strategies have advantages not only from an engineering point of view, but from a theoretical point of view as well. The problem is formalized in [LMR88] as follows. Initially, there are k packets in the network, each is assigned a path of length d, and no edge in the network is to be traversed by more than c packets. The scheduling algorithm has to forward the packets along the assigned paths.

The first result of [LMR88] is a proof that there exists a schedule that terminates in $O(d+c)$ time. However, it seems that determining this schedule requires a complex *centralized* computation, relying on global information. The paper provides also some *randomized* distributed protocols for the problem. These protocols are simple, online and local (in the sense discussed above). The first applies to arbitrary sets of paths and requires $O(c + d\log|V|)$ time (where $|V|$ is the number of nodes), and the second completes the routing in $O(c+l+\log|V|)$ time, and applies to the case when the paths are leveled with l levels. (Informally, a set of paths is *leveled* if the nodes of the network can be partitioned into levels in such a way that each edge of the paths connects two consecutive levels.)

In contrast with both types of algorithms considered in [LMR], in this paper we consider the complexity of *deterministic distributed* algorithms. In fact, we concentrate on a class of very simple on-line routing algorithms, termed *greedy* algorithms. These are algorithms that will always forward some message over each link whenever they can (i.e., whenever there is a message waiting to be forwarded on that link). The class of greedy policies is very natural [Ko78], and in fact, all routing policies used in practical systems of which we are aware fall in this class. Greedy algorithms differ in the rule employed locally for deciding on the packet to be forwarded, in the case that more than one packet awaits to be sent over the same link.

In the sequel we present several results concerning the behavior of greedy scheduling algorithms. We first look at some restricted path classes. To begin with, in Section 3 we show that for a *leveled* set of paths, the time required for routing k packets on paths of length d by any greedy algorithm is at most $d + k - 1$. This holds even for an arbitrary (yet greedy) scheduling policy, or put another way, even when an adversary is permitted to select the next packet to be forwarded in each step. We actually establish an appropriate extension of this result, applying to the more general case where packets start at different times and traverse routes of different lengths. The time by which the i'th packet arrives its destination is bounded by $\tau_i^A + d_i + k - 1$, where d_i is the length of the ith packet's route and τ_i^A is the start time of the i'th packet. This implies the same result for the natural case of *unique subpaths*. (A collection of routes enjoys the *unique subpaths* property if for every pair of nodes u and v, whenever two different routes both have a segment connecting u and v then these two segments are identical.)

In Section 4 we consider the class of *shortest paths*, where each route is required to be a *shortest path* between its source and destination. For this class, we present a strategy that requires at most $d + k - 1$ time. Again, an appropriate extension holds for the case of varying route lengths (specifically, the arrival of the i'th packet at its destination occurs within $d_i + k - 1$ time units). This strategy is based on advancing the packet that has progressed the least so far. We conjecture that the same bound is true for any greedy algorithm.

We then turn to general route classes, where the set of paths is not restricted in any way. In contrast with the special cases discussed above, we show in Section 5 that greedy algorithms might behave badly for an *arbitrary* set of paths. This is true even when we consider natural greedy schedulers, like fixed priority, FIFO, or preferring the packet that traversed the minimum (or maximum) distance so far. We show that in such a case the time may be $\Omega(d\sqrt{k} + k)$. These negative results hold even for the case where both $k = \Theta(|V|)$ and $d = \Theta(|V|)$. This strengthens the counter-examples given in [LMR88] for the case of long routes and a large number of packets.

2 Model

We view the communication network as a directed graph, $G = (V, E)$, where an edge (u, v) represents a bidirectional link connecting the processors u and v. We assume synchronous communication, i.e., the system maintains a global clock, characterized by the property that a packet sent at time t is received by time $t + 1$.

Next let us define formally the routing problem and its relevant parameters. The input to the problem is a collection $\mathcal{P}$ of k packets p_i and k associated routes ρ_i, $1 \leq i \leq k$. Packet p_i is originated at node A_i, its destination is B_i, and it is transmitted along the route ρ_i. We deal with node-simple (or, loop-free) routes. The length of the route ρ_i is d_i, and we denote $d(\mathcal{P}) = \max_i\{|d_i|\}$.

Two packets are said to *collide* at time t if they are currently waiting at the same node to be sent over the same link. The scheduling algorithm has to decide at each time t which packet to forward at time t. (Note that the paths are fixed, and hence the algorithm has no choice with respect to the edges that a packet traverses.) Let τ_i^A denote the time at which packet p_i was sent from its originator A_i, and let τ_i^B denote the arrival time of p_i at its destination B_i. Let T_i denote the time elapsing from τ_i^A until τ_i^B, i.e., $T_i = \tau_i^B - \tau_i^A$. The *schedule time* of $\mathcal{P}$ is $T(\mathcal{P}) = \max_i \{T_i\}$.

Some of our results apply only to special path types. Below we characterize these route classes.

A set of paths $\mathcal{P}$ is leveled if there exists an assignment $level : V \to [1, \cdots, |V|]$, such that for each path $\rho = (v_1 \ldots v_l)$, $level(v_j) = level(v_{j-1}) + 1$. A directed graph is leveled if there exists an assignment $level$, such that for every directed edge (u, v), $level(u) + 1 = level(v)$. In a leveled directed graph, every set $\mathcal{P}$ of routes is leveled.

The path ρ_i is a *shortest path* if its length equals the distance between its endpoints A_i and B_i. A set of paths $\mathcal{P}$ is *shortest* if every path $\rho_i \in \mathcal{P}$ is a shortest path.

A set of paths $\mathcal{P}$ has the *unique subpaths* property if for every pair of nodes u and v, all the subpaths connecting them in any path of $\mathcal{P}$ are identical; that is, if both the routes ρ_i and ρ_j go through u and v, then the segments of the paths connecting u and v are identical.

Finally, let us formally define the concept of greedy packet scheduling algorithms. A *greedy algorithm* is an algorithm satisfying the property that at each time unit, the set of packets that are forwarded is maximal, i.e., if there are messages waiting to be forwarded on some link then one of these messages is forwarded. Note that this includes also an algorithm that selects the message to be forwarded next on each link arbitrarily from among the waiting messages (or alternatively, allows an adversary to decide which packet will be sent next). The only restriction on the adversary is that it must send some packet out of the available ones, on every link.

3 Leveled routing

In this section we prove our first result, concerning greedy scheduling on leveled paths.

Theorem 3.1 *Let $\mathcal{P}$ be a set of k leveled paths. Then for any greedy algorithm used for routing $\mathcal{P}$,*

1. *every packet p_i arrives within $T_i \leq d_i + k - 1$ time units, and*

2. *the algorithm has schedule time $T(\mathcal{P}) \leq d(\mathcal{P}) + k - 1$.*

Proof: The intuition behind the proof arises from considering, for any given time during the execution, the levels that are *occupied* at that time (i.e., that contain at least one packet). We observe that at every time unit there is some progress, in the sense that either the number of occupied levels grows, or the lowest occupied level (the one whose number is the smallest) becomes unoccupied.

Let us first define some terminology. For each packet p_i, let $level(p_i, t)$ denote the number of the level where p_i resides at time t. A level L is said to be *occupied* at time t if there exists a packet p_i such that $level(p_i, t) = L$.

We adopt the convention that at any time $t > \tau_i^B$, $level(p_i, t)$ is incremented by one. This can be thought of as if the packet continues progressing indefinitely along some path ρ'_i extended from the destination B_i and dedicated to it, and hence never collides afterwards. This does not restrict generality in any way, since such an extension ρ'_i of the packet's route has no influence on the routes of other packets, and the arrival time of the packet is still considered to be τ_i^B, the time it has reached its original destination B_i.

Consider the collection $\mathcal{L}(t)$ of occupied levels at time t. We break this collection into "blocks" of consecutive levels (separated by unoccupied levels). We define the following parameters for each packet p_i:

- $B(p_i, t)$ is the *block* of p_i at time t (i.e., the block containing $level(p_i, t)$).

Suppose that $B(p_i, t) = \{L, L+1, \ldots, H\}$. Then

- $min(p_i, t) = L - 1$ is the maximal level that is smaller than $level(p_i, t)$ and is not occupied at time t.

- $max(p_i, t) = H + 1$ is the minimal level greater than $level(p_i, t)$ that is not occupied at time t.

- $width(p_i, t) = |B(p_i, t)| = max(p_i, t) - min(p_i, t) - 1 \ (= H - L + 1)$ is the number of levels in p_i's block, $B(p_i, t)$.

The following potential function is defined per packet p_i:

$$\Phi(p_i, t) = min(p_i, t) + width(p_i, t).$$

We are interested in the changes in this potential from one step to the next. Consequently, let us denote

$$\begin{aligned}
\Delta_\Phi(p_i, t, t') &= \Phi(p_i, t') - \Phi(p_i, t), \\
\Delta_{min}(p_i, t, t') &= min(p_i, t') - min(p_i, t), \\
\Delta_{width}(p_i, t, t') &= width(p_i, t') - width(p_i, t),
\end{aligned}$$

hence
$$\Delta_\Phi(p_i, t, t') = \Delta_{min}(p_i, t, t') + \Delta_{width}(p_i, t, t'). \tag{1}$$

The proof goes along the following lines. We show that $\Delta_\Phi(p_i, t, t+1) \geq 1$, and therefore $\Delta_\Phi(p_i, \tau_i^A, \tau_i^B) \geq T_i$. We then show that $\Delta_\Phi(p_i, \tau_i^A, \tau_i^B) \leq k + d_i - 1$. Combining the two claims implies that $T_i \leq k + d_i - 1$.

Claim 3.2 $\Delta_{width}(p_i, t, t+1) \geq 0$ for every $t \geq 0$.

Proof: Since the algorithm is greedy, we are guaranteed that if the levels $L, L+1, \ldots, H$ are occupied at time t, then the levels $L+1, \ldots, H+1$ are occupied at time $t+1$. Also, $L \leq level(p_i, t) \leq H$ implies $L \leq level(p_i, t+1) \leq H+1$, and therefore $L+1, \ldots, H+1 \in B(p_i, t+1)$. This implies that $width(p_i, t)$ is monotonically non-decreasing with t, and the claim follows. ∎

Unfortunately, $min(p_i, t)$ may decrease with time. We now show that the sum of the two parameters, $\Phi(p_i, t)$, is monotonically increasing.

Claim 3.3 $\Delta_\Phi(p_i, t, t+1) \geq 1$ for every $t \geq 0$.

Proof: As time progresses from t to $t+1$, $min(p_i, t)$ can either grow by 1, remain the same or decrease. We analyze these cases one by one. First assume that $min(p_i, t+1) = min(p_i, t) + 1$. Then $\Delta_{min}(p_i, t, t+1) = 1$, and by Claim 3.2, $\Delta_{width}(p_i, t, t+1) \geq 0$, implying the claim by Eq. (1).

Next, suppose $min(p_i, t+1) = min(p_i, t)$. This implies that level $min(p_i, t) + 1$ is occupied at time $t+1$, and hence $width(p_i, t)$ grows by at least one (since $max(p_i, t)$ also becomes occupied at time $t+1$). Thus $\Phi(p_i, t+1) - \Phi(p_i, t) \geq 1$.

The remaining case is that $min(p_i, t+1) = min(p_i, t) - x$, where $x > 0$. There is only one way for that to happen, namely, there has to be a packet in level $min(p_i, t) + 1$ that did not progress, and the block $B(p_i, t)$ is "joined" by the block just below it. But in this case, the decrease of $min(p_i, t)$ by x is offset by an increase of $width(p_i, t)$ by $x + 1$. This holds since there are x more levels below $min(p_i, t)$ in the united block $B(p_i, t+1)$, and $max(p_i, t)$ becomes occupied. Hence $width(p_i, t+1) - width(p_i, t) \geq x + 1$. By Eq. (1), the net change in $\Phi(p_i, t)$ satisfies

$$\Delta_\Phi(p_i, t, t+1) = \Delta_{min}(p_i, t, t+1) + \Delta_{width}(p_i, t, t+1) \geq (-x) + (x+1) = 1.$$

∎

Corollary 3.4 $\Delta_\Phi(p_i, \tau_i^A, \tau_i^B) \geq T_i$.

This corollary is complemented by the following claim, bounding the increase in Φ from above.

Claim 3.5 $\Delta_\Phi(p_i, \tau_i^A, \tau_i^B) \leq d_i + k - 1$.

Proof: Consider a packet p_i whose origin A_i is at level
$$L_A = level(p_i, \tau_i^A)$$
and whose destination B_i is at level
$$L_B = level(p_i, \tau_i^B) = level(p_i, \tau_i^A) + d_i.$$

Initially, $min(p_i, \tau_i^A) \geq L_A - width(p_i, \tau_i^A)$. On the other hand, upon arrival at the destination, $min(p_i, \tau_i^B) \leq L_B - 1$. Hence
$$\Delta_{min}(p_i, \tau_i^A, \tau_i^B) \leq (L_B - 1) - (L_A - width(p_i, \tau_i^A)) = d_i + width(p_i, \tau_i^A) - 1.$$

Since $width(p_i, \tau_i^B) \leq k$,
$$\Delta_{width}(p_i, \tau_i^A, \tau_i^A + T) \leq k - width(p_i, \tau_i^A).$$

The claim now follows by Eq. (1) above. ∎

Combining Corollary 3.4 and Claim 3.5, we get
$$T_i \leq \Delta_\Phi(p_i, \tau_i^A, \tau_i^B) \leq d_i + k - 1.$$

This completes the proof of Part (1) of the Theorem. Part (2) follows immediately from Part (1). ∎

The natural class of paths with the unique subpaths property can be analyzed using the above theorem.

Corollary 3.6 *Let $\mathcal{P}$ be a set of k paths satisfying the unique subpaths property. Then for any greedy algorithm used for routing $\mathcal{P}$,*

1. *every packet p_i arrives within $T_i \leq d_i + k - 1$ time units, and*

2. *the algorithm has schedule time $T(\mathcal{P}) \leq d(\mathcal{P}) + k - 1$.*

Proof: We prove that the delay suffered by any packet p_i is no greater than in an execution on a leveled graph (with the same k and d_i). Note that the subgraph induced by the route of any particular packet p_i plus all the packets it encounters, is leveled. One obstacle is that every other packet p_j may suffer some delay as a result of collisions before its route joins that of p_i. (A similar event can also happen after they part, but this need not interest us, since it does not affect p_i.) This problem can be mended by extending the path of any such packet p_j in the induced leveled graph, before it meets the path of p_i. Notice that we have not changed the length of the route of packet p_i. Thus, by Part (1) of Theorem 3.1 the delay suffered by p_i in the unique subpaths case is the same as the one in the leveled paths case we have constructed. ∎

4 Shortest path routing

In this section we consider a scheduling algorithm for the case in which each route ρ_i in the set $\mathcal{P}$ uses a shortest path from its origin to its destination. We shall assume that all packets start at the same time, i.e., $\tau_i^A = 0$ for $1 \leq i \leq k$. For every time $0 \leq t \leq T_i$, let $d_i(t)$ denote the distance traversed by p_i by time t (note that in particular, $d_i(T_i) = d_i$). If p_i and p_j "collide" at time t, the algorithm resolves the collision by preferring p_i iff

$$d_i(t) < d_j(t) \text{ or } (d_i(t) = d_j(t) \text{ and } i < j).$$

We refer to this algorithm as the *Min Went* algorithm. The rest of this section is devoted to proving the following theorem.

Theorem 4.1 *If the set of paths $\mathcal{P}$ consists of shortest paths and $\tau_i^A = 0$ for $1 \leq i \leq k$ (i.e., all the packets start at the same time) then the Min Went scheduling algorithm guarantees*

1. *Every packet p_i arrives at time $T_i \leq d_i + k - 1$.*

2. *the schedule time is $T(\mathcal{P}) \leq d(\mathcal{P}) + k - 1$.*

We begin the proof by pointing out the following trivial fact regarding the relationship between packets in consecutive collisions.

Fact 4.2 *If p_i and p_j collide twice (at times t_1 and t_2), then the relation between $d_i(t)$ and $d_j(t)$ remains the same.*

Proof: The claim follows from the fact that $d_i(t_2) - d_i(t_1) = d_j(t_2) - d_j(t_1)$, since otherwise, one of the routes is not a shortest path. ∎

Definition 4.3 *Given an execution of the algorithm the collision relation C is defined as the collection of all triples $\langle p_i, p_j, t \rangle$ such that at time t packets p_i and p_j collide (i.e., they are at the same node, waiting for the same edge), and p_i wins the collision resolution and gets to use the edge (at time t).*

Since only one packet can go on a specific edge at a time t we can deduce the following fact.

Fact 4.4 *For every p, t there is at most one triple $\langle p', p, t \rangle$ in C.*

Consider some packet p_{i_0}. If this packet is never delayed, then $T_{i_0} = d_{i_0}$ and we are done. Hence suppose the packet was delayed along its route. We now define a *delay sequence* for p_{i_0} on the run. Let t_0 be the last time that packet p_{i_0} was delayed. (Note that such a time exists since the run is finite; a bound of $T(\mathcal{P}) \leq k \cdot d(\mathcal{P})$ on the scheduling time of any algorithm is trivial.) Namely, there is a triple $\langle p_{i_1}, p_{i_0}, t_0 \rangle$ in C, and there is no such triple for p_{i_0} in later times $t > t_0$. (Recall that by Fact 4.4 there is only one such p_{i_1}.) Let t_1 be the last time p_{i_1} was delayed before time t_0. Namely, there is a triple $\langle p_{i_2}, p_{i_1}, t_1 \rangle$ and no such triple for p_{i_1} in any time between t_1 and t_0. Continue the sequence in this way until reaching a packet p_{i_ℓ} was not delayed prior to time $t_{\ell-1}$.

It is convenient to define also $t_{-1} = T_{i_0}$ (the arrival time of p_{i_0}) and $t_\ell = 0$ (the start time of p_ℓ).

We get a sequence $\mathcal{DS}$ of triples

$$\mathcal{DS} = \langle p_{i_1}, p_{i_0}, t_0 \rangle, \langle p_{i_2}, p_{i_1}, t_1 \rangle, \ldots, \langle p_{i_\ell}, p_{i_{\ell-1}}, t_{\ell-1} \rangle,$$

where $T_{i_0} = t_{-1} > t_0 > t_1 > \ldots > t_{\ell-1} > t_\ell = \tau_{i_\ell}^A = 0$.

Lemma 4.5 $T_{i_0} \leq d_{i_0} + \ell$.

Proof: By definition of the relation C, we have the inequalities

$(X_j) \quad d_{i_{j+1}}(t_j) \leq d_{i_j}(t_j), \qquad$ for $j = 0, 1, \ldots, \ell - 1$.

For $j = 0, \ldots, \ell$ let θ_j denote the segment of the route ρ_{i_j} traversed by p_{i_j} between the times t_j and t_{j-1}, and let

$$\Delta_j = |\theta_j| = d_{i_j}(t_{j-1}) - d_{i_j}(t_j).$$

Substitute this definition in the inequalities (X_j) to get

$(Y_j) \quad d_{i_{j+1}}(t_{j+1}) + \Delta_{j+1} \leq d_{i_j}(t_j), \qquad$ for $j = 0, \ldots, \ell - 1$.

We also have

$(Y_{-1}) \quad d_{i_0}(t_0) + \Delta_0 = d_{i_0}(t_{-1}) = d_{i_0}$.

Summing the inequalities (Y_j) for $j = -1, 0, \ldots, \ell - 1$, we get

$$\Delta_0 + \Delta_1 + \ldots + \Delta_{\ell-1} + \Delta_\ell \leq d_{i_0} \qquad (2)$$

We also construct a chain of equalities for the times involved in these collisions. Since packet p_{i_j} (for $0 \leq j \leq \ell - 1$) was delayed at time t_j but never delayed since that time until time t_{j-1}, we have

(Z_j) $t_{j-1} = t_j + 1 + \Delta_j,$ for $j = 0, \ldots, \ell - 1.$

We also have

(Z_ℓ) $t_{\ell-1} = d_{i_\ell}(t_{\ell-1}) = \Delta_\ell + t_\ell.$

Combining the equalities (Z_j) for $j = 0, \ldots, \ell$ we get

$$T_{i_0} = t_{-1} = \Delta_0 + \Delta_1 + \ldots + \Delta_{\ell-1} + \Delta_\ell + t_\ell + \ell. \tag{3}$$

Combining Eq. (2) and (3) with the assumption that $t_\ell = \tau_{i_\ell}^A = 0$, we get that

$$T_{i_0} \leq d_{i_0} + \ell,$$

and the lemma follows. ∎

In order to complete the proof of the theorem, it therefore remains to bound the length of the sequence $\mathcal{DS}$. This is done by proving the following claim.

Lemma 4.6 *The packets p_{i_j} appearing in the triples of the sequence $\mathcal{DS}$ are all distinct.*

Proof: The main idea of the proof is to show that if there is a packet that appears twice in $\mathcal{DS}$, then the path traverse by this packet is not the shortest. This fact is established by constructing an alternative path, using the route segments traversed by the packets in $\mathcal{DS}$, and showing that this alternative path is shorter.

The proof is by contradiction. Assume that some packet occurs twice in the sequence, for instance $p_{i_m} = p_{i_r}$ (or, $i_m = i_r$,) for $m > r$. By the structure of the sequence, every two consecutive packets are distinct, so necessarily $m \geq r + 2$. This means that the sequence contains a subcycle

$$\langle p_{i_{r+1}}, p_{i_r}, t_k \rangle, \langle p_{i_{r+2}}, p_{i_{r+1}}, t_{r+1} \rangle, \ldots, \langle p_{i_{m-1}}, p_{i_{m-2}}, t_{m-2} \rangle, \langle p_{i_m}, p_{i_{m-1}}, t_{m-1} \rangle = \langle p_{i_r}, p_{i_{m-1}}, t_{m-1} \rangle$$

where $t_r > t_{r+1} > \ldots > t_{m-1}$, and $m \geq r + 2$. (See Figure 1.)

We argue that among the inequalities (X_j), for $r \leq j \leq m - 1$, at least one of the inequalities is strict. Otherwise, all the collision resolutions in the cycle were made on the basis of indices, so $i_r = i_m < i_{m-1} < \ldots < i_{r+1} < i_r$; contradiction.

It follows that among the corresponding inequalities (Y_j), for $r \leq j \leq m - 2$, plus (X_{m-1}), at least one is strict.

Combine these inequalities in chain. Recalling that at least one of the inequalities is strict, we get

$$d_{i_m}(t_{m-1}) + \Delta_{m-1} + \ldots + \Delta_{r+1} < d_{i_r}(t_r). \tag{4}$$

Finally, let $\bar{\theta}$ denote the segment of the route ρ_{i_r} traversed by p_{i_r} between the times t_{m-1} (when it won) and t_r (when it lost), and let $\bar{\Delta} = |\bar{\theta}| = d_{i_r}(t_r) - d_{i_r}(t_{m-1})$. We get that

$$\Delta_{m-1} + \ldots + \Delta_{r+1} < d_{i_r}(t_r) - d_{i_r}(t_{m-1}) = \bar{\Delta},$$

or in other words, the segment $\bar{\theta}$ of ρ_{i_r} is not shortest; contradiction to the assumption that all the paths in $\mathcal{P}$ are shortest paths. ∎

Corollary 4.7 *The sequence $\mathcal{DS}$ is of length $\ell \leq k - 1$.* ∎

Combining this corollary with Lemma 4.5 completes the proof of Part (1) of the theorem. Part (2) follows immediately. ∎

5 Greedy algorithms in the general case

The purpose of this section is to demonstrate the fact that, unlike the case of leveled routes, for general route classes not every greedy algorithm delivers the messages fast. Moreover, even the specific greedy algorithm used in Section 4 is not good enough in the general case. In fact, other "reasonable" greedy algorithms have bad performance too. Some of the complications of the following proof result from the fact that we strive to make it hold also for the case of "long" ($\Theta(|V|)$) routes and "many" ($\Theta(|V|)$) packets. We start with analyzing the following *fixed priority* algorithm.

Theorem 5.1 *There exist a graph G and a collection of paths $\mathcal{P}$ whose schedule time under the fixed priority algorithm is $T(\mathcal{P}) = \Omega(d\sqrt{k})$ even for $d, k = \Omega(|V|)$.*

Proof: Define the graph $G = (V, E)$ to be the union of the subgraphs $G^0, L^1, ..., L^x, C, S$, where x is a parameter to be determined later. See Figure 2. Intuitively, G^0 is the "main" route, while the purpose of the subgraphs L^i is to generate delays.

The nodes of the subgraph G^0 are

$$V(G^0) = \{1, \cdots, x\} \times \{1, \cdots, \frac{n}{x}\} \times \{\text{in}, \text{out}\},$$

and the edges are

$$E(G^0) = \{(v_{<i,j,\text{out}>}, v_{<i,j+1,\text{in}>}) : 1 \leq i \leq \frac{n}{x}, 1 \leq j \leq x-1\} \cup$$

$$\{(v_{<i,j,\text{in}>}, v_{<i,j,\text{out}>}) : 1 \leq i \leq x, 1 \leq j \leq \frac{n}{x}\} \cup$$

$$\{(v_{<i,x,\text{out}>}, v_{<i+1,1,\text{in}>}) : 1 \leq i \leq \frac{n}{x} - 1\}.$$

The graph L is a straight line of y nodes and $y-1$ edges, i.e., $V(L) = \{l_1, \cdots, l_y\}$ and $E(L) = \{(l_i, l_{i+1}) : 1 \leq i \leq y-1\}$. We require that $y \geq x$, where the exact value of y will be determined later. For $1 \leq i \leq x$, the subgraph L^i is a copy of the graph L with superscript i.

The set C of edges connects nodes in G^0 to nodes in L^i. The first node in L^i, node l_1^i, is connected to a node $v_{<i,j,\text{in}>}$, for $1 \leq j \leq \frac{n}{x}$. The last node in L^i, node l_y^i, is connected to the "next" node on G^0, that is, to node $v_{<i,j,\text{out}>}$, for $1 \leq j \leq \frac{n}{x}$. Formally,

$$C = \{(v_{<i,j,\text{in}>}, l_1^i) : 1 \leq i \leq x, 1 \leq j \leq \frac{n}{x}\} \cup \{(l_y^i, v_{<i,j,\text{out}>}) : 1 \leq i \leq x, 1 \leq j \leq \frac{n}{x}\}$$

The component S includes the start node s, and one edge $(s, v_{<1,1,\text{in}>})$.

When the algorithm starts, all $n = k$ packets are in node s. We denote the *identifier* of a packet by a pair of numbers $[i,j]$, where $1 \leq i \leq \frac{n}{y}$ and $1 \leq j \leq y$. We define a complete order between the packet identifiers by comparing them lexicographically. When comparing the identifiers of two packets, the packet with the smaller identifier has higher priority. The packets are grouped into *batches* according to their first identifier field, setting $B(i) = \{[i,j] \mid 1 \leq j \leq y\}$. The packets of each batch $B(i)$ must all traverse the same route ρ_i.

The routes of all the packets start with the edge $(s, v_{<1,1,\text{in}>})$. The route ρ_1 (of batch $B(1)$) traverses L^1, returns to node $v_{<1,1,\text{out}>}$, from there to $v_{<1,2,\text{in}>}$ and traverses L^2, etc. Formally, define the edge sets

$$R_j = \bigcup_{1 \leq i \leq x} \left(\{(v_{<j,i,\text{in}>}, l_1^i)\} \cup L^i \cup \{(l_y^i, v_{<j,i,\text{out}>})\}\right)$$

and

$$M_j = \{(s, v_{<1,1,\text{in}>})\} \cup \bigcup_{1 \leq i \leq x-1} \{(v_{<j,i,\text{out}>}, v_{<j,i+1,\text{in}>})\}.$$

Then ρ_j is composed of the edges of $M_1 \cup \cdots \cup M_{j-1} \cup R_j$.

Consider what happens until the packets of batch $B(1)$ arrive at their destination. Note that until time y, only packets of $B(1)$ move, since the others wait for them to be done with edge $(s, v_{<1,1,\text{in}>})$.

At time $y+1$, the packet $[2,1]$ arrives at $v_{<1,1,\text{in}>}$, and at time $y+2$ at node $v_{<1,1,\text{out}>}$. However, at that time the first packet of $B(1)$, i.e., $[1,1]$, also arrives at node $v_{<1,1,\text{out}>}$ (from node l_y^1). Thus, by time $2y+2$ (on which the first batch traverses the edge $(v_{<1,1,\text{out}>}, v_{<1,2,\text{in}>})$), all the packets of $B(2)$ are still delayed at node $v_{<1,1,\text{out}>}$.

It takes packet $[2,1]$ $y+2$ time units to traverse the subpath from s to $v_{<1,1,\text{out}>}$. The same argument as above can be repeated for every L^i. Therefore, when packet $[2,1]$ arrives at $v_{<2,1,\text{in}>}$ the time is $x(y+2)$. At this time the packets of $B(1)$ arrive their destination. A similar argument as above shows that when the jth batch has traversed $M_1, \cdots, M_{j-1}$, the time is $jx(y+2)$. Thus the total time is $\Omega(\frac{n}{y}xy)$. The theorem follows by choosing $x = y = \lfloor \sqrt{n} \rfloor$. (Note that $|V| = \Theta(n) = \Theta(k)$.) ∎

A similar, although somewhat more complicated proof can be constructed for other greedy algorithms, specifically the Min Went policy of Section 4, the analogous *Max Went* policy, or the *FIFO policy*, namely, the algorithm that resolves a collision between two packets in node v by sending the first to have arrived at node v (breaking ties by packet identifier numbers).

Theorem 5.2 *There exist a graph G and a collection of paths $\mathcal{P}$ whose schedule time under the the Min Went, Max Went and FIFO policies is $T(\mathcal{P}) = \Omega(d\sqrt{k})$ even for $d, k = \Omega(|V|)$.*

Proof: The proof is obtained by simple variations on the construction in the proof of Theorem 5.1.

For the Max went policy, observe that the schedule constructed in the proof of Theorem 5.1 obeys the Max Went policy.

For the FIFO policy we modify the graph, by changing the set C to

$$C = \{(v_{<i,j,\text{in}>}, l_1^i) : 1 \le i \le x\} \cup \{(l_k^i, v_{<i,j,\text{out}>}) : 1 \le i \le x, 1 \le k \le y\}.$$

The routes of the packets also change. The kth packet in the jth batch, i.e. packet $[k, j]$, moves from l_{y-k+1}^i to $v_{<i,j,\text{out}>}$, for $1 \le i \le x$. Formally, let

$$R_{k,j} = \bigcup_{1 \le i \le x} \{(v_{<j,i,\text{in}>}, l_1^i), (l_1^i, l_2^i), \ldots, (l_{y-k}^i, l_{y-k+1}^i), (l_{y-k+1}^i, v_{<j,i,\text{out}>})\}$$

and now the route of packet $[k, j]$ is $M_1 \cup \ldots \cup M_{j-1} \cup R_{k,j}$.

Note that the whole first batch arrives at node $v_{1,1,\text{out}}$ at the same time. Hence the FIFO policy too resolves the collisions (with the second batch) in favor of the first batch. Similarly, the resolution of the conflicts in this case is the same as before.

For the Min Went Policy the construction is similar but somewhat more complicated, due to the fact that the packets "switch priorities" during the execution; whenever a packet uses one of the bypasses L^i, its priority reduces. This requires us to organize the connections between the delay loops and the segments of G^0 in a more careful way, according to the priorities upon leaving the current segment. We omit the details of this construction. ∎

Acknowledgment

It is a pleasure to thank Amotz Bar-Noy for working with us in the early stages of this research.

References

[BGGJP85] A. E. Baratz, J.P. Gray, P.E. Green, J.M. Jaffe, and D.P. Pozenski, SNA Networks of Small Systems, *IEEE Trans. on Comm.* **sac-3**, May 1985, 416–426.

[BG87] D. Bertsekas and R. Gallager, *Data Networks*, Prentice Hall, Englewood Cliffs, NJ, 1987.

[CGK88] Israel Cidon, Inder Gopal and Shay Kutten, New Models and Algorithms for Future Networks. *Proc. 7th Annual ACM Symp. on Principles of Distributed Computing*, Toronto, Canada, August 1988, 74–89.

[Ko78] Hisashi Kobayashi, *Modeling and Analysis*, Addison-Wesley, 1978.

[LMR88] T. Leighton, B. Maggs, and S. Rao, Universal Packet Routing Algorithms, *Proc. 29th IEEE Symp. on Foundations of Computer Science*, White Plains, NY, October 1988, 256–269.

[Mar82] James Martin, *SNA: IBM's Networking Solution*, Prantice Hall, Englewood Cliffs, NJ, 1982

[MRR80] J. McQuillan, I. Richer and E.C. Rosen, The New Routing Algorithm for the ARPANET, *IEEE Trans. on Commun.* **com-28**, May 1980, 711–719.

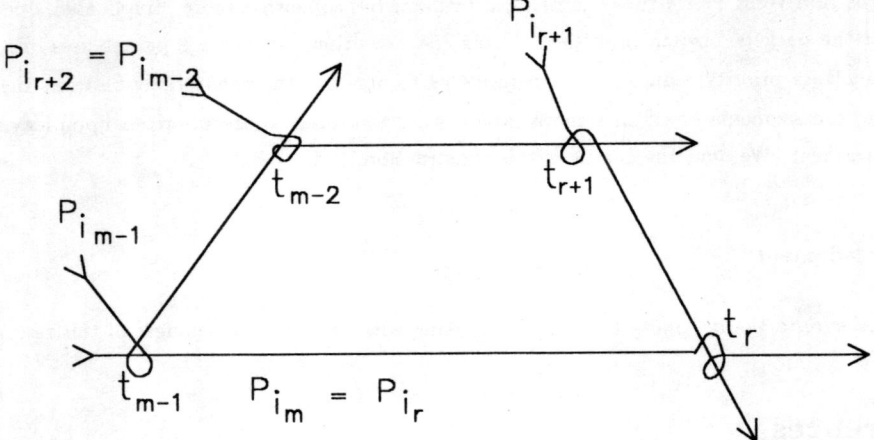

Figure 1 - Time Ordered Cycle

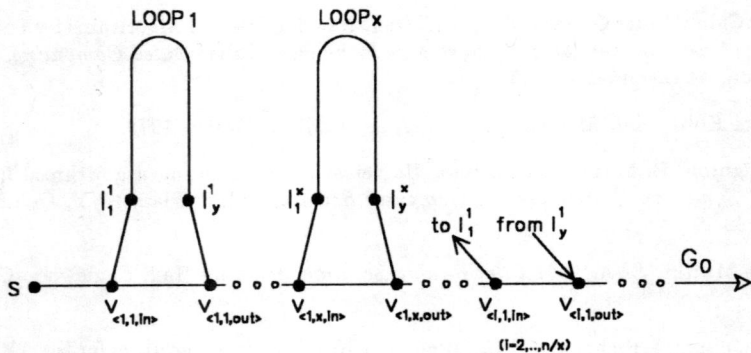

Figure 2 - Counter Example for Arbitrary Routes

Optimal Computation of Global Sensitive Functions in Fast Networks

Israel Cidon, Inder Gopal, and Shay Kutten

IBM T. J. Watson Research Center
Yorktown Heights, NY 10598

Abstract

The common practice in distributed algorithms research was to assume that the computation time is zero, and assign a cost only to the communication. This was justified at the times that the communication was the bottleneck of a network. However, lately there has been a dramatic increase in the capacity of communication links so that we can no longer make this assumption.

In this paper we find optimal algorithms for distributed computation of "global sensitive" functions in the case that the computation time is some given P, and the communication time is some given C. (Roughly speaking, a global sensitive function is one that depends on each of its inputs.)

1. Introduction

The common practice in distributed algorithms research was to assume that the computation time is zero, and assign a cost only to the communication. (See e.g. [GHS83]) This was justified at the times that the communication was the bottleneck of a network. However, lately there has been a dramatic increase in the capacity of communication links so that we can no longer make this assumption.

In [CGK88] we have investigated the complement assumption, i.e. that of zero communication cost, with a charge only for the computation. However, if the network covers long distances, then propagation delays may still be significant. In addition, the speed of the computation may increase in the future. Thus it makes sense to analyze distributed computations when the ratio between the computation time and the communication time is given as a parameter.

In this paper we make the first step, by analyzing the distributed computation of "global sensitive" functions in the case that the computing time is some given P, and the communication time is some given C. Roughly speaking, a global sensitive function is one that depends on every one of its inputs, though a more formal definition will be given later. Most "natural" functions are global sensitive. In addition to their theoretical importance they arise in many applications, e.g. call set-up [ACGKK90], directory search, parallel computing and more.

An optimal solution is presented for the above problem.

2. Model and Problem Definition

The communication network is represented by a graph (V, E) where V is the set of nodes and E is the set of bidirectional communication links. We denote $|V|$ by n, and assume, without loss of generality, that the nodes are numbered $1, 2,..., n$. In this paper we assume that the graph is complete.

Messages are assumed to arrive within finite but unbounded time. For purposes of time complexity analysis, we count computation and communication delays separately. (In the terms of [CGK88] The computation delay is the delay of the software, or *system call complexity* and the communication delay is the delay of the switching hardware including the links' propagation delays.) In particular, we assume that the communication delays are upper bounded by C time units and the computation delays are upper bounded by P time units. (Note that these upper bounds are introduced for the purpose of the time complexity analysis and the algorithm operates correctly even if delays are unbounded.) Thus we define

- Time Complexity - The time complexity is then the maximal time which may elapse from the algorithm starting point to its end under the assumption that the hardware and software delays are upper bounded as described above.

We assume that each node i maintains some *independent* input value I_i which is drawn from some finite alphabet. A distributed algorithm is triggered at time 0 at all nodes. The algorithm terminates eventually at node 1. Upon termination, node 1 should have computed the correct value of some function $f(I_1, I_2,..., I_n)$. f is defined for all vectors of length equal or smaller than n.

The function f is defined to be:
1. Associative - $f(X, Y, Z) = f(X,f(Y,Z)) = f(f(X,Y),Z)$
2. Commutative - $f(X,Y) = f(Y,X)$

In addition f is **global sensitive in the weak sense**. This means that there exit at least one input vector $\bar{I} = (I_1, I_2,..., I_n)$ (which is called **a global sensitive input vector**) such that for all j, $(1 \le j \le n)$, if we define $\bar{I}_j$ to be equal to $\bar{I}$ except for the single input value I_j, then there exists some selection of I_j such that $f(\bar{I}) \ne f(\bar{I}_j)$

A global sensitive function **in the strong sense** was defined [ALSY90]. There, the function must be different for **any** two input vectors that differ in one entry value. Our definition of global sensitive in the weak sense captures a larger class of functions including the basic Boolean functions (OR, AND) and cases in which some input values are not distinguishable.

3. Properties of Optimal Protocols

During the execution of the algorithm messages are sent. We distinguish between messages that can affect the final result and messages that cannot. A *Causal Message* is defined in the following recursive way. It is either received by node 1 before it terminates or it is received by some other node before that node sends a causal message. Any other message is defined to be a non-causal message. It is clear from the definition that a causal message sent over a link cannot be preceded by a non-causal message as we assume FIFO reception. If a message is received at a node before a causal message has been received then, this previous message must be causal as well. Lemma 1 shows that we can essentially ignore or suppress any non-causal message. We assume the operation of an arbitrary correct algorithm that computes the function f.

Lemma 1: Consider any arbitrary execution of the algorithm then, we can delay any non-causal message at a node or a link for any arbitrary amount of time without changing the algorithm's results or its termination time.

Proof: The proof follows immediately from the fact that non-causal messages cannot be received before a causal one at any node. Since causal messages cannot be sent after the reception of a non-causal message and since node 1 makes its final decision only due to reception of causal messages the lemma holds.

Lemma 2: Assume that a function f is computed. Then, there exist at least one execution of the (correct) distributed algorithm in which **every** node sent at least one causal message whose information depends on its local input value.

Proof: Let $\bar{I}$ be a global sensitive input vector and assume an execution under this input vector. Define the termination time as T. By using lemma 1 we construct a new feasible execution of the algorithm by delaying all non-causal messages that have been sent before T, to be received after time T. Since we are dealing with an asynchronous network, this execution is still a feasible one. We now focus on the new execution.

Let A be the set of nodes that have sent a causal message (we include node 1 in Λ as well), and let $B = A^c$. If B is empty then the proof is completed. If B is not empty than no message sent from B is received at any node in A before T. This also means that the input values of nodes in B cannot affect the computation results before T. Now assume the operation of the algorithm with a different input value for a single node $j \in B$. Since input values of nodes in Λ are exactly as before, and no message from nodes in B are received until time T, repeating the exact execution of the algorithm at these node is a feasible execution. This means that no matter how the input value I_j is chosen, node 1 terminates with the same result. This contradicts the correctness of the algorithm.

Following the results of lemma 2, for every correct algorithm (in particular optimal ones) there are executions in which each node must send at least one causal message. From now on we will focus on these particular executions.

Let us further assume that message delays do not depend on their contents. Then the following Lemma 3 can be applied.

Definition: A *tree based algorithm* is an algorithm where the function is computed over a predefined rooted tree (the same tree for all possible input vectors). The computation is done in the following: At initialization, all tree leaves send their input values to their parents. Each node (which is not a leaf) waits until it has received a message from all its children and then computes the partial function of its subtree and forward the result to its parent.

Since we assumed that message delays do not depend on their contents, a particular tree based algorithm will have the same worst case time-cost for all functions f and all possible input vectors. We are looking for optimal algorithms in the worst case (over all executions and possible input vectors). Therefore, in order to prove the optimality of a tree based algorithm, it is sufficient to show that for any arbitrary correct algorithm that compute an arbitrary function f there exists an input vector and a tree based protocol which is at least efficient as the original algorithm (only for this particular input vector).

Lemma 3: For any function f and any correct algorithm that computes f there exist a tree-based algorithm with worst case time and message delays better or equal to that algorithm.

Proof: Consider some execution of the algorithm for a global sensitive input vector. Consider the last causal message that is sent by each node. Since this is the last message then, the causal path of these messages define a tree rooted at node 1. (From the definition of causal messages this paths must define a connected structure. Since we are dealing with the last causal message, this structure cannot have any cycles.) Similarly, each node must wait for all its children last causal messages before it sends its message. Otherwise, some of these messages will be non-causal. This implies that this last causal messages process is equivalent to the operation of a tree based algorithm.

Now let us execute the tree-based algorithm over the above defined tree (for all input vectors). This algorithm is equivalent in its structure to the last causal messages sent by the original algorithm for a specific input vector. Therefore, the tree based algorithm just introduced, has worst case message and time cost which is bounded by a particular execution of the original algorithm (not necessarily the worst one). This implies that its time and message delays are bounded by the worst case delay of the original one.

Corollary 1: There exists a single tree-based algorithm which is worst-case optimal for all functions f.

4. Optimal Tree-based Algorithm

We assume that there are two worst-case delays associated with each message. We define C to be the worst case communication delay and P the worst case system call reception delay. The communication cost, reflects the time elapsed from starting the message transmission until it is received at the other hand. (This can be further split into transmission and propagation delays.) The system call delay is the time it takes to process the received message and accomplish the computation steps that are needed. We also assume that the initialization of the algorithm at some node (if it is not via a message reception) takes one system call - P.

We define an optimal (t,P,C) tree, as the rooted tree with the maximum number of nodes over which a tree-based algorithm can compute any function f and terminates no later than t. Fixing P and C, we denote this optimal tree by $OT(t)$ and the number of nodes in $OT(t)$ by $S(t)$. It is clear that the size of the optimal tree is a non-decreasing function of t. C and P were defined as worst case times. It is easy to prove that increasing any message delay in the tree-based algorithm never decrease the overall time to complete an execution. Therefore, we can assume that in the worst case situation messages encounter exactly the maximum allowable delays (P and C).

Let us consider an optimal tree OT(t). The root of this tree is receiving causal messages from its children. It is able to process all of them until time t. Let us observe the last message it processes. This message must be received no later than (t-P) otherwise it cannot terminate at t. This in turn implies that the message must have been sent before (t-P-C). In addition, assuming that messages are being processed in FIFO order, the root must have received and processed all other messages by that time (t-P). It is also clear that:

$$S(t) = \begin{cases} 0 & t < P \\ 1 & t < 2P + C \end{cases} \qquad (1)$$

Let X and Y be rooted trees. We define the operation $\overleftarrow{\cup}$ as an operation that get as input two rooted trees and output a new tree in the following way. The rooted tree Z which is the result of $Z = X \overleftarrow{\cup} Y$ is constructed by augmenting the root of Y (we add a directed edge) as a child of the root X.

Since $S(t)$ is a non decreasing function of t, this lead us to the following recursive structure of the tree ($t \geq 2P + C$).

$$OT(t) = OT(t - P) \overleftarrow{\cup} OT(t - C - P) \tag{2}$$

This also translates to the numerical equation:

$$S(t) = S(t - P) + S(t - C - P) \tag{3}$$

Both the tree structure and size can be solved using the recursions defined by (2) and (3) respectively and by the initial conditions (1). One has to compute $OT(t - i(C + P) - jP)$ for all i,j such that $t - i(C + P) - jP \geq 2P + C$. The simplest way to do it is by computing the values of t, to be considered, ordering them from low to high and computing the structures and the values according to this order. Basically we can restrict the computation to discrete values of time that are in the form $iP + jC$. Other times t can be truncated to the highest value that still satisfied $t \geq iP + jC$.

5. Examples

1) New model - $C = 0$, $P = 1$. We can assume only integer values of time. The equations take the following simple structure:

$$S(t) = \begin{cases} 0 & t < 1 \\ 1 & t < 2 \end{cases} \tag{4}$$

$$OT(k) = OT(k - 1) \overleftarrow{\cup} OT(k - 1) \tag{5}$$

This structure is a binomial tree [SCH81]. Solving for the tree size results in

$$S(k) = 2S(k - 1) = 2^{k-1} \tag{6}$$

2) Old model - $C = 1$, $P = 0$. It is easy to see that the recursion is blowing up. This is because using a star configuration we can add any number of nodes to the structure in order to get any tree size for $t = 1$.

3) $C = 1$, $P = 1$

$$S(k) = \begin{cases} 0 & k < 1 \\ 1 & k < 3 \end{cases} \tag{7}$$

$$OT(k) = OT(k - 1) \overleftarrow{\cup} OT(k - 2) \tag{8}$$

This structure is growing slower than a binomial tree. Solving for the tree size results in

$$S(k) = S(k - 1) + S(k - 2) \tag{9}$$

Via Z transform it is easy to derive the equation:

$$S(Z) = \sum_{k=1}^{\infty} S(k)Z^k = \frac{-Z}{Z^2 + Z - 1} \tag{10}$$

Taking the inverse transform results in:

$$S(k) = \frac{\left((1+\sqrt{5})^k - (1-\sqrt{5})^k\right)}{\sqrt{5}\,(2)^k} \tag{11}$$

Note that these are the well known Fibonacci numbers.

Computing the optimal time for a given size

Since the recursion is applied in the order of increasing value of time, one should just computes it for the discrete values of time until $S(t) \geq n$. As was mentioned before we should use only values of time that are in the form of $iP + jC$ where i and j are integers. Since in the tree-based algorithm there are only $n-1$ messages sent clearly: $i \leq n, j \leq n$. Therefore, there are at most n^2 possible number of points that should be computed.

6. Conclusion

Our model is a generalization of both the model in [CGK88] and the model of telephone networks of [SCH81].

This paper also clarifies a certain point regarding the models for fast networks [CGK88]. In that paper we assumed that nodes do not know in advance the path to some remote nodes. However, if node v does know (or learned) the route to some other node u (and no topological change has disconnected that route) then v could have sent a message that reached node u costing only one system call (i.e. P). This might lead to the (wrong) conclusion that once each node knows how to directly reach all other nodes, the difference between the old model and the new class of models diminishes. The current paper demonstrates that even in complete networks, charging for the computation does make the model different than the traditional model.

An interesting open problem is to generalized the problem investigated here for other graphs.

References

[ACGKK90], A.Awerbuch, I. Cidon, I.S. Gopal, M. Kaplan and S. Kutten, "Distributed Control for PARIS," to appear in *Proceedings of the 9th ACM Symposium on Principles of Distributed Computing*, Quebec City, Canada, August 1990.

[ALSY90] Y. Afek, G.M. Landau, B. Schieber and M. Yung, "The Power of Multimedia: Combining Point-to-Point and Multiaccess Networks", *Information and Computation* vol. 84 no. 1 january 1990, pp. 97-118.

[CGK88] I.Cidon, I.S.Gopal, S.Kutten, "New Models and Algorithms for Future Networks," *Proceedings of the 7th Annual ACM Symposium on Principles of Distributed Computing* Toronto, Ontario, Canada, August 1988, pp. 75-89.

[SCH81] Slater, P.J., E.J. Cockayne, and S.T. Hedetniemi, "Information dissimination in Trees," *SIAM J. on Computing* 10 (1981), pp. 692-701.

[GHS83] Gallager, R.G., P.M. Humblet and P.M. Spira, "A Distributed Algorithm for Minimum- Weight Spanning Trees", ACM Transactions on Programming Languages and Systems, January 1983, Vol. 5, No. 1, pp. 67-77.

Efficient Mechanism for Fairness and Deadlock-Avoidance in High-Speed Networks

Yoram Ofek   Moti Yung

IBM T. J. Watson Research Center
P.O.Box 704, Yorktown Heights, NY 10598

Extended Summary

abstract

High-speed network is a new environment, motivated largely by the advance of fiber-optics technology. Such a network cannot operate in conditions which may cause deadlock or may starve certain network nodes since these phenomena result in long waiting times and invocations of delaying recovery procedures which imply (by definition) very low speed.

On the other hand, in order to accelerate the network algorithms, such networks are likely to have the following characteristics: 1. network traffic has higher priority than input traffic (to guarantee fast uniform routing services at intermediate nodes independent of local demand), and 2. packets are removed at destination unlike traditional broadcast approach in local-area network (to avoid unnecessary communication resource wasting). These two characteristics imply the possibility of overloading a node with network traffic which implies an inherent lack of fairness that cannot be solved by a global mutual exclusion mechanism such as the traditional token (as in a token ring).

In this work we present a general control mechanism for fair access to communication links of a network with an arbitrary topology. In this abstract motivated by high-speed operation, we assume that the network nodes operate in real time, i.e. known links delay. (The work can be translate to other asynchronous models.) The proposed control mechanism operates over a spanning tree that was constructed for this purpose in the network. The fairness mechanism implies deadlock-freeness. Furthermore, it guarantees a stronger condition which is equifairness (equal opportunity is given to every node at each control cycle).

The mechanism regulates the network access globally and can work much faster than previous ones. It is optimal as its time is proportional to the network diameter (previous mechanisms were proportional to the network's size). This is crucial since a shorter control cycle unloads hot spots in a faster rate. Furthermore, the mechanism is self-balancing as it adapts itself to the relative speed of the various parts of the network and it also automatically tolerates one failure per cycle.

As a result, the mechanism can be part of a high-speed network architecture in which, on one hand a node can try to transmit asynchronously, without reservation, as much as it can only by observing the state of its adjacent links, and on the other hand the network access and flow control will ensure no loss, fair access to the network and no deadlocks.

In addition, a **clock-driven self-stability** property of the algorithm will ensure that local state information and a reliable local clock will be sufficient to recover from any faulty state within a single control cycle.

1 Introduction

High speed asynchronous network is a new environment motivated by the increased speed and bandwidth of the communication media. In such an environment fast service is required both at the intermediate node routing as well as at the network access process. In a recent work such an architecture was characterized and the operation principles required or highly desired in such an architecture were suggested [OY90a].

One principle of operation presented is the fact that deadlock or starvation is anathema to the fast service requirement of a high-speed network implying that deadlock avoidance is an important operation principle. (By deadlock we mean either of input accessing the network, or of internal traffic; we concentrate from now and on, on the input network access process). A stronger requirement than deadlock-freeness is fairness, which guarantees the eventual performance of any network input access— it provides progress of the operation and does not deny service from any node (namely, it is starvation-free). We claim that in the new environment a stronger requirement is needed. This is **equifairness** which means that in every control cycle along an infinite (non-faulty) execution, each node gets an equal opportunity to transmit (its predetermined quota) (see [Fr] for a much more formal treatment of the notion).

Next we elaborate on **the cause of unfairness** in the network access process. First, in high-speed networks nodes make decisions based on local information only. Then unfairness may be caused because demand for both network input access from the node and internal network traffic flow are competing for the same resources (buffers, links). The resulting unfair case is typically an outcome of a combination of the following two primary reasons:

1. Network traffic priority. A priority is given to the traffic already in the network over external traffic from the node into the network. (This is necessary in order to minimize the packet loss at intermediate nodes in the network while providing the service in a timely uniform fashion.)

2. Destination packet removal. Packets are removed from the network by their destinations (rather then being broadcast to the entire network). Furthermore, since nodes make decisions based on local information, some nodes may see more idle links than others and be able to generate more traffic into the network. Since network traffic has priority, this may cause other nodes to be deprived of access to the network.

We remark that the above two reasons seem to be required. First, destination packet removal is desirable for spatial reuse (concurrency) which gives the network much higher throughput. Second, local (distributed) decision making is necessary, since operation based on global information requires the network to freeze for a while (in order to maintain consistency), but this prevents high-speed changes which usually occur in our environment.

Next, we present **our result** and contrast it with previous related designs. In previous work [CO89a, CO89b] it was shown how to ensure fair use of communication links on a ring network with spatial reuse. The fairness is quite strong (equifairness) as it guarantees that in every control cycle each node has a chance to satisfy its local conditions. The control is done by a signal which is passed along the ring whose task is to regulate the traffic according to fairness conditions. The work has been further adapted to a general network by embedding a ring structure in a general connected topology and using the virtual embedded ring to manage the fairness mechanism [OY90a].

The drawback of the previous solution is the fact that the control cycle was linear in the network size (as each edge of the virtual ring is traversed in a cycle by a unique signal which regulates the fairness in the network). The regulating cycle has to visit the network globally (to ensure equifairness), but we would like to have a short turn around time between each visit. The length of the control cycle is measured as the worst case time between two consecutive receptions of the control signal, assuming each node passes the signal in one unit of time. (This is, of course, an assumption made for the analysis only, in reality the delay of the signal is a function of the actual load).

In this work, we present an equifairness mechanism for a spanning tree that can be part of an arbitrary topology and the control cycle can be proportional to the graph's diameter. Most existing networks has a diameter substantially shorter than their size, and time proportional to the diameter is optimal when we want the regulating activity to be global and to take care of each and every node in the network.

The new mechanism has further advantages. It has a built-in fault-tolerance and is self-balancing in the sense that slow parts of the network have less influence on the global rate of the control cycle. The extreme case of self-balancing is when one failure occurs and one part of the network is not active, the mechanism automatically wakes this part up. This part which failed has infinite delay, so the control cycle converges to it (in its attempt to equalize the delay), this will automatically prevent problems in the "one failure" model which is the most realistic one. Another property of the mechanism is self-stabilization based on the local clocks and knowledge of the network delays, i.e. a real-time or clock-driven self-stabilized property.

In more details, the fairness algorithm presented here is designed to achieve the following:

1. **Equifair access and Deadlock-freeness.** The network interface that controls the media access has a built-in fairness mechanism. All nodes will have equal opportunity to transmit packets into the network over their adjacent links. The flow control mechanism in the node interface (which makes sure messages in transit will proceed) together with the fairness property ensure that deadlocks will not occur.

2. **Self-balancing and Fault-tolerance.** The algorithm which manages the control signal propagates in the network. It consists of a basic cycle which first propagates towards the tree leaves, and then converges according to the relative speed to the weighted middle point on the tree (weights are the speeds along links). One-failure per cycle is tolerated automatically as an extreme case of the above self-balancing property.

3. **Clock-driven Self-Stabilization.** In addition, the algorithm uses local information and time-out mechanism to stabilize into a legal starting state. (We present a stabilization based on time-out in a clocked controlled (real-time) model.)

Many works have dealt with deadlock prevention and avoidance in computer networks [AIP90, MM79, MS80, TU81, BT84, G81, Obe82, CM82, Gop84, CJS87, JS89, AKP90]. As mentioned above, we believe that in the context of high-speed environment the strictly stronger condition of equifairness is necessary. Fair access to resources was traditionally assured by mutual exclusion as in a token-ring network.

The rest of the paper we present in Section 2 the background on the system basic mechanisms, then we present the algorithm in Section 3, its correctness and properties are given in Section 4, while Section 5 discusses extensions and conclusions.

2 The System Basic Model and Mechanisms

In this section we describe the basic structure and mechanisms of the system, that will help in understanding the model of the fairness algorithm. Since the main context of this work is the *MetaNet* architecture [OY90a, OY90b], we describe the system in this specific context. It is done for clarity of the presentation and to motivate this work, never-the-less the algorithm itself can be implemented in many other environments.

The topology of the network is arbitrary and all links are bidirectional. Each node has one or more full-duplex ports which are connected to full-duplex links, as shown in Figure 1. On the receiving side of each link there is a link buffer (LB), which can store one maximum size packet. The switch is part of the network interface and it can transfer packets from the LBs to the host or the out-going links, and to transfer packets from the host to the out-going links.

2.1 The Node Model

The **host interface**, as shown in Figure 1, is the connection of the network-interface with the outside world (which is the end-node). It contains the output buffers to the network and the input buffers from the network. As part of the control algorithm the following is done. When the node has a packet to transmit it puts it in the output buffer and it decides over what output link the packet should be sent. If the link is available the packet can be sent to the link via the switch (provided that the fairness condition, that will be described later, has not been violated). When a packet destined to the node arrives it will be sent to the input buffer via the switch.

The **switch** has input and output links and has the following realizable properties:

- Only one input link can forward a packet to an output link at any given time. This means that if there is another packet that should be sent over a busy link this packet will be stored in the link buffer (LB).

- A packet that should be sent via an output link which is not busy will start to be transmitted before the whole packet has arrived to the input link, i.e., the packets "cut-through" the switch from input to output.

- The switch can forward multiple packets to the host interface at the same time, and the host interface sends several packets at the same time via the switch to different output links.

- The throughput of the switch is larger than the sum of the bandwidths of all the processor-input links, and the switch can serve all the input links at the same time.

- The switch provides two additional important properties that are discussed in more details in [OY90a]. One is self-routing, the destination information in the packet header is sufficient for reaching the destination (not necessarily on the same route each time), and the other is lossless, packets cannot be lost in the network because of asynchronous traffic congestion.

2.2 The Link and its Control Signal Model

A **full-duplex link** is viewed as two directed links. For a link to be operational, it is necessary that both directions are operational, since the fairness algorithm requires both directions. Thus, in case one direction is faulty, the other direction is declared faulty as well.

The traffic into the network is regulated and controlled by a mechanism that uses **hardware massages or control signals**. These signals use the same physical medium as the data but there transfer via the link is transparent to the data. The exchange of these signals together with the fairness algorithm is performed by the fairness controller, as shown in Figure 1.

The fairness controller has three parts (as shown in Figure 1): (i) queues for storing the different types of control messages, (ii) timer that is the time-reference of the fairness algorithm, and (iii) Fairness-FSM (Finite State Machine) that executes the fairness algorithm and uses the control signal queues and timer as parameters.

The control signals can be realized by redundant or unused serial codewords. As a result, each control signal is short and can be sent in the middle of a data packet in a way that does not damage the data packets which they preempt (i.e., they have non-destructive preemptive resume priority), as illustrated in Figure 2. We will define three specific control signals in the next section.

For the efficient operation of the fairness algorithm no error recovery protocol is performed on the control signals. Only error detecting is performed on the signal, and as a result, it is only possible to determine reliably that a control signal has arrived, but not that it has been sent. The network interface will not be able to determine whether or not a faulty control signal has been sent. Therefore, a lost signal will not be directly recovered and the fairness algorithm will have to take a special care of this event. As it will be shown, the algorithm can tolerate a single signal loss in every fairness cycle. If more than one signal has been lost, any node in the network can start to send the fairness control signals.

The link delay is assumed to be known to the fairness controller. This fact is acceptable in high-speed environment, furthermore it can be simulated by a hand-shake protocol. The maximum control signal transfer delay, T_{LINK}, is the delay to sent a signal from a fairness-FSM on one side of a link to a control queue on the other side, as shown in Figure 2. This delay comprises of the link propagation delay and some small internal delays (no queuing delays). The knowledge of T_{LINK} is used for stabilizing the fairness algorithm in a very efficient manner.

3 The Fairness Algorithm

In the description of the following algorithm we assume that a full-duplex spanning tree has been constructed. The following fairness algorithm does not have to be aware of the actual structure of the tree, but it must be a tree (without cycles). As a result, when the algorithm runs the tree can be modified by adding or deleting nodes and edges without stopping the algorithm, that is a dynamic tree maintenance algorithm can be used [ACK90].

The fairness algorithm operates directly on the tree, but the result is equivalent to the *MetaRing* fairness algorithm [CO89b] in the sense that in every round or cycle of the algorithm each node can transmit a **predefined quota** on its out-going links. The fairness in the *MetaRing* is based on a control signal, SAT, that circulates in the ring, as shown in Figure 3. In every round of the SAT each node can transmit a predefined quota. (In the *MetaRing* there is a one SAT signal in each direction, which regulates the traffic in the other direction, see Figure 3.) In the following algorithm the fairness is achieved by **broadcasting and merging** of multiple control signals on a spanning tree.

The simplest way to implement an algorithm with a broadcast and merge phases is to elect a root node as a tree leader [GHS83], and to start a PIF (broadcast and echo) cycle on the tree. It is done by sending control signals to all the leaves, these signals are then merged back to the root node. These signals do not traverse the links of the tree unconditionally, they are forwarded only when a predefined condition is satisfied. When the root receives the signal from all its children it will start a new cycle.

What are the drawbacks of the above naive approach? A mechanism which depends on a static root node is more likely to fail and requires more complex recovery mechanism than a completely distributed one. In addition the convergence of any cycle depends on the slowest node and is not adjustable. The algorithm may also get into a deadlock when the tree is disconnected, which we want to prevent and continue operation on the connected components. Therefore, our goal is to present a more sophisticated algorithm that can, for example, tolerate automatically any single failure, and has a dynamic merging point (self-balancing). The following is an informal discussion of that algorithm.

3.1 Informal Description

In the following algorithm the node can be in one of two states: PARENT (P) or CHILD (C). The algorithm has two basic phases that are performed in a continuous cycles over the given spanning tree:

Broadcast phase. The broadcast phase starts from a root edge, as shown in Figure 4, when a control signal, called SAT-BRD, is sent down the tree to all the nodes until it reaches the leaves; (i.e., a node after receiving a SAT-BRD from its parent node will forward it to all its children.)

Merge phase. A leaf node after receiving a SAT-BRD will return to its parent a control signal, called SAT-MRG. These signals are forwarded from all the leaves of the tree until they reach the root edge, as shown in Figures 5 and 6. (A node will forward a SAT-MRG to its parent on the tree after receiving SAT-MRG from all its children.) After the nodes on the two sides of the root edge receive the SAT-MRG on all its links a new broadcast phase starts. Then the node will send SAT-BRD to all its children and will send back a signal call MRG-ACK (merge acknowledge) to the other node on the root edge. The MRG-ACK is redundant and will be used to start a broadcast phase only if a SAT-MRG or SAT-BRD were lost.

The continuous broadcast and merge operations can be viewed as a continuous PIF, which terminates on two adjacent nodes that can be different from cycle to cycle, i.e., the termination of the merge phase is dynamic.

The control signals, SAT-BRD and SAT-MRG, are not transferred unconditionally on the tree. These signals has a fairness and flow control meaning and they indicate that the node that has sent this signal is SATisfied. Satisfied means that the node has had its fair share use of links or other network resources. Usually, it indicates that the node had its fair share access on the link the control signal is received on; i.e., for SAT-MRG the node is satisfied on all the links to its children, and for SAT-BRD the node had a satisfied access on the current link to its parent.

As a result of the fairness requirements the broadcast and merge operations are asynchronous but have bounded delays, since the maximum amount of data each node can transmit before becoming satisfied is finite. This time bound is used for detecting missing broadcast and/or merge signals after TIME-OUTs are expired. The recovery procedure from missing control signals is performed **locally**, by **initiating a broadcast phase**, and the algorithm is guaranteed to stabilize within one cycle.

3.2 Notations, Conditions, States and Time-bounds for Node i

We use the following notations:

- d^i. The tree degree of node i.

- $\vec{L}^i$. This is the vector of all the tree links of node i, $\{L_1^i, L_1^i, \ldots\ldots, L_d^i\}$.

- $\vec{L}_{MRG}^i$. This is a local variable that lists the links from which node i already received the SAT-MRG signals.

- $Q_{L_j^i}^{min}$. This is the minimum quota node i should send over link L_j^i in order to be SATisfied.

- $Q_{L_j^i}^{max}$. This is the maximum quota node i can send over link L_j^i before receiving a new quota ($Q_{L_j^i}^{max} \geq Q_{L_j^i}^{min}$).

- $q_{L_j^i}$. This is the actual number of packets node i already sent over link L_j^i.

Note that the minimum and maximum quotas are not necessarily the same, and can be changed dynamically in time or as a result of some set-up procedure.

Operation Conditions:

We define two operating conditions: one for determining when the node is SATisfied on a link, and the other for determining when the node can transmit on a link, providing that the link is not busy.

- **Satisfied on link L_j^i.** Node i is satisfied on link L_j^i if between successive control signals received on this link it has sent at least $Q_{L_j^i}^{min}$ on it ($q_{L_j^i} \geq Q_{L_j^i}^{min}$) or if it has nothing to send on L_j^i.

- **Transmission on link L_j^i.** Node i can send a packet on link L_j^i, if $q_{L_j^i} \leq Q_{L_j^i}^{max}$ and the link is idle.

States of node i:

The node can be in one of two states:

1. **CHILD.** The node enters this state after forwarding SAT-MRG to its current parent on the tree.

2. **PARENT.** The node enters this state after forwarding SAT-BRD to all its current children on the tree.

States of a full-duplex link:

A full-duplex link is considered to be operational when both directions are operational. It is assumed that a data link control (DLC) protocol informs the algorithm on the link state which can be either UP (operational) or DOWN (not operational).

Time bounds:

The algorithm operation has two time bounds, a local time-bound and a global time-out. (As mentioned above, the model is that the node is clock-driven, which replaces global information gathering in a traditional network.)

- **Global time-out** is the maximum time a node can be at the same state. The time that node i can be in the parent state is bounded by T^i_{PARENT} and in the child state it is bounded by T^i_{CHILD}. Therefore, each time node i changes its state the time-out timer is being reset ($TIMER(t)^i = 0$). When a failure occur (e.g., loss of two or more control signals) the timer will expire and the node will restart the algorithm. It is assumed that when a time-out occurs, all the nodes are satisfied. After time-out the node will change its state to PARENT and will send SAT-BRD to all its neighbors. Notice that the bound can be computed from local bounds on the tree.

- **Local time-bound** is the maximum propagation delay to transfer a control signal from one node to its neighbor. This time-bound, T_{LINK}, will be used in order to ensure that after a failure a broadcast and merge phases will not cross one another. As was mentioned before, this will ensure the return to a stable steady-state within one round after the failure has occurred.

3.3 Formal Description

The algorithm has two parts, one for node i in the PARENT state and the other for node i in the CHILD state. These two parts use six procedures and the general structure of the algorithm is shown in Figure 7. The names near the arrows of Figure 7 correspond to statements in the algorithm.

The network interface of each node has two message queues, one for SAT-BRD messages and the other for SAT-MRG and MRG-ACK messages. These messages are stored together with the name of the link they were received on.

We first describe the algorithm performed by the node in a state of **PARENT** when it is part of a broadcast tree and is waiting in normal operation state to merge signals from all its subtrees. The idea is that when all but one subtrees are merged into the node it decides to change its status and merge to the remaining subtree.

A. PARENT Loop for Node i:
{ While in **PARENT** state do the following loop: }

- **P1.** If SAT-MRG queue is not empty, receive SAT-MRG on link L_j^i, then add L_j^i to $\vec{L}_{MRG}^i$.

- **P2.** If the cardinality of $\vec{L}_{MRG}^i$ is $d^i - 1$, then call **Legal Merge Procedure**. {merge from all but one neighbors has been received}:

- **P3.** If SAT-BRD queue is not empty, receive SAT-BRD on link L_j^i, then call **Broadcast Recovery Procedure** with a parameter L_j^i. else, if time-out has occurred ($TIMER(t)^i \geq T_{PARENT}^i$), then call **Broadcast Recovery Procedure**.

- **P4.** If SAT-MRG queue is not empty, receive MRG-ACK on link L_j^i, then discard this message.

- **P5.** If link L_j^i goes DOWN, call **PARENT Omission Procedure**.

- **P6.** repeat loop.

The following is the algorithm performed at the node in a **CHILD** state. The node monitors its parent link (the link on which the SAT-MRG was sent). It waits for either a SAT-BRD signal or SAT-MRG or MRG-ACK in order to change its state to PARENT.

B. CHILD Loop for Node i:
{ While in **CHILD** state do the following loop: }

- **C1.** If receive SAT-MRG from parent link L_k^i, send MRG-ACK on link L_k^i and call **Legal Broadcast Procedure**. {merge acknowledge is sent on the ROOT EDGE, merge phase ends and broadcast phase starts }

- **C2.** If received SAT-BRD from parent link L_k^i, and $TIMER(t)^i > T_{LINK}$ then call **Legal Broadcast Procedure**, else discard signal. {SAT-BRD signal is discarded to avoid phase crossing}

- **C3.** If received MRG-ACK from parent link L_k^i, then call **Legal Broadcast Procedure**. {The signal MRG-ACK means that merge phase terminates, then start broadcast phase}

- **C4.** If SAT-BRD queue is not empty, receive SAT-BRD on non-parent link L_j^i, or if time-out has occurred $(TIMER(t)^i \geq T_{CHILD}^i)$, then call **Broadcast Recovery Procedure**. {Illegal event has occurred}

- **C5.** If SAT-MRG queue is not empty, receive SAT-MRG on non-parent link L_j^i, then discard this message. {Illegal signal}

- **C6.** If link L_j^i goes DOWN, call **CHILD Omission Procedure**.

- **C7.** If link L_j^i goes UP, call **CHILD Join Procedure**.

- **C8.** repeat loop.

The following merge procedure is used for changing state from PARENT to CHILD, and to ensure that the node is satisfied on its children's links.

The following procedure is used for ensuring that the node is satisfied on all the links to its children, and then changes its state to CHILD and sends SAT-MRG signal to its current parent.

Legal Merge Procedure.

- M1. If SATisfied on all links in $\vec{L}^i_{MRG}$, then SAT-MRG is sent to $L^i_k = \vec{L}^i - \vec{L}^i_{MRG}$, else wait until SATisfied and then send SAT-MRG to L^i_k.
  { L^i_k becomes the link to the current parent node }

- M2. $q_{L^i_j} = 0$ for all L^i_j in $\vec{L}^i_{MRG}$.
  { the transmission quota on all the merged links is renewed }

- M3. $\vec{L}^i_{MRG} = \emptyset$.

- M4. $TIMER(t)^i = 0$. { reinitialize the timer }

- M5. **Phase crossing elimination** - discard all SAT-BRD messages from link L^i_k currently in the SAT-BRD queue. { The merge phase has priority over the broadcast phase, and therefore, these SAT-BRD messages are discarded }

- M6. Change state to CHILD, goto **C1**.

The following broadcast procedure is used for changing state from CHILD to PARENT, and to ensure that the node is satisfied on its parent's link.

Legal Broadcast Procedure.

- B1. If SATisfied on L^i_k, then send SAT-BRD to all $L^i_j \in \vec{L}^i - L^i_k$,
  s.t. a SAT-MRG has not received on L^i_j into the SAT-MRG queue, { in order to ensure phase crossing elimination }
  else
  wait until SATisfied and then send SAT-BRD to all $L^i_j \in \vec{L}^i - L^i_k$,
  s.t. a SAT-MRG has not received on L^i_j into the SAT-MRG queue, { in order to ensure phase crossing elimination }

- B2. $TIMER(t)^i = 0$. { reinitialize the timer }

- B3. $q_{L^i_k} = 0$ { the transmission quota on link L^i_k is renewed }

- B4. Change state to PARENT, goto **P1**.

In the broadcast recovery procedure, which occurs after a time-out, the satisfied conditions are not checked. Thus, the recovery broadcast phase covers the tree within the propagation delay across the diameter of the network. This procedure is the basis for the clock-driven self-stabilization property. It ensures recovery from any initial and transient faults provided that the local clocks are functional. Together with the phase crossing elimination the procedure stabilizes this algorithm in one control cycle, which is important in a high speed environment.

Broadcast Recovery Procedure.

- R1. If time-out has occurred, send SAT-BRD to all $L_k^i \in \vec{L^i}$,
  s.t. a SAT-MRG has not received on L_k^i into the SAT-MRG queue, { in order to ensure phase crossing elimination }
  else { SAT-BRD was received on L_j^i }
  send SAT-BRD to $L_k^i \in \vec{L^i} - L_j^i$,
  s.t. a SAT-MRG has not received on L_k^i into the SAT-MRG queue, { in order to ensure phase crossing elimination }

- R2. $TIMER(t)^i = 0$. { reinitialize the timer }

- R3. For all $L_k^i \in \vec{L^i}$, $q_{L_k^i} = 0$. { the transmission quotas are renewed }

- R4. $\vec{L_{MRG}^i} = \emptyset$.

- R5. Empty the SAT-BRD queue.

- R6. Goto **P1** (PARENT state).

The child omission procedure is called by a node in the CHILD state. The omission of a link from the algorithm does not interfere with the normal operation of the algorithm. If the tree is disconnected each part will continue to perform the algorithm independently.

CHILD Omission Procedure

- CO1. $\vec{L^i} = \vec{L^i} - L_j^i$.

- CO2. $d^i = d^i - 1$.

- CO3. If L_j^i is parent link do:
  { The node becomes root }

  - Send SAT-BRD to all $L_j^i \in \vec{L^i}$,
    s.t. a SAT-MRG has not received on L_j^i into the SAT-MRG queue, { in order to ensure phase crossing elimination }
  - $TIMER(t)^i = 0$. { reinitialize the timer }
  - Change state to PARENT, goto **P1**.

- CO4. If L_j^i is NOT parent link, then goto **C1** (CHILD state).

The parent omission procedure is called by a node in the PARENT state. As a result of this procedure the node can return to the PARENT algorithm or change state to CHILD via the legal merge procedure.

PARENT Omission Procedure

- PO1. $d^i = d^i - 1$.

- PO2. Remove L^i_j from $\vec{L^i_{MRG}}$ and from $\vec{L^i}$.

- PO3. If L^i_j is NOT in $\vec{L^i_{MRG}}$ do:

  - If the cardinality of $\vec{L^i_{MRG}}$ is $d^i - 1$, then call **Legal Merge Procedure**, else goto **P1** (PARENT state).

- PO4. If L^i_j is in $\vec{L^i_{MRG}}$,
  goto **P1** (PARENT state).

A link can join the algorithm on the tree only when the node is in the CHILD state. The control signals will start to be transferred after the node completes the join operation and changes its state to PARENT.

CHILD Join Procedure.

- J1. $d^i = d^i + 1$.

- J2. Add L^i_j to the $\vec{L^i}$.

- J3. All signals received on this link are ignored until the transition to PARENT state.
  { Only in the next broadcast phase the link joins the algorithm }

- J4. Goto **C1** (CHILD state).

4 Correctness and Properties

In this section we first outline the proof of correctness and stability of the algorithm and then we present its basic properties.

4.1 Stability Proof

In the following proofs we refer to the algorithm statements by their line numbers. We start by arguing about the behavior of the algorithm in the regular fault-free case.

Lemma 1 (Leaf Merge): *A leaf node in the CHILD state after receiving SAT-BRD will change its state to PARENT, and then it will change its state back to CHILD.*

Proof: The leaf changes its state to PARENT once it is satisfied on its only link. This will happen since the amount of traffic in each control cycle is bounded by the quota. A leaf in the

PARENT state changes its state back to CHILD, since it has no children and it does not have to wait for any SAT-MRG signal. See the parent algorithm (P1), the cardinality of $\vec{L}_{MRG}^i$ of the leaf node is always $d^i - 1$ (i.e., zero), as a result, the **Legal Merge Procedure** is called and since the leaf is automatically SATisfied it changes its state to CHILD. A SAT-MRG signal is sent on the link to the PARENT, and the merge phase starts.

Lemma 2 (Cycle Convergence): *If no failures occur, all SAT-MRG signals are transferred reliably, the merge phase will terminate at two adjacent nodes, i.e., on the ROOT EDGE (see Figure 6). These two nodes will be in the CHILD state and will receive SAT-MRG from one another, and as a result, both will broadcast SAT-BRD away from one another.*

Proof: The node always forward the SAT-MRG on the link that it has not received SAT-MRG. We can show that two adjacent nodes that are ready to forward SAT-MRG will send them to one another, s.t. the broadcast phase will start correctly. This must happen since if one of the nodes forwards the SAT-MRG to a third node we will get a contradiction to the fact that the SAT-MRG from the second adjacent has not arrived yet. Regardless of the order of the two SAT-MRG's, the two end-points of the ROOT EDGE recognize the situation and broadcast out of this edge to the rest of the sub-tree starting at the node (and they also send each other MRG-ACK).

Lemma 3 (Broadcast Cover): *The broadcast phase started at the ROOT EDGE will cover the tree, provided that no failures occur.*

Proof: Immediately from the tree structure.

Lemma 4 (Phase Crossing Elimination): *A broadcast phase cannot cross a merge phase, that is, if a SAT-MRG signal is sent on a link in one direction a SAT-BRD in the opposite direction will be ignored.*

Proof: In other words, we have to show that a broadcast cannot progress down a sub-tree, while a merge progresses up in the tree. SAT-BRD signal is sent in B1, R1 and CO3 with the provision that SAT-MRG has not received on that link. After a SAT-MRG is sent on a link (M1) all SAT-BRDs received on that link are discarded for the duration of T_{LINK}. After this time interval the SAT-MRG reaches the other node which will ensure that any later broadcast will cancel this SAT-MRG signal.

Note again that in slow networks the **phase crossing elimination** can be achieved by using hand-shakes. In high-speed networks hand-shakes that should be performed regularly will significantly decrease the effectiveness of our mechanism.

The lemmas can be summarized as follows:

Theorem 1 (Normal Operation): *A cycle that starts from the leaf nodes will continue to perform merge and broadcast phases continuously, as long as the are no signal loses.*

The proof for this theorem follows immediately from the first three lemmas, and the complete details are deferred to the full paper.

Theorem 2 (Stability): *When the algorithm starts at an arbitrary point (e.g., after it stops when two or more control signals were lost in the same cycle), a merge will converge on a single ROOT EDGE and all nodes are in the CHILD state. (Thus, a normal operation can continue immediately).*

The proof of this theorem is based on two mechanisms in the algorithm: (i) if single or multiple time-out occurs the broadcast phase will cover the entire tree and a then a uniform merge phase will converge to the root edge and (ii) if as a result of some faulty conditions a merge and broadcast phases are progressing on the tree at the same time, they will not cross one another (lemma 4). Furthermore, the merge phase stops the broadcast phase and since all merge phases have started from the leaves if the algorithm does not stop it will converge to the root edge, and normal operation will resume. (The complete details of the induction on the tree and case analysis showing that the process converges to a root edge are deferred to the full paper.)

4.2 Fairness and Basic Properties

The following are properties of the algorithm.

Theorem 3 (Equifairness): *The algorithm ensures equifair access to the tree links in each round.*

Detailed proof is deferred to the full paper. The idea is that each fault-free cycle gives opportunity to every node to transmit its quota on every link.

Corollary 1 *The traffic into the network is deadlock-free.*
A packet at the head of an output queue will enter the network as a result of the fairness property.

Theorem 4 (Self-balancing): *The merge phase will converge to a root edge in the network, such that, the delays on both subtrees (of the root edge) are equal.*

Detailed proof is differed to the full paper.

Corollary 2 (Built-in Fault Tolerance): *The algorithm can tolerate any single control signal loss in a cycle, and the algorithm will stop only if at least two control signals were lost in a cycle.*

5 Conclusion and Discussion

We have presented a self-stabilized algorithm for fairness on a spanning tree. The algorithm is completely distributed and no node is designated as a root or a leader in the tree. This fact resulted in a couple of highly desirable properties: the algorithm can tolerate any single loss of a control signal in every cycle, and therefore, if the signal loss probability in a cycle of 1 millisecond is 10^{-6}, then the probability that a cycle will be stopped (after two intermittent failures occur) is 10^{-12}, which gives a very long expected time between failure. In addition, component of the tree will continue operation after disconnection.

The elimination of merge and broadcast phases crossing stabilized the algorithm. This operation is equivalent to symmetry breaking, as it provides knowledge of the cycle and provides an effect which can be achieved by using ID numbers on the phase messages.

The mechanism gives external flow-control which provides all the nodes in this network with equal opportunity or fairness for accessing their adjacent links (equifairness), and as an immediate result deadlocks are automatically avoided. A possible useful extension for this work is to use the same mechanism for internal flow-control in order to reduce internal congestions and to improve network's delays.

The application of this work is widely varied and it ranges from parallel architecture switches to local and wide area networks. In multi-stage switches (parallel machines) our mechanism can be used for regulating the access into the switch and for congestion avoidance if the switch uses internal buffers between stages. The more obvious application is for fairness and flow-control mechanism in the *MetaNet* architecture (in particular it can be used to provide deadlock-free internal flow-control on threads, see [OY90a]). Another possible use is as a real-time feedback from the network to an input rate control mechanism such as the "leaky bucket" [Tur86]. In this application the rate is adaptive and depends on the cycle time of the broadcast-merge phases. When the cycle time is short the rate is higher than in the case of a longer cycle.

References

[AIP90] C. Arbib, G. F. Italiano, A. Panconesi, "Predicting deadlock in store-and-forward networks," *Networks*, to appear.

[AKP90] B. Awerbuch, S. Kutten and D. Peleg. Efficient Deadlock-free Routing, INFOCOM-91 (to appear).

[ACK90] B. Awerbuch, I. Cidon and S. Kutten. Dynamic Tree Maintenance, Found. of Comp. Scie (FOCS) 90., IEEE.

[BT84] G. Bracha and S. Toueg. A distributed algorithm for generalized deadlock detection. In *Proc. 3rd ACM Symp. on Principles of Distributed Computing*, pages 285–301. ACM, August 1984.

[CO89a] I. Cidon and Y. Ofek, "Distributed Fairness Algorithm for Local Area Networks with Concurrent Transmissions," the 3rd International Workshop on Distributed Algorithms, Nice, September 1989, IBM Research Report RC 15051, October 1989.

[CO89b] I. Cidon and Y. Ofek, "MetaRing - A Full-Duplex Ring with Fairness and Spatial Reuse," IBM Research Report RC 14961, September 1989, also INFOCOM'90.

[Fr] N. Francis, *Fairness*. Springer Verlag, New York.

[CJS87] I. Cidon, J. Jaffe, and M. Sidi. Distributed store-and forward deadlock detection and resolution algorithms. *IEEE Trans. on Commun.*, COM-35:1139–1145, May 1987.

[CM82] K.M. Chandi and J. Misra. A distributed algorithm for detecting resource deadlocks in distributed systems. In *Proc. 1st ACM Symp. on Principles of Distributed Computing*, pages 157–164. ACM, August 1982.

[G81] K.D. Günther. Prevention of deadlocks in packet-switched data transport systems. *IEEE Trans. on Commun.*, COM-29:512–524, May 1981.

[GHS83] R.G. Gallager, P.A. Humblet, and P.M. Spira. A distributed algorithm for minimum weight spanning trees. *ACM Trans. on Programming Lang. and Syst.*, 5:66–77, 1983.

[Gop84] I.S. Gopal. Prevention of store-and-forward deadlock in computer networks. *IEEE Trans. on Commun.*, COM-33:1258–1264, Dec. 1985.

[JS89] J.M. Jaffe and M. Sidi. Distributed deadlock resolution in store-and forward networks. *Algorithmica*, 4, 1989.

[LiRo83] M. T. Liu and D. M. Rouse, "A Study of Ring Networks," *Proc. IFIP WG6.4/University of Kent Workshop on Ring Technology Based Local Area Networks*, September 1983, pp. 1-39.

[MM79] D.A. Menascoe and R. Muntz. Locking and deadlock detection in distributed databases. *IEEE Trans. on Software Eng.*, SE-5:195–202, 1979.

[MS80] P.M. Merlin and P.J. Schweitzer. Deadlock avoidance in store-and-forward networks i: Store and forward deadlock. *IEEE Trans. on Commun.*, COM-28:345–352, March 1980.

[Obe82] R. Obermarck. Distributed deadlock detection algorithm. *ACM Trans. on Database Syst.*, 7:187–208, 1982.

[OY90a] Y. Ofek and M. Yung, "Principles for High Speed Network Control: loss-less and deadlock-freeness, self-routing and a single buffer per link," *ACM PODC'90*, pp. 161-165.

[OY90b] Y. Ofek and M. Yung, "Lossless Asynchronous Broadcast-and-Feedback on the MetaNet Achitecture," INFOCOM'91 (to appear).

[Ros86] F. E. Ross, "FDDI - a Tutorial," *IEEE Communication Magazine*, Vol. 24, No. 5, May 1986, pp. 10-17.

[SLCG] M. Sidi, W. Z. Liu, I. Cidon and I. Gopal, "Congestion Avoidance through Input Rate Regulation," *GLOBCOM'89*, Dallas Texas, 1989.

[TU81] S. Toueg and J.D. Ullman. Deadlock-free packet switching networks. *SIAM J. on Comput.*, 10:594–611, 1981.

[Tur86] J. Turner, "New Directions in communications (or Which Way to the Information Age?)", *IEEE Communications Magazine*, October 1986, Vol. 24, No. 10.

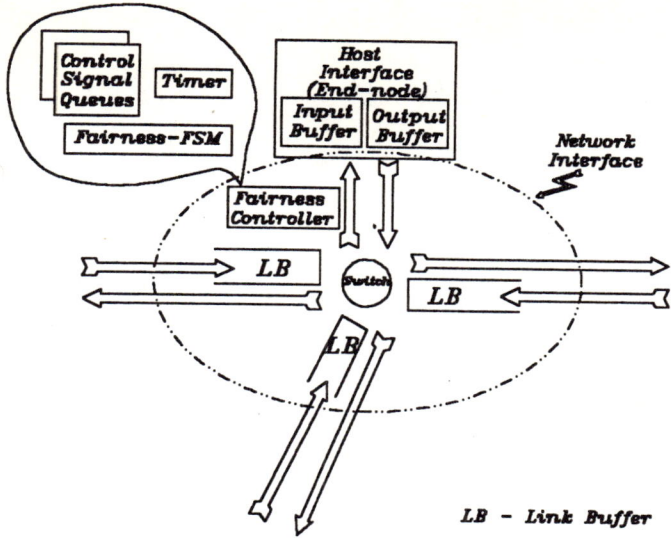

Figure 1.   Node Basic Structure

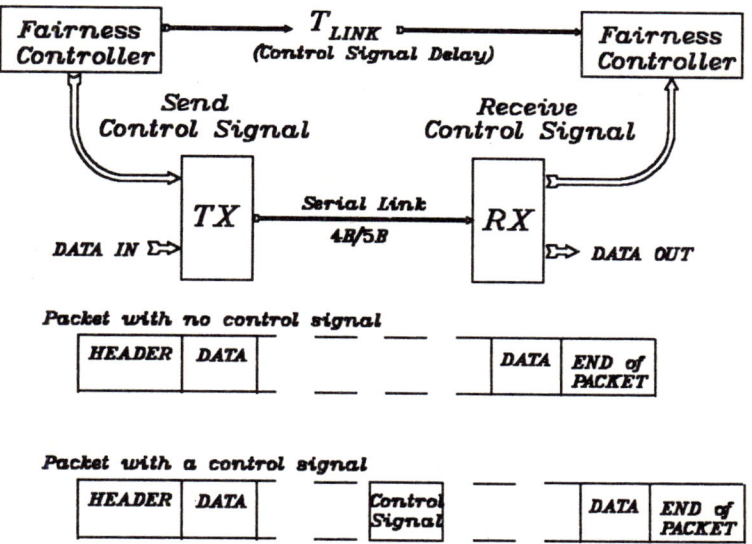

Figure 2.   The Preemptive/Nondestructive Control Signal Mechanism

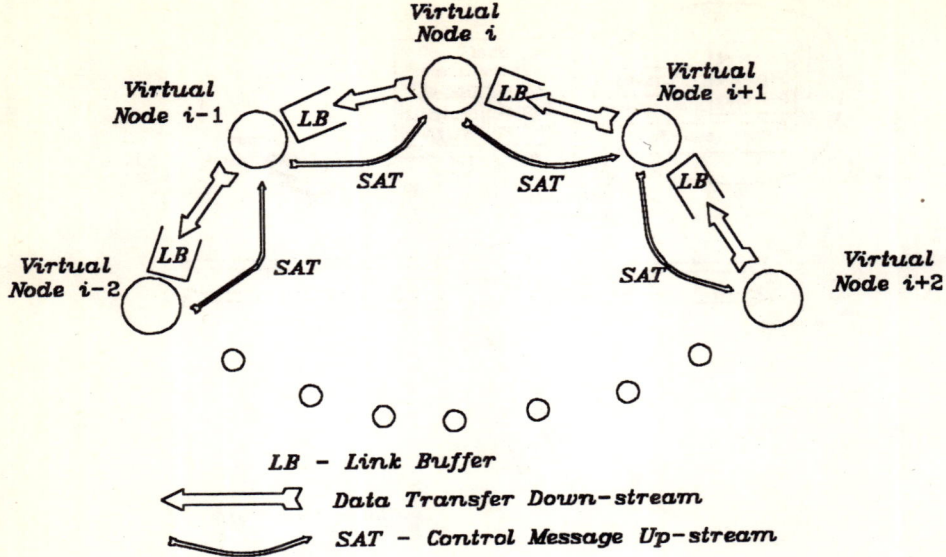

Figure 3. The MetaRing Fairness Mechanism

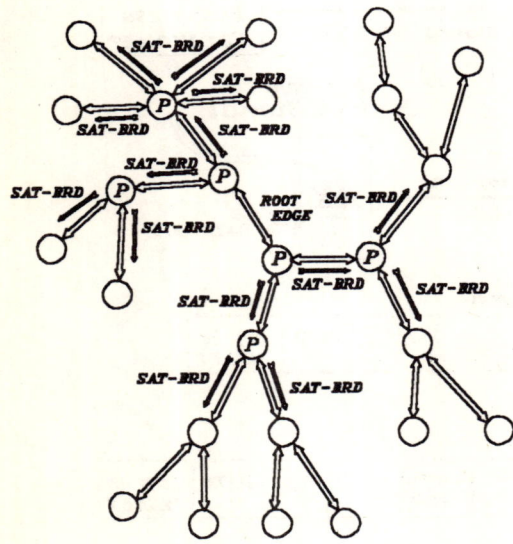

Figure 4. The Broadcast Phase

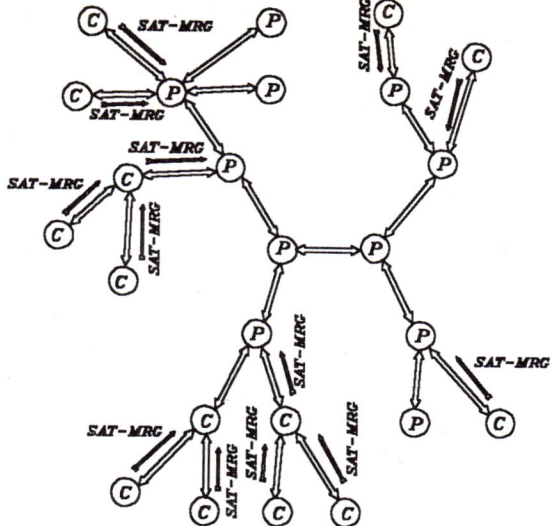

Figure 5. The Merge Phase

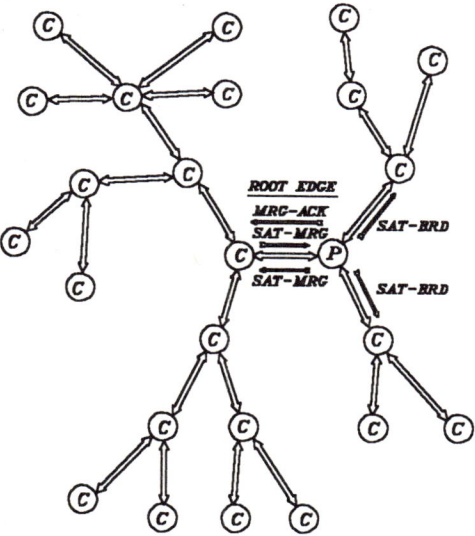

Figure 6. The Merge Termination with MRG-ACK

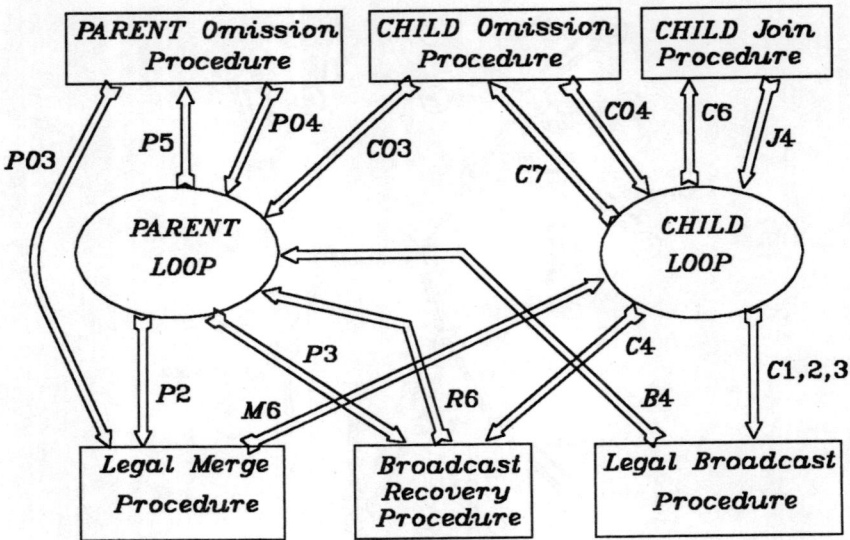

Figure 7. The Algorithm Structure

Strong Verifiable Secret Sharing
Extended Abstract

Cynthia Dwork
IBM Almaden Research Center

Abstract

Verifiable secret sharing has proven to be a powerful tool in the construction of fault-tolerant distributed algorithms. Many algorithms for VSS exist in the literature. These are of two types: small-error and error-free. In the small-error solutions, there is a small probability either that the dealer has not properly distributed the secret or that the faulty players can figure out the secret before reconstruction. In the error-free solutions neither of these can occur. However, the error-free solutions of which we are aware have a small weakness: the faulty processors can force a correct dealer to publicly reveal so much information that *every correct* processor learns the secret prematurely. This occurs despite the fact that no faulty processor learns anything at all about the secret. We overcome this weakness with no increase in the number of processors while remaining error-free.

1 Introduction

The t-resilient Verifiable Secret Sharing problem (t-VSS) was defined by Chor, Goldreich, Micali, and Awerbuch [4]. A solution to t-VSS is a protocol for a distributed system of n processors allowing a distinguished *dealer* processor to irreversibly commit to a secret value which can be reconstructed with certainty at a later time, even if the dealer attempts to block the reconstruction. Specifically, if the dealer is nonfaulty until reconstruction then no set F of up to t faulty processors can learn anything about the secret before reconstruction, and the value reconstructed will indeed be the secret value of the dealer. Moreover, the members of F cannot prevent reconstruction of the committed secret, even if the dealer is in F.

VSS and its applications have been widely studied in the literature [1, 2, 3, 6, 10, 12]. In particular, Ben-Or, Goldwasser, and Wigderson, and independently Chaum, Crepeau, and Damgard [2, 3] showed that in a system of $n \geq 3t + 1$ processors, where t is an upper bound on the number of faulty processors, any function of n inputs can be computed in such a way that not only can the faulty processors not disrupt the computation, but they cannot learn any additional information about the inputs of the nonfaulty processors than that implied by their own inputs and the output of the function. Both these papers use t-resilient VSS to share out the inputs to the function.

The VSS protocols in [2, 10] differ from other VSS protocols in that they are error-free; secrecy is perfect (information-theoretic) and reconstruction never fails.[1] However, both these error-free VSS protocols suffer from one weakness: the faulty processors can force a correct dealer to publicly reveal so much information that *every correct* processor learns the secret prematurely. This occurs despite the fact that no faulty processor learns anything at all about the secret. This weakness is easily overcome by increasing the number of processors (as a function of t). We originally conjectured such an increase necessary. In any VSS protocol the processors must exchange information in order to verify that a secret is correctly dealt out. It therefore seemed plausible that even in an execution of a distribution protocol in which the dealer is not faulty, the faulty processors could "lie" during verification about what they received from the dealer, causing nonfaulty processors to object and forcing the dealer to reveal previously secret information sent to the faulty processors. We show this is not the case. We obtain a VSS protocol requiring no additional processors and only moderate additional communication and computation costs above those of the protocols in [2, 10], with the additional property that no coalition of up to t curious but otherwise nonfaulty processors prematurely learns the secret.

Dolev, Dwork, Waarts, and Yung [6] defined a generalization of t-VSS, in which there are two threshholds: a secrecy threshold ℓ and a disruptor threshold d. (ℓ, d)-VSS considers the case in which there are ℓ faulty processors, *of which* only d are disruptors. As in t-VSS, the faulty processors should learn nothing about the secret before reconstruction, and the disruptors cannot prevent reconstruction. When $\ell = d = t$ this is precisely t-VSS.

We define the (g, ℓ, d)-*Strong Verifiable Secret Sharing* problem $((g, \ell, d)$-SVSS) to be (ℓ, d)-VSS with the added property that if the dealer remains nonfaulty until reconstruction then for all coalitions G of up to g nonfaulty but curious processors not containing the dealer, G learns nothing about the secret before reconstruction. In other words, for all such coalitions G, no fact about the secret that is not initially implicitly known (in the sense of Halpern and Moses [11]) to G before execution of the protocol becomes implicitly known to G before reconstruction. This does not rule out the possibility that G gains some probabilistic information about the secret. However, we also require that the protocol messages sent by the nonfaulty processors contain absolutely no information about the secret. Thus, any "information" gained is obtained from the faulty processors. Since anything the faulty processors say is suspect, such "information" can be probabilistic at best. A precise definition appears in the next section. An equivalent formulation considers a model of computation in which there are only faulty and pure processors (pure processors never gossip). The additional requirement beyond that of (ℓ, d)-VSS in this case is that no coalition of g pure processors together gain enough information to learn anything about the secret.

---

[1] In some protocols secrecy is imperfect because it relies on public key cryptography; in others, reconstruction may fail because verification is performed using interactive proof systems techniques.

The motivation is that even if the pure processors were a little less pure, and some g of them were to pool their information, they still would be unable to learn anything about the secret. It turns out that the number of processors needed for Strong Verifiable Secret Sharing is a function of d and $\max\{\ell, g\}$. We therefore set $g = \ell$ and keep the discussion down to two parameters. We call this version of the problem (ℓ, d)-Strong Verifiable Secret Sharing $((\ell, d)$-SVSS).

There are several situations in which it is desirable to separate the "nonfaulty" gossips from the faulty processors. We give two scenarios.

Scenario 1: Simultaneous Exchange of Secrets. Countries A and B wish to simultaneously exchange secret passwords (to be used in verifying adherence to a treaty). They wish the exchange to take place at a specific future date. Each of them distributes its password using SVSS to the United Nations General Assembly, to be reconstructed at the time of exchange. A and its client states wish to learn B's password early. B and its clients want to abrogate the treaty and prevent reconstruction of B's password.

Scenario 2: Competition for Design of Public Works. Engineer E has developed a bridge design that, at current costs, can hold load Z at \$1 per square foot. Engineer E wishes to publicize and time-stamp her result, but does not wish to publicize her techniques early because she is afraid some other engineer will improve upon her results and win the contract. She therefore commits to her proof using SVSS, to be reconstructed at the submission deadline. Engineer D, also competing for the bridge contract, wishes to destroy E's time-stamp, without helping anyone (except himself) to learn E's techniques. Finally, Professor L, who is not larcenous but is intellectually curious, wants to know Engineer E's technique.

It is known that (ℓ, d)-VSS requires $\ell + 2d + 1$ processors, with or without a broadcast channel, and this bound is tight [6]. We present an error-free protocol for (ℓ, d)-SVSS requiring no additional processors.

2 Definitions

We consider a completely synchronous distributed system of n processors, $\{p_0, \ldots, p_{n-1}\}$, connected by a complete network of perfectly secure channels. Our definition of protocol is the standard one. Let E be a finite field with a primitive nth root of unity such that $|E| > n$. We assume the secrets are always elements of our finite field E. There is a fixed underlying probability distribution Π on messages in E. *Perfect (information-theoretic) secrecy* says essentially that the best one can do at guessing a secret is to guess an element with maximal probability according to Π. We say a polynomial is of degree ℓ if it is of degree *at most ℓ*.

Given a protocol $\mathcal{P}$, a processor executing $\mathcal{P}$ is a *disruptor* if it does not follow $\mathcal{P}$ correctly. It can misbehave by not choosing random values according to the specified distribution, by failing to send specified messages, possibly sending arbitrary messages in their stead, and by making erroneous state transitions. A processor is a *gossip* if it follows $\mathcal{P}$ correctly but in addition sends extraneous messages labelled as gossip messages. Since the messages are labelled as extraneous, they are never confused with protocol messages. A processor is *pure* if it is neither disruptive nor a gossip. A processor is *faulty* if it is a disruptor, or if it sends a gossip message to a non-gossip, or if it sends a gossip message to another faulty processor. We discuss the intuition for these definitions after defining VSS and SVSS. We assume the existence of an *adversary* that selects processors to be made faulty and controls the behavior of the faulty processors. We think of the adversary as a (possibly probabilistic) algorithm that takes as input private random choices and the messages received by compromised processors and, based on these inputs, chooses which additional processors to compromise, as well as the behavior of all compromised processors.

A solution to the (ℓ, d)-*Verifiable Secret Sharing* problem $((\ell, d)$-VSS$)$ is a pair of protocols $(\mathcal{P}_1, \mathcal{P}_2)$, called the *distribution* and *reconstruction* protocols, respectively. Let p_0 be a dinstinguished *dealer* processor with secret input s. If in $\mathcal{P}_1$ and $\mathcal{P}_2$ combined there are at most ℓ faulty processors (disruptors and faulty gossips), *of which* at most d are disruptors, then the following properties are required.

1. If p_0 remains pure until the beginning of execution of $\mathcal{P}_2$, then no set of up to ℓ faulty processors has any information (in the information-theoretic sense) about the secret s. ($\mathcal{P}_2$ may begin long after $\mathcal{P}_1$ is completed, so the end of $\mathcal{P}_1$ is not the same as the beginning of $\mathcal{P}_2$.)

2. Whether or not p_0 is pure during $\mathcal{P}_1$, the value obtained by executing $\mathcal{P}_2$ is completely determined by the end of $\mathcal{P}_1$, and remains unchanged regardless of the behavior of the disruptive processors.

3. If p_0 is nondisruptive throughout execution of $\mathcal{P}_1$ then the value obtained when $\mathcal{P}_2$ is executed is s, the true secret input of p_0.

When $\ell = d$ this is precisely the usual definition of d-resilient Verifiable Secret Sharing. The change is the addition of the category of "faulty gossips." These only affect secrecy and do not affect reconstructability. This point is subtle, and comes from the fact that we require perfect reconstructability.

Let $Q(\cdot)$ be a predicate. Let ξ be an execution of $\mathcal{P}$ in which p_0 is pure until execution of $\mathcal{P}_2$ and p_0 has input s satisfying $Q(\cdot)$. We say a coaliton of processors L *prematurely learns* $Q(\cdot)$ in ξ if the members of L do not have *implicit knowledge*[2] before execution of $\mathcal{P}_1$ that the dealer's input satisfies $Q(\cdot)$, but before $\mathcal{P}_2$ they do have implicit knowledge that the secret satisfies $Q(\cdot)$.

A protocol solves the (ℓ, d)-*Strong Verifiable Secret Sharing* problem $((\ell, d)$-SVSS$)$, $\ell \geq d$, if in addition to solving (ℓ, d)-VSS it satisfies the following conditions:

1. If p_0, with input s, remains pure until the beginning of execution of $\mathcal{P}_2$, then no set L of up to ℓ nonfaulty gossips prematurely learns anything about s.

2. If p_0, with input s, remains pure until the beginning of $\mathcal{P}_2$, then the protocol messages received by L from the nondisruptive processors contain no information about s.

---

[2]Implicit knowledge was defined by Halpern and Moses [11]. Roughly speaking, the definition says that at time k in ξ, the group L has implicit knowledge of a fact ϕ if and only if ϕ holds in all executions of the protocol in which the views of the members of L is identical to their views at time k of ξ.

Remark 1 *As mentioned in the Introduction, there are really three parameters to Strong Verifiable Secret Sharing: ℓ, the upper bound on the number of faulty processors, d, the upper bound on the number of disruptors, and g the upper bound on the number of nonfaulty gossips (that is, the size of the coalition L). However, since the (tight) bound on processors needed for SVSS is a function of $\max\{g, \ell\}$ and d, in this paper we let $g = \ell$.*

Since this definition is delicate, we give an equivalent formulation. Suppose it were guaranteed that the system contains only pure and disruptive processors. Fix an execution of an (ℓ, d)-SVSS protocol, and consider the histories of all pure processors until just before execution of $\mathcal{P}_2$. Then the first additional requirement says that if p_0 is pure at this point then, for every set L of ℓ pure processors, the histories of the members of L do not jointly have enough information to unambiguously determine anything about the secret not implicit in Π. Thus, even if members of L were to "gossip" among themselves (in the colloquial sense), they would not prematurely learn anything about the secret. For this reason we formally define the notion of gossip. Intuitively, a gossip is not quite the same as a "faulty" or "malicious" processor, unless it collaborates with a malicious processor. We therefore differentiate faulty gossips (those whose gossip may eventually be heard by the disruptors) from nonfaulty gossips (those gossiping only among themselves). The second additional requirement says that any probabilistic information gained by L necessarily comes from unreliable sources.

Finally, a protocol solves (ℓ, d)-*Perfect SVSS* if it solves (ℓ, d)-VSS and in addition secrecy with respect to the nonfaulty gossips is perfect. More formally, for all adversaries $\mathcal{A}$, for all legal messages m and m', for all L of up to ℓ nonfaulty gossips, and for every view $\mathcal{V}$ of L before $\mathcal{P}_2$ in an execution in which the dealer remains pure until $\mathcal{P}_2$,

$$\Pr\left[\mathcal{V} \mid \mathcal{A}, m\right] = \Pr\left[\mathcal{V} \mid \mathcal{A}, m'\right],$$

where the notation "$\Pr\left[\mathcal{V} \mid \mathcal{A}, m\right]$" denotes the probability, taken over the random choices of all processors and the adversary $\mathcal{A}$, that L has view $\mathcal{V}$, given that the adversary is $\mathcal{A}$ and that the dealer's secret is m.

Remark 2 *An alternative definition for (ℓ, d)-Perfect SVSS would be to take the probabilities over the choice of $\mathcal{A}$, and condition only on the secret. However, we can think of no reasonable probability distributions on adversaries.*

3 Strong Verifiable Secret Sharing

Dolev, Dwork, Waarts, and Yung have shown that error-free (ℓ, d)-VSS requires $\ell + 2d + 1$ processors, even in the presence of a broadcast channel [6]. They also observed that the t-VSS protocol of Ben-Or, Goldwasser, and Wigderson appearing in [2] is easily modified to achieve this bound. Henceforth, *Protocol VSS* refers to the Ben-Or, Goldwasser, and Wigderson protocol, modified to handle differing secrecy and resiliency thresholds. We repeatedly use Protocol VSS as a subroutine for our SVSS protocol. The secrecy and resiliency threshholds may vary, and will always be mentioned explicitly.

In the following we assume the existence of a broadcast channel. This assumption is not necessary because, since $n \geq \ell + 2d + 1 \geq 3d + 1$, there are enough processors in the system to run d-resilient Byzantine agreement. However, it simplifies presentation of the protocol.

We briefly review some aspects of Protocol VSS with secrecy threshhold ℓ and resiliency threshhold d. Let ω be a primitive n-th root of unity in our finite field E. Throughout this paper, we say a polynomial is of degree ℓ if it is of degree at most ℓ. The dealer selects a random polynomial $f(x, y)$ of degree ℓ in both variables, so that $f(0, 0) = s$ (the secret). To each player p_i the dealer sends two polynomials, the *principal polynomial* $f_i(x) = f(x, \omega^i)$ and the *checking polynomial* $g_i(y) = f(\omega^i, y)$. (This can be done by sending to p_i the $\ell + 1$ coefficients of each of the polynomials.) The dealer treats itself uniformly, giving itself a share of the secret and participating in verification as does any other processor.

Intuitively, the ith *share* of the secret s is $f_i(0)$. The vector of shares is a codeword in a d-error correcting Reed-Müller code. The rest of the information sent to p_i is checking information, used in checking that the secret was properly dealt out. Note that if the dealer behaved correctly then for all $0 \leq i, j \leq n-1$, $f_i(\omega^j) = f(\omega^i, \omega^j) = g_j(\omega^i)$. To verify that the secret was dealt out correctly, each p_i sends $f_i(\omega^j)$ to p_j, who checks that the value received matches p_j's checking polynomial at ω^i. Let us assume p_j is nondisruptive. If p_0 and p_i are also nondisruptive then the verification of this point succeeds. Otherwise p_j broadcasts a request for clarification from p_0.

The following lemma says that if there are "enough" processors and there are no requests for clarification, then the secret shared out enjoys the additional secrecy condition required in Strong Verifiable Secret Sharing.

Lemma 3.1 *Let $n = \ell + 2d + 1$. Consider an execution ξ of Protocol VSS with secrecy parameter ℓ and resiliency parameter d in a system of n processors. If p_0 is pure in ξ until the beginning of $\mathcal{P}_2$ and if there are no requests for clarification during execution of $\mathcal{P}_1$, then in ξ no coalition L of up to ℓ nonfaulty gossips prematurely learns anything about the secret before execution of $\mathcal{P}_2$ begins. Moreover, the protocol messages received by L from the nondisruptive processors contain no information about s.*

Proof: Let s be the secret distributed during ξ. Let $Q(\cdot)$ be an arbitrary nontrivial predicate, and let L be an arbitrary coalition of up to ℓ nonfaulty gossips in ξ. Since the bivariate polynomial $f(x, y)$ chosen by the dealer is of degree ℓ in both variables, the material received by L from the dealer contains no information about $s = f(0, 0)$. Moreover, since the execution is without request for clarification, the remaining protocol messages in $\mathcal{P}_1$ also give no information about s to L. Thus, there exists a failure-free execution, ξ', of Protocol VSS, *possibly with a different adversary*, in which the dealer is pure, has secret s', the protocol messages received by the members of L are exactly as in ξ, and $Q(s')$ does *not* hold. Since the dealer is pure in ξ', there exists a third execution of Protocol VSS, ξ'', possibly with a third adversary, in which the dealer again has secret s', and sends exactly the same messages as in ξ', but the nonprotocol messages of the faulty processors in ξ'' are exactly as in ξ. Now, the members of L receive exactly the same messages in ξ as in ξ'', but in the first of these executions the secret (s) satisfies $Q(\cdot)$, while in the second the secret (s') does not satisfy $Q(\cdot)$. Thus, by definition, the members of L do not learn $Q(\cdot)$ in ξ. ∎

We present two protocols. The first protocol is slow; the second is a time optimization of the first, but has higher communication costs. Our approach in the slow protocol is to have the system repeatedly try to generate a well shared out random pad using as a subroutine Protocol VSS with secrecy and resiliency parameters ℓ and d, respectively. Specifically, each processor p_i shares out a random value s_i using Protocol VSS. If the system succeeds, that is, if in all the executions of the subroutine there are no requests for clarification, the pad is the sum of all the s_i. All players send their shares of the pad to the dealer, who reconstructs the pad P and broadcasts the sum $v = s + P$ (all arithmetic is in E). In this case, since the pad P is well shared out it can be reconstructed later, and the disruptors cannot change the value of the pad during reconstruction. Thus the dealer is irreversibly committed to $v - P$. Moreover, by Lemma 3.1 and the fact that there is at least one pure processor, the pad satisfies the strong secrecy requirements for SVSS. If the system does not succeed in generating the random pad then the protocol will guarantee that for some agreed upon pair (p_i, p_j) of processors, *every* processor knows that at least one of these two is faulty. In this case, every processor from then on ignores both p_i and p_j, and the system tries again to generate a random pad. Thus, the remaining number of active processors is $\ell + 2(d-1) + 1$, of which only $d-1$ may be disruptors. The attempt to generate the pad is iterated, this time using secrecy and resiliency parameters ℓ and $d-1$, respectively, in Protocol VSS. Clearly, after d iterations of this procedure no disruptors remain. The next iteration of the loop is guaranteed to succeed.

For the fast protocol we "parallelize" the iterations of this loop. Thus, the two protocols differ only in the procedure for generating a random pad. Similar techniques for collapsing loop iterations were used in [6] to quickly achieve perfectly secure message transmission in a general network. The new results require an extension of those techniques, while demonstrating the power of the basic approach.

Theorem 3.1 $\ell + 2d + 1$ *processors are necessary and sufficient for (ℓ, d)-Strong Verifiable Secret Sharing.*

Proof: Let $n = \ell + 2d + 1$. We need one technical point before describing the slow algorithm. Recall that in Protocol VSS the ith *share* of the secret s is $f_i(0)$. The vector of n shares is a codeword in a d-error correcting Reed-Müller code with design distance $n - \ell = 2d + 1$. However, at some point the number of active processors may be $n' = n - 2(d-d') = \ell + 2d' + 1$, in which case the information available for reconstructing the secret may only be $\ell + 2d' + 1$ components of this codeword, of which at most d' will be erroneous. In essense, since the codeword is of length n, in addition to the d' errors the error correction procedure must handle $e = n - n' = 2(d - d')$ erasures. Fortunately, the necessary design distance to correct d' errors and e erasures is precisely $2d' + e + 1 = 2d + 1$.

(ℓ, d)-Strong Verifiable Secret Sharing

Generation of a Random Pad

Initialization (performed by all processors):
$n' = n$;
$d' = d$;
$done = false$;
$Active = \{p_0, \ldots, p_{n-1}\}$;

Main Loop: While $done = false$ DO:

Phase 0:
$p_i \in Active$:
    $Conflicts = \emptyset$;

Phase 1:
$p_i \in Active$:
    Choose s_i uniformly at random from E;
    Perform the dealing phase of Protocol $VSS(s_i)$, with resilience d'
    and secrecy ℓ, sending shares only to the n' processors in $Active$;

Phase 2:
$p_j \in Active$:
    For each $p_i \in Active$ DO:
        For each $p_k \in Active$, exchange checking information for $VSS(s_i)$ with p_k;
        If an inconsistency is discovered
            then broadcast "(j,k) discrepancy in VSS_i"; fi
   OD

A *credible discrepancy* is a message of the form "(j, k) discrepancy in VSS_i," broadcast by p_j, where $\{p_i, p_j, p_k\} \subseteq Active$.

Phase 3:
$p_j \in Active$:
    If no credible discrepancy broadcast in previous phase then set $done := true$; fi

If there was at least one credible discrepancy broadcast during Phase 2, then let i, j, k be the lexicographically least triple such that p_j broadcast "(j, k) discrepancy in VSS_i". Note that the correct processors are in agreement about i, j, k.

The remaining Phases are executed only if $done = false$.

Phase 4:
p_i (i is as described in the previous paragraph):
    Broadcast all the private information of p_j and p_k;
    Let α denote this information;

Phase 5:
$p_y \in Active$:
    If α is syntactically incorrect
        then $Conflicts := Conflicts \cup \{(i,j)\}$; fi

$p_x \in \{j, k\}$ (j and k are as described in the paragraph above):
    If α is syntactically correct and
    α conflicts with information received from p_i during Phase 1
        then broadcast "$Conflict(i,x)$" fi

Phase 6:
$p_x \in Active$:
    $Conflicts := Conflicts \cup \{(a,b) | \text{"}Conflict(a,b)\text{"}$ was broadcast in Phase 5 $\}$;
    If neither (i,j) nor (i,k) is in $Conflicts$
        then $Conflicts := Conflicts \cup (j,k)$; fi

Phase 7:
$p_x \in Active$:
    Let (a,b) be the lexicographically least pair in $Conflicts$;
    Delete p_a and p_b from Active;
    $d' := d' - 1$;
    $n' := n' - 2$;

OD (End of Main Loop)

On exiting the Main Loop, for every $p_i, p_j \in Active$, p_i holds both a principal polynomial and a checking polynomial for s_j, where s_j is the secret shared by p_j in the final iteration of the Main Loop. Let us call these $f_{ij}(x)$ and $g_{ij}(y)$, respectively. Note that even if $p_i \notin Active$, these polynomials are well defined, but p_i does not know what they are. The polynomials f_{ij}, for $p_i \in Active$, are sufficient for p_i to compute its share of $PAD = \sum_{p_j \in Active} s_j$. However, we will use both sets of polynomials to generate shares of PAD for the processors no longer in $Active$.

Reactivation of Inactive Processors:

Let

$$F_i(x) = \sum_{p_j \in Active} f_{ij}(x) \tag{1}$$

$$G_i(y) = \sum_{p_j \in Active} g_{ij}(x). \tag{2}$$

Each $p_i \in Active$ computes $F_i(x)$ and $G_i(y)$. For each $p_m \notin Active$, p_i sends to p_m the points $F_i(\omega^m)$ and $G_i(\omega^m)$. The n-tuple $(F_0(\omega^m), \ldots, F_{n-1}(\omega^m))$ is a codeword. Processor p_m receives only n' pieces of the codeword, of which up to d' may be in error. Using the errors and erasures correcting procedure, p_m computes the codeword. The degree ℓ polynomial interpolating this codeword is $G_m(y)$. Similarly, p_m computes the codeword $(G_0(\omega^m), \ldots, G_{n-1}(\omega^m))$, from which it obtains $F_m(x)$.

To complete the distribution protocol, every processor p_i sends $F_i(0)$, its share of the random pad, to p_0. Using the erasures and errors correction procedure, p_0 computes the codeword and from this obtains the value of PAD. It then broadcasts $v = PAD + s$. Processor p_i's share of the secret s is $v - F_i(0)$. If p_0 fails to broadcast a value, its input is simply taken to be PAD itself. Each processor already has a share of PAD, so nothing else must be done.

To reconstruct the secret, each processor broadcasts its share. The shares are interpolated using the error correction procedure for d errors. This completes the description of the slow protocol.

We now list the main steps in proving correctness of this protocol. We say that an iteration of the loop *succeeds* if there is no credible discrepancy in the iteration (in which case the loop terminates). Consider a particular iteration of the loop that does not succeed. Then at the end of Phase 6 of this iteration there is at least one pair in $Conflicts$.

Lemma 3.2 *For each pair* $(a,b) \in Conflicts$, *at least one of* p_a *and* p_b *is a disruptor.*

Proof: A pair is added to $Conflicts$ in Phases 4-6. Let "(j,k) discrepancy in VSS_i" be a credible discrepancy broadcast in Phase 2 of some iteration of the Pad Generation protocol. The protocol calls for processor p_i to broadcast some information α in response to this complaint. By "syntactically correct" we mean that α consists of a principal and checking polynomial for each of p_j and p_k, and that they agree on the necessary points.

If (i,j) is added to $Conflicts$ because α is syntactically incorrect, then p_i is a disruptor.

If p_j broadcasts "$Conflict(i,j)$" then either the supposed private information of p_j contained in α differs from the information p_j received in Phase 1, in which case p_i is a disruptor, or not, in which case p_j is a disruptor. A similar argument applies to the case in which p_k broadcasts $Conflict(i,k)$.

Suppose neither p_j nor p_k broadcasts a conflict message in Phase 5, but α is syntactically correct. In this case, (j,k) is added to $Conflicts$. Let α_j (respectively, α_k) denote the supposed private information of p_j (respectively, p_k) broadcast by p_i in Phase 4. Let β_j

(β_k) be the private information actually sent by p_i to p_j (p_k) in Phase 1. If $\alpha_j = \beta_j$ and $\alpha_k = \beta_k$ and α is syntactically correct, then there is no discrepancy in the initial information. Since p_j broadcast the credible discrepancy "(j,k) discrepancy in VSS_i", either p_k sent the wrong checking information to p_j, in which case p_k is a disruptor, or p_j incorrectly claimed a discrepancy, in which case p_j is a disruptor. On the other hand, if $\alpha_j \neq \beta_j$ (respectively, $\alpha_k \neq \beta_k$), then p_j (respectively, p_k) should have broadcast a conflict message in Phase 5. Since, by assumption, neither did so, at least one of p_j and p_k is a disruptor. ∎

Corollary 3.1 *The Main Loop satisfies the invariant $n' = \ell + 2d' + 1$, where d' is an upper bound on the number of disruptors remaining in Active.* ∎

This corollary, together with the fact that $\ell + 2d' + 1$ processors suffice for (ℓ, d')-VSS and the technical point about errors and erasures made at the beginning of this proof, says that in the next iteration of the loop we can use (ℓ, d')-VSS among the remaining processors.

If in some iteration of the loop there are no disruptors in *Active*, then the iteration will succeed. Since in each unsuccessful iteration some disruptor is removed, if there are at most d disruptors, there will eventually be a successful iteration.

Lemma 3.3 *On completion of a successful iteration of the main loop, there is a bivariate polynomial P of degree ℓ in both variables such that $P(0,0) = PAD$ and for all $p_i \in$ Active, $F_i(x) = P(x, \omega^i)$ and $G_i(y) = P(\omega^i, y)$, where the F_i and G_i are as defined Equations 1 and 2.* ∎

Corollary 3.2 *For each $p_m \notin$ Active, the polynomials F_m and G_m computed during the Reactivation Step satisfy $F_m(x) = P(x, \omega^m)$ and $G_m(y) = P(\omega^m, y)$.* ∎

Lemma 3.4 *Let PAD be generated in a successful iteration of the Main Loop. Then PAD is a uniformly distributed random element of E.* ∎

This completes our outline of the key steps in proving the algorithm correct. The lower bound follows immediately from the lower bound in [6] for (ℓ, d)-VSS. In particular, it is shown there that to reconstruct in the presence of d disruptors, the secret must be completely determined by the shares held by any $n - 2d$ players. Thus, to be secure against coalitions of ℓ faulty players requires $\ell < n - 2d$, whence $n \geq \ell + 2d + 1$. ∎

Remark 3   1. Note that the lower bound, since it applies to ordinary (ℓ, d)-VSS (as opposed to Strong VSS), does not even rely on the fact that we want some secrecy with respect to the nonfaulty gossips. In fact, the bound holds even if secrecy is only required in executions in which no processor is faulty during the distribution protocol. From this we immediately obtain an actual lower bound of $\max\{\ell, g\} + 2d + 1$, for (g, ℓ, d)-SVSS, where g is an upper bound on the number of nonfaulty gossips, ℓ is an upper bound on the number of faulty players, and d is an upper bound on the number of disruptors.

2. The fact that the secret must be determined by any $n-2d$ shares explains why we cannot solve (ℓ,d)-SVSS simply by increasing the degree of the polynomial to $\ell + g$, without increasing the number of processors. Were we to increase the degree then, since $\ell+g+1$ shares would then be needed to obtain the secret, the total number of processors would have to be at least $\ell + g + 2d + 1$.

We briefly describe the fast pad generation protocol. Each processor shares $n^3 + 1$ independent random values chosen uniformly from E. We say the mth valued shared by p_j is p_j's *input to the mth pad*. Checking information is exchanged for all of these at once. Discrepancy messages are issued as in Phase 2 of the Pad Generation protocol. Each credible discrepancy gives rise to a triple of processors. The triples of the kth pad are the processors named in the discrepancy messages generated in the dealing of the inputs to the kth pad (e.g., (i,j,k), if the credible discrepancy "(j,k) discrepancy in VSS_i" is broadcast). A pigeonhole argument shows there is at least one pad P whose triples are contained in the union of the triples of the remaining pads R. The triples of the pads in R are resolved to pairs (conflicts), as done in Phases 4-6 of the slow protocol. Each processor orders the *Conflicts* list of pairs lexicographically. Let (x,y) be the first conflict on the ordered list. Each processor *processes* the pair (x,y) as follows. All communication from both p_x and p_y in generating Pad P is ignored; p_x and p_y are removed from *Active*; all pairs containing either x or y are removed from *Conflicts* (these additional pairs are not "processed," they are simply deleted). This procedure, which requires no communication, is iterated until the *Conflicts* list is empty. The resulting random pad is the sum of the inputs to P of the processors remaining in *Active* when the *Conflicts* list becomes empty.

Lemma 3.5 *At most $2d$ processors are removed from Active.*

Proof: A processor p_y is deleted from active only if some pair (y,z) or (z,y) is processed. By Lemma 3.2, every conflict contains at least one disruptor. For each disruptor p_x, at most one conflict containing x is processed. Thus, at most d pairs are processed. ∎

Lemma 3.6 *For every triple of P at least one member of the triple is not in Active at the end of the fast protocol.*

Proof: Let (i,j,k) be a triple of P. Then by choice of P, (i,j,k) is a triple of some pad S other than P. Since every triple of S is resolved to a pair, at least one of $(i,j),(j,k),(i,k)$ is in *Conflicts*. Without loss of generality, let $(i,j) \in Conflicts$. Then for some a at least one of $(i,a),(a,i),(j,a)$ or (a,j) is processed, and both members of a processed pair are removed from *Active*. ∎

Lemma 3.7 *Let p_i,p_j,p_k be in Active at the end of the fast protocol. If p_j and p_k are nondisruptive then the checking information p_j and p_k received in $VSS(s_i)$ agrees, where s_i is p_i's input to pad P.*

Proof: Suppose the checking information does not agree. Then (i, j, k) would be a triple of P. By Lemma 3.6, one of p_i, p_j, and p_k would not be in *Active* at the end of the fast protocol, contradicting the hypothesis of the lemma. ∎

Thus, at the end of the fast protocol, pad P has no remaining conflicts. The value of P is the sum of the secrets shared out by the processors remaining in *Active* after the *Conflicts* list is empty.

We next show that obtaining perfect secrecy with respect to coalitions of ℓ nonfaulty processors requires more than $\ell + 2d + 1$ processors. Recall from Section 2 that there are really three parameters: the number g of nonfaulty gossips, the upper bound ℓ on the number of faulty processors, and the upper bound d on the number of disruptors. Now, $\max\{\ell, g\} + 2d + 1$ processors are necessary and sufficient for (ℓ, d)-SVSS. We therefore simply took $g = \ell$ for our discussion. In the case of Perfect Strong Verifiable Secret Sharing, we must be a little more careful.

Theorem 3.2 *Any protocol for (ℓ, d)-Perfect SVSS requires $g + \ell + 2d + 1$ processors, and this bound is tight.*

Proof: Our protocol for $(g + \ell, d)$-SVSS yields perfect secrecy with respect to any g nonfaulty gossips because in this case the nonfaulty gossips and the faulty processors together hold only $g + \ell$ shares of the secret. Since the bivariate polynomials used in generating the random pad in our protocol for $(g + \ell, d)$-SVSS are of degree $g + \ell$ in each variable, these shares contain absolutely no information (in the information-theoretic sense) about the free term.

For necessity, suppose $n = g + \ell + 2d$ processors suffice, and consider an adversary $\mathcal{A}$ that always compromises the ℓ processors $p_1 \ldots p_\ell$ (recall, the dealer is p_0) and causes the compromised processors always to truthfully announce their shares in gossip messages. As mentioned earlier, results of Dolev, Dwork, Waarts, and Yung, show that any $n - 2d$ shares must be enough to reconstruct the secret [6]. Thus, if $n = g + \ell + 2d$ then $g + \ell$ shares suffice. Clearly, for any two legal messages m and m', with this adversary the views of the g nonfaulty gossips when the secret message is m do not occur with the same probability that they do when the message is m'. Thus, the information-theoretic definition of secrecy is violated. ∎

4 Recent Results

The original goal in the research leading to this work was to understand precisely what must occur during verification. It seemed plausible that the loss of secrecy occurring during verification in the protocol of Ben-Or, Goldwasser, and Wigderson was unavoidable with only $3t + 1$ processors, but this proved not to be the case. Unfortunately, the resulting $3t + 1$ processor algorithm for Strong VSS yields no insight into verification. However, by focussing on the quality of secrecy with respect to coalitions of nonfaulty gossips it *is* possible to separate a variant of Unverified Secret Sharing, where there are no requirements if the dealer is faulty, from its Verified Secret Sharing analogue. These results are described in [7]. Finally, in [7] it is shown how to extend the secret computation results of Ben-Or, Goldwasser, and Wigderson in [2] to obtain *strongly* secret computation.

References

[1] D. Beaver, and S. Goldwasser, Multiparty Computation with Faulty Majority, *Proc. 30th Symp. on Foundations of Comp. Science*, pp. 468-473, 1989.

[2] M. Ben-Or, S. Goldwasser, and A. Wigderson, Completeness Theorems for Non-Cryptographic Fault-Tolerant Distributed Computation, *Proc. 20th Symp. on Theory of Computing*, pp. 1-10, 1988.

[3] D. Chaum, C. Crepeau, and I. Damgard, Multiparty Unconditionally Secure Protocols, *Proc. 20th Symp. on Theory of Computing*, 11-19, 1988.

[4] B. Chor, S. Goldwasser, S. Micali, and B. Awerbuch, Verifiable Secret Sharing and Achieving Simultaneity in the Presence of Faults, *Proc. 26 Symp. on Foundations of Computing*, pp. 383-395, 1985.

[5] B. Chor, and E. Kushilevitz, A Zero-One Law for Boolean Privacy, *Proc. of the 21st Annual ACM Symp. on Theory of Computing*, pp. 62-72, 1989.

[6] D. Dolev, C. Dwork, O. Waarts, and M. Yung, Perfectly Secure Message Transmission, IBM RJ 7496, 1990; *To Appear, Proc. 31st IEEE Symposium on Foundations of Computer Science*, 1990.

[7] Cynthia Dwork, Unverified, Verified, and Strong (Un)Verified Secret Sharing, *manuscript*, 1990.

[8] Ran El-Yaniv, Interactive Consistency in Constant Expected Time, *Masters' Thesis, Department of Mathematics and Computer Science, Hebrew University*, 1988.

[9] P. Feldman, Optimal Algorithms for Byzantine Agreement, *PhD Thesis*, Department of Mathematics, MIT, 1988.

[10] P. Feldman, and S. Micali, Optimal Algorithms for Byzantine Agreement, *Proc. 20th Symp. on Theory of Computing*, pp. 148-161, 1988.

[11] J. Halpern and Y. Moses, Knowledge and Common Knowledge in a Distributed Environment, *Proc. 3rd ACM Symposium on Principles of Distributed Computing*, pp. 50-61, 1984; also, *IBM RJ 4421*, revised June, 1989.

[12] T. Rabin, and M. Ben-Or, Verifiable Secret Sharing and Multiparty Protocols with Honest Majority, *Proc. 21st Symp. on Theory of Computing*, pp. 73-85, 1989.

WEAK CONSISTENCY AND PESSIMISTIC REPLICA CONTROL

Alain SANDOZ   André SCHIPER

Département d'Informatique
Ecole Polytechnique Fédérale de Lausanne
CH-1015 Lausanne (Switzerland)
e-mail: sandoz@elma.epfl.ch

Abstract

In distributed systems replication is used to enhance availability and performance. Concurrent access to copies on different sites must be synchronized so transactions remain serializable. The main difficulty is the possibility of a partition of the network due to site or communication failures. Several protocols have been designed to synchronize transactions running in different components. Most pessimistic algorithms restrict access to a unique component per object and impose mutual consistency of copies. In this paper we show that this is not necessary for pessimistic control. We present a forwarding strategy for missing updates and a method to globally order conflicting transactions in a partitioned system. This enables consistent views of *objects* in minority components, logical conflicts in different components *and* one-copy serializability. We present an algorithm based on these ideas which achieves higher availability than other pessimistic protocols. This leads us to define a new concept for control as opposed to the traditional adaptation/recovery paradigm for replica control.

Keywords: distributed systems, replication, serializability, pessimism, weak consistency.

1. Introduction

Replicating objects in a distributed system can enhance availability and global performance. For example, availability of an object is less dependent on single-site failures. Communication overhead due to lengthy research is limited. Thus managing replicated objects has been studied extensivly [8] and is still an area of active research [2,6,13,14,15]. The advantages however have a price: maintaining data consistency and correct execution of transactions. This is equivalent to ensuring one-copy serializability (1-SR) [4] of transactions. Execution of concurrent transactions must be equivalent to a serial schedule of the same operations as if they were accessing single-copy objects.

The main problem is the partitioning of the underlying network into several unlinked components. Transactions running in one component can read or write copies there, but cannot access copies in other components. If this happens, identity of copies in the whole network cannot be maintained unless, either all the copies of an object are in the same component, or updates on this object are disabled altogether. In the perspective of maintaining high availability, this is too restrictive. Different methods have been designed to allow some degree of divergence between copies in the system while ensuring 1-SR of transactions.

Methods differ according to their global strategy, i.e. synchronizing transactions before (*pessimistic* control) or after (*optimistic* control [7]) they access objects. Which strategy is best for a given application depends on a tradeoff between high availability of objects and the cost of a possible rollback of transactions. Considering the cost of control when the system is *not* partitioned is also important. Pessimistic strategies strongly limit availability in order to ensure that committed transactions will never be backed out. This is generally achieved (1) by restricting access to at most one network-component per object and (2) by (explicitly or implicitly) maintaining mutual consistency of copies in that component.

The objective of this paper is to show that neither of these two conditions is necessary for pessimistic control. We first develop a missing-update information forwarding strategy [9] enabling consistent minority-reads in a connected network. Next, we extend this scheme to order conflicting transactions in different components and prove the correctness of the method. We propose an algorithm for replica control based on these ideas. Finally we suggest a method to keep the algorithm running smoothly without the burden of a recovery protocol when network components reconnect. This leads us to shed the old distinction between adaptation and recovery protocols, since the effects of partitions (divergence of copies) are considered a matter of life even when no failures occur in the system.

The content of the paper is as follows. In the next section we examine some related work. In section 3 we describe our model of the distributed system and give some definitions along with several examples. In section 4 we recall the graph theoretical setting of 1-SR and formally justify our approach. In section 5 we develop each of the points stated above. This is followed by some comments on the generality of the method and the meaning of recovery in this context.

2. Related work

The first formal definition of one-copy serializability was given in [3] and later placed in the general context of distributed concurrency control in [4]. An important survey of replication techniques was given in [8]. In that paper classifications for pessimistic vs. optimistic and syntactic vs. semantic algorithms were established. In [9] several necessary conditions were found for correct pessimistic processing of transactions, as well as some conditions which we will find to be overly restrictive. In particular, forwarding missing updates to *quorums* of copies of objects accessed by transactions depending on these updates. This is strongly related to mutual consistency of copies, which we claim is not necessary for correct scheduling. In [15] a formal definition of pessimism is stated. In that paper, it is proved that such pessimistic algorithms enforce 1-SR of transactions. A dynamic voting scheme is also proposed to optimize existence of a majority component for objects in a partitioned system and achieve better availability. In [1] a *group paradigm for concurrency control* is defined and shown to generalize most existing replica control methods. In [13,14] efficient dynamic voting schemes have been proposed that allow some consistent access to old copies in a partitioned system. In this sense, their motivation is similar to ours. The results are however not comparable: in some cases our algorithm definitely allows for better availability; also, some cases in [13] cannot be resolved by our algorithm and we have not examined its compatibility with the use of multiple versions [14]. This last paper and several others [2,18] define consistent views in terms of *sites* and network evolution. We define a consistent view in terms of accessed copies, which enables us to take a low-cost approach to updates by writing a minimal set of copies enforcing consistency. Recently, the notion of *update transport* has been proposed in [20]. Our motivation for forwarding updates in background mode is strongly related to this idea.

3. Model

3.1. The distributed system

A distributed system (DS) is a set of nodes connected by a network of bidirectional communication links. Initially the system is connected, but both nodes and links can break down, possibly partitioning sites into subsets unable to communicate over component boundaries. The model of node failures is "crash-failure" [17]. In particular, nodes do not send false information after failing. Partitions in the network are eventually repaired after finite time. This delay is however not bounded *a priori*.

Objects in the DS can be simple or replicated. A simple object resides at one site. A replicated object (RO) is represented in the system by a set of copies. Copies of an object are unique at sites and do not migrate. We do not consider multiple versions as in [14]. Replication can be partial: sites might not have a copy of some replicated objects.

Operations on objects are conducted by transactions [4]. A transaction executes *logical reads* and/or *logical writes* on objects. These operations are physically carried out on subsets of copies. Each site controls concurrent access to local copies. For each transaction T, a distinguished site, the transaction coordinator, or home site $h(T)$ centrally controls the global effects of the transaction. We do not consider nested transactions [16]. An object read (resp. written) by a transaction T is in the read-set $R(T)$ (resp. write-set $W(T)$) of T. We always consider $W(T) \subseteq R(T)$. This condition is quite natural and not so restrictive. It will simplify the statement and understanding of our algorithm. We note $R_\varphi(T)$ (resp. $W_\varphi(T)$) the set of copies read (resp. written) by T. When a transaction cannot reach all copies of an object in the write-set, we say the update is *incomplete*. Transactions have all the good properties enabling correct execution if the system were not replicated [10]. Transactions are two-phased and release their locks upon termination, there exists a commit protocol (for example 2PC) and deadlocks are handled. Transactions either commit, making their output permanent in the whole system, or abort, leaving no trace at all. A transaction that should but cannot access an object for any reason is aborted altogether. This model emphasizes the fact that we are not concerned in this paper with concurrency control, but only with consistency of access to replicated objects.

3.2. Replication

The problem of achieving global consistency of transactions accessing objects in different components of a network is illustrated in figure 1. The system is composed of two sites: s_1 and s_2, initially each with identical copies x_i and y_i (i=1,2) of objects X and Y. Suppose communication breaks down between the nodes and a transaction T_i runs at each site s_i:
- T_1 reads x_1 and y_1 and writes x_1;
- T_2 reads x_2 and y_2 and writes y_2.

These transactions cannot together have executed correctly: when the system is repaired and their results become global, neither one can be considered to have run before the other. They are not serializable.

One approach to this problem is to treat inconsistencies after they are detected. This is optimistic control, for which it is assumed that such situations seldom occur and that possibly backing out some transactions is cheaper than not running them at all. Both transactions above will be allowed to run to commit-point and then be suspended until the network is repaired. After repair, all inconsistencies are detected and (in this example) either T_1 or T_2 is aborted [7].

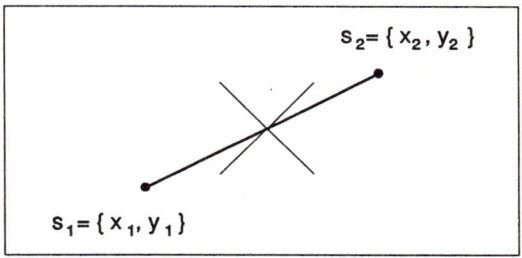

Fig. 1: Simple partitioned system.

The alternative is to avoid the problem by assuming the worst case situation: pessimistic control. Consistency is guaranteed through severe limitations on availability. In most pessimistic approaches, both transactions would be aborted for lack of sufficient information on a majority[1] of copies of X and Y in each component. In this paper we consider a more general approach that always allows one of these transactions to commit without delay when it reaches its commit-point. The justification for pessimism is correctness at the possible expense of availability. To further set the context of the paper, we borrow the following definition of pessimism from [15]:

---

[1] We define "majority" in a broad sense below.

Definition: a pessimistic replica control protocol ensures strong consistency (called *SC-pessimistic*) if, for any replicated object X the following three conditions hold (1) at any moment there exists at most one component in which object X can be read or written. This component is called the *majority-* or *distinguished-component* for X, noted $M(X)$. (2) All copies of X in $M(X)$ are identical. (3) As the network evolves due to failures and repairs, successive majority-components for object X overlap, i.e. if $M(X)$ exists, it contains the most recent version of X.

Any protocol following this definition, as do most pessimistic schemes implicitly, ensures 1-SR. Mutual consistency of copies in $M(X)$ is a necessary condition. For example, in the *general-quorum* algorithm of [12], write-quorums intersect, as well as write-quorums with read-quorums. This is equivalent to mutual consistency of copies in a distinguished component, since only the most recent copies can be read or written over.

In what follows we consider the notion of majority in a very general sense, not restricting our method to voting or quorums. We assume the existence of a mechanism ensuring that for every object X and every site s, s can determine (1) if it is in a connected group of nodes containing the latest version of X and (2) if it can lock enough copies such that a transaction updating X must access at least one of them. Any site satisfying these conditions is in a majority group for X. A set of sites which are not in a majority group for an object is called a *minority-group*. If a read is executed on a copy in a majority (resp. minority) group, it is a *majority* (resp. *minority*) *-read*. In the latter case, there is no way to know if the copy is the most recent version of the accessed object.

SC-pessimistic algorithms have several drawbacks: if partitions occur recurrently, there might not exist a majority component in the system for an object. Moreover, read operations are usually expensive because they must access several copies to get hold of a majority. So most recent developments, such as *dynamic-voting algorithms* [6,15], focus on optimizing the *existence* of a majority component. This is aimed at curtailing the first weakness. These algorithms however follow traditional access-rules: read or write on X are allowed only in $M(X)$ where copies are mutually consistent. We consider this to be overly restrictive, as shown in the following two examples:

Example 1: suppose we define in figure 1 $M(X) = \{s_1\}$ and $M(Y) = \{s_2\}$. Following the definition above, T_1 and T_2 must be aborted, because neither one runs in a majority component for all objects it accesses. Suppose now that we introduce a total order "$<_o$" on replicated objects such that $X <_o Y$, and that we weaken the read-access rule as follows: T_i can read object O in component C_i iff
- either $C_i = M(O)$;
- or else O is smaller ($<_o$) than any RO incompletely updated by T_i or by any transaction on which T_i depends.

This rule enables T_2 to read copy x_2 and commit during the partition, whereas T_1 must still be aborted. Control is pessimistic even though strong consistency is not enforced. This idea is developed in [19].

Example 2: in figure 2 the system is composed of six sites $s_1,...,s_6$ containing three objects X, Y and Z. Replica control is *optimistic*. Transaction managers time-out requests unresponded to, so they cannot make the difference between slow communication and a partition in the network. Though the system is connected, link (s_1,s_4) happens to be overloaded, making some communication delays very long.

Suppose transaction T_2 running at site s_4 requests a majority of object X, but the request cannot get over the slow link. Thus, the optimistic protocol running at s_4 will consider the system to be partitioned and allow T_2 to access copies x_4 and x_5. On the other hand, transaction T_1 running on s_2 might be able to read Y on s_5 and s_6 *and* update X on sites s_1, s_2 and s_3. Both transactions reach commit point and T_1 commits whereas T_2 is suspended. When communication speeds up, site s_4 runs the optimistic recovery protocol. If no inconsistency is detected, delayed transaction T_2 will commit, leading to a weakly consistent, but correct, execution in the connected network.

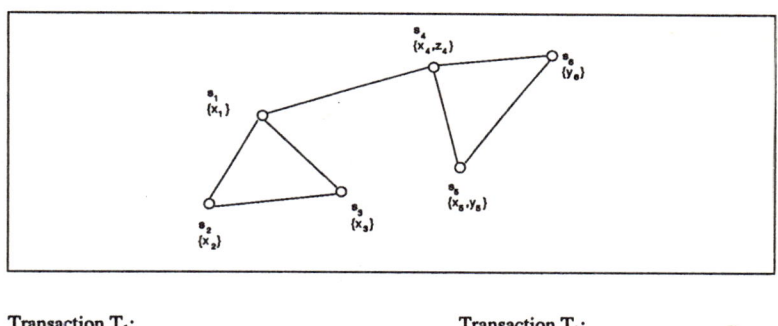

Transaction T_1:

$h(T_1) = s_2$
$T_1: <r[X,Y], w[X]>$
$\Rightarrow r_1[x_1,x_2,x_3,y_5,y_6], w_1[x_1,x_2,x_3]$

Transaction T_2:

$h(T_2) = s_4$
$T_2: <r[X,Z], w[Z]>$
$\Rightarrow r_2[x_4,x_5,z_4], w_2[z_4]$

Fig. 2: Quasi-partitioned network.

Mutual consistency[2] inside a component is generally admitted as a desirable condition in optimistic protocols. However, in this example, we see that it is not a necessary condition. This leads to the feeling that it is not the case either for pessimistic approaches.

These examples suggest a twofold approach for enhancing availability of pessimistic algorithms: (1) when a majority is not available, find a weaker condition for reads which still enforces global consistency of transactions; (2) find the weakest possible condition for reading old copies in a majority component while guarantying 1-SR inside that component. These ideas are developed in the following sections.

---

[2] When speaking of copies in a context related to components, mutual consistency generally means equality; in the context of transactions, it means that objects are seen in a consistent manner, i.e. copies *seem* to be equal and transactions to run serially. Hence the terminology of "restoring mutual consistency" [7] for recovery in the optimistic protocol.

Weak consistency and 1-SR can be useful in a number of situations, in particular when using large archival distributed database systems. Consider for example a bank which sends customers a log of transactions on their account at the beginning of each trimester. If account information is replicated at different branches, activity on thousands of accounts at once is likely to slow down communication when a majority of copies is requested for each computation. On the other hand, the most recent information on the account is not needed: only information dated before the end of the past quarter is necessary. So access to old copies can be used to enhance performance, without impairing consistency.

4. One-copy serializability and pessimism

In any database, transactions should execute serially as far as the user is aware. In a distributed system the network and replication should also be transparent. When a transaction T_i executes a logical read $r_i(X)$ or write $w_i(X)$ on an object, physical operations $r_i(x_j)$ (resp. $w_i(x_j)$), $j=j_1,...j_n$, are executed on a subset of copies. Two transactions conflict if they access one copy in common, at least one operation being a write. This implicitly orders the two transactions at the copy's resident site, since concurrency control is enforced there. To order transactions globally, the notions of *log*, *serial log*, and *serialization graph of a log* formalize the notion of correctness [4] in a non-replicated system. Acyclicity of the serialization graph guaranties serializability of corresponding transactions. In the replicated case, other conflicts arise when a transaction T_1 reads some copy x_{j1} of object X and copies different than x_{j1} are updated by another transaction T_2. This happens for example during partition when transaction T_1 reads an "old" copy of X.

Definition: the *one-copy serialization graph* $1\text{-}SG(\sigma)$ of a set σ of transactions is defined as (σ, E), where E is a set of directed edges. For an edge $s=(T_1,T_2) \in E$, we note $T_-(s)=T_1$ and $T_+(s)=T_2$. Edges in E represent conflicts between transactions and are such that:
  (1) E contains all direct conflicts on physical objects;
  (2) for every replicated object X in the system, $1\text{-}SG(\sigma)$ induces a total order $<_x$ on the transactions writing X;
  (3) if T_i, T_j, and T_k are three transactions such that
      a) there exists a copy x_a of an object X in the system such that $w_i(x_a)$ precedes $r_j(x_a)$;
      b) there is no transaction T_n such that $w_i(x_a)$ precedes $w_n(x_a)$, and $w_n(x_a)$ precedes $r_j(x_a)$;
      c) $T_i <_x T_k$ (i.e. T_k updates X after T_i);
      then there is a path in $1\text{-}SG(\sigma)$ from T_j to T_k.

This last point means that, if T_j "reads-X-from" T_i and T_k updates X after T_i, then T_j must be considered as having executed before T_k. If there exists an acyclic one-copy serialization graph for σ, then σ is 1-SR. A one-copy serial schedule is given by any topological sort of the graph. Conversely, if we can build a partial order on σ and there exists a $1\text{-}SG(\sigma)$ with which this order is compatible, then the graph is acyclic and σ is 1-SR.

We will use the following general method to design and prove our access-control algorithm. For a set σ of transactions, we consider the $1\text{-}SG(\sigma)$ described below. We show that our algorithm partially orders σ in a manner compatible with this graph. So the 1-SG(σ) is acyclic and induces a one-copy serial schedule. As in traditional pessimistic schemes, we first require that *updating* an object is possible only in a majority component. Since successive majority components for an object overlap, successive updates conflict directly and no edge must be added in 1-SG to ensure condition (2). So the 1-SG we consider is characterized by the edges introduced to satisfy condition (3). Given T_i, T_j, and T_k as in the definition, we add an edge joining T_j directly to T_k. We call this edge a *serialization edge* (noted *se*). This notion is illustrated in the following examples.

Example 1: When applying an *SC-pessimistic* strategy, serialization edges simply do not occur in the 1-SG. During system partition, access to objects is limited in such a manner that the conflict graph of transactions in each component is an acyclic 1-SG. Moreover, subsets of σ proper to each component do not have any outgoing *se* joining a transaction in another component. This is because each component is a majority group for all available objects, so no minority read is possible. Thus the global graph is also acyclic.

Example 2: In [9] missing-update information is forwarded to transactions which depend on an incomplete update. This is used implicitly to prevent any cycle with a unique serialization edge in the graph. Such a cycle could occur in the following situation. During a partition, a transaction T_1 makes an update on a majority, but not all, of the copies of an object X (say x_1 is not updated); then the network reconnects and a transaction T_2 which depends on T_1 reads copy x_1, introducing an *se* from T_2 to T_1 and a cycle in the 1-SG. To prevent this, T_1 runs in failure-mode and posts missing-update information to all depending transactions, which will be aware of which copies of X not to read.

Our approach to weak consistency and serializability is the following:
- We transport missing-update information along paths in the conflict graph. This information is stored with copies. We state a simple condition on missing-update information associated to copies accessed by a minority-read. In this way we prevent cycles containing only one serialization edge.

A cycle with one *se* reflects an inconsistency strongly related to the notion of physical time in the system. Since other edges than the only *se* represent physical conflicts, $T_-(se)$ which executes an old read actually executes after $T_+(se)$. This condition on time is not true for minimal cycles with two or more *se*'s, which translate *logical* conflicts not detectable with missing-update information.
- We state a second condition (see *example 1* of §3.2) for minority-reads on copies of objects that might have been updated in disconnected parts of the network. This prevents cycles with two or more *se*'s in the 1-SG.

The combined approach guaranties that there are no cycles with one or more serialization edge in the 1-SG, thus leading to 1-SR of transactions. Since the conditions for access are weaker than the majority requirement, the result is better availability. Mutual consistency of copies within a given component is not necessary. Moreover, the algorithm will not need to construct any part of the 1-SG.

Since paths in the conflict graph translate physical conflicts between transactions, information on potential *logical* conflicts will be stored at copies. Reading and updating this information takes place when the copies are accessed. Control information used by a transaction T is of two types: (1) information on incomplete updates of objects actually accessed by T and (2) information on incomplete updates and minority-reads executed on other objects by transactions preceding T. A transaction manager possessing this information on all objects in R(T) can decide if a minority-read is consistent. A majority-read is always possible.

5. Controlling weakly consistent copies
5.1. Forwarding missing-update information

The objective of this paragraph is to present a strategy for posting information along paths in the conflict graph of a connected network-component. This information concerns incomplete updates in the component. Missing-update information (noted *mui*) is stored at copies accessed by transactions to be forwarded to transactions depending on them. When a transaction accesses a copy, it may pick up some missing-update information stored there, complete it with some of its own, and store it at copies on which further depending transactions might conflict. For the moment we assume *mui* is stored at a majority of copies of objects read or written. This restriction is the main drawback of [9][3], since it is equivalent to the majority requirement. For now it enables transactions running in the same component to be serialized, but the restriction will be released in §5.2.

We first describe in a general manner the notion of forwarding missing-update information. We note t_mui(T) the missing-update information of which a transaction T is aware. The first point, which is fairly clear, is that if T updates copy x_i it should store t_mui(T) at x_i. This is because any new operation on x_i will conflict, and thus depend, on the update. For now we note $c\_Wmui(x_i,T)$ this information stored at x_i by T writing x_i. When T only reads x_i, t_mui(T) should not be forwarded to every transaction which next accesses the copy, because a transaction can read x_i without ever depending on T. On the other hand, t_mui(T) must still be stored at x_i, because the next access to x_i could be a write, and thus conflict with T. We note $c\_Rmui(x_i,T)$ the *mui* stored at x_i by a transaction T only reading the copy. If we note $c\_Rmui(x_i) = \cup_T c\_Rmui(x_i,T)$ and $c\_Wmui(x_i) = \cup_T c\_Wmui(x_i,T)$, we have the following access patterns for missing update information:
- for a transaction T only reading x_i, T reads $c\_Wmui(x_i)$, completes t_mui(T), and overwrites $c\_Rmui(x_i)$ with t_mui(T);
- for a transaction T writing x_i, T reads $c\_Rmui(x_i)$, completes t_mui(T), and overwrites $c\_Wmui(x_i)$ and $c\_Rmui(x_i)$ with t_mui(T);
- $c\_Rmui(x_i)$ is always a superset of $c\_Wmui(x_i)$.

This is illustrated by figure 3 in which transactions $T_1,..,T_4$ access copy x_i. Transaction T_1 updates x_i and deposits $t\_mui(T_1)$ at the copy. Transaction T_2 reads x_i and accesses some other objects from which it gathers *mui*. As T_2 conflicts with T_1, it must consider $t\_mui(T_1)$ as relevant information. This holds also for T_3. However, even though T_3 executes after T_2, they do not conflict. So $t\_mui(T_2)$ is not relevant to T_3. Moreover, forwarding $t\_mui(T_2)$ to T_3

---

[3] This is due to the fact that missing-update information must be forwarded to *quorums* of copies of entities *read* by a transaction, thus requiring a quorum of copies for reads and writes if a transaction runs in failure mode.

would serialize T_3 after T_2, which is too restrictive. So T_2 and T_3 read *mui* from $c\_Wmui(x_i)$ and deposit $t\_mui(T_2)$ and $t\_mui(T_3)$ in $c\_Rmui(x_i)$. When transaction T_4 updates x_i, it reads $c\_Rmui(x_i)$ and inherits all missing-update information from transactions T_1, T_2, T_3 and transactions preceding them. It then overwrites $c\_Wmui(x_i)$ and $c\_Rmui(x_i)$ with $t\_mui(T_4)$.

For simplicity, we suppose the system contains N objects, simple or replicated. Objects are mapped on the interval $1..N$ by the function $X \rightarrow o\_cn(X)$, for *object class number* of X. This arbitrary numbering implicitly gives a total order of replicated objects. To each copy x_i of X is associated a version number $v(x_i)$. Value $v(x_i)=n$ means copy x_i was overwritten by the nth update of X, which is consistent because transactions updating X are all physically ordered. The version number of object X is $v(X)=\sup_i(v(x_i))$. For the moment we suppose that every transaction updates at least one object.

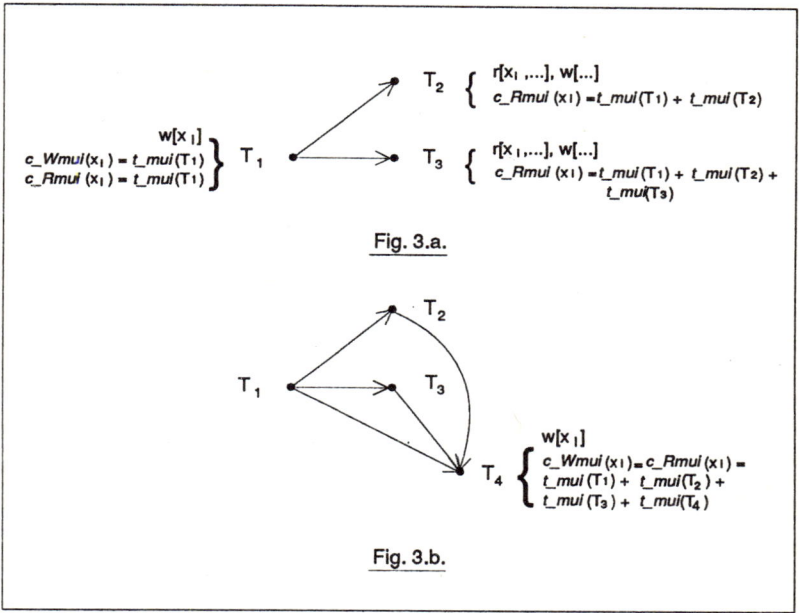

Fig. 3: Principle of forwarding missing-update information.

Consider the use of $v(X)$ as *mui*. If a transaction is aware of $v(X)$, it cannot read a copy x_i with $v(x_i)<v(X)$. So only one version number must be kept for X at any copy as missing-update information. To take advantage of this we introduce the following notation, i.e. vectors of version numbers (noted *vvn*, $vvn \in N^N$). Each entry $vvn[o\_cn(Y)]$ (also noted $vvn[Y]$) is a version number of Y or has the special value *nil* which is defined to be strictly smaller than any version number in the system. The binary relation "$<_v$" on vectors of version numbers is defined by: $vvn_1 <_v vvn_2$ iff $\forall X, vvn_1[X] \leq vvn_2[X]$. The supremum of vectors $vvn_1,..,vvn_k$ is given by: $\forall X, \sup(vvn_1,..,vvn_k)[X] = \sup(vvn_1[X],..,vvn_k[X])$.

A transaction depends on version $v(X)$ if it accesses a copy x_i with $v(x_i) = v(X)$ or if it conflicts with a transaction depending on $v(X)$. So we use *vvn*'s as missing-update

information. At each copy x_i in the system are stored two vectors, the *copy-write* and *copy-read* dependency vectors, noted $c\_Wdv(x_i)$ and $c\_Rdv(x_i)$. Suppose copy x_i has been written by transaction T. These dependency vectors are defined by: $\forall Y$,
- $c\_Wdv(x_i)[Y] = v(Y)$ if T writes Y; or
- $c\_Wdv(x_i)[Y] = \sup(v(y_i)$, T depends on $y_i)$ if T depends on any copy of Y; or else
- $c\_Wdv(x_i)[Y] = nil$.

In other words, $c\_Wdv(x_i)[Y]$ is the highest version number of a copy of Y accessed by T or by a transaction on which T depends, or *nil* if there is none. Similarly, $c\_Rdv(x_i)[Y]$ is the highest version number of a copy of Y on which depends a transaction which reads x_i after update by T, or *nil* if there is none. Both $c\_Wdv(x_i)[X]$ and $c\_Rdv(x_i)[X]$ are equal to $v(x_i)$.

An example of c_Wdv and c_Rdv associated to three simple objects X, Y, and Z is given in figure 4 in which these objects are accessed by transactions $T_1,..,T_4$ (after initialization to version 0 by transaction T_0).

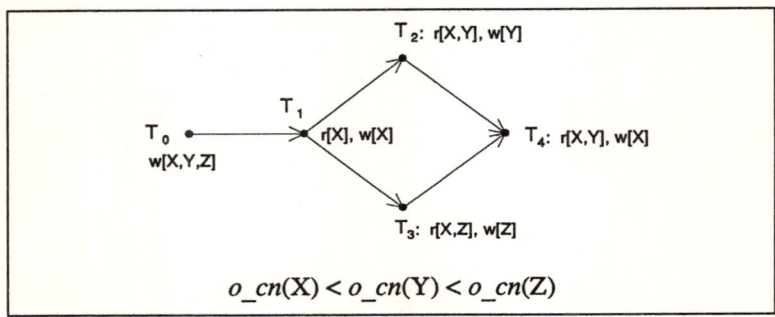

	$c\_Wdv(X)$	$c\_Rdv(X)$	$c\_Wdv(Y)$	$c\_Rdv(Y)$	$c\_Wdv(Z)$	$c\_Rdv(Z)$
T_0	(0,*nil*,*nil*)	(0,*nil*,*nil*)	(*nil*,0,*nil*)	(*nil*,0,*nil*)	(*nil*,*nil*,0)	(*nil*,*nil*,0)
T_1	(1,*nil*,*nil*)	(1,*nil*,*nil*)	(*nil*,0,*nil*)	(*nil*,0,*nil*)	(*nil*,*nil*,0)	(*nil*,*nil*,0)
T_2	(1,*nil*,*nil*)	(1,1,*nil*)	(1,1,*nil*)	(1,1,*nil*)	(*nil*,*nil*,0)	(*nil*,*nil*,0)
T_3	(1,*nil*,*nil*)	(1,1,1)	(1,1,*nil*)	(1,1,*nil*)	(1,1,1)	(1,1,1)
T_4	(2,1,1)	(2,1,1)	(1,1,*nil*)	(2,1,1)	(1,1,1)	(1,1,1)

Fig. 4: Vectors associated to a copy in the distributed system.

Remarks: it is clear from the definition of read- and write-dependency *vvn*'s that correct replica control must enforce condition $c\_Wdv(x_i) <_v c\_Rdv(x_i)$ for every copy x_i in the DS. Also, relation $<_v$ is reflexive, antisymmetric and transitive, and thus defines an order on *vvn*'s.

Our objective is to associate *vvn*'s to transactions in such a manner to order them consistently with the 1-SG. For a transaction T the *transaction dependency vector* of T, noted $t\_dv(T)$ is defined by:
- for $X \in W(T)$, $t\_dv(T)[X] := v(X)$;
- for $X \in R(T)$, $X \notin W(T)$, $t\_dv(T)[X] := \sup(\{c\_Rdv(y_i)[X], y_i \in W_\varphi(T)\} \cup \{c\_Wdv(y_i)[X], y_i \in R_\varphi(T)\})$.

Given the rules for forwarding *mui* in §5.1 we have, if T commits:
- for $x_i \in W_\varphi(T)$, $c\_Wdv(x_i) := c\_Rdv(x_i) := t\_dv(T)$;
- for $x_i \in R_\varphi(T)$, $x_i \notin W_\varphi(T)$, $c\_Rdv(x_i) := t\_dv(T)$.

Forwarding dependency vectors to majorities of copies in this manner clearly ensures 1-SR inside a connected component. This is only a restatement of [9] and contributes nothing to the general theory.

5.2. Minority-reads in the partitioned network

We now consider forwarding *mui* only to the copies actually accessed, even in the case of minority reads. We are thus confronted with the problem of synchronizing logical conflicts in a partitioned network.

Recall the definition of serialization edges: existence of an *se* (T_i,T_j) in 1-SG means that transaction T_i in component C_i has read an object X that was updated by T_j in component C_j, possibly with $i \neq j$. Consequently (1) C_j was a majority component for X and (2) not all copies of X were reached when this update took place. The transaction manager of T_j takes notice of (2) and transmits this knowledge down all paths in the conflict graph. To realize this, we consider using object class numbers as *mui*. If transaction T writes copy x_i, we associate to x_i as missing-update information the smallest o_cn of a replicated object incompletely updated by T or by a transaction on which T depends, or ∞ if there is none. This number is called the *copy write maximum class number* and is formally defined by:

$c\_Wmcn(x_i) =$
  $\inf(\ \{\ o\_cn(Y),\ T\ \text{wrote}\ x_i\ \text{and depends on an incomplete update of}\ Y\ \} \cup \{\infty\}\ )$.

In the same manner we associate to x_i the smallest o_cn of an incompletely updated RO on which depends a transaction reading x_i or also ∞ if there is none:

$c\_Rmcn(x_i) =$
  $\inf(\ \{\ o\_cn(Y)\ /\ \exists T_1,\ T_1\ \text{reads}\ x_i\ \text{and depends on an incomplete update of}\ Y\ \}\ ) \cup \{\infty\}$.

Finally we associate to a transaction T a transaction class number, noted $t\_cn(T)$, defined by:

$t\_cn(T) :=$
  $\inf(\ \{o\_cn(O),\ O\ \text{incompletely updated by}\ T\} \cup \{c\_Wmcn(x_i),\ x_i \in R_\varphi(T)\} \cup \{c\_Rmcn(x_i),\ x_i \in W_\varphi(T)\}\ )$.

Using dependency vectors and class numbers associated to copies and transactions, we can state the conditions for a transaction T reading a copy o_i of object O to be consistent:

(a) T reads a majority of copies of O, and o_i is the most recent version;
(b) or else, $o\_cn(O) < t\_cn(T)$;
(c) or else, $o\_cn(O) = t\_cn(T)$ and $v(o_i) = t\_dv(T)[O]$.

5.3. Globally ordering transactions

Definition: the binary relation "$<_T$" between transactions is defined by: $T_1 <_T T_2$ iff any one of conditions (1)-(3) is true:
  (1): $t\_cn(T_1) > t\_cn(T_2)$

(2): $t\_cn(T_1) = t\_cn(T_2) = \infty$ and $t\_dv(T_1) <_v t\_dv(T_2)$
(3): $\exists O \in RO \ o\_cn(O) = t\_cn(T_1) = t\_cn(T_2)$ and
   [$t\_dv(T_1)[O] < t\_dv(T_2)[O]$ or
   ($t\_dv(T_1)[O] = t\_dv(T_2)[O]$ and $t\_dv(T_1) <_v t\_dv(T_2)$)]

Theorem: consider a set σ of transactions running in the distributed system. If transactions in σ follow the rules of §5.1 and §5.2, then the restriction of relation $<_T$ to σ is reflexive, transitive and antisymetric and this order is compatible with 1-SG(σ).

Proof: *Relation $<_T$ is an order on committed transactions*. Reflexivity and transitivity are straightforward. To show antisymetry, consider two committed transactions T_1 and T_2 such that $T_1 <_T T_2$ and $T_2 <_T T_1$. Then $t\_cn(T_1) = t\_cn(T_2)$, and so $t\_dv(T_1) <_v t\_dv(T_2)$ and $t\_dv(T_2) <_v t\_dv(T_1)$. So for every O in *RO* we must have $t\_dv(T_1)[O] = t\_dv(T_2)[O]$. Recall that every committed transaction updates at least one object, incrementing its version number, and that updates on the same object always conflict on at least one copy. So $t\_dv(T_1)[O] = t\_dv(T_2)[O]$, for all O, implies $T_1 = T_2$.

Relation $<_T$ is compatible with 1-SG(σ). We must show that if there exists a path in 1-SG(σ) between two transactions T_1 and T_2, then $T_1 <_T T_2$. Because $<_T$ is transitive, it suffices to consider only the case of an edge between T_1 and T_2 in 1-SG(σ).

Suppose this edge between T_1 and T_2 is *direct*: then both T_1 and T_2 access a physical object o_i in common, at least one access being a write. If T_1 writes, then $c\_Wdv(o_i)$ read by T_2 is equal to $t\_dv(T_1)$. If T_2 writes, then $t\_dv(T_1) = c\_Rdv(o_i)$ written by T_1 is smaller ($<_v$) than $t\_dv(T_2)$. In both cases we have $t\_dv(T_1) <_v t\_dv(T_2)$. Reasoning in the same manner on $c\_Wmcn(o_i)$ and $c\_Rmcn(o_i)$, we get $t\_cn(T_2) \le t\_cn(T_1)$. Thus $T_1 <_T T_2$.

Suppose the edge is a *serialization edge*. Then T_1 reads an old copy o_i of an object O that T_2 updated incompletely (say T_2 wrote at least copy o_j, i≠j, so $v(o_i)<v(o_j)$) and thus $t\_cn(T_2) \le o\_cn(O)$. Note first that $t\_cn(T_1) < t\_cn(T_2)$ is not possible because if T_1 can make a minority read on O, then $o\_cn(O) \le t\_cn(T_1)$. So the only case to consider is $o\_cn(O) = t\_cn(T_1) = t\_cn(T_2)$. But then we have $t\_dv(T_1)[O] = v(o_i) < v(o_j) = t\_dv(T_2)[O]$ by the last rule at the end of sect. 5.2. So here also $T_1 <_T T_2$. ◊

5.4. The replica control protocol

An algorithm controlling access to replicated objects by a transaction T with the rules described above must ensure the three following points (1) gather *mui* at copies accessed by T; (2) verify conditions on minority-reads to allow T to commit; and (3) construct $t\_mui(T)$ and forward it to accessed copies upon commit.

Only the third needs some detail in case several different copies of the same object O are in $R_\varphi(T)$. If the access is a majority-read, $t\_dv(T)$ and $t\_cn(T)$ are forwarded to all accessed copies of O as is the latest version of O to all old copies in the majority, if any. If the access is a minority-read, then $h(T)$ can choose any subset of copies to read (containing one with the highest version number or deliberately only older ones), as long as the conditions on $t\_cn(T)$, which depends on inf($c\_Wmcn(o_i)$, $o_i \in R_\varphi(T)$), are respected; $t\_dv(T)$, $t\_cn(T)$ and the most recent accessed version are then forwarded to these copies. This ensures that for any copy o_i, $c\_Rdv(o_i)[O] = c\_Wdv(o_i)[O] = v(o_i)$.

5.5. Example

Consider again *example 2* of §3.2. Define for every object more than half the total number of copies as a majority. All version numbers of copies are equal to zero at initialization. Objects X, Y, and Z are ordered by $o\_cn(X) < o\_cn(Y) < o\_cn(Z)$.

Transaction T_1 reads all copies of Y and incompletely updates X on copies x_1, x_2, x_3. Following the rules above,
- $t\_cn(T_1) = o\_cn(X)$ and $t\_dv(T_1) = (1,0,nil)$ are posted at copies x_1, x_2, x_3, y_5, y_6;
- $c\_Wdv(x_i) = c\_Rdv(x_i) = c\_Rdv(y_j) = t\_dv(T_1)$ and $c\_Wmcn(x_i) = c\_Rmcn(x_i) = c\_Rmcn(y_j) = t\_cn(T_1)$.

Transaction T_2 reads copies of x_4, x_5 and completely updates Z on copy z_4. Similarly as above,
- $t\_cn(T_2) = \infty$ and $t\_dv(T_2) = (0,nil,1)$ are posted at copies x_4, x_5, z_4;
- $c\_Wdv(z_4) = c\_Rdv(z_4) = c\_Rdv(x_i) = t\_dv(T_2)$ and $c\_Wmcn(z_4) = c\_Rmcn(z_4) = c\_Rmcn(x_i) = t\_cn(T_2)$.

Both T_1 and T_2 can commit immediately. Consider a third transaction T_3 which reads copy x_4 and updates y_5 and y_6. Then,
- $t\_cn(T_3) = o\_cn(X)$, which does not yet impair minority-read of X on x_4 because condition (c) of §5.2 might still be true; however,
- $\max\{c\_Rdv(o_i)[X], o_i \in W_\varphi(T_3)\} = c\_Rdv(y_j)[X] = 1$, from which it follows that T_3 must be aborted, because $v(x_4)=0$ and T_3 depends on a transaction (here T_1) having written version 1 of X.

5.6. Read-only transactions

We assumed that all transactions wrote at least one object. In fact transactions can also be *read-only*, leaving no trace in the system. Read-only transactions need not be serialized relatively to transactions which will depend on them (*read-write* conflicts). Moreover this would introduce undesirable restrictions on further transactions. However, they must be correctly serialized relatively to transactions on which they depend. So we consider a read-only transaction as a transaction aborting at commit-point. To reach commit-point, a read-only transaction must respect the access rules given above. Once it commits, we do not forward any information concerning it to copies read (just as if it had aborted). This enables us to include this particular case in the general scheme without modification.

5.7. Forwarding updates in background mode

We have considered updates of RO's to be executed on any subset of copies ensuring that the notion of majority is consistent. This allows for incomplete updates and for old reads in the system. However, an incomplete update on object O by transaction T inhibits minority-reads on copies of objects X with $o\_cn(X) > o\_cn(O)$ by transactions depending on T. If these restrictions are left free to proliferate, it is likely that the availability in the system will soon become as bad as in traditional pessimistic schemes. So it is necessary to design some mechanism to bring old copies up to date. We have described how this is partially realized by transactions reading subsets of different copies of the same object in §5.4. This however does not relieve the constraints due to c_Wmcn's of these copies.

We propose the following method, which has not yet been formally proven, to solve this problem. Assume the existence of a fault-tolerant broadcast facility [11]. This mechanism ensures that messages broadcasted to a set of nodes eventually arrive at destination, maybe after arbitrary delay due to partitions or other failures. Using this, we define an update on an object O as comprising the following two steps:

- commit the update to a sufficient subset of copies consistently with the notion of majority in the system;
- if the update was incomplete, broadcast it in background mode to all unaffected copies.

The second point is totally transparent to the writing transaction T which has already committed. When the new version of O arrives at a copy o_i, the copy is updated as are $c\_Wmcn(o_i)$ and $c\_Wdv(o_i)$ respectively with the values of $t\_cn(T)$ and $t\_dv(T)$. When all copies have been brought up to the newest version, and if no incomplete update of O has taken place in the meanwhile, completion of the second phase is signaled to all sites. These in turn must release access restrictions due to the earlier incomplete update of O. For copy x_i such that $c\_Wmcn(x_i) = o\_cn(O)$, this means updating $c\_Wmcn(x_i)$ to the smallest value $o\_cn(Y)$ such that $o\_cn(Y) > o\_cn(O)$ and x_i depends on an incomplete update of Y. This can be done by associating one more item of missing-update information to each copy, namely, a vector $c\_muv(x_i) \in \{0,1\}^N$ such that

$c\_muv(x_i)[Y]=1 \Leftrightarrow x_i$ depends on an incomplete update of Y.

Forwarding this information to transactions follows the same dual strategy as described in §5.1. To optimize research of copies with $c\_muv(x_i)[Y]=1$, we can maintain on each site s a vector $s\_muv(s)$ such that $s\_muv(s)[Y]$ is the list of all copies $x_i \in s$ with $c\_muv(x_i)[Y]=1$.

When update of O is completed on site s, the site looks its local entry $s\_muv(s)[O]$ up and updates $c\_Wmcn(x_i)$ of every copy x_i in this list to o_cn of the first object Y such that $o\_cn(Y) > o\_cn(O)$ and $c\_muv(x_i)[Y]=1$. This releases restrictions on minority reads of x_i which were due to the incomplete update of O.

6. Conclusions

We have shown in this paper that pessimistic replica control doesn't need to enforce mutual consistency of accessible copies. Our approach is equivalent to considering a failure-prone network as one having arbitrarily long communication delays. In such an asynchronous context, the notion of partition is only a special case of the communication model, so the classical notion of adaptation during partition and recovery after repair [5] is not necessary. Our algorithm takes advantage of this situation by considering a pessimistic control protocol allowing reads on old copies and using an efficient (i.e. fault-tolerant) broadcast protocol to eventually bring old copies up to date.

Our approach to consistency control is syntactic. However, semantic knowledge on transactions can be used in several manners to enhance performance of the method. In [19] we described how choosing the order on replicated objects can do this. Another way is to choose on line the best (i.e. most performing) conditions on reads. This can mean reading any old copy if possible, or choosing the first good available copy if some old ones are inconsistent, or finding a read-quorum of copies to ensure reading of a new copy, or aborting the transaction if the desirable alternative is not possible, but the next highest one in the list is considered too expensive.

Compared to traditional pessimistic algorithms, this one is surprisingly complicated. This is due to the generality of our conditions on pessimistic processing. Traditional algorithms impose overly restrictive (and thus easily implemented) conditions for serializability. Ours imposes very weak conditions on accessibility, i.e. conditions close to necessary and sufficient conditions for correct scheduling, which is hard to implement. Note however that necessary and sufficient conditions on accessibility for pessimistic control do not exist. This is because any pessimistic multiple-se ordering algorithm can only handle a fixed class of orders on serialization edges, thus limiting availability in a strictly too restrictive manner in respect to the transactions that might execute in the system.

References

[1] A.El Abbadi, S.Toueg: "The Group Paradigm for Concurrency Control Protocols", ACM-SIGMOD Conf. on Management of Data 6/1988, pp.126-134.

[2] A.El Abbadi, S.Toueg: "Maintaining Availability in Partitioned Replicated Databases", ACM Trans. on Database Systems, 14:2, 6/1989, pp.264-290.

[3] P.A.Bernstein, N.Goodman: "The Failure and Recovery Problem for Replicated Databases", ACM 2nd Annual Symp. on Principles of Distributed Computing 1983, pp.114-122.

[4] P.A.Bernstein, V.Hadzilacos, N.Goodman: "Concurrency Control and Recovery in Database Systems", Addison-Wesley 1987.

[5] B.K.Bhargava: "Concurrency Control and Reliability in Distributed Systems", chap. 1, van Nostrand-Reinhold, 1987.

[6] D.Davcev: "A Dynamic Voting Scheme in Distributed Systems", IEEE-TSE, 15:1, 1/1989, pp.93-97.

[7] S.B.Davidson: "Optimism and Consistency in Partitioned Distributed Database Systems", ACM Trans. on Database Systems, 9:3, 9/1984, pp.456-481.

[8] S.B.Davidson, H.Garcia-Molina, D.Skeen: "Consistency in Partitioned Networks", ACM Computing Surveys, 17:3, 9/1985, pp.341-370.

[9] D.L.Eager, K.C.Sevcik: "Achieving Robustness in Distributed Database Systems", ACM Trans. on Database Systems, 8:3, 9/1983, pp.354-381.

[10] K.P.Eswaran, J.N.Gray, R.A.Lorie, I.L.Traiger: "The Notions of Consistency and Predicate Locks in a Database System", Comms. of the ACM, 19:11, 11/1976, pp.624-633.

[11] H.Garcia-Molina, B.Kogan: "An implementation of Efficient Broadcast using an Unreliable Multicast Facility", 7th IEEE Symposium on Reliable Distributed Systems, 10/1988, pp.101-111.

[12] D.K.Gifford: "Weighted Voting for Replicated Data", 7th ACM Symp. on Operating Systems Principles, 1979, pp.150-162.

[13] C.-L. Huang, V.O.K. Li: "Missing-Partition Dynamic Voting Scheme for Replicated Database Systems", 9th IEEE ICDCS, 1989, pp.579-586.

[14] C.-L. Huang, V.O.K. Li: "Regeneration-Based Multiversion Dynamic Voting Scheme for Replicated Database Systems", 10th IEEE ICDCS, Paris 1990, pp.370-377.

[15] S.Jajodia, D.Mutchler: "A Pessimistic Consistency Control Algorithm for Replicated Files which Achieves High Availability", IEEE-TSE, 15:1, 1/1989.

[16] J.E.B.Moss: "Nested Transactions, an Approach to Reliable Distributed Computing", MIT Press, 1985.

[17] G.Neiger, S.Toueg: "Automatically Increasing the Fault-Tolerance of Distributed Systems", ACM PODC 1989, pp.248-262.

[18] B.M.Oki, B.H.Liskov: "Viewstamped Replication: A New Primary Copy Method to Support Highly-Available Distributed Systems", 7th ACM Symp. on Principles of Distributed Computing, 8/1988, pp.8-17.

[19] A.Sandoz: "Achieving High Availability in a Replicated File System by Dynamically Ordering Transactions", 10th IEEE ICDCS, Paris, 6/1990, pp.432-439.

[20] M.Singhal: "Update Transport: A New Technique for Update Synchronization in Replicated Database Systems", to appear in IEEE-TSE, 12/1990.

Localized-Access Protocols for Replicated Databases*

D. Agrawal   A. El Abbadi

Department of Computer Science
University of California
Santa Barbara, CA 93106

Abstract

In this paper, we present two protocols for efficient execution of transactions in replicated databases. Transactions are executed at a single site thus avoiding communication overhead and distributed commitment, which are required by most other replica control protocols. In the first protocol, data accessibility at a site can be dynamically reconfigured using special transactions, which are executed on demand. In the second protocol, data accessibility is reconfigured by migrating ownership of individual objects in the database. The two protocols present trade-offs with respect to atomicity, resiliency, and data availability. The approach of local execution of user transactions improves response time, eliminates the need for distributed commit protocols, and accommodates database heterogeneity.

1 Introduction

Data replication is traditionally used for increasing data availability and fault-tolerance in distributed systems. When a data object is replicated on several sites, the object may still be available for reading or writing even after failures have occurred. Thus data replication may increase data availability and hence increase the fault-tolerance of the database. On the other hand, replication also holds the promise of reducing communication costs incurred when a transaction can access locally stored data. If data is stored at the site where a transaction is initiated, operations can be executed locally, thus reducing both communication costs and the response time. However, local execution is

---

*This research is supported by the NSF under grant numbers CCR-8809387 and IRI-8809284.

often not possible since several sites may need to be accessed for replica synchronization. In this paper, we assume that sites are *fail-stop* [SS82] and communication links may fail to deliver messages. Combinations of such failures may lead to *network partitioning* [DGMS85], We propose two schemes that employ replicated data primarily to eliminate remote data access by transactions, resulting in better performance in distributed databases.

Most replicated data management protocols have focussed on increasing data availability and fault-tolerance, often at the expense of communication costs. Several protocols are based on the quorum approach proposed by Gifford [Gif79] where both read and write operations must access several copies of an object, thus requiring communication between several sites. Few attempts have, however, been made to use replication to improve performance in distributed databases. The virtual partitions, the dynamic quorum, and the views protocols [ESC85, Her87, ET89a] address the issue of communication costs by always allowing a read operation to be executed by accessing a single copy. Write operations, on the other hand, must access several copies. The viewstamped replication protocol proposed by Oki and Liskov [OL88] extends the basic primary site approach [Sto79], and allows both read and write operations to be executed at a single site. However, before a transaction can commit it must communicate with a set of sites to inform them of its results. This protocol therefore reduces the overall response time of transactions, but does not effect the communication requirements. Kogan and Garcia-Molina proposed Bakunin networks [KGM87] where read operations can be executed locally at any site while write operations must be executed locally at specific, predefined sites. This approach reduces the communication costs of executing transactions and results in low response time. It, however, requires pre-analysis of the transactions in the database to determine which sites can support write operations.

More recently, several protocols have been proposed to eliminate the problem of distributed commitment that arises in distributed databases. Tam and Hsu [TH90] have used the idea of tokens to localize the access of data objects in distributed databases. Tokens are used to migrate data among the sites while guaranteeing data coherency. Special token transactions reliably manage the transitions of tokens. A special unilateral protocol is used to commit these token transactions. Soparkar and Silberschatz [SS90] have proposed a more radical approach to execute transactions locally by partitioning the values of data objects among several sites. This technique, however, is restrictive since it is only applicable to replicated databases that permit data-value partitioning.

In this paper, we employ replication to reduce communication costs as well as to maintain data availability in distributed databases. We develop two localized-access replica control protocols that allow user transactions to execute at the sites where they are initiated. This approach results in three major benefits. First, user transactions require low communication costs, thus improving their response time. Second, distributed commit protocols are not required and therefore resources are not blocked unnecessarily due to failures. Third, since user transactions are executed and synchronized locally, the need for a distributed concurrency control and recovery protocols is eliminated. This reduces the complexity of these protocols, since user transactions do not have to access any data stored at remote sites. Hence, the localized-access replica control protocols can be easily integrated in a heterogeneous environment with autonomous concurrency control and recovery protocols at different sites.

Our first protocol localizes access of data objects by executing a special system transaction. By executing these system transactions, data accessibility is localized while allowing old information to be read at other sites. The second protocol localizes access on a per object basis. This results in a more decentralized protocol with weaker atomicity requirements. Both protocols ensure serializability; however, they provide varying degrees of data availability and require different levels of atomicity. The paper is organized as follows. In the next section we present the transaction based localized-access replica control protocol. The object based localized-access replica control protocol is described in Section 3. The paper concludes with a discussion of our results.

2 The Transaction based Localized-Access Protocol

A distributed replicated database is a collection of objects where each object is implemented by a set of copies residing on different sites. We use the notation x_s to denote a copy of an object x residing on site s. Users interact with the database by invoking a transaction, which is a partially ordered sequence of read and write operations that are executed atomically on the objects. To be consistent, a replicated object should be *one-copy equivalent*, i.e., it should behave as if each object has only one copy in so far as the users can tell. A *replica control protocol* is one that ensures that the database is one-copy equivalent. A database system should also ensure *serializability* [EGLT76], i.e., if operations of transactions are interleaved, the system behaves as if all the transactions were executed in some serial order. A *concurrency control protocol* is one that ensures serializability [EGLT76, Ree78, KR81, AE90]. A replicated database system is correct if it is *one-copy serializable* [BG87], i.e., it ensures serializability and is one-copy equivalent.

2.1 Overview of the Protocol

In this protocol, each site, s, has an *owner set*, $OwnerSet[s]$, which is a set of all objects that may be accessed (read or written) by transactions executing at s. Also, each site s has a *read-only set*, $ReadOnlySet[s]$, which contains objects that may only be read by transactions executing at s. All user transactions are executed *locally*, i.e., they read and write the copies that reside on the site where they are issued. To execute a read operation on an object x, a transaction executing at site s must first determine whether x is accessible locally, i.e., it is in $ReadOnlySet[s]$ or in $OwnerSet[s]$. Similarly, for write operations on x, x must be in $OwnerSet[s]$. If the object is accessible, the operation is executed locally by using the local concurrency control protocol. Otherwise, the transaction is aborted. Different sites may use different concurrency control protocols to ensure serializability. Note that the two sets are treated as any other object in the database and therefore a transaction uses the local concurrency control protocol to access them.

When a transaction aborts due to the inaccessibility of an object x at a site s, it submits a request for modifying $OwnerSet[s]$ to include x. All such requests are collected by site s and periodically it initiates a special transaction to update $OwnerSet[s]$ and $ReadOnlySet[s]$. The transactions that update these sets are called *system transactions*. Note that if the read and write sets of user transactions are known *a priori*, unnecessary aborts of user transactions can be avoided by scheduling system transactions more efficiently.

2.2 System Transactions

We now describe how the owner and read-only sets are reconfigured by executing system transactions. In order to implement system transactions, additional information is maintained at every site. At a site s, each copy x_s of an object x has associated with it an *access identifier*, $AccessId[x_s]$, as well as a *temporary access identifier*, $TempId[x_s]$. Furthermore, each system transaction T_s has a transaction identifier $Tid[T_s]$, which is associated with all messages it sends. A transaction identifier is a unique timestamp and all transaction identifiers form a total order.

Initiator s	Participant p
$Tid[T_s] = timestamp;$ $\forall x \in NewOwnerSet[s]:$ $\quad TempId[x_s] = Tid[T_s];$ $send(T_s, NewOwnerSet[s])$ to all participants;	$receive(T_s, NewOwnerSet[s]);$ $GrantedSet[p] = \{x \mid x \in NewOwnerSet[s] \land$ $\quad Tid[T_s] > Max(AccessId[x_p], TempId[x_p])\};$ /* Execute reads according to local cc */ $ValOfGrantedSet[p] =$ $\quad \{\langle x_p, AccessId[x_p]\rangle \mid x \in GrantedSet[p]\};$ $ReadOnlySet[p] = ReadOnlySet[p] \bigcup$ $\quad (OwnerSet[p] \cap GrantedSet[p]);$ $OwnerSet[p] = OwnerSet[p] \setminus GrantedSet[p];$ $send(GrantedSet[p], ValOfGrantedSet[p]);$
$receive(GrantedSet[p], ValOfGrantedSet[p])$ $\quad$ from all participants $p;$ $OwnerSet[s] = \{x \mid$ all copies of x were read $\};$ $ReadOnlySet[s] = \emptyset;$ $\forall x \in OwnerSet[s]:$ $\quad$ /* Execute writes according to local cc */ $\quad$ Let x_q be the copy with the highest access id; $\quad x_s$ is copied from $ValOfGrantedSet[q];$ $\quad AccessId[x_s] = Tid[T_s];$	

Figure 1: Execution of a System Transaction T_s at site s

The execution of a system transaction T_s initiated at site s is illustrated in Figure 1. T_s determines its transaction identifier $Tid[T_s]$, which is greater than any of the access identifiers of copies at site s. $Tid[T_s]$ is used to guarantee global serializability of system transactions. The new owner set, which is to be installed, is maintained in $NewOwnerSet[s]$. The transaction updates $TempId[x_s]$ to $Tid[T_s]$ for all objects that are member of $NewOwnerSet[s]$. Next, T_s sends requests to all sites including itself to localize the access of all objects in $NewOwnerSet[s]$, i.e., to get their permission to allow access to those objects at site s.

When a participant p receives a request from system transaction T_s, it constructs a set $GrantedSet[p]$, which contains all objects whose access can be transferred to s. An object x is included in $GrantedSet[p]$ if $Tid[T_s]$ is greater than both $AccessId[x_p]$ and $TempId[x_p]$. This is required to enforce the global synchronization of system transactions, which may be executing concurrently. Note that $GrantedSet[p]$ is a subset of $NewOwnerSet[s]$. For each object x in $GrantedSet[p]$, T_s reads copy x_p by using the local concurrency control protocol. Also, if x is in $OwnerSet[p]$, the transaction removes x from $OwnerSet[p]$ and includes it in $ReadOnlySet[p]$. This ensures that user transactions at different sites do not execute conflicting writes on the same object. Read operations, however, may read old data.

After receiving $GrantedSet[p]$ from all participants p, T_s constructs and installs the newly configured owner set, $OwnerSet[s]$ and the new read only set, $ReadOnlySet[s]$. If all copies of x were read successfully, x is added to $OwnerSet[s]$. $ReadOnlySet[s]$ is initialized to the empty set. Finally, for all objects x in $OwnerSet[s]$ the system transaction updates the local copy x_s by the most recent copy written and sets $AccessId[x_s]$ to $Tid[T_s]$. The most recent copy for an object x is copy x_q such that $AccessId[x_q]$ is the maximum among all copies read by T_s.

2.3 Analysis of the Protocol

In order to argue the correctness, we describe the *group paradigm* [ET89b], which was developed for proving the correctness of concurrency control and replica control protocols for replicated databases. The paradigm hierarchically divides the problem of achieving one-copy serializability by introducing the notion of a *group*, which is a set of transactions. Each group g has associated with it a *local policy*, P_g, which ensures the correct serialization of all transactions in that group, e.g., a local policy may require transactions to use a specific owner set with a specific concurrency control protocol. We use $<_g$ to refer to the serialization order of the transactions in group g given by P_g. Furthermore, a *global policy*, P, is used to ensure a total order $<$ on all groups, so that when combined with the serialization order of transactions in each group, it results in the serializability of all transactions. In [ET89b], it is shown that to ensure the overall serializability of all user transactions, the total order $<$ must satisfy the following condition. If t_j reads the value of x that was written by another transaction t_i, where $t_i \in g_i$, $t_j \in g_j$, and $g_i \neq g_j$ then

1. $g_i < g_j$.

2. There is no g_k with a transaction that writes x such that $g_i < g_k < g_j$.

3. In the serialization order, $<_{g_i}$, of g_i determined by local policy P_{g_i}, t_i is the last transaction to write x in g_i.

We now use the group paradigm to argue the correctness of the transaction based localized-access protocol. A *group* is defined to be any set of transactions that use owner and read-only sets installed by a single system transaction at a site. This system transaction is uniquely identified by its transaction identifier. Since all transactions use a correct concurrency control protocol, hence each group is serializable, thus satisfying the local policy condition. We now argue that the global policy is satisfied, i.e., there exists a total order on all groups that satisfies conditions 1, 2, and 3. The total order $<$ is defined by the transaction identifiers associated with system transactions installing the owner sets used by transactions in the group.

User transactions either read values written by other user transactions executing at the same site, or by a system transaction installing the owner set containing the objects being read. Since the global policy only restricts transactions reading values written in other groups, we only have to consider system transactions. A system transaction, T_s, requesting to read a copy is rejected by all copies with access identifiers greater than $Tid[T_s]$. This ensures that a system transaction always reads a value written in a group associated with a system transaction with a lower transaction identifier, thus satisfying condition 1.

A system transaction, T_s that reads an object x removes it from the owner set and includes it in the read-only set of the owner site. This ensures that user transactions executing write operations at those sites (with access identifiers less than $Tid[T_s]$), and which follow T_s's local read operations in the serialization order are aborted. Read operations can still be executed at previous owner sites since they do not violate condition 2. Note, however, this data may be old (stale data), and transactions in this group may read the stale data and can modify any objects still in the owner set. System transaction T_s reads the copy with the highest access identifier. Furthermore, since any site that allows x to be accessed must read all copies, any later system transaction that succeeds in installing an owner set including x must have a transaction identifier greater than $Tid[T_s]$, and thus condition 2 is satisfied. Finally once a system transaction reads a copy of x written in group g, no user transaction can write that copy in g, and hence the copy read by T_s is the final value written in that group. Hence, condition 3 is satisfied.

We now analyze the communication costs incurred by the above protocol. If the total number of system transactions executed is M and there are n sites in the system then the communication cost will be $2Mn$. User transactions execute locally, hence, incur no communication costs. This is in contrast to the quorum protocol [Gif79] where each individual transaction incurs communication costs. Let N be the total number of user transactions and k be the average number of objects accessed per transaction. If we only consider the communication costs of accessing k objects and ignore the cost of committing the transactions, the communication costs in the quorum protocol is $2kN\frac{n}{2}$ (where quorums are chosen as a majority). Thus, the transaction based localized-access protocol has a communication cost advantage over the quorum protocol as long as the number of system transactions is less than or equal to $\frac{kN}{2}$. In general, kN is expected to be a large number for database systems and therefore the localized-access protocol will be useful in such environments. In the worst case, if each user transaction results in a system transaction, the localized-access protocol will require less communication as long as user transactions access at least two objects.

One of the main problems in designing fault-tolerant distributed database protocols is the blocking of transactions in the presence of failures. A distributed transaction is *blocked* if some of its components can neither commit nor abort. Skeen has shown that there exists no non-blocking distributed commit protocol [Ske82]. In our protocol, system transactions are the only transactions that access distributed data. However, the semantics of these transactions does not require a distributed commit protocol. Neither the participants nor the initiator of a system transaction may suffer from blocking. In particular, a system transaction can decide to commit unilaterally without waiting for all the responses. A premature commit decision may only result in a smaller owner set. For simplicity, we require that a system transaction initiated at s reads all copies of an object x, in order to include it in $OwnerSet[s]$. However, several optimizations can be incorporated that will require x to be read only at the previous owner site.

3 The Object based Localized-Access Protocol

In this section, we present a protocol where object accessibility is localized by dynamically transferring the ownership on a per object basis. The protocol is based on the notion of cache-coherency which is used as the correctness criteria in shared virtual memory systems [LH89]. Unlike the transaction based protocol, this protocol does not require a single distributed system transaction to update all owned copies at the site. Also, it does not require a site to receive messages from all sites when requesting ownership transfers. This results in a more resilient protocol. When used in conjunction with a concurrency control mechanism, e.g., two phase locking, this protocol guarantees one-copy serializability. Also, unlike the previous protocol, user transactions in this scheme are never permitted to read stale data.

3.1 Overview of the Protocol

The database model is the same as described in the previous section. We consider a distributed database in which the information about objects is categorized in two components: *data* and *control information*. The degree of data replication changes dynamically during system execution. However, control information for an object is fully replicated. If a site is an *owner* of an object x, then user transactions at that site can access (read and write) x. The control information keeps track of the possible owner of an object. Thus, for an object x at site s, the following variables are stored: $x_s.data$, $x_s.owner$, and $x_s.time$. At $x_s.time$, which is a logical time [Lam78], $x_s.owner$ became the owner of x.

As before, each site s has associated with it an owner set, $OwnerSet[s]$ but no read-only set.

User transactions use the local concurrency control mechanism to ensure serializability. When a user transaction executes at a site s and issues an operation on object x, the operation is executed if x is in $OwnerSet[s]$. Otherwise, the transaction initiates a request at site s to transfer ownership of x to s. The transaction may abort if the ownership of some of the objects, which it attempts to access, cannot be transferred to the local site.

3.2 Transferring the Ownership of Objects

We now describe how ownership is transferred between different sites on a per object basis. Similar approaches have been used to solve the distributed mutual exclusion problem [CR83, SK85, Ray89, BAA89] and the data-migration problem [TH90]. Unlike our scheme, the token or privilege based distributed mutual exclusion algorithms do not explicitly handle failures. Also, these algorithms do not use the notion of logical time [Lam78] to solve the distributed mutual exclusion problem. Tam and Hsu [TH90] also use the notion of tokens for migrating data but maintain the ownership information at a centralized locator. Our approach differs from the above solutions in that we use the combined notion of tokens (or ownership) and logical time to localize data access in a fault-tolerant manner.

The transfer ownership operation is illustrated in Figures 2 and 3. Figure 2 illustrates the execution of the protocol when no adverse failures are encountered during its execution. The algorithm in Figure 3 is initiated only when the normal mode of ownership transfer times out possibly due to site and/or communication failures. We assume the existence of a logical clock [Lam78] that captures the causality of communication events. Note that the region of code bracketed within "$\ll$" and "$\gg$" is executed atomically at a site.

In the normal mode, the initiator site s sends a request to transfer the ownership of object x to itself. This request is sent to $x_s.owner$, say p, along with $x_s.time$. Site s assumes that p is the current owner of x; if this assumption is incorrect, s's request will be forwarded to another site. In general, if a receiving site p is not the current owner and $x_p.time$ is larger than the time included in the request by the immediate sender, p forwards the transfer request to $x_p.owner$. Assume site p receives a request initiated at site s forwarded by sender p'. If p is the current owner and x is not being used by local transactions at p (e.g., x is not locked), p responds by transferring ownership to s and updating its control information accordingly. That is, $x_p.owner$ is set to s and $x_p.time$ is set to the current logical time at p. Finally, $x_p.data$ is sent to s. On the other hand, if $x_p.time$ is smaller than the time in the forwarded request from p' then p' had transferred the ownership of x to p but due to failures p is unaware of it. In this case, p fetches $x_{p'}.data$ which is the current value of x, and transfers the ownership to s as described above. We will argue the correctness of this algorithm in the next subsection.

initiator s	participant p
$send(Tid, s, s, x, x_s.time)$ to $x_s.owner$	
	$receive(Tid, initiator, p', x, time)$;
	IF $x_p.owner \neq p \wedge time < x_p.time$ THEN
	/* Forward the request */
	$send(Tid, initiator, p, x, x_p.time)$ to $x_p.owner$;
	ELSE IF $x_p.owner = p$ THEN
	/* Transfer Ownership to Initiator*/
	1. $\ll x_p.owner = initiator$;
	$x_p.time = LogicalClock_p; \gg$
	$send(Tid, x_p.data)$ to $intiator$;
	ELSE IF $x_p.owner \neq p \wedge time > x_p.time$ THEN
	• Get $x_{p'}.data$;
	/* Transfer Ownership to Initiator */
	2. $\ll x_p.owner = initiator$;
	$x_p.time = LogicalClock_p; \gg$
	$send(Tid, x_p.data)$ to $intiator$;
	ENDIF;
$receive(Tid, x_s.data) \longrightarrow$	
3. $\ll x_s.owner = s$;	
$x_s.time = LogicalClock_s; \gg$	
‖ TIMEOUT $\longrightarrow$	
/* execute the failure mode (Figure 3) */	

Figure 2: Normal Mode of the ownership transfer algorithm initiated at site s

The above algorithm has a drawback that if a site in the chain of forwarded requests is down or if a forwarded request is not delivered, the ownership transfer will not complete. If this happens, the initiator of the transfer request times out and executes the failure mode of the algorithm illustrated in Figure 3. The initiator broadcasts a request for transfer of ownership to all sites. If the current owner p receives the request it transfers ownership to the initiator. The initiator completes the ownership transfer by updating its local variables. All other sites reply with their owner and time information for object x. The initiator may not receive the ownership information if the current owner is down or the last transfer of ownership was incomplete due to site or communication failures. If the current owner is down then the transfer operation is aborted. On the other hand, if the last ownership transfer was incomplete due to failures, the initiator waits to hear from two sites u and v such that $x_u.owner$ points to v and $x_v.time$ is smaller than $x_u.time$. As will be argued later, the current data resides at u. Initiator s gets the data from u and, u transfers ownership from v to s, i.e., sets $x_u.owner$ to s. If no such sites respond then the ownership transfer is aborted.

3.3 Proof of Correctness

In this section, we argue the correctness of the object-based localized-access protocol by demonstrating that there is at most one site that can access each object at any time. Initially we assume that for each object x there is exactly one owner s and for all sites p, $x_p.owner = s$ and $x_p.time < x_s.time$ $(p \neq s)$.

Theorem 1 The following is an invariant in the system:

$$[(\exists \text{ exactly one } u \ni x_u.owner = u) \Rightarrow (\not\exists v, w \ni x_v.owner = w \wedge x_v.time > x_w.time)]$$

$$\wedge$$

$$[(\not\exists u \ni x_u.owner = u) \Rightarrow (\exists \text{ exactly one pair } v, w \ni x_v.owner = w \wedge x_v.time > x_w.time)]$$

Proof. The invariant initially holds since there is only one owner, all other sites point to the owner, and the timestamp associated with all other sites is less than the owner's timestamp. Next we show that the invariant holds during the execution of the normal

initiator s	participant p
$send(Tid', s, x)$ to all sites	
	$receive(Tid', s, x)$
	IF $x_p.owner = p$ THEN
	/* Transfer ownership to Initiator s */
	4. $\ll x_p.owner = initiator;$
	$x_p.time = LogicalClock_p; \gg$
	$send(Tid', x_p.data)$ to s;
	ELSE
	$send(Tid', x_p.owner, x_p.time)$ to s;
Wait for a reply from the Owner $\longrightarrow$	ENDIF;
/* copy $x_s.data$ from the reply */	
5. $\ll x_s.owner = s;$	
$x_s.time = LogicalClock_s; \gg$	
|| Wait for replies from sites $u, v \ni$	
$x_u.owner = v$ AND $x_u.time > x_v.time \longrightarrow$	
6. • u transfers ownership to s;	
• s gets data and ownership from u;	
|| TIMEOUT $\longrightarrow$	
Abort;	

Figure 3: Failure Mode of the ownership transfer algorithm initiated at site s

mode and the failure mode of ownership transfer algorithm. In both figures, the statements that effect the invariant are numbered. We argue that the execution of these statements leaves the invariant true.

Statement 1 atomically invalidates the antecedent of the first conjunct. We need to show that the consequent of the second conjunct holds. Exactly one pair p and s are created such that $x_p.owner = s$ and $x_p.time > x_s.time$. The latter is true because the algorithm uses logical clocks and $x_s.time$ has not yet been updated.

From the invariant and the boolean condition, Statement 2 is executed when there is exactly one pair p and p' such that $x_{p'}.owner = p$ and $x_{p'}.time > x_p.time$. In this case, from the invariant we can conclude that there is no owner of x. Hence, the antecedent of the second conjunct is true. After the execution of statement 2, the pair p and p' is atomically replaced by a new pair p and s such that $x_p.owner = s$ and $x_p.time > x_s.time$. Thus the invariant still holds since exactly one pair p and s is created.

Finally, when statement 3 is executed the receipt of the message transfers the ownership from p to s. Hence there is no owner implying that there exists only one pair p and s such that $x_p.owner = s$ and $x_p.time > x_s.time$ (i.e., after statement 1). However, after executing statement 3, the pair is replaced since $x_s.time$ is updated and s becomes the owner. This ensures that the antecedent of the first conjunct holds, and since the pair is eliminated the consequence of the same conjunct holds too.

A similar analysis can be used for the failure mode of the algorithm to show that the invariant holds. □

From the above theorem, at most one copy of an object can be accessed at all times, the protocol ensures data coherency and hence one-copy equivalence. Since objects are accessed using the local concurrency control mechanism, one-copy serializability is guaranteed.

We now argue that if a site requests transfer of ownership, it will succeed if there are no failures. From Theorem 1, we can assert that either there exists a site with the ownership of an object or there exists a pair of sites with incomplete transfer of ownership. In the first case, the ownership can be easily transferred. In the second case, the pair v, w is such that v contains the most recent version of data and w is the rightful owner (although w does not know it). The requesting site transfers the data and ownership from v and completes the transfer successfully.

The protocol is resilient to a large class of failures, e.g., site failures, lost messages, out-of-order messages. In the normal mode of the protocol, a site will be successful in acquiring the ownership as long as all the sites in its request forwarding chain are able to forward its request. Note that site failures after the request has been forwarded do not effect the algorithm. In the best case, the ownership may be available locally while in the worst case it may require n messages, where n is the total number of sites in the system. If the forwarding chain is broken due to failures, the failure mode is initiated. In this mode, the initiating site broadcasts its request to all sites. Since the requesting site needs to hear either from the owner or from a pair of specific sites, the protocol can tolerate the failure of all other sites. The message cost for the failure mode is always n messages. For this reason, the broadcast mode should be used only as the last alternative.

4 Discussion

In this paper we propose two different approaches for the design of replica control protocols that allow transactions to execute locally in a distributed environment. The two approaches for localizing access to distributed data represent contrasting alternatives. The first approach is based on executing system transactions that require all ownership transfers to be performed atomically in the entire system. The advantage of this protocol is that it permits higher data availability since old information can still be read at non-owner sites. The second approach transfers ownership on a per object basis and need only be executed incrementally for a single object. This protocol only needs local atomicity and provides higher resiliency to failures. The first protocol fails to transfer ownership of an object if any copy of that object is inaccessible. The second protocol fails only when the owner of the object is inaccessible or if any one of a pair of sites, which were involved in an incomplete transfer, is unavailable. When either of these protocols is unable to localize the access to an object, user transactions attempting to access that object cannot execute.

Local execution of user transactions results in three major benefits: low response time of user transactions, eliminating the need for distributed commit protocols, and supporting heterogeneity of concurrency control and recovery protocols. Response time is minimal if all the objects being accessed by a transaction are owned at the local site. However, this is at the expense of response time of those transactions which result in system transactions or ownership transfer requests for localizing the access of remote data. The most important aspect of our approach is that it eliminates the need for distributed commitment. Distributed commit protocols have high communication overhead and still may block due to partition failures [Ske82]. Blocking is undesirable in distributed databases since it results in loss of data availability and wasted resources. In both our approaches, the system transactions and the ownership transfer requests do not block even when failures occur. In heterogeneous databases, global synchronization of distributed transactions is difficult because different sites may use different concurrency control and recovery mechanisms. Using the localized-access protocols, user transactions are executed locally, thus completely eliminating any interaction between user transactions executing on different sites.

References

[AE90]   D. Agrawal and A. El Abbadi. Locks with Constrained Sharing. In *Proceedings of the Ninth ACM Symposium on Principles of Database Systems*, pages 85–93, April 1990.

[BAA89]   J.M. Bernabéu-Aubán and M. Ahamad. Applying a Path-Compression Technique to Obtain an Efficient Distributed Mutual Exclusion Algorithm. In *Proceedings of the Third International Workshop on Distributed Algorithms*, pages 33–44, September 1989.

[BG87]   P. A. Bernstein and N. Goodman. A Proof Technique for Concurrency Control and Recovery Algorithms for Replicated Databases. *Distributed Computing, Springer-Verlag*, 2(1):32–44, January 1987.

[CR83]   O. Carvalho and G. Roucairol. On Mutual Exclusion in Computer Networks. *Communications of the ACM*, 26:146–147, February 1983.

[DGMS85]   S. B. Davidson, H. Garcia-Molina, and D. Skeen. Consistency in partitioned networks. *ACM Computing Surveys*, 17(3):341–370, September 1985.

[EGLT76] K. P. Eswaran, J. N. Gray, R. A. Lorie, and I. L. Traiger. The Notion of Consistency and Predicate Locks in Database System. *Communications of the ACM*, 19(11):624–633, November 1976.

[ESC85] A. El Abbadi, D. Skeen, and F. Cristian. An Efficient Fault-Tolerant Protocol for Replicated Data Management. In *Proceedings of the Fourth ACM Symposium on Principles of Database Systems*, pages 215–228, March 1985.

[ET89a] A. El Abbadi and S. Toueg. Maintaining Availability in Partitioned Replicated Databases. *ACM Transaction on Database Systems*, 14(2):264–290, June 1989.

[ET89b] A. El Abbadi and S. Toueg. The Group Paradigm for Concurrency Control Protocol. *IEEE Transactions on Knowledge and Data Engineering*, pages 376–386, September 1989.

[Gif79] D. K. Gifford. Weighted Voting for Replicated Data. In *Proceedings of the Seventh ACM Symposium on Operating Systems Principles*, pages 150–159, December 1979.

[Her87] M. Herlihy. Dynamic Quorum Adjustments for Partitioned Data. *ACM Transactions on Database Systems*, 12(2):170–194, June 1987.

[KGM87] B. Kogan and H. Garcia-Molina. Update Propagation in Bakunin Data Networks. In *Proceedings of the Sixth ACM Symposium on Principles of Distributed Computing*, pages 13–26, August 1987.

[KR81] H. T. Kung and J. T. Robinson. On Optimistic Methods for Concurrency Control. *ACM Transactions on Database Systems*, 6(2):213–226, June 1981.

[Lam78] L. Lamport. Time, Clocks, and the Ordering of Events in a Distributed System. *Communications of the ACM*, 21(7):558–565, July 1978.

[LH89] K. Li and P. Hudak. Memory Coherence in Shared Virtual Memory Systems. *ACM Transactions on Computer Systems*, 7(4), November 1989.

[OL88] B. Oki and B. Liskov. Viewstamped Replication: A New Primary Copy Method to Support Highly-Available Distributed Systems. In *Proceedings of* the Seventh ACM Symposium on Principles of Distributed Computing, pages 8–17, August 1988.

[Ray89] K. Raymond. A Tree-Based Algorithm for Distributed Mutual Exclusion. *ACM Transactions on Computer Systems*, 7(1):61–77, February 1989.

[Ree78] D. P. Reed. Naming and Synchronization in a Decentralized Computer System. Technical Report MIT-LCS-TR-205, Massachusetts Institute of Technology, Cambridge, Massachusetts, September 1978.

[SK85] I. Suzuki and T. Kasami. A Distributed Mutual Exclusion Algorithm. *ACM Transactions on Computer Systems*, 3(4):344–349, November 1985.

[Ske82] D. Skeen. *Crash Recovery in a Distributed Database Systems*. PhD thesis, Department of Electrical Engineering and Computer Science, University of California at Berkeley, 1982.

[SS82] R. Schlichting and F. B. Schneider. Fail-Stop Processors: An Approach to Designing Fault-Tolerant Computing Systems. *ACM Transactions on Computer Systems*, 1(3):222–238, August 1982.

[SS90] N. Soparkar and A. Silberschatz. Data-value Partitioning and Virtual Messages. In *Proceedings of the Ninth ACM Symposium on Principles of Database Systems*, pages 357–367, April 1990.

[Sto79] M. Stonebraker. Concurrency Control and Consistency in Multiple Copies of Data in Distributed INGRES. *IEEE Transactions on Software Engineering*, 3(3):188–194, May 1979.

[TH90] V. O. Tam and M. Hsu. Token Transactions: Managing Fine-Grained Migration of Data. In *Proceedings of the Ninth ACM Symposium on Principles of Database Systems*, pages 344–356, April 1990.

WEIGHTED VOTING FOR OPERATION DEPENDENT MANAGEMENT OF REPLICATED DATA[1]

Mirjana Obradovic   Piotr Berman

Department of Computer Science
333 Whitmore Laboratory
Penn State University
University Park, PA 16802

Abstract

We consider the problem of finding an optimal static pessimistic replica control scheme. It has been recognized that operation mix plays an important role in finding optimal schemes. We demonstrate that *voting* provides the highest possible availability for fully connected networks and Ethernet systems for the cases of one or two operations. We introduce a technique for reducing the number of operations considered in the analysis. Using this technique we extend the above results to all cases of three operations.

1 Introduction

In distributed systems, replication of data is a way to improve the reliability and performance. When several copies of the same data items are stored at different sites (nodes), read/write operations may be possible in spite of site or link failures, even if they lead to network partitions. The performance may improve when we access the local copy of a data item, rather than a copy at a remote site. Therefore, we need to study mechanisms (protocols) that ensure the consistency of replicated data.

Several protocols for controlling data replication have been proposed so far (for an excellent survey see [4]). Here we are interested exclusively in *pessimistic* protocols which assure the data consistency without making any application dependent assumptions.

The most popular pessimistic protocol is *voting* [7,14]: every node is assigned a number of votes while every operation must collect certain quorum of votes in order to be performed. In the simplest case, the quorum is equivalent to the majority of votes. The main advantage of the voting scheme is its easy implementation.

Voting is an example of a *static* scheme, because partition groups allowed to perform an operation are declared in advance. Static schemes are simple to implement and have

---

[1]This work was partially supported by AFOSR contract 87-0400 and NSF grant CR 8805978.

potentially lower overhead, since nodes react only to transaction requests, rather than to changes in the status of the network, and they always react to the requests in the same manner. Therefore, even when dynamic schemes offer better availability, it is worthwhile to identify the optimal static scheme, evaluate its performance and decide whether the gains of a dynamic scheme sufficiently justify the increased complexity of the protocol. Besides, static protocols may serve as components of dynamic protocols.

This paper focuses on *operation dependent* schemes, in which a decision whether to perform an operation depends also on its type. These schemes may yield higher availability, assuming that the operation mix is known in advance.

Discussion of the static pessimistic schemes is based on the assumption that we know the following parameters of a system with n nodes:

- set Ω of m operations performed on data;

- operation mix, i.e. the relative frequencies f_i given for all $\omega_i \in \Omega$;

- set of integrity constraints (intersection constraints that groups of nodes allowed to perform particular operations must satisfy);

- description of the network with node and communication reliabilities.

Our problem is to identify an optimal static scheme (i.e. a scheme that satisfies the integrity constraints and provides the highest possible availability for performing operations). Here parameters n and m characterize the size of the problem. Initial studies mostly discussed the case $m = 1$ (operation independent schemes).

It has been widely accepted that *coteries* (proposed by Garcia-Molina and Barbara [5,6]) provide the most general framework for analyzing static schemes. In this paper we use an essentially equivalent concept of an *acceptance set* (proposed by Tang and Natarajan [12]) which is the set of partition groups allowed to perform an operation on data items. This concept can be easily extended to systems with many operations.

The problem of finding an optimal static pessimistic scheme was studied by several authors [1,3,4,12,15]. To simplify the analysis, most of these studies discuss models in which communication links never fail. Tong and Kain [15] proposed an elegant algorithm for computing the optimal vote assignment for the fully connected networks for the operation independent case. Later, Cheung, Ahamad and Ammar [3] proposed an exhaustive search approach for computing optimal vote assignments for the same type of networks, but in the operation dependent case. Tang and Natarajan [12] formulated the problem of finding an optimal acceptance set as a 0-1 linear programming problem. The latter algorithms have exponential running time which makes them impractical for a large number of nodes. Therefore, it is of interest to identify important classes of systems for which optimal static pessimistic schemes can be found and evaluated in polynomial time.

For the operation independent case we have shown [10] that the vote assignments computed by the algorithm of Tong and Kain [15] are not merely optimal vote assignments, but also optimal as static pessimistic schemes, both for fully connected networks with perfect links and for the Ethernet systems.

In this paper we introduce a technique which allows us to extend these results for fully connected networks for various operation dependent schemes with many operations. In particular, we show that for all cases with 2 or 3 operations voting is optimal, and we show how our method can handle an example with 4 operations. Also, we extend some of these results for Ethernet systems.

One could view every operation as a combination of reads and writes, and reduce the problem to at most two operations. However, such a simplistic approach may preclude us from finding an optimal solution. The reason is as follows: the knowledge of the purpose of various update operations allows us to relax the integrity constraints, which in turn admits more solutions, some of them may offer a higher availability. Therefore, it is of uttermost importance to analyze the model with more than two operations.

The rest of the paper is organized as follows. In Section 2 the model and the notation are described. Section 3 discusses optimal static schemes for two operations. Section 4 deals with the problem of computing optimal static schemes in general. In section 5 extensions of the previous results for Ethernet systems are described. Finally, in Section 6, conclusions and open problems for further research are given.

2 Definitions

A network consists of n nodes $V = \{1, 2, \ldots, n\}$. Nodes store separate copies of the data that can be transmitted through communication links. We consider two types of networks: *fully connected* (in which there is a link between every two nodes) and *Ethernet systems* (in which nodes communicate through an Ethernet-like medium).

In the first case the link failure rates are assumed to be negligible i.e. they are not taken into account. This assumption simplifies the analysis: otherwise merely computing the probability that a subset of nodes remains connected is an NP-hard problem (see Rosenthal [11]). Later we will present an extension for the case of an Ethernet system where we take into account the failure rate of the Ethernet.

For every node i, value p_i represents the steady state probability that the node is operational. We say that nonempty set G ($G \subseteq V$) is a partition group (or simply, a group) if G is a maximal set of nodes such that all nodes of G are operational and can communicate with each other. In the fully connected networks a group consists of all operational nodes (since we assume that links do not fail). In Ethernet systems, when the Ethernet is operational, a group consists of all operational nodes. In the situation when the Ethernet is down, each operational node forms a separate group.

For nonempty set G, $E(G)$ denotes the event that G is a group. We also use the shorthand $\Pr(G)$ for $\Pr(E(G))$.

A static *acceptance scheme* $\mathcal{S} = (\mathcal{S}_1, \ldots, \mathcal{S}_m)$ can be described by defining for each operation ω_i the *acceptance set* $\mathcal{S}_i$ containing all groups allowed to perform ω_i (accepted for ω_i). For each operation ω_i we compute *availability* of the system for that operation (denoted $A(\omega_i)$) which represents the probability that the network is available for operation ω_i:

$$A(\omega_i) = \Pr\left(\bigcup_{G \in \mathcal{S}_i} E(G) \right).$$

Availability $A_\mathcal{S}$ of acceptance scheme $\mathcal{S}$ is computed as

$$A_\mathcal{S} = \sum_{i=1}^{m} f_i A(\omega_i)$$

(f_i's define the operation mix discussed before).

The events $E(G)$ are disjoint for fully connected networks. Hence, for such networks

$$A(\omega_i) = \sum_{G \in \mathcal{S}_i} \Pr(G) \ .$$

In this case, the probability of G being a group is

$$\Pr(G) = \left(\prod_{i \in G} p_i\right) \prod_{i \notin G}(1 - p_i) = \left(\prod_{i=1}^{n}(1 - p_i)\right) \prod_{i \in G} \frac{p_i}{1 - p_i} \ .$$

A replica control scheme $\mathcal{S}$ may be required to satisfy an integrity constraint $\mathcal{I}$. In general, $\mathcal{I}$ is a conjunction of several intersection constraints of the form $\mathcal{S}_i \sqcap \mathcal{S}_j \equiv$ (for every $G \in \mathcal{S}_i$ and every $H \in \mathcal{S}_j$) $G \cap H \neq \emptyset$.

3 Read-Write Acceptance Schemes

In [10] we have discussed optimal acceptance schemes with single acceptance set for all operations (operation independent schemes). In this section we consider acceptance schemes for the set of two operations with separate acceptance sets. This special case will be important in the general discussion continued in the next section. A good illustration is $\Omega = \{\text{read, write}\}$. If timestamps are used, then for each data item the maximum timestamp of reads and maximum timestamp of writes can be recorded. Consequently, to ensure the consistency (one-copy serializability) it is enough to require that every group from a read acceptance set $\mathcal{R}$ intersects with every group from a write acceptance set $\mathcal{W}$; we denote this requirement $\mathcal{R} \sqcap \mathcal{W}$. (A conflict between two write operations is resolved according to the *Thomas' Write Rule* [2,14]).

Given the relative frequency f for read operation, we can express the availability of *read-write acceptance scheme* $\mathcal{S} = (\mathcal{R}, \mathcal{W})$ in a fully connected network as

$$A_{\mathcal{R},\mathcal{W}} = f \sum_{G \in \mathcal{R}} \Pr(G) + (1 - f) \sum_{G \in \mathcal{W}} \Pr(G) \ .$$

In general, identifying an optimal read-write acceptance scheme constitutes a complex problem. Usually, it requires that $\mathcal{R} \neq \mathcal{W}$. In the context of fully connected networks, we have shown that it can be defined via voting, moreover we have provided the first efficient algorithm to find the vote assignment for this case [10]. The following theorem is needed in the optimality proof of the vote assignment. Conditions (1) and (2) are sufficient but not necessary. (We use $\overline{G}$ as a shorthand for $V - G$.)

Theorem 1 $(\mathcal{R}, \mathcal{W})$ *is an optimal read-write scheme satisfying $\mathcal{R} \sqcap \mathcal{W}$ if it satisfies the following conditions:*

(1) for every $G \subseteq V$ either $G \in \mathcal{R}$ or $\overline{G} \in \mathcal{W}$;

(2) for every $G \subseteq V$ we have $G \in \mathcal{R}$ iff $f \Pr(G) \geq (1 - f) \Pr(\overline{G})$.

Proof (sketch): Because groups from $\mathcal{R}$ must intersect all groups from $\mathcal{W}$, at most one of the inclusions $G \in \mathcal{R}$ and $\overline{G} \in \mathcal{W}$ may be true. The first inclusion (if true) contributes $f \Pr(G)$ to the availability $A_{\mathcal{R},\mathcal{W}}$; the second (if true) contributes $(1-f)\Pr(\overline{G})$. Condition (1) implies that at least one of these contributions is made, condition (2) implies that chosen contribution is larger one. $\square$

Now, let us consider a vote assignment $v : \{1, \ldots, n\} \to \mathbf{R}$ with a quorum of r votes for read operation and a quorum of w votes for write operation (where $r + w = \sum_{i=1}^{n} v(i)$). In this case acceptance scheme $(\mathcal{R}, \mathcal{W})$ can be defined as $\mathcal{R} = \{G|v(G) \geq r\}$ and $\mathcal{W} = \{G|v(G) > w\}$; thus $(\mathcal{R}, \mathcal{W})$ automatically satisfies condition (1). In [10] we proposed a linear time algorithm for computing vote assignment v and quorums r and w, so that condition (2) is satisfied as well. For the rest of the paper we assume that node probabilities are greater than 0.5 (other cases can be handled trivially).

Algorithm 1

for $i := 1$ **to** n **do** $v(i) := \log \dfrac{p_i}{1 - p_i}$;

$sum := \sum_{i=1}^{n} v(i)$;

$r := \dfrac{1}{2}(sum - \log \dfrac{f}{1-f})$;

$u := $ minimal $v(i)$;

$r := \max(r, u)$;

$r := \min(r, sum)$;

$w := sum - r$;

One should note that optimal availability can be also achieved by $\mathcal{R} = \{G|v(G) > r\}$ and $\mathcal{W} = \{G|v(G) \geq w\}$; in the proof we merely substitute the occurrences of $<$ with $\leq$ and vice versa.

In the case when version number are used, every two groups belonging to the write acceptance set must intersect (to ensure the correct updates of version numbers). We denote this requirement $\mathcal{W} \sqcap \mathcal{W}$. When relative frequency of read operation f is at least 0.5, the optimal solution is provided by the vote assignment and quorums of Algorithm 1 (write quorum has a value that ensures intersection of every two write groups). However, when $f < 0.5$ the resulting scheme does not satisfy the same constraint for write operation. It turns out that in this case the optimal availability is achieved when read and write acceptance sets are equal.

Theorem 2 *If the frequency rate of read operation f is less than 0.5 then read-write acceptance scheme $(\mathcal{R}, \mathcal{W})$ defined as $\mathcal{R} = \mathcal{W} = \{\ G \mid \Pr(G) > \Pr(\overline{G})\ \} \cup \{\ G \mid \Pr(G) = \Pr(\overline{G}) \ \&\ 1 \in G\ \}$ is an optimal acceptance scheme satisfying $\mathcal{R} \sqcap \mathcal{W}$ & $\mathcal{W} \sqcap \mathcal{W}$.*

Proof: The read acceptance set $\mathcal{R}$ satisfies the following two conditions:

(1) for every $G \subseteq V$ either $G \in \mathcal{R}$ or $\overline{G} \in \mathcal{R}$;

(2) for every $G \subseteq V$ if $G \in \mathcal{R}$ then $\Pr(G) \geq \Pr(\overline{G})$.

We will show that these conditions imply that $(\mathcal{R}, \mathcal{W})$ is optimal.

For a given set of groups $\mathcal{S}$ let $C_\mathcal{S}$ be defined as

$$C_\mathcal{S}(G) = \begin{cases} 1, & G \in \mathcal{S} \\ 0, & \text{otherwise.} \end{cases}$$

The availability of read-write acceptance scheme $(\mathcal{R}, \mathcal{W})$ is $A_{\mathcal{R},\mathcal{W}} = \sum_{G \in \mathcal{R}} \Pr(G)$. Let $(\mathcal{R}', \mathcal{W}')$ be another read-write acceptance scheme satisfying $\mathcal{R} \sqcap \mathcal{W}$ & $\mathcal{W} \sqcap \mathcal{W}$. Then, because of (1), $A_{\mathcal{R}',\mathcal{W}'}$ is equal to

$$\sum_{G \in \mathcal{R}} \left\{ f \left[C_{\mathcal{R}'}(G) \Pr(G) + C_{\mathcal{R}'}(\overline{G}) \Pr(\overline{G}) \right] + (1-f) \left[C_{\mathcal{W}'}(G) \Pr(G) + C_{\mathcal{W}'}(\overline{G}) \Pr(\overline{G}) \right] \right\}.$$

Let $G \in \mathcal{R}$. At most one of $C_{\mathcal{W}'}(G)$ and $C_{\mathcal{W}'}(\overline{G})$ is 1. Also, if $C_{\mathcal{R}'}(G) = C_{\mathcal{R}'}(\overline{G}) = 1$ then $C_{\mathcal{W}'}(G) = C_{\mathcal{W}'}(\overline{G}) = 0$. Since $\Pr(G) \geq \Pr(\overline{G})$ and $f < 0.5$, we have

$$\Pr(G) \geq f \left[C_{\mathcal{R}'}(G) \Pr(G) + C_{\mathcal{R}'}(\overline{G}) \Pr(\overline{G}) \right] + (1-f) \left[C_{\mathcal{W}'}(G) \Pr(G) + C_{\mathcal{W}'}(\overline{G}) \Pr(\overline{G}) \right].$$

Therefore, $A_{\mathcal{R},\mathcal{W}} \geq A_{\mathcal{R}',\mathcal{W}'}$. □

Obviously, if $\Pr(G) \neq \Pr(\overline{G})$ for every group G, then vote assignment $v(i) = \log \frac{p_i}{1-p_i}$ and quorums $r = w = \frac{1}{2} \sum_{i=1}^{n} v(i)$ correctly define $\mathcal{R}$ and $\mathcal{W}$. However, if there is a group G such that $\Pr(G) = \Pr(\overline{G})$ (and consequently $v(G) = v(\overline{G})$) a small adjustment of votes is necessary to break ties in favor of groups containing node 1. Following algorithm (due to Tong and Kain [15]) has the tiebreaking part that solves the problem.

Algorithm 2

for $i := 1$ to n do $v(i) := \log \frac{p_i}{1-p_i}$;
if there exists G s.t. $v(G) = v(\overline{G})$ then begin
    $d := \min\{ |v(G) - v(\overline{G})| \mid v(G) \neq v(\overline{G}) \}$;
    $v(1) := v(1) + \frac{1}{2}d$
end;
$r := w := \frac{1}{2} \sum_{i=1}^{n} v(i)$;

In general, breaking ties takes exponential time. The algorithm checks a condition which is an exponentially long disjunction and then it minimizes over an exponentially large set. At the first look, this appears unavoidable. For example, the condition is an NP-complete version of the knapsack problem. However, in [10] we propose a 'practical approach' for computing optimal vote assignments. This method uses approximation and resolves the tiebreaking part in linear time with no practical degradation in the resulting availability.

4 Acceptance Schemes with Several Operations

In this section we will examine the optimal solutions (acceptance schemes) for the set of m operations $\Omega = \{\omega_1, \ldots, \omega_m\}$. An acceptance scheme is an m-tuple $S = (S_1, \ldots S_m)$ where S_i is an acceptance set for operation ω_i. In order to ensure the consistency of data, these acceptance sets must satisfy some constraints. For example, in the previous section, we had to assume that every group from the read acceptance set must intersect with every group from the write acceptance set. We are not going to discuss here why particular constraints are required, as they may vary depending on the implementation (we have seen that using timestamps instead of version numbers imposes less restrictive constraints). Rather, we will assume some constraints and search for an optimal acceptance scheme that satisfies them. We assume that an integrity constraint $\mathcal{I}$ is a conjunction of some intersection constraints defined in Section 2.

Definition 1 *Given an integrity constraint $\mathcal{I}$, we say that S_i is* replaceable *by S_j if simultaneous substitution of all occurrences of S_i by S_j does not produce an intersection constraint that is not in $\mathcal{I}$. S_i and S_j are* replaceable *iff S_i is replaceable by S_j and vice versa.*

Optimal solution for a problem with $m = 1$ and the constraint $S_1 \sqcap S_1$ was presented in [10] (optimal acceptance scheme is defined by vote assignment of Algorithm 2). The case $m = 2$ was discussed in the previous section. Two constraints were considered: $S_1 \sqcap S_2$ and $S_1 \sqcap S_1$ & $S_1 \sqcap S_2$. We will show that many problems of larger size can be reduced to problems with $m \leq 2$. Below we formulate our most frequently used reduction method.

Lemma 1 *Let $S = (S_1, \ldots, S_m)$ be an acceptance scheme satisfying constraint $\mathcal{I}$ in which S_i and S_j are replaceable. Then S' obtained from S by substituting S_j by S_i also satisfies $\mathcal{I}$.*

Proof: Follows directly from Definition 1. □

This lemma shows that for the optimization purposes operations with replaceable acceptance sets can be considered as a single operation (with the frequency rate equal to the sum of respective frequency rates); such identification reduces parameter m of the problem.

We will show that voting provides optimal acceptance schemes when $m = 3$. To examine all possible integrity constraints we will represent them as *integrity constraint graphs* with nodes $\{1, 2, 3\}$. In such a graph edge (a, b) stands for the intersection constraint $S_a \sqcap S_b$. We will discuss only connected graphs (for disconnected graphs the reduction to problems with smaller m is trivial, as each component can be solved separately). We divide these graphs into two groups: *chain structures* (Figure 1) and *cycle structures* (Figure 2). For every case we will show how to obtain an optimal acceptance scheme through voting.

If not mentioned otherwise, the vote assignment is given according to the formula $v(i) = \log \dfrac{p_i}{1 - p_i}$. Let t_i be a quorum (threshold) for operation ω_i. If $t_i > \dfrac{1}{2} \sum_{j=1}^{n} v(j)$ or

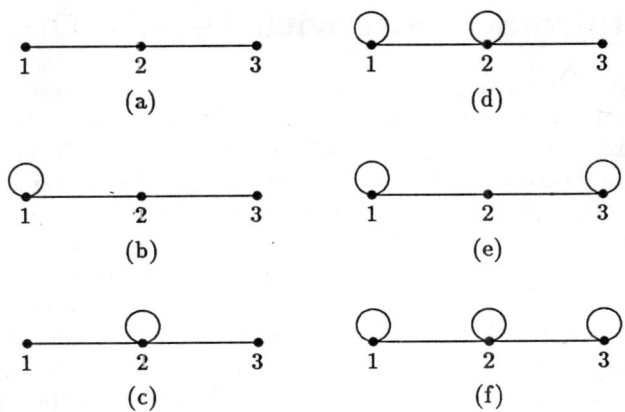

Figure 1: Chain Structures

if $t_i = \frac{1}{2}\sum_{j=1}^{n} v(j)$ and $S_i \sqcap S_i$ we define S_i to be $\{G|v(G) > t_i\}$. Otherwise, S_i is equal to $\{G|v(G) \geq t_i\}$. This definition is consistent with our solutions from the previous section.

Theorem 3 *The voting scheme provides the highest possible availability for fully connected networks for any case of three operations.*

Proof:
Cases (a), (c), (e) and (f): Acceptance sets S_1 and S_3 are replaceable, so the problem reduces to $m = 2$.

Case (d): We divide operations into two groups: $\{\omega_1\}$ and $\{\omega_2, \omega_3\}$. These cases are solved separately and resulting vote assignments are v and u respectively. If $v = u$, then $S_1 \sqcap S_1$ and $S_2 \sqcap S_2$ (which are assured) imply $S_1 \sqcap S_2$. Otherwise, we know that $v(1) > u(1)$ because v was adjusted to break ties in favor of groups containing node 1. In order to get the unified solution, we have to make $v(1)$ and $u(1)$ equal, taking care not to change S_1, S_2 or S_3. This can be done by $v(1) = u(1) = \log\frac{p_1}{1-p_1} + d$ where d is positive but sufficiently small so that acceptance sets are not changed.

Case (b): The availability of the acceptance scheme is equal to $A = f_1 A(\omega_1) + f_2 A(\omega_2) + f_3 A(\omega_3)$. If $f_2 \geq f_1 + f_3$ we will look at operation ω_2 as two operations ω_2' and ω_2'' with respective frequencies $f_2' = \frac{f_1 f_2}{f_1 + f_3}$ and $f_2'' = \frac{f_2 f_3}{f_1 + f_3}$. Optimization for two groups of operations $\{\omega_1, \omega_2'\}$ and $\{\omega_2'', \omega_3\}$ provides the same vote assignments and thresholds for ω_2' and ω_2'' (because the frequencies in these two groups are proportional). Suppose that $f_2 < f_1 + f_3$. If $f_2 \leq f_3$ then we divide operations into two groups: $\{\omega_1\}$ and $\{\omega_2, \omega_3\}$ and apply case (d). Otherwise, if $f_2 > f_3$ we will view ω_2 as two operations ω_2' and ω_2'' with frequencies $f_2' = f_2 - f_3$ and $f_2'' = f_3$. Since $f_1 > f_2'$, the optimal availability is achieved when $S_1 = S_2 = S_3$ is defined via vote assignment of Algorithm 2.

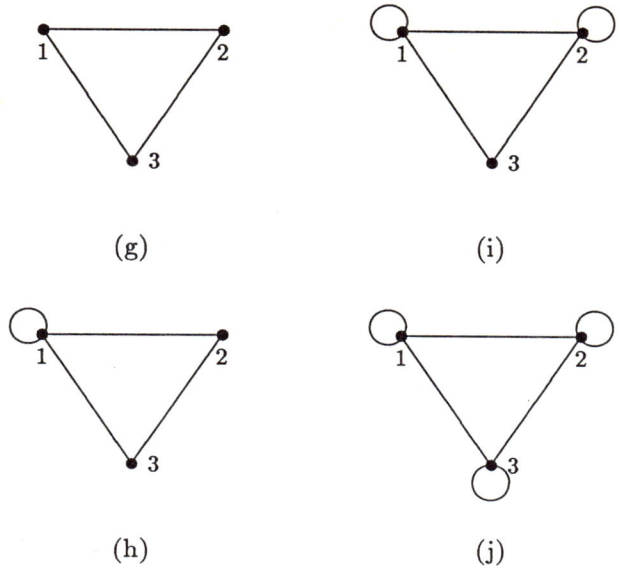

Figure 2: Cycle Structures

In the remaining part of the proof we shall say that an acceptance scheme $S = (S_1, \ldots, S_m)$ is a *maximal acceptance scheme satisfying constraint* $\mathcal{I}$ if adding one more group to one of S_i's would violate $\mathcal{I}$. Following the definition of availability, it is clear that one of the maximal acceptance schemes achieves the highest availability.

Case (g): Without loss of generality, let $f_1 \geq f_2 \geq f_3$. Let $S = (S_1, S_2, S_3)$ be a maximal acceptance scheme satisfying this integrity constraint. Then $S' = (S_1 \cup (S_2 \oplus S_3), S_2 \cap S_3, S_2 \cap S_3)$ also satisfies it[2]. Comparing respective availabilities we have:

$$\begin{aligned}
A_S &= f_1 \sum_{G \in S_1} \Pr(G) + f_2 \sum_{G \in S_2} \Pr(G) + f_3 \sum_{G \in S_3} \Pr(G) \\
&\leq f_1 \sum_{G \in S_1} \Pr(G) + f_1 \sum_{G \in S_2 \oplus S_3} \Pr(G) + (f_2 + f_3) \sum_{G \in S_2 \cap S_3} \Pr(G) \\
&= A_{S'}
\end{aligned}$$

The last equation holds since $(S_2 - S_3) \cap (S_3 - S_2) = \emptyset$. Also, $S_1 \cap (S_2 - S_3) = \emptyset$ and $S_1 \cap (S_3 - S_2) = \emptyset$; the latter two hold because acceptance scheme S is maximal. (For example, if $G \in S_1 \cap (S_2 - S_3)$ then $(S_1, S_2, S_3 \cup \{G\})$ also satisfies the integrity constraint.) So, the problem is to find optimal S_1 and S_2 with respect to the constraint $S_1 \sqcap S_2$ & $S_2 \sqcap S_2$.

Case (h): Without loss of generality we may assume that $f_2 \geq f_3$. Let $S = (S_1, S_2, S_3)$ be a maximal acceptance scheme satisfying the integrity constraint. Then $S' = (S_1, S_2 \cup$

---
[2] $S_2 \oplus S_3$ is a shorthand for $(S_2 - S_3) \cup (S_3 - S_2)$.

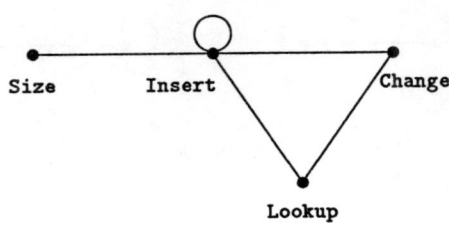

Figure 3: Integrity Constraint Graph of Example 1

S_3, S_1) also satisfies it. It is easy to prove that, since S is maximal, $S_1 = S_2 \cap S_3$. Comparing the availabilities we have

$$
\begin{aligned}
A_S &= f_1 \sum_{G \in S_1} \Pr(G) + f_2 \sum_{G \in S_2} \Pr(G) + f_3 \sum_{G \in S_3} \Pr(G) \\
&\leq f_1 \sum_{G \in S_1} \Pr(G) + f_2 \sum_{G \in S_2} \Pr(G) + f_2 \sum_{G \in S_3 - S_2} \Pr(G) + f_3 \sum_{G \in S_2 \cap S_3} \Pr(G) \\
&= A_{S'}
\end{aligned}
$$

Therefore, as in (g) the problem reduces to $m = 2$.

Cases (i) and (j): Reduction to $m = 2$ is possible since S_1 and S_2 are replaceable. □

As we see, when there are at most three operations the optimal availability is achievable through voting. The reduction can be extended further, for larger sets of operations, as shown by the next example of a set of 4 operations (due to Herlihy [8]).

Example 1 A *Directory* stores pairs of values, where one value (the key) is used to retrieve the other (the item).
    Insert = Operation(k: Key, i: Item) signals (Present)
inserts a new pair in the directory, signaling if the key is already present.
    Change = Operation(k: Key, i: Item) signals (Absent)
alters the item associated with the given key, signaling if the key is absent.
    Lookup = Operation(k: Key) returns(Item) signals (Absent)
returns item associated with the given key, signaling if the key is absent.
    Size = Operation() returns(Int)
returns the number of pairs currently in the directory.

The integrity constraint graph is given in Figure 3.

Without loss of generality, we may assume that $f_{\text{Change}} \geq f_{\text{Lookup}}$. As in case (h) it is easy to prove that we may request $S(\text{Insert}) = S(\text{Lookup})$ and still obtain an optimal solution. This identification reduces the problem to case (e) from the proof of Theorem 3.

5 Ethernet Systems

So far, our discussion of optimal schemes has assumed negligible link failure rates. Now, we will show how to extend some of the previous results for the case of an Ethernet system where we take into account the failure rate of the Ethernet-like medium.

An Ethernet system consists of nodes connected through an Ethernet-like medium. Every two nodes can communicate directly when the medium is operational. There is no communication when the medium is down. The reliability of the medium is denoted by e. For a nonempty set G, let the probabilities that G is a group in the fully connected network with perfect links and in the Ethernet system (with the same node probabilities), be denoted by $\Pr_F(G)$ and $\Pr_E(G)$, respectively. Then

$$\Pr_E(G) = \begin{cases} e \Pr_F(G) + (1-e)p_i & \text{if } G = \{i\}, \\ e \Pr_F(G) & \text{otherwise.} \end{cases}$$

Tong and Kain [15] proposed an algorithm for computing the optimal vote assignment for Ethernet systems for the operation independent case. We have shown in [10] that this algorithm also provides an optimal acceptance scheme. Now, we will describe another algorithm which provides the solution for two operations (read-write acceptance scheme) with the only constraint $\mathcal{R} \sqcap \mathcal{W}$.

Let us consider the availability $A_{\mathcal{R},\mathcal{W}}$ of a maximal read-write acceptance scheme $(\mathcal{R}, \mathcal{W})$ in the Ethernet system with reliability e of the medium. If $(\mathcal{R}, \mathcal{W})$ is a maximal acceptance scheme, then both $\mathcal{R}$ and $\mathcal{W}$ are closed for supersets (i.e. if $G \in \mathcal{R}$ then for every $H \supseteq G$, $H \in \mathcal{R}$). We distinguish the following cases:

Case 1: No singleton group belongs to $\mathcal{R} \cup \mathcal{W}$. Then the availability $A_{\mathcal{R},\mathcal{W}}$ is equal to the availability of this scheme in the fully connected network (with the same node probabilities) multiplied by e.

Case 2: There exists a singleton group $\{i\} \in \mathcal{R} \cap \mathcal{W}$. Then $(\mathcal{R}, \mathcal{W})$ is defined as

$$\mathcal{R} = \mathcal{W} = \{ G \mid i \in G \}.$$

Therefore $A_{\mathcal{R},\mathcal{W}} = p_i$.

Case 3: Set $U = \{i | \{i\} \in \mathcal{R}\}$ has more than one element. Then $\mathcal{R} = \{ G \mid G \cap U \neq \emptyset \} \cup \mathcal{R}_1$ and $\mathcal{W} = \{ H \cup U \mid H \in \mathcal{W}_1\}$ where $(\mathcal{R}_1, \mathcal{W}_1)$ is a read-write acceptance scheme defined on the network with nodes $\bar{U}$. The availability $A_{\mathcal{R},\mathcal{W}}$ is equal to

$$f[1 - \prod_{i \in U}(1-p_i)] + ef\left(\prod_{i \in U}(1-p_i)\right)\sum_{H \in \mathcal{R}_1}\Pr_1(H) + e(1-f)\left(\prod_{i \in U}p_i\right)\sum_{H \in \mathcal{W}_1}\Pr_1(H)$$

where $\Pr_1(H)$ is the probability of group H in the fully connected network with nodes $\bar{U}$. Maximizing $A_{\mathcal{R},\mathcal{W}}$ is equivalent to maximizing

$$f' \sum_{H \in \mathcal{R}_1}\Pr_1(H) + (1-f')\sum_{H \in \mathcal{W}_1}\Pr_1(H)$$

where
$$f' = \frac{f \prod_{i \in U}(1-p_i)}{f \prod_{i \in U}(1-p_i) + (1-f)\prod_{i \in U} p_i}$$

and thus we can apply Algorithm 1 to find vote assignment v_1 and thresholds r_1 and w_1. Then $(\mathcal{R}, \mathcal{W})$ is defined through vote assignment

$$v(i) = \begin{cases} r_1 & i \in U \\ v_1(i) & i \in \overline{U}. \end{cases}$$

and thresholds
$$r = r_1, \quad w = w_1 + |U|r_1.$$

If $\mathcal{R}_1$ contains groups with just one node, those nodes should be placed into U and the analysis repeated.

Case 4: Symmetric to Case 3 with respect to $\mathcal{R}$ and $\mathcal{W}$.

Thus, search for an optimal acceptance scheme reduces to search for the optimal acceptance schemes in these four cases and comparison of obtained availabilities. Case 1 can be handled by Algorithm 1 from Section 3. For Case 2 we simply choose i to be the node with the highest probability. Cases 3 and 4 require checking availabilities for many possible sets U. Importantly, the following two lemmas help us reduce the number of possibilities to only $n-1$.

Lemma 2 *Let $(\mathcal{R}, \mathcal{W})$ be a maximal acceptance scheme with $\mathcal{W}$ containing a singleton group and $f > 0.5$. Then the availability of $(\mathcal{W}, \mathcal{R})$ is not lower than the availability of $(\mathcal{R}, \mathcal{W})$.*

Proof: Let $\{i\} \in \mathcal{W}$ and $G \in \mathcal{R}$. Then $i \in G$ and therefore $G \in \mathcal{W}$. Consequently $\mathcal{R} \subseteq \mathcal{W}$. Therefore $A(\text{read}) \leq A(\text{write})$ and

$$A_{\mathcal{R},\mathcal{W}} = fA(\text{read}) + (1-f)A(\text{write}) \leq fA(\text{write}) + (1-f)A(\text{read}) = A_{\mathcal{W},\mathcal{R}}. \quad \square$$

Lemma 2 allows us to consider only one of the cases 3 and 4.

Lemma 3 *Let $(\mathcal{R}, \mathcal{W})$ be a maximal acceptance scheme such that $\{i\} \in \mathcal{R}$ for some node i. If $p_i < p_j$ for some node j, then the acceptance scheme obtained by swapping nodes i and j does not have lower availability.*

Proof: Let $(\mathcal{R}', \mathcal{W}')$ be a read-write acceptance scheme obtained from $(\mathcal{R}, \mathcal{W})$ by swapping nodes i and j. If $\{j\} \in \mathcal{R}$ then $(\mathcal{R}', \mathcal{W}') = (\mathcal{R}, \mathcal{W})$. Let us assume that $\{j\} \notin \mathcal{R}$. For a set of groups $\mathcal{S}$ define $\mathcal{S}_i = \{G \mid i \in G \ \& \ j \notin G\}$, $\mathcal{S}_j = \{G \mid i \notin G \ \& \ j \in G\}$, $\mathcal{S}_{ij} = \mathcal{S} - \mathcal{S}_i - \mathcal{S}_j$. Obviously, $\mathcal{R}_{ij} = \mathcal{R}'_{ij}$. If $G \in \mathcal{R}_j$ then $((G - \{j\}) \cup \{i\}) \in \mathcal{R}_i$. So, the only difference in availability concerning read operation is with groups such that $G \in \mathcal{R}_i$ but $((G - \{i\}) \cup \{j\}) \notin \mathcal{R}_j$. Since $p_i < p_j$, new groups increase the availability and so $A(\text{read}) \leq A'(\text{read})$. For write the analysis is straightforward because $\mathcal{W}_j = \emptyset$. $\square$

Without loss of generality, we can assume that $p_1 \geq p_2 \geq \ldots \geq p_n$. Then Case 3 reduces to checking availabilities with U of the form $\{1, \ldots, k\}$ where $k = 2, \ldots, n$.

In all these cases the optimal acceptance schemes can be represented via vote assignments. The only limitation of this approach is the running time of the algorithm which is exponential in n because of the computation of availabilities in cases 1 and 3. However, by using the same approximation techniques as in [10], we are able to compute in polynomial time the acceptance scheme that is arbitrarily close to the optimal.

The analysis with the constraint $\mathcal{R} \sqcap \mathcal{W}$ & $\mathcal{W} \sqcap \mathcal{W}$ is similar except that Case 4 does not have to be considered because it does not satisfy the second intersection constraint.

Theorem 4 *Voting scheme provides the highest possible availability for Ethernet systems for any case of two operations.*

6 Conclusion

Our study of static pessimistic replica control protocols has shown that adapting to the operation mix is important for increasing the availability. Also, it has been shown that voting scheme provides the optimal availability in many cases.

It is encouraging that the algorithms developed for operation dependent case with two operations extend to many cases with several operations. The reduction methods described here are quite elementary and they do not increase significantly the cost of optimization. It is worthwhile to identify other situations where such reductions are possible. Ring networks should provide an interesting test case. Our recent result states that voting provides optimal availability for the operation independent case for this type of networks.

The *dynamic voting* protocols proposed by Jajodia and Mutchler [9] and Paris and Long [13] are examples of dynamic pessimistic schemes for managing replicated data. Dynamic voting offers better solution than static voting for fully connected networks with equal node probabilities under the assumption that updates (write operations) are frequent. Therefore, we might expect that optimization of a vote assignment in dynamic voting leads to an even better replica control protocol.

Another interesting problem is to estimate the discrepancy between the availability achieved through voting and through an unrestricted static scheme. So far, we have not found an example with a significant discrepancy.

References

[1] M Ahamad, M.H. Ammar. *Performance Characterization of Quorum-Consensus Algorithms for Replicated Data.* IEEE Conf. on Reliability in Distributed Software and Database Systems, 1987, pp. 161-167.

[2] A. Bernstein, V. Hadzilacos, N. Goodman. *Concurrency Control and Recovery in Database Systems.* Addison-Wesley, 1987.

[3] S.Y. Cheung, M. Ahamad, M.H. Ammar. *Optimizing Vote and Quorum Assignments for Reading and Writing Replicated Data.* Proc. of Fifth Int. Conference on Data Engineering 1989, pp. 271-279.

[4] S.B. Davidson, H. Garcia-Molina, D. Skeen. *Consistency in Partitioned Networks.* ACM Computing Surveys, Vol.17, No.3, Sep. 1985, pp. 341-370.

[5] H. Garcia-Molina, D. Barbara. *Optimizing the Reliability Provided by Voting Mechanisms.* Proc. of Fourth International Conference on Distributed Computing Systems, Oct. 1984, pp. 340-346.

[6] H. Garcia-Molina, D. Barbara. *How to Assign Votes in a Distributed System.* Journal of ACM, Vol. 32, No. 4, Oct. 1985, pp. 841-860.

[7] D.K. Gifford. *Weighted Voting for Replicated Data.* Proc. of Seventh ACM Symposium on Operating System Principles, Dec. 1979, pp. 150-162.

[8] M. Herlihy. *A Quorum-Consensus Replication Method for Abstract Data Types.* ACM Transactions on Computer Systems, Vol. 4, No. 1, Feb. 1986, pp. 32-53.

[9] S. Jajodia, D. Mutchler. *Dynamic Voting.* Proc. of ACM SIGMOD Int. Conf. on Management of Data, 1987, pp. 227-238.

[10] M. Obradovic, P. Berman. *Voting as the Optimal Static Pessimistic Scheme for Managing Replicated Data.* Proc. of Ninth IEEE Symposium on Reliable Distributed Systems, 1990, pp. 126-135.

[11] A. Rosenthal. *Computing the Reliability of a Complex Network.* SIAM J. Appl. Math., vol. 32, pp. 384-393, 1977.

[12] J. Tang, N. Natarajan. *A static Pessimistic Scheme for Handling Replicated Databases.* ACM SIGMOD Int. Conf. on Management of Data 1989.

[13] J.-F. Paris, D. Long. *Efficient Dynamic Voting Algorithms.* Proc. of Fourth Int. Conference on Data Engineering, Feb. 1988, 268-275.

[14] R.H. Thomas. *A Majority Consensus Approach to Concurrency Control for Replicated Databases.* ACM Transactions on Database Systems, Vol. 4, No. 2, June 1979, pp. 180-209.

[15] Z. Tong, R.Y. Kain. *Vote Assignments in Weighted Voting Mechanisms.* Proc. of Seventh Symposium on Reliable Distributed Systems, 1988, pp. 138-143.

Wakeup under Read/Write Atomicity

Prasad Jayanti   Sam Toueg
Department of Computer Science
Upson Hall, Cornell University
Ithaca, New York 14853
e-mail(prasad@cs.cornell.edu, sam@cs.cornell.edu)

Abstract

We study the wakeup problem in an asynchronous system of anonymous processes that are identically coded. Informally, a protocol solves wakeup if, in all infinite fair runs of the protocol, every process eventually learns that every process has executed at least one step and is therefore 'awake'[FMRT90]. We study this problem in the context of two models of shared memory communication. In the first model, each word of shared memory may be read and written by every process. In the second model, each process has a register which only that process may write, but all processes may read. We show that wakeup is not solvable in the first model. For the second model, we derive matching lower and upper bounds on the amount of shared memory necessary to solve wakeup. We also study the feasibility of leader election and consensus in the above two models.

1 Introduction

Self-stabilization refers to the ability of a distributed system to eventually start behaving correctly even if the initial states of the processes and the shared memory are arbitrary. Following the pioneering work of Dijkstra[Dij74], there has been a lot of research in this area. Although self-stabilization adds fault tolerance to non-terminating protocols such as mutual exclusion, it is not appropriate, as noted by Fischer *et al* [FMRT90], for terminating protocols such as wakeup, consensus, and leader election. To see this, note that there will be no system execution if the initial state of each process is a termination state. Fischer *et al* have, therefore, proposed an alternate model of interest to terminating protocols[FMRT90]. In their model, the system consists of identically programmed anonymous asynchronous processes communicating through shared memory. Unlike self-stabilizing systems, the program counter of each process points to the appropriate instruction initially. The initial state of shared memory, however, is arbitrary as in self-stabilizing systems. This gives a process the flexibility to start its participation in the protocol at any time it chooses without any global synchronization[1].

In this paper, we study primarily the *wakeup* problem in the model outlined above. Leader election and consensus are the other problems studied. Informally, a wakeup protocol is one in which, in every infinite fair run, each process eventually learns that every other process has woken up and begun participating in the protocol. The non-triviality condition is the obvious one: process p cannot learn that process q has woken up unless q has taken at least one step of the protocol.

The wakeup problem was first defined and studied in [FMRT90]. In the communication model of [FMRT90], there is a single shared variable in the entire system and processes use this shared variable for communication. It is assumed that each process may do a *test-and-set* operation on this variable, and that this operation is atomic. Our first result shows that, if in the above model one assumes only read/write atomicity, even a weaker version of the wakeup problem becomes unsolvable.

We then consider the wakeup problem in an alternate model of communication: we assume that there is a communication register associated with each process and that this register may be written only by that process, but may be read by all processes. This is a Single-Writer-Multi-Reader (SWMR) model in contrast to the earlier described Multi-Writer-Multi-Reader (MWMR) model. We present a terminating wakeup protocol when each communication register is a three-valued regular register[2]. Following that, we prove the optimality of this protocol by showing that wakeup is not solvable if each communication register is only a boolean.

In this paper, we also consider two other fundamental problems: leader election and consensus (see [Ita90] and [CIL87] for further references). We show that leader election is not solvable in both SWMR and MWMR models. Consensus, on the other hand, is solvable in SWMR but not in MWMR. Impossibility of consensus in MWMR is not due to process failures since they are not allowed in our model. It is interesting to contrast this result with [CIL87] in which consensus is proved impossible in the presence of failures, even if the model is not symmetric.

The paper is organized as follows. In section 2, we describe the model and specify the wakeup problem. In section 3, we prove that even the weakest form of wakeup is not solvable in the MWMR model. In section 4, we focus on the SWMR model. First, we present a terminating wakeup protocol when each communication register is a three-valued regular register. Then we show that wakeup is not solvable when each communication register is only a boolean. We present results on consensus and leader election in section 5. We conclude in section 6.

---

[1]This is in contrast to most models in which each process p participating in the protocol must synchronize with the process q responsible for the initialization of shared memory such that this initialization by q precedes the steps of p.

[2]A register r is *regular* if the value returned by every read operation is either the value written into r by the most recent write operation that completed before that read operation started or the value written into r by some write operation overlapping with that read operation.

2 Model and Wakeup Specification

In our model, we borrow several definitions from [FLP85]. We consider an asynchronous system of N non-faulty processes ($N \geq 2$) in which processes have no names. For ease of exposition, however, we refer to the processes in this paper by the unique names $p_1, p_2, \ldots, p_N$. Each process p_i has a *wakeup register* w_i, a *termination register* t_i, and an unbounded amount of internal storage. The registers w_i and t_i, together with the program counter and internal storage, comprise the *internal state* of p_i. Initial state prescribes starting values to program counter, wakeup register, termination register and internal storage. In particular, the values in the wakeup and termination registers must be both zero in the initial state. A state in which wakeup register has the value 1 is distinguished as a *realization state*, and a state in which termination register has the value 1 is distinguished as a *termination state*. Process p_i acts according to a transition function. The transition function cannot change the value in the wakeup register once the process has reached a realization state. Further, the transition function can change neither the internal state nor the shared memory (see below) once the process has reached a termination state. We assume that processes are all identically coded: each process has the same transition function and starts in the same initial state. A *protocol* is a specification of an initial state and a transitive function. We study the following two models of communication between processes.

1. Processes communicate by reading and writing the words of shared memory M. Each process may read and write every word of M. We refer to this model of communication as the *Multi Writer Multi Reader* (MWMR) model.

2. Each process p_i has a communication register c_i associated with it. Only p_i may write c_i, but every process may read c_i. We refer to this model of communication as the *Single Writer Multi Reader* (SWMR) model. In this model, $c_1, c_2, \ldots, c_N$ comprise the shared memory. We require that, for any i and j ($i \neq j$), there be no correlation between the order in which p_i perceives its neighbors (the remaining $N - 1$ processes) and the order in which p_j perceives its neighbors. We formalize this notion as follows: each process p_i has a *local-ordering* α_i such that $\alpha_i[0] = i$, and $\langle \alpha_i[1], \alpha_i[2], \ldots, \alpha[N-1] \rangle$ is an arbitrary permutation of $\langle 1, 2, \ldots, i-1, i+1, \ldots, N \rangle$.

The two models are illustrated in figure 1 when there are only two processes in the system.

A *configuration* consists of the internal state of each process, together with the values in the words of shared memory. If P is a protocol, then an *initial configuration* in P is one in which each process is in the initial state specified by P and the words of shared memory contain arbitrary values. A *step* takes one configuration to another and consists of a primitive step by a single process. Let C be a configuration. A step s in the MWMR model is of one of the following three types.

1. (p_i, r, j, v): p_i reads the value in $M[j]$ (the jth word of shared memory M) as v; then depending on p_i's internal state in C and on v, p_i enters a new internal state dictated by the transition function.

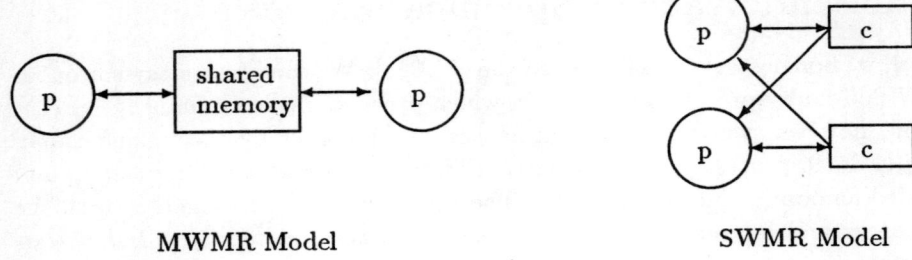

MWMR Model　　　　　　　　　　　SWMR Model

Figure 1: The two communication models

2. (p_i, w, j, v): p_i writes the value v in $M[j]$; then depending on p_i's internal state in C, p_i enters a new internal state dictated by the transition function.

3. $(p_i, *)$: p_i takes this step. This step does not change the configuration.

Steps 1 and 2 above can be applied to C only if p_i's internal state in C is not a termination state, while step 3 can be applied to C only if p_i's internal state in C is a termination state.

A step s in the SWMR model is of one of the following three types.

1. (p_i, r, j, v): p_i reads the value in $c_{\alpha_i[j]}$ as v; then depending on p_i's internal state in C and on v, p_i enters a new internal state dictated by the transition function.

2. (p_i, w, v): p_i writes the value v in c_i; then depending on p_i's internal state in C, p_i enters a new internal state dictated by the transition function.

3. $(p_i, *)$: p_i takes this step. This step does not change the configuration.

Steps 1 and 2 above can be applied to C only if p_i's internal state in C is not a termination state, while Step 3 can be applied to C only if p_i's internal state in C is a termination state.

If s is a step and C is a configuration to which s can be applied, then $s(C, P)$ denotes the configuration that results when s is applied to C using the protocol P. When the protocol P being referred to is clear from the context, we write $s(C)$ for $s(C, P)$. A *run* from C in P is a sequence of steps that can be applied, in turn, starting from C using the protocol P. If $R = s_1, s_2, \ldots, s_n$ is a finite run from C in P, then $R(C, P)$ denotes $s_n(s_{n-1}(\cdots s_1(C)\cdots))$. When the protocol P is understood, we write $R(C)$ for $R(C, P)$. A *schedule* is a sequence of process names. Given a schedule $S = p_{i_1}, p_{i_2}, \ldots$, a configuration C, and a protocol P, we say that $R = s_1, s_2, \ldots$ is the run from C that corresponds to S in P if s_j is a step by p_{i_j} that can be applied to the configuration $s_{j-1}(s_{j-2}(\cdots s_1(C)\cdots))$ in P. Note that such an R exists and is unique. If $R = s_1, s_2, \ldots$ is a run, then $S = p_{i_1}, p_{i_2}, \ldots$ is the schedule that corresponds to R if, for all j, p_{i_j} takes the step s_j. An infinite schedule S is *fair* if the name of each process occurs infinitely many times in S.

If R_1 is a finite sequence and R_2 is a sequence, we use $R_1.R_2$ to denote the sequence obtained by the concatenation of R_1 and R_2, and R_1^z to denote $R_1.R_1.\cdots.R_1$ where R_1 is repeated z times.

A run R from C is a *weak wakeup run* with respect to p_i iff (i) there is a finite prefix R' of R such that p_i is in a realization state in $R'(C)$, and (ii) for each prefix R'' of R, if p_i is in a realization state in $R''(C)$, then R'' includes a step of p_k for some $k \neq i$. A run R from C is a *wakeup run* iff (i) for each process p_i, there is a finite prefix R_i of R such that p_i is in a realization state in $R_i(C)$, and (ii) for each prefix R' of R, and for each process p_i, if p_i is in a realization state in $R'(C)$, then R' includes a step of each process. A run R from C is a *terminating wakeup run* iff R is a wakeup run and for each process p_i, there exists a finite prefix R_i of R such that p_i is in a termination state in $R_i(C)$.

A protocol P is a *weak wakeup protocol* iff, for each initial configuration C_0 and for each infinite fair schedule S, if R is the run from C_0 that corresponds to S in P, then there is a p_i such that R is a weak wakeup run with respect to p_i. A protocol P is a *(terminating) wakeup protocol* iff, for each initial configuration C_0 and for each infinite fair schedule S, if R is the run from C_0 that corresponds to S in P, then R is a (terminating) wakeup run.

Note that weak wakeup, wakeup, and terminating wakeup protocols form a hierarchy: a terminating wakeup protocol is also a wakeup protocol, and a wakeup protocol is also a weak wakeup protocol.

3 Wakeup in MWMR model

In this section, we show that no weak wakeup protocol exists for the MWMR model. This is because there is no process p_i that can distinguish between the run R that corresponds to the schedule $S = \langle p_1, p_2, \ldots, p_N \rangle^\infty$ and the run R_i that corresponds to the schedule $S_i = \langle p_i \rangle^\infty$. The formal arguments are as follows.

Lemma 3.1 *Let C be a configuration in which every process is in the same state, and let P be a protocol. For all $z > 0$, let S^z ans S_i^z denote the schedules $\langle p_1, p_2, \ldots, p_N \rangle^z$ and $\langle p_i \rangle^z$ respectively. Let R^z and R_i^z be the runs from C that correspond to S^z and S_i^z respectively in P. Then, the following hold.*

1. *the value in each word $M[j]$ of shared memory M is the same in $R^z(C)$ and $R_i^z(C)$.*

2. *all processes are in the same internal state in $R^z(C)$; this state is the same as the state of p_i in $R_i^z(C)$.*

Proof The lemma follows from an easy induction on z if we use the fact that all processes have identical transition functions. □

Theorem 3.1 *There is no weak wakeup protocol in the MWMR model.*

Proof Suppose P is a weak wakeup protocol in the MWMR model. Let C be an initial configuration. Let S^z, S_i^z, R^z, and R_i^z be as in lemma 3.1. Since S^∞ is an infinite fair schedule, there must exist a p_i such that R^∞ is a weak wakeup run with respect to p_i. This implies, by claim 2 of lemma 3.1, that R_i^∞ is also a weak wakeup run with respect to p_i. But R_i^∞ cannot be a weak wakeup run with respect to p_i since no process other than p_i takes a step in R_i^∞. Thus we have a contradiction. □

4 Wakeup in SWMR model

In this section, we provide a terminating wakeup protocol when each communication register is a three-valued register. Each of these need only be a *regular register*, in the terminology of [Lam86]. This is less stringent than requiring them to be *atomic registers*. Following this, we show that a wakeup protocol does not exist when each communication register is a boolean register, even if atomic.

4.1 Wakeup protocol

We assume that the communication register c_i of p_i is a three-valued regular register. The algorithm for p_i is given in figure 2. The vectors *Awake* and *InitialReading* are local to each process.

```
{Initialization}
for j := 1 to N do
   Awake[j] := false;

for j := 1 to N do
   InitialReading[j] := c_j;

c_i := c_i + 1;

repeat
   for j := 1 to N do
      if ¬Awake[j] then
         if InitialReading[j] ≠ c_j then
            Awake[j] := true;
until (∀j Awake[j]);

c_i := c_i + 1;
```

Figure 2: Wakeup protocol in SWMR model

Theorem 4.1 *The protocol in figure 2 is a terminating wakeup protocol.*

Proof The only subtle point is to show that the protocol is dead-lock free. Visualizing the steps of the different processes on a global time line, it can be seen that there is no deadlock in the protocol. The proof is omitted. □

4.2 Impossibility result

The impossibility result stated as theorem 4.2 holds irrespective of the number of processes N in the system. Here we prove the theorem for the specific case when $N = 2$. The proof for the general case is in [JT90].

Throughout this subsection, we assume that p_1 and p_2 are the only processes in the system, and c_1 and c_2 are booleans. We denote the state of p_i ($1 \leq i \leq 2$) in a configuration C by $state_i(C)$, and the value in c_i in a configuration C by $c_i(C)$. We say C is a *homogenous configuration* if $state_1(C) = state_2(C)$ and $c_1(C) = c_2(C)$. If C is a homogenous configuration, then $state(C)$ denotes the state of p_1 or p_2 in C, and $c(C)$ denotes the value of c_1 or c_2 in C. The additional symbols used in this section are as follows: P is a wakeup protocol; C_0 is the homogenous initial configuration such that $c(C_0) = 0$; for all $z \geq 0$, R^z, R_1^z, and R_2^z denote the runs from C_0 that correspond to the schedules $\langle p_1, p_2 \rangle^z$, p_1^z, and p_2^z respectively in P.

Lemma 4.1 *There exists an $\alpha \geq 1$ such that (i) $\forall j \geq \alpha$ $c_1(R_1^j(C_0)) = 1$ and (ii) $\forall j \geq \alpha$ $c_2(R_2^j(C_0)) = 1$.*

Proof Since p_1 and p_2 are identically coded, either both (i) and (ii) are true or both are false. We assume that the lemma is false and proceed to derive a contradiction by constructing an infinite fair schedule σ such that the run π from C_0 that corresponds to σ in P is not a wakeup run. The algorithm below builds this schedule σ incrementally. At each point in the algorithm, π denotes the run from C_0 that corresponds to σ in P, and C denotes $\pi(C_0)$.

1. $\sigma = NULL$ {initialization}

2. Note that $c_1(C) = c_2(C) = 0$. Further, p_1 does not see p_2 take a step in the run π. Therefore, if the lemma is false, there must exist a $j \geq 1$ such that if π' is the run from C_0 that corresponds to $\sigma' = \sigma.p_1^j$ in P, then $c_1(\pi'(C_0)) = 0$.
   Let $\sigma = \sigma.p_1^j$.

3. Note that $c_1(C) = c_2(C) = 0$. Further, p_2 does not see p_1 take a step in the run π. Therefore, if the lemma is false, we have $c_2(\pi'(C_0)) = 0$ where π' is the run from C_0 that corresponds to $\sigma' = \sigma.p_2^j$ and j is the same as in step 2.
   Let $\sigma = \sigma.p_2^j$.

4. Go to step 2.

Clearly, the infinite schedule σ constructed by the above algorithm is fair. However, no finite prefix π' of π is a wakeup run since neither process sees the other take a step in the run π'. This implies that P is not a wakeup protocol, a contradiction. □

Lemma 4.2 Let $C_1 = R_1^\alpha(C_0)$ and $C_2 = R_2^\alpha(C_0)$ where α is in lemma 4.1. For all $z \geq 0$, let $\bar{R}_2^z$ denote the run from C_1 that corresponds to the schedule p_2^z in P and $\bar{R}_1^z$ denote the run from C_2 that corresponds to the schedule p_1^z in P. Then there exists a β such that (i) $\forall j \geq \beta \ c_2(\bar{R}_2^j(C_1)) = 1$ and (ii) $\forall j \geq \beta \ c_1(\bar{R}_1^j(C_2)) = 1$.

Proof Since p_1 and p_2 are identically coded, (i) implies (ii). We assume that (i) is false and proceed to derive a contradiction by constructing an infinite fair schedule σ such that the run π from C_0 that corresponds to σ in P is not a wakeup run. The algorithm below builds this schedule σ incrementally. At each point in the algorithm, π denotes the run from C_0 that corresponds to σ in P, and C denotes $\pi(C_0)$.

1. $\sigma = p_1^\alpha$ where α is as in lemma 4.1.

2. Note that $c_1(C) = 1$ and $c_2(C) = 0$. Further, p_2 does not see p_1 take a step in the run π. Therefore, if the lemma is false, there must exist a $j \geq 1$ such that if π' is the run from C_0 that corresponds to $\sigma' = \sigma.p_2^j$ in P, then $c_2(\pi'(C_0)) = 0$.

   Let $\sigma = \sigma.p_2^j$.

3. Note that $c_1(C) = 1$ and $c_2(C) = 0$. Further, p_1 does not see p_2 take a step in the run π. Therefore, if π' is the run from C_0 that corresponds to $\sigma' = \sigma.p_1$ in P, then, by lemma 4.1, $c_1(\pi'(C_0)) = 1$.

   Let $\sigma = \sigma.p_1$.

4. Go to step 2.

Clearly, the infinite schedule σ constructed by the above algorithm is fair. However, no finite prefix π' of π is a wakeup run since neither process sees the other take a step in the run π'. This implies that P is not a wakeup protocol, a contradiction. $\square$

Lemma 4.3 Let $\sigma' = p_2^{f-1}.p_1^\alpha.p_2$ where α is the same number as in lemma 4.2, and f is such that $c_2(R_2^f(C_0)) = 1$ and for all $j < f$, $c_2(R_2^j(C_0)) = 0$. Let π' be the run from C_0 that corresponds to σ' in P, and $C' = \pi'(C_0)$. For all $z \geq 0$, let Φ_1^z denote the run from C' that corresponds to p_1^z in P. Then, there exists a $\delta' \geq 0$ such that for all $j \geq \delta'$, $c_1(\Phi_1^j(C')) = 1$.

Proof First note that the existence of f is guaranteed by lemma 4.1. Assuming that the lemma is false, we proceed to derive a contradiction by constructing an infinite fair schedule σ such that the run from C_0 that corresponds to $\sigma'.\sigma$ in P is not a wakeup run. We build σ incrementally through the following algorithm. At each point in the algorithm, π denotes the run from C' that corresponds to σ in P, and $C = \pi(C')$.

1. If the lemma is false, there exists a $j \geq 1$ such that $c_1(\Phi_1^j(C')) = 0$.

   Let $\sigma = p_1^j$

2. Note that $c_1(C) = 0$ and p_2 does not see p_1 take a step in the run π. Therefore, if $\bar{\pi}$ is the run from C' that corresponds to $\sigma.p_2^{\alpha-f}$ (α is as in lemma 4.1), then, by lemma 4.1, $c_2(\bar{\pi}(C')) = 1$.

   Let $\sigma = \sigma.p_2^{\alpha-f}$.

3. Note that $c_2(C) = 1$ and the process p_1 does not see p_2 take a step in the run π. Therefore, if the lemma is false, there must exist a j such that if $\bar{\pi}$ is the run from C_0 that corresponds to $\sigma.p_1^j$ in P, then $c_1(\bar{\pi}(C')) = 0$.

   Let $\sigma = \sigma.p_1^j$

4. Note that $c_1(C) = 0$. Further, process p_2 does not see p_1 take a step in the run $\pi'.\pi$. Therefore, if $\bar{\pi}$ is the run from C' that corresponds to $\sigma.p_2$ in P, then, by lemma 4.1, $c_2(\bar{\pi}(C')) = 1$.

   Let $\sigma = \sigma.p_2$

5. Go to step 3.

Clearly, the infinite schedule $\sigma'.\sigma$ is fair. However, no finite prefix $\bar{\pi}$ of the run $\pi'.\pi$ from C_0 is a wakeup run since p_2 does not see p_l take a step in the run $\bar{\pi}$. This implies that P is not a wakeup protocol, a contradiction. □

Lemma 4.4 *Let $\sigma'' = p_1^\alpha.p_2^\beta$ where α and β are as in lemmas 4.1 and 4.2. Let π'' be the run from C_0 that corresponds to σ'' in P, and $C'' = \pi''(C_0)$. For all $z \geq 0$, let Φ_1^z denote the run from C'' that corresponds to p_1^z in P. Then, there exists a $\delta'' \geq 1$ such that for all $j \geq \delta''$, $c_1(\Phi_1^j(C'')) = 0$.*

Proof Assuming that the lemma is false, we proceed to derive a contradiction by constructing an infinite fair schedule σ such that the run from C_0 that corresponds to $\sigma''.\sigma$ in P is not a wakeup run. We build σ incrementally through the following algorithm. At each point in the algorithm, π denotes the run from C'' that corresponds to σ in P, and $C = \pi(C'')$.

1. {Initialization}
   $\sigma = NULL$

2. Note that $c_2(C) = 1$ and the process p_1 does not see p_2 take a step in the run $\pi''.\pi$. Therefore, if the lemma is false, there must exist a j such that if $\bar{\pi}$ is the run from C'' that corresponds to $\sigma.p_1^j$ in P, then $c_1(\bar{\pi}(C'')) = 1$.

   Let $\sigma = \sigma.p_1^j$

3. Note that $c_1(C) = 1$. Further, process p_2 does not see p_1 take a step in the run π. Therefore, if $\bar{\pi}$ is the run from C'' that corresponds to $\sigma.p_2$ in P, then, by lemma 4.2, $c_2(\bar{\pi}(C'')) = 1$.

   Let $\sigma = \sigma.p_2$

4. Goto step 2.

Clearly, the infinite schedule $\sigma''.\sigma$ is fair. However, no finite prefix $\bar{\pi}$ of the run $\pi''.\pi$ from C_0 is a wakeup run since p_2 does not see p_1 take a step in the run $\bar{\pi}$. This implies that P is not a wakeup protocol, a contradiction. □

Theorem 4.2 *There is no wakeup protocol when each communication register is a boolean.*

Proof The proof given here is for the particular case when the number of processes N in the system is 2. The proof for the general case can be found in [JT90]. Let π', π'', C', and C'' be as in lemmas 4.3 and 4.4. For all $z \geq 0$, let $\dot{\Phi}_1^z$ denote the run from C' that corresponds to p_1^z in P, and $\ddot{\Phi}_1^z$ denote the run from C'' that corresponds to p_1^z in P. By lemma 4.4, there is a δ'' such that for all $k \geq \delta''$, $c_1(\ddot{\Phi}_1^k(C'')) = 0$. This, together with the fact that the runs π' and π'' are indistinguishable to p_1, implies that $c_1(\dot{\Phi}_1^k(C')) = 0$ for all $k \geq \delta''$. However, by lemma 4.3, there exists a δ' such that $c_1(\dot{\Phi}_1^j(C')) = 1$ for all $j \geq \delta'$. Clearly, δ' and δ'' cannot both exist, a contradiction. □

5 Related results

We consider two related problems: *leader election* and *consensus*. A leader election protocol is a protocol for N processes where each process has a write-once output register. In every infinite fair run, there must exist a finite prefix π of that run such that exactly one process writes a 1 in its output register and every other process writes a 0 in its output register in the run π.

A consensus protocol is a protocol for N processes where each process has a read-only input register and a write-once output register. In every infinite fair run, there must exist a finite prefix π of that run such that every process writes the same value (known as the decision-value) into its output register in that run π. This decision-value must be the value of the input register of some process.

We show that leader election is not solvable in both MWMR and SWMR models. Consensus, on the other hand, is solvable in the SWMR model but not in the MWMR model.

5.1 Leader election

Theorem 5.1 *There is no leader election protocol in the MWMR model or in the SWMR model.*

Proof Let P be a leader election protocol and C_0 be an intial configuration. In the SWMR model, we require C_0 to be such that each communication register has the same value in C_0. For all $z \geq 0$, let R^z denote the run from C_0 that corresponds to the schedule $\langle p_1, p_2, \ldots, p_N \rangle^z$ in P. Since all processes are in the same state in C_0 and are identically programmed, it is easy to verify that all processes are in the same state in $R^z(C_0)$. Therefore, for all $z \geq 0$, either every process has written a 1 in the output register in the run R^z or no process has written a 1 in the run R^z. This contradicts that P is a leader election protocol. □

5.2 Consensus

Lemma 5.1 *Let P be a consensus protocol in the MWMR model, C be any initial configuration, and p_i be any process. For all $z \geq 0$, let R_i^z be the run from C that corresponds to the schedule p_i^z in P. Then, there exists an $\alpha \geq 1$ such that p_i decides the value v in its input register in the run R_i^α.*

Proof For all $z \geq 0$, let R^z denote the run from C that corresponds to the schedule $\langle p_1, p_2, \ldots, p_N \rangle^z$ in P. Suppose the input register of each process holds the same value v. By lemma 3.1, all processes are in the same state in $R^z(C)$ and this state is the same as the state of p_i in $R_i^z(C)$. Therefore, if the lemma is false, it follows that either no process decides in the run R^∞ or every process decides some $v' \neq v$. Either way, we have a contradiction to P being a consensus protocol. □

Theorem 5.2 *There is no consensus protocol in the MWMR model.*

Proof Suppose there is a consensus protocol P in the MWMR model. Let C_0 be an initial configuration, and for all $1 \leq i \leq N$, let v_i be the value in the input register of p_i. We show by induction that, for all i ($0 \leq i \leq N$), there exist $\alpha_1, \alpha_2, \ldots, \alpha_i$ such that if π_i is the run from C_0 that corresponds to $p_1^{\alpha_1} . p_2^{\alpha_2} \ldots p_i^{\alpha_i}$ in P, then each process $p_k \in \{p_1, p_2, \ldots, p_i\}$ decides v_k in the run π_i. The basis, when $i = 0$, is trivial. For the induction step, first note that p_{i+1} cannot distinguish between $C_i = \pi_i(C_0)$ and an initial configuration C in which the shared memory is in the same state as in $\pi_i(C_0)$. Therefore, by lemma 5.1, there exists $\alpha_{i+1} \geq 1$ such that p_{i+1} decides v_{i+1} in the run π' from C_i that corresponds to the schedule $p_{i+1}^{\alpha_{i+1}}$; equivalently, p_{i+1} decides v_{i+1} in the run π_{i+1} from C_0 that corresponds to the schedule $p_1^{\alpha_1} . p_2^{\alpha_2} \ldots p_{i+1}^{\alpha_{i+1}}$. This completes the induction step. Therefore, if two processes p_k and p_l have different input values in C_0, then they decide on different values in the run π_N. This contradicts that P is a consensus protocol. □

In contrast to the above result, consensus is possible in the SWMR model. A minor enhancement to the wakeup protocol in figure 2 leads to a consensus protocol. This protocol is easy to obtain and we leave it to the reader.

Theorem 5.3 *There is a consensus protocol in the SWMR model.*

6 Conclusion

In the model of [FMRT90], Fischer *et al.* proved that consensus and leader election are both reducible to wakeup. In our multiple-writer multiple-reader model, which is the model of [FMRT90] without the test-and-set assumption, none of these three problems is solvable. However, in the single-writer multiple-reader model, wakeup and consensus are solvable, but not leader election. Thus, we showed that the reducibility of consensus and leader election to wakeup is model dependent. Another contribution of this paper is the derivation of matching lower and upper bounds on the amount of shared memory for the wakeup problem in the single-writer multiple-reader model.

References

[CIL87]   Benny Chor, Amos Israeli, and Ming Li. On processor coordination using asynchronous hardware. In *The 16th ACM Symposium on Principles of Distributed Computing*, pages 86–97, August 1987.

[Dij74]   Edsger Dijkstra. Self-stabilizing systems in spite of distributed control. *CACM*, 17:643–644, 1974.

[FLP85]   Michael Fischer, Nancy Lynch, and Michael Paterson. Impossibility of distributed consensus with one faulty process. *JACM*, 32(2):374–382, 1985.

[FMRT90] Michael Fischer, Shlomo Moran, Steven Rudich, and Gadi Taubenfeld. The wakeup problem. In *Twenty-second Annual Symposium on Theory of Computing*, May 1990.

[Ita90]   Alon Itai. On the computation power needed to elect a leader. In *4th International Workshop on Distributed Algorithms*, September 1990.

[JT90]   Prasad Jayanti and Sam Toueg. Optimal wakeup and related problems. Technical report, Cornell University, Dept. of Computer Science, Cornell University, Ithaca, NY, 1990. In preparation.

[Lam86]   Leslie Lamport. On interprocess communication, parts i and ii. *Distributed Computing*, 1:77–101, 1986.

Time and Message Efficient Reliable Broadcasts*

Tushar Deepak Chandra   Sam Toueg
Department of Computer Science
Upson Hall, Cornell University
Ithaca, New York 14853
e-mail (chandra@cs.cornell.edu, sam@cs.cornell.edu)

Abstract

This paper describes the first Reliable Broadcast algorithms that are simultaneously efficient in both time and messages. These algorithms tolerate *crash* and *omission* failures. Each Reliable Broadcast takes $O(f)$ time and $O(fn)$ messages, where f is the number of processes that actually fail during this broadcast and n is the total number of processes. In other words, each additional process that fails during a broadcast can increase the broadcast time by at most a constant, and the number of messages by at most $O(n)$. The algorithm tolerant of *crash* failures requires $f+2$ rounds. The one for *general-omission* failures requires $2f+3$ rounds.

1 Introduction

Reliable Broadcast is a fundamental problem of fault-tolerant distributed computing. Informally, Reliable Broadcast requires that when a process *broadcasts* a message, all correct processes consistently *decide* on the value of that message. In particular, the following formulation of this problem has been extensively studied [PSL80, LSP82].

Validity: If a correct process *broadcasts* a message m, then all correct processes must eventually *decide* m.

Agreement: If a correct process *decides* m, then all correct processes must also *decide* m.

Eventual Decision: If a process *broadcasts* a message, then all correct processes eventually *decide* on some value.

The first solutions to this problem were expensive in both time and message complexity. Algorithms tolerant of t faulty processes usually required $O(t)$-time for each broadcast. With *early-stopping* algorithms, the broadcast time was later reduced to $O(f)$, where f is the number of processes that actually fail during the broadcast [Ske82, DRS86]. However,

---
*This work was supported by NSF grant number CCR-8901780.

early-stopping algorithms were still expensive in messages [Had83, PT84, DR85, Per85, PT86, Ezh87]. For example, [PT86]'s early-stopping algorithm for general-omission failures requires only $f+2$ rounds[1] but requires $O(fn^2)$ messages (where n is the total number of processes).

Some Reliable Broadcast algorithms were designed to minimise the number of messages rather than the message delivery time [Bra82, DS83, Web89]. For example, the algorithm in [Bra82] requires only $O(n+t\sqrt{t})$ messages, but takes $O(t)$ rounds. Other algorithms were optimised for the failure-free case. These algorithms minimise the number of messages when no failures actually occur [Web89, AWH90, HH90].

In this paper, we present the first Reliable Broadcast algorithms that are *both* time and message efficient: each broadcast completes in $O(f)$-time with $O(fn)$ messages. In other words, each additional process that fails during a broadcast can increase the broadcast time by at most a constant, and the number of messages by at most $O(n)$. Our algorithms are designed to tolerate *crash, send-omission*, and *general-omission* failures, respectively. For general-omission failures, the performance of our algorithm is close to the known lower bounds of $\Omega(f)$ rounds and $\Omega(ft+n)$ messages per broadcast[2].

Most of the theoretical research on Reliable Broadcast (including the above algorithms) assumes that all processes have a priori knowledge of the initial time of each broadcast. In many systems, this information is not available: indeed, any process can *spontaneously* broadcast a message at *any* time [CASD85]. For such systems, we define the *Spontaneous Reliable Broadcast* problem, and present a solution that is tolerant of crash failures.

The Reliable Broadcast problem does not impose any restriction on the decision of faulty processes. However, for "benign" failures such as omission failures (where processes do not change state arbitrarily or lie), it is desirable *and* feasible to place some restrictions on the decision of faulty processes. The *Uniform Reliable Broadcast* problem [Nei88, GT89] is the Reliable Broadcast problem with the additional requirement that all processes that reach a decision, including the faulty ones, decide on the same value. Thus the Uniform Reliable Broadcast problem strengthens the *agreement* condition to:

Uniform Agreement: If a process (correct or faulty) *decides* m, then all correct processes must also decide m.

Uniform Reliable Broadcast algorithms can be used to solve many other important problems, such as the *distributed commit* problem [Ske82, CW86]. In fact, it is the Uniform Agreement requirement that ensures that either all processes commit or all abort [Had86]. Several of our algorithms actually solve the Uniform Reliable Broadcast problem.

The rest of this paper is organised as follows. In Section 2 we describe the models of failure. In Section 3, we outline the central idea of our algorithms. In Sections 4, 6 and 7 we describe algorithms tolerant of crash, send-omission and general-omission failures respectively. Section 5 describes a Spontaneous Reliable Broadcast algorithm for crash

---

[1] Informally, a round is an interval of time sufficient for a process to broadcast a message, to receive all messages sent to it, and to process those messages.

[2] We recently improved the performance of that algorithm to match these bounds.

failures. In Section 8 we present some lower bounds for general-omission failures and Section 9 discusses and summarises the results.

2 Model and Definitions

We assume a system of n processes that can communicate through reliable links in a fully connected point-to-point network. Processes have unique ids in the range $[1, n]$ which are known a priori to all processes. The computation proceeds in synchronous rounds. Informally, a round is an interval of time where processes first send messages (according to their states), wait to receive messages sent by other processes in the same round, and then change their states accordingly[3]. We consider the following types of process failures:

Crash failures: A process may fail by halting prematurely. Until it halts, it behaves correctly[4].

Send-omission failures: A process may fail not only by crashing, but also by omitting to send some of the messages that it should send.

General-omission failures: A process may fail by halting or by omitting to send or receive messages.

3 Outline of Algorithms

The algorithms in this paper use the rotating coordinator paradigm [Rei82, CM84]. A subset of $t+1$ processes cyclically become coordinators for a constant number of rounds each. The sender is the first coordinator and its id is 1. When a process becomes a coordinator, it determines a "consistent" decision value and tries to impose it on the remaining processes. Our algorithms ensure that when a correct process becomes the coordinator, it will succeed in enforcing agreement on the message broadcast. Since at most f coordinators can be faulty during an execution of the algorithm, agreement is achieved in $O(f)$ rounds. Moreover, in each round, most of the messages are to or from the coordinator; thus the number of messages sent is $O(n)$ per round, and agreement is reached with $O(fn)$ messages.

Each process p maintains a variable $estimate_p$ that represents p's current estimate of the final decision value. Processes can be in one of two states: *undecided* or *decided*. A process p *decides* v when it sets its variables $decision_p$ to v, and $state_p$ to *decided*. Our algorithms ensure that if a correct process p decides v (for some v), then all correct processes eventually decide v.

---

[3]In Section 5 we relax this assumption.

[4]If a process p crashes in round r then any subset of events on p in round r could fail to occur. It is possible to weaken this assumption and derive better algorithms.

{*Initialisation*}
$estimate_p \leftarrow \begin{cases} m & \text{if } p \text{ is the sender } (m \text{ is sender's value}) \\ \bot & \text{otherwise} \end{cases}$
$state_p \leftarrow undecided$
{*End Initialisation*}

For $c \leftarrow 1, 2, ..., t+1$ **do**
{*Processor c becomes the coordinator for three rounds*}
Round 1: All *undecided* processes p send *request* to c
         if c does not receive any *request*s then it skips rounds 2 and 3
Round 2: c broadcasts $estimate_c$
         All *undecided* processes p that receive $estimate_c$ set $estimate_p \leftarrow estimate_c$
Round 3: c broadcasts *decide*
         All *undecided* processes p that receive a *decide* do
             $decision_p \leftarrow estimate_p$
             $state_p \leftarrow decided$
od

Algorithm 1a: Reliable Broadcast with crash failures

4 Reliable Broadcast for Crash Failures

Algorithm 1a tolerates crash failures. This algorithm takes $3f + 3$ rounds to achieve decision. In Appendix A we indicate how this can be improved to $f + 2$ rounds. Each coordinator becomes "active" for three rounds. In the first round *undecided* processes send a request for "help" to the current coordinator c (an *undecided* coordinator "sends" a request to itself). If the current coordinator c does not receive any request, it skips rounds 2 and 3. If c receives a request, it broadcasts $estimate_c$ in round 2, and *decide* in round 3. Note that due to the crash failure assumption, if c begins to broadcast *decide*, then it must have successfully sent $estimate_c$ to all. Thus, *decide* is sent only if all processes receive $estimate_c$: all future coordinators are guaranteed to have the same message estimate, and will eventually force all processes to decide on it. The proof outline of the correctness of algorithm 1a is as follows[5].

Let T be the round in which the first *decide* message is received by any process. Let p be the coordinator that sent this *decide*, and let $estimate_p$ be the message p broadcast in round $T - 1$. We can show that all correct processes eventually decide $estimate_p$.

Lemma 1.1: At round $T - 1$, all processes q which did not crash received $estimate_p$ and set $estimate_q$ to $estimate_p$.

---

[5] The complete proof is given in [CT90].

Lemma 1.2: If c is a coordinator after p, and c sends $estimate_c$, then $estimate_c = estimate_p$.

Lemma 1.3: If coordinator c is correct, all processes which have not crashed decide by the end of round $3c$.

Theorem 1: Algorithm 1a solves the Reliable Broadcast problem in the presence of crash failures. The correct processes decide by round $3f + 3$ after sending at most $O(fn)$ messages.

It is easy to show (from Lemma 1.2) that Algorithm 1a actually solves the Uniform Reliable Broadcast problem.

We can easily improve the time complexity of Algorithm 1a by merging rounds 1 and 2. In round 1, any process which has not decided sends the coordinator a *request*. Furthermore, if the coordinator c has not decided, it broadcasts $estimate_c$ (note that if c is decided at this point, then all surviving processes must have the same estimate as c). In round 2, c sends *decide* if it received a *request* in round 1. With this modification, the correct processes decide by round $2f + 2$.

Further improvements are possible using pipelining. So far we only allowed a single coordinator to be active at a time. We can speed up the algorithm by pipelining its execution so that coordinator $i + 1$ starts only one round after coordinator i (while i is still active). Thus coordinator c starts in round c. The resulting algorithm achieves decision in $f + 2$ rounds. See Appendix A for details.

5 Spontaneous Reliable Broadcast - Crash Failures

Algorithm 1a relies on the following assumptions:

1. Execution proceeds in synchronous rounds.

2. All processes know *a priori* which process initiated a broadcast and in which round.

3. The identities and the order of all the coordinators is common knowledge.

These assumptions preclude the use of Algorithm 1a in an environment where any process can spontaneously initiate a broadcast at any time, and where dynamic failures prevent the use of a static agreement on the coordinators. A Reliable Broadcast algorithm designed for this setting is called a *Spontaneous Reliable Broadcast* algorithm. Algorithm 1b is a Spontaneous Reliable Broadcast algorithm that we derived from Algorithm 1a. Algorithm 1b overcomes the limitations of Algorithm 1a follows:

1. Assumption 1 is replaced with the assumption that processes have synchronised clocks[6], and that communication delay is bounded by a constant δ.

---

[6]This is to simplify the presentation of the algorithm. However, approximately synchronised clocks are sufficient.

{Initialisation}
$(estimate_p, coord\text{-}list_p) \leftarrow \begin{cases} (m, \text{ list of } t \text{ processes}) & \text{if } p \text{ is the sender} \\ (\bot, \bot) & \text{otherwise} \end{cases}$
$state_p \leftarrow undecided$
{End Initialisation}

If p is the sender then
    broadcast $(estimate_p, coord\text{-}list_p, 0)$
    broadcast *decide*

cobegin
☐ When p first receives an estimate   *{Say estimate came from c}*
    $(estimate_p, coord\text{-}list_p, coord\text{-}index_p) \leftarrow (estimate_c, coord\text{-}list_c, coord\text{-}index_c)$
    $start\text{-}time_p \leftarrow local\text{-}time$ *{Decide should come by $start\text{-}time_p + 2\delta$}*
    from time = $start\text{-}time_p + 2\delta$
        repeat at intervals of 3δ
            if $state_p = decided$ then **exit repeat**
            else
                $coord\text{-}index_p \leftarrow coord\text{-}index_p + 1$
                send $(request, estimate_p, coord\text{-}list_p, coord\text{-}index_p)$ to $coord\text{-}list_p[coord\text{-}index_p]$
    forever
☐ When p first receives a *decide*
    $decision_p \leftarrow estimate_p$
    $state_p \leftarrow decided$
☐ When p first receives a *request*   *{Say request came from q}*
    $(estimate_p, coord\text{-}list_p, coord\text{-}index_p) \leftarrow (estimate_q, coord\text{-}list_q, coord\text{-}index_q)$
    broadcast $(estimate_p, coord\text{-}list_p, coord\text{-}index_p)$
    broadcast *decide*
coend

Algorithm 1b: Spontaneous Reliable Broadcast for crash failures
(version of Algorithm 1a)

2. Assumption 2 is removed. Any process can initiate a broadcast at any time. However, since there is no a priori knowledge of who broadcasts and when, correct processes may never become aware of a broadcast initiated by a process that crashes. Therefore, we must replace the *eventual decision* condition with:

   Uniform Decision: If any correct process decides, then all correct processes eventually decide.

   The resulting specification allows the correct processes to completely ignore a broadcast initiated by a faulty process. Similar specifications have been studied in [CM84, SGS84, CASD85, GT89].

3. Assumption 3 is eliminated. In Algorithm 1b, the initiator of a broadcast chooses a sequence of future coordinators and includes this sequence in its initial broadcast[7]. This list of coordinators is also piggybacked on subsequent messages related to this broadcast.

6 Reliable Broadcast for Send-Omission Failures

Algorithms 1a and 1b do not tolerate send-omission failures. For example a faulty coordinator c could first omit to send $estimate_c$ to the next coordinator, and then send *decide* to one correct process p. Thus p decides on $estimate_c$ while the next coordinator, unaware of this estimate, can make undecided processes decide on $\bot$. This leads to disagreement. To correct this problem, we add an extra round in which processes that did not receive $estimate_c$ send a NACK to c. If c receives any NACK, it does not broadcast *decide*.

However, even with this modification, disagreement is possible. For example, a faulty coordinator c omits to send $estimate_c$ to a faulty process c' which fails to send a NACK to c. c does not receive any NACKs and thus proceeds to send some *decides*. Then c' becomes the new coordinator without having received $estimate_c$. At this point it is possible that some correct process decided on $estimate_c$ while other correct processes are still undecided and rely on c' for a decision value.

To solve this problem, a request message from an undecided process p now includes $estimate_p$ with an associated *coordinator id*. This is the *id* of the coordinator that sent this estimate to p. An undecided coordinator c considers all the requests that it receives, and sets $estimate_c$ to the estimate with the largest associated coordinator id. The resulting algorithm (Algorithm 2) takes $4f + 4$ rounds to decide. In Appendix A we indicate how this can be improved to $2f + 1$ rounds.

The proof outline of Algorithm 1 is as follows. Let T be the round in which the first *decide* is received by any process. Let p be the coordinator that sent this *decide* and

---

[7]Note that the initiator of a broadcast can decide the resiliency of that broadcast: the length of the sequence of coordinators chosen determines the maximum number of coordinator crashes that can be tolerated.

{*Initialisation*}
$(estimate_p, coord\text{-}id_p) \leftarrow \begin{cases} (m, 0) & \text{if } p \text{ is the sender } (m \text{ is sender's value}) \\ (\perp, -1) & \text{otherwise} \end{cases}$
$state_p \leftarrow undecided$
{*End Initialisation*}

For $c \leftarrow 1, 2, ..., t+1$ **do**
{*Processor c becomes the coordinator for four rounds*}
Round 1: All *undecided* processes p send $(request, estimate_p, coord\text{-}id_p)$ to c
      if c does not receive any *request* then it skips rounds 2 to 4
      else
            $estimate_c \leftarrow estimate_p$ with largest $coord\text{-}id_p$
Round 2: c broadcasts $(estimate_c, c)$
      All *undecided* processes p that receive $(estimate_c, c)$ do
            $(estimate_p, coord\text{-}id_p) \leftarrow (estimate_c, c)$
Round 3: All *undecided* processes p that did not receive $estimate_c$ in round 2
      send NACK to c
Round 4: If c does not receive a NACK then c broadcasts *decide*
      else c HALTS   {*The coordinator detects its own failure*}
      All *undecided* processes p that receive *decide* do
            $decision_p \leftarrow estimate_p$
            $state_p \leftarrow decided$
od

Algorithm 2: Reliable Broadcast with send-omission failures

---

let $estimate_p$ be the message p broadcast in round $T-2$. We can show that all correct processes eventually decide $estimate_p$.

Lemma 2.1: By the end of round $T-2$, all correct processes q receive $estimate_p$ and set $estimate_q$ to $estimate_p$.

Lemma 2.2: Suppose a correct process q sets $(estimate_q, coord\text{-}id_q)$ in "round 2" of the algorithm to some value (v, r). Then, until q decides, all the coordinators can only send v as their estimate.

Lemma 2.3: If coordinator c is correct, all correct processes decide by the end of round $4c$.

Lemma 2.4: All correct processes which decide must decide on the same value.

Theorem 2: Algorithm 2 solves the Reliable Broadcast problem in the presence of send-omission failures. The correct processes decide by round $4f + 4$ after sending at most $O(fn)$ messages.

7 Reliable Broadcast for General-Omission Failures

Algorithm 2 tolerates any number of send-omission failures (i.e., for any $t < n$). To tolerate general-omission failures we used a "translation" technique from [NT90] which requires $n > 2t$. Thus, the resulting algorithm (Algorithm 3), tolerates up to $\lfloor \frac{n-1}{2} \rfloor$ general-omission failures, and actually solves the Uniform Reliable Broadcast problem[8].

Informally, running algorithm 2 in a system with general-omission failures does not work for the following three reasons:

1. A faulty coordinator could fail to receive a NACK, and thereby send a *decide* when it should not. This problem is remedied by the translation mechanism: essentially the NACK mechanism is replaced with $n - t$ positive ACKs.

2. A faulty coordinator that is activated by a $(request, estimate_p, coord\text{-}id_p)$ may fail to receive a $(request, estimate_q, coord\text{-}id_q)$ with $coord\text{-}id_p < coord\text{-}id_q$, where q is a correct process. To solve this problem, an activated coordinator c broadcasts a *probe* asking all processes p to send $(estimate_p, coord\text{-}id_p)$. The coordinator must receive at least $n - t$ responses before it updates $estimate_c$ and broadcasts it.

3. A faulty process may continuously fail to receive *decide* messages and thus successively send *requests* to all coordinators, thereby activating all of them. This results in too many messages. To overcome this problem, we introduce a technique that prevents a faulty process from activating more than one correct coordinator. So at most $2f + 1$ coordinators will be activated, resulting in $O(fn)$ messages. The technique works as follows. An activated coordinator c selects one of the processes which woke it up, called the *requester*. Any process p that decides, relays its decision value to the *requester*. If later p becomes a coordinator, it ignores any request from this *requester*.

Algorithm 3 tolerates upto $\lfloor \frac{n-1}{2} \rfloor$ failures. A single coordinator executes every 7 rounds. The correct processes take $7f + 6$ rounds to decide. In Appendix A we indicate how this can be improved to $2f + 3$ rounds.

The proof outline of Algorithm 3 is as follows. If $Q_c \neq \emptyset$, we say that coordinator c is *active*. Let T be the round in which the first *decide* is received by any process. Let p be the coordinator that sent this *decide* and let $estimate_p$ be the message p broadcast in round $T - 2$. We can show that all correct processes eventually decide $estimate_p$.

Lemma 3.1: By the end of round $T - 2$, at least $n - t$ processes q receive $estimate_p$ and set $estimate_q$ to $estimate_p$.

---

[8]The complete version of the paper will describe an algorithm that solves the Reliable Broadcast problem (without the uniform agreement condition) for more than $\lfloor \frac{n-1}{2} \rfloor$ failures.

{*Initialisation*}

$(estimate_p, coord\text{-}id_p) \leftarrow \begin{cases} (m, 0) & \text{if } p \text{ is the sender } (m \text{ is sender's value}) \\ (\bot, -1) & \text{otherwise} \end{cases}$

$state_p \leftarrow undecided$
$finishedset_p \leftarrow \emptyset$
{*End Initialisation*}

For $c \leftarrow 1, 2, ..., t + 1$ do
{*Processor c becomes the coordinator for seven rounds*}
Round 1: All *undecided* processes p send *request* to c
  Let $Q_c = \{q \mid c \text{ received a request from } q \wedge q \notin finishedset_c\}$
  If $Q_c = \emptyset$ then c skips rounds 2 to 7
  else *requester* $\leftarrow$ an element of Q_c
Round 2: c broadcasts *probe*
Round 3: All processes p that receive a *probe* send $(answer, estimate_p, coord\text{-}id_p)$ to c
Round 4: If c receives $\geq n - t$ answers then
    $estimate_c \leftarrow estimate_p$ with largest $coord\text{-}id_p$
    c broadcasts $(estimate_c, c)$
  else c skips rounds 5 to 7   {*The coordinator detects its own failure*}
  All *undecided* processes p that receive $(estimate_c, c)$ do
    $(estimate_p, coord\text{-}id_p) \leftarrow (estimate_c, c)$
Round 5: All processes p that received $(estimate_c, c)$ send an ACK to c
Round 6: If c receives $\geq n - t$ ACKs then c broadcasts $(decide, requester)$
  All *undecided* processes p which received $estimate_c$ and *decide* do
    $decision_p \leftarrow estimate_p$
    $state_p \leftarrow decided$
Round 7: All processes p that received $estimate_c$ and *decide* do
    Add *requester* to $finishedset_p$
    send $(decide, estimate_p)$ to *requester*
  If *requester* is *undecided* and it receives $(decide, estimate_p)$ for some p
    $decision_{requester} \leftarrow estimate_p$
    $state_{requester} \leftarrow decided$
od

Algorithm 3: Reliable Broadcast with general-omission failures

Lemma 3.2: If $t+1$ processes receive $estimate_c$ from coordinator c, all future coordinators which send out their estimate, send out $estimate_c$.

Lemma 3.3: If coordinator c is correct, all correct processes decide by the end of round $7c-1$.

Lemma 3.4: All processes which decide, decide on the same value.

Lemma 3.5: At most $f+1$ correct coordinators become active.

Theorem 3: Algorithm 3 solves the Reliable Broadcast problem in the presence of general-omission failures. The correct processes decide by round $7f+6$ after sending at most $O(fn)$ messages.

In fact we can show that the *total* number of messages sent in the system (including those sent by *faulty* processes) is $O(fn)$. It is easy to see that when the current coordinator is correct but not active, correct processes do not send any messages. Let $c_{correct}$ be the number of correct processes which become active coordinators and c_{faulty} be the number of faulty coordinators. From Lemma 3.5, $c_{correct} \leq f+1$. The number of messages sent by correct coordinators is bound by:

$$c_{correct} O(n) \leq O(fn) \text{ messages}$$

The number of messages sent by the $n-f$ correct processes while they are not coordinators is bound by:

$$(n-f) * (c_{correct} + c_{faulty}) * O(1) \leq O(fn) \text{ messages}$$

Each faulty processes sends at most $O(n)$ messages during the algorithm. Thus the total number of messages sent in the algorithm is $O(fn)$.

From Lemma 3.4, Algorithm 3 solves the Uniform Reliable Broadcast problem.

8 Lower bounds for general-omission failures

Dolev et. al. showed that any Reliable Broadcast algorithm for general-omission failures requires $\frac{t^2}{2}$ messages in the worst case [DR85]. With a minor modification we can extend this result to *early-stopping* algorithms as follows[9]:

Theorem 4: For any algorithm A which solves the Reliable Broadcast problem tolerant of general-omission failures and for any f, there exists a run of A with f failures in which correct processes send $\frac{ft}{4}$ messages.

The following result from [NT90] shows that Algorithm 3 is optimal in the number of failures that it tolerates.

---

[9]The proof will be found in the complete version of the paper.

Failure Model	Number of rounds (for pipelined version)
Crash	$f+2$
Send-Omission	$2f+1$
General-Omission	$2f+3$

Table 1

Theorem 5: Any Uniform Reliable Broadcast algorithm for general-omission failures requires $n > 2t$.

9 Summary of results

The Reliable Broadcast protocols in this paper require $O(f)$ time and $O(fn)$ messages. In Table 1, we list the exact number of rounds they require for each model of failure.

In the complete version of this paper we will present an algorithm which matches the $\Omega(f)$ time and $\Omega(ft+n)$ messages lower bounds for general-omission failures. We will also describe a Reliable Broadcast algorithm that tolerates more than $\lfloor \frac{n-1}{2} \rfloor$ general-omission failures.

Acknowledgements

We would like to thank Navin Budhiraja, Ajei Gopal, Prasad Jayanti, Keith Marzullo, Gil Neiger, Pat Stephenson and the distributed systems group at Cornell University for their critical comments. Navin Budhiraja proposed a simplified version of Algorithm 1a which took $O(n^2)$ messages to reach agreement.

Appendix A - Pipelining and other Improvements

Some of the initial rounds can be avoided for the first coordinator. For crash and send-omission failures, the first round can be skipped and for general-omission failures, the first 3 rounds can be omitted. This gives us algorithms which take $3f + 2$, $4f + 3$ and $7f + 3$ rounds to reach decision for crash, send-omission and general-omission failures ($n > 2t$) respectively. For the send-omission algorithm, a process can decide the moment its initial value changes i.e., process p can decide the moment $coord\text{-}id_p \neq 0$. With this change, the send-omission algorithm can decide in $4f + 1$ rounds.

Further speed-ups are possible using pipelining. In the pipelined versions, in each round there are many active coordinators - each one at a different stage of the algorithm. A brief description of this pipelining scheme and its performance follows:

Crash failures: When process i begins its second round as coordinator, process $i + 1$ begins its first round. It can be shown that the correctness of the algorithm is preserved and the system decides by round $f + 2$.

Send-omission failures: Coordinator $i + 1$ starts when coordinator i begins its third round. With this, the system decides in $2f + 1$ rounds.

General-omission failures ($n > 2t$): As in the send-omission case, two successive coordinators can be run with a gap of two rounds between them. With this modification decision is achieved in $2f + 3$ rounds. Note that to achieve decision in $2f + 3$ rounds, the first coordinator must skip its first 3 rounds and the second coordinator must skip its first.

We can reduce the number of messages for the general-omission failure algorithm by a constant. This is achieved by having decided coordinators skip rounds 2 through 5 when they receive a message in round 1. From Lemma 3.4, all processes that decide in Algorithm 3 decide on the same value. Thus the coordinator can broadcast its decision value in round 6 without worrying about the estimates of other processes.

References

[AWH90] Eugene Amdur, Sam Weber, and Vassos Hadzilacos. On the message complexity of binary byzantine agreement under crash failures. 1990. Submitted to Distributed Computing.

[Bra82]   Gabriel Bracha. Personal communication. 1982.

[CASD85] Flaviu Cristian, Houtan Aghili, H. Raymond Strong, and Danny Dolev. Atomic broadcast: From simple message diffusion to Byzantine agreement. In *Proceedings of the Fifteenth International Symposium on Fault-Tolerant Computing*, pages 200–206, June 1985. A revised version appears as IBM Research Laboratory Technical Report RJ5244 (April 1989).

[CM84]   J. Chang and N. Maxemchuk. Reliable broadcast protocols. *ACM Transactions on Computer Systems*, 2(3):251–273, August 1984.

[CT90]   Tushar Deepak Chandra and Sam Toueg. Time and message efficient reliable broadcasts. Technical Report 90-1094, Department of Computer Science, Cornell University, May 1990.

[CW86]   Brian A. Coan and Jennifer L. Welch. Transaction commit in a realistic fault model. In *Proceedings of the Fifth ACM Symposium on Principles of Distributed Computing*, pages 40–51, August 1986.

[DR85]   Danny Dolev and Rüdiger Reischuk. Bounds on information exchange for Byzantine agreement. *Journal of the ACM*, 32(1):191–204, January 1985.

[DRS86]   Danny Dolev, Rüdiger Reischuk, and H. Raymond Strong. Early stopping in Byzantine agreement. Technical Report RJ5406, IBM Research Laboratory, December 1986.

[DS83]   C. Dwork and D. Skeen. The inherent cost of nonblocking commitment. In *Proceedings of the 2nd Annual ACM Symposium on Principles of Distributed Computing*, pages 1–11, August 1983.

[Ezh87]   Paul D. Ezhilchelvan. Early stopping algorithms for distributed agreement under fail-stop, omission, and timing fault types. In *IEEE 1987 Sixth Symposium on Reliability in Distributed Software and Database Systems*, pages 201–212, Computing Laboratory, The university, Newcastle upon Tyne, England, 1987. IEEE computer society press.

[GT89]   Ajei Gopal and Sam Toueg. Reliable broadcast in synchronous and asynchronous environments (preliminary version). In J.-C. Bermond and M. Raynal, editors, *Proceedings of the Third International Workshop on Distributed Algorithms*, volume 392 of *Lecture Notes on Computer Science*, pages 110–123. Springer-Verlag, September 1989.

[Had83]   Vassos Hadzilacos. Byzantine agreement under restricted types of failures (not telling the truth is different from telling lies). Technical Report 18-83, Department of Computer Science, Harvard University, 1983. A revised version appears in Hadzilacos's Ph.D. dissertation [Had84].

[Had84]   Vassos Hadzilacos. *Issues of Fault Tolerance in Concurrent Computations.* PhD thesis, Harvard University, June 1984. Department of Computer Science Technical Report 11-84.

[Had86]   Vassos Hadzilacos. On the relationship between the atomic commitment and consensus problems. Workshop on Fault-Tolerant Distributed Computing, March 17-19, 1986, Pacific Grove, CA. (Proceedings to be published in a volume, edited by Brabara Simons, of the Springer-Verlag Series "Lecture Notes on Computer Science"), 1986.

[HH90]   Vassos Hadzilacos and Joseph Y. Halpern. Message and bit-optimal protocol for byzantine agreement. 1990. To appear.

[LSP82]   Leslie Lamport, Robert Shostak, and Marshall Pease. The Byzantine generals problem. *ACM Transactions on Programming Languages and Systems*, 4(3):382–401, July 1982.

[Nei88]   Gil Neiger. *Techniques for Simplifying the Design of Distributed Systems.* PhD thesis, Cornell University, August 1988. Department of Computer Science Technical Report 88-933.

[NT90]   Gil Neiger and Sam Toueg. Automatically increasing the fault-tolerance of distributed algorithms. *Journal of Algorithms*, 11(3):374–419, September 1990.

[Per85]   Kenneth J. Perry. *Early Stopping Protocols for Fault-Tolerant Distributed Agreement.* PhD thesis, Cornell University, February 1985. Department of Computer Science Technical Report 85-662.

[PSL80]   M. Pease, R. Shostak, and Leslie Lamport. Reaching agreement in the presence of faults. *Journal of the ACM*, 27(2):228–234, April 1980.

[PT84]   Kenneth J. Perry and Sam Toueg. An authenticated Byzantine generals algorithm with early stopping. Technical Report 84-620, Department of Computer Science, Cornell University, June 1984.

[PT86]   Kenneth J. Perry and Sam Toueg. Distributed agreement in the presence of processor and communication faults. *IEEE Transactions on Software Engineering*, 12(3):477–482, March 1986.

[Rei82]   Rüdiger Reischuk. A new solution for the Byzantine general's problem. Technical Report RJ 3673, IBM Research Laboratory, November 1982.

[SGS84]   Fred B. Schneider, David Gries, and Richard D. Schlichting. Fault-tolerant broadcasts. *Science of Computer Programming*, 4(1):1–15, April 1984.

[Ske82]   Dale Skeen. *Crash Recovery in a Distributed Database System.* PhD thesis, University of California at Berkeley, Department of EECS, 1982.

[Web89]   Samuel Weber. Bounds on the message complexity of Byzantine agreement. Master's thesis, University of Toronto, October 1989.

Early–Stopping Distributed Bidding and Applications
(Preliminary Version)

Navin Budhiraja*   Ajei Gopal[†]   Sam Toueg[‡]

Department of Computer Science
Cornell University
Ithaca, NY 14853
{navin, ajei, sam}@cs.cornell.edu

Abstract

We define the problem of *Distributed Bidding* and derive efficient solutions for several models of process failures. Our algorithms are *early–stopping*: their performance gracefully degrades as the number of processes that actually fail increases. We use our Distributed Bidding algorithms to derive the first known *early-delivery Atomic Broadcast* algorithms that deliver messages in time proportional to the number of processes that actually fail during the broadcast, rather than to the maximum number of faulty processes that the algorithm tolerates.

1 Introduction

In *Distributed Bidding*, a process solicits a bid for some object from all processes, and all processes must eventually agree on the bid submitted by each process. In this paper we formally define the fault–tolerant version of this problem, and present efficient solutions for process failures ranging from crash to arbitrary failures.

The Distributed Bidding problem can be solved using an "off-the-shelf" *Reliable Broadcast* algorithm (such as [LSP82]): a process reliably broadcasts the object, and all processes reply by reliably broadcasting their bids for that object. Although this simple solution can be speeded up using an *early–stopping* Reliable Broadcast algorithm (such as [DRS82,PT84]), it is still too slow.

---
*Partially supported by the Defense Advanced Research Projects Agency (DoD) under NASA Ames grant number NAG 2-593, Contract N00140-87-C-8904
[†]Partially supported by an IBM Graduate Student Fellowship.
[‡]Partially supported by NSF grant CCR-8901780.

Our Distributed Bidding algorithms complete within approximately aD time, where a is the number of processes that actually fail during the bidding, and D is the maximum link delay. That is, our algorithms are *early–stopping*, as their performance gracefully degrades with increase in the number of processes that actually fail during execution of the algorithm.

Distributed Bidding can be used in the design of many fault–tolerant algorithms. For example, we use our Distributed Bidding algorithms to derive the first known *early-delivery Atomic Broadcast* algorithms that deliver messages in time proportional to the number of processes that actually fail during the broadcast, rather than to the maximum number of faulty processes that the algorithm tolerates. Previous solutions guaranteed early–delivery only in failure–free executions [GSTC90,CS87].

In contrast to most early–stopping algorithms [DRS82,PT86,TPS87], we avoid the "round" model of computation. In particular we do not make the unrealistic assumption that all the processes have *a priori* knowledge of the initiation time of each Distributed Bidding: any process can initiate a bid request at any time.

The next two sections present the model and the formal definition of Distributed Bidding. Section 4 overviews our results, Section 5 presents the Distributed Bidding algorithms for several models of failure, and Section 6 describes our early–delivery Atomic Broadcast algorithms. In the Appendix, there is a brief summary of some simulation results for our implementation of early–delivery Atomic Broadcast. Due to space considerations, we omit all proofs from this preliminary version of the paper. We only include the statements of the main lemmas and theorems.

2 Model

We consider a distributed system that consists of a completely connected point-to-point network. Processes may be either *correct* or *faulty*. A process subject to *crash* failures may prematurely halt execution. A process subject to *send–omission* failures may crash or may occasionally fail to send a message during execution. A process subject to *general–omission* failures may crash or may occasionally fail to send or receive a message during execution. There is no constraint on the behavior of a process subject to *arbitrary* failures.

We assume non-faulty links. The link delay between any two correct processes is at most D, and at least d. To simplify the presentation, we assume that the clocks of all correct processes are perfectly synchronized. (However, our algorithms can be easily modified for the case that clocks are only synchronized to within ϵ of each other.)

Let $\mathcal{P}$ be the set of all processes in the system, and let $n = |\mathcal{P}|$. Let f denote the *maximum* number of faulty processes that the algorithm must tolerate, and a be the number of processes that *actually fail* during Distributed Bidding.

3   Distributed Bidding

If a process p, the *initiator*, requests a bid for an object m_p, then correct processes should eventually *decide* on an *outcome vector* that contains the bids of all processes. Formally:

- *Validity:* If a correct process p initiates Distributed Bidding for m_p, then every correct process q eventually *decides* on some outcome vector $V_q^{m_p}$ for m_p.

- *Integrity:* If a correct process q *decides* on outcome vector $V_q^{m_p}$ for m_p, then this vector contains a bid for m_p from each correct process—*i.e.*, for all correct processes r, $V_q^{m_p}[r] = \beta_r(m_p)$, where $\beta_r(m_p)$ denotes process r's bid for m_p.

- *Agreement:* If a correct process q *decides* on outcome vector $V_q^{m_p}$ for m_p, then all correct processes eventually *decide* on the same outcome vector for m_p—*i.e.*, for all correct processes r, $V_q^{m_p} = V_r^{m_p}$.

- *Termination:* If a correct process q bids for m_p, then q eventually *decides* on an outcome vector $V_q^{m_p}$ for m_p, or q eventually *decides* NIL for m_p.

Validity implies that a correct process may decide NIL only if the initiator p is faulty. Note that this specification permits some correct processes to decide NIL although other correct processes neither bid nor decide. However, if any correct process decides on an outcome vector, then *Agreement* forces all correct processes to do the same.

4   Goals, Techniques and Results

In most systems, failures seldom occur. Thus, our goal is to derive Distributed Bidding algorithms that are fast in the likely case that no failures, or very few failures occur during bidding. In addition, in many systems the *expected* link delay is much closer to d (the minimum link delay) than to D (the maximum link delay) [Cri89, HK89]. Our algorithms take advantage of this property by terminating faster when actual message transmission times are small.

Our Distributed Bidding algorithms consist of two stages. First, the initiator requests a bid for an object m_p by disseminating m_p to all processes—this is called the *request* stage. Second, each process disseminates its bid for m_p, and all correct processes agree on these bids—the *bidding* stage.

A simple solution to Distributed Bidding is to implement these two stages using an "off-the-shelf" Reliable Broadcast algorithm. In the request stage, the initiator reliably broadcasts the object. The subsequent bidding stage consists of n concurrent reliable broadcasts, one broadcast for each process's bid. This scheme takes about $(f+f)D$ time to terminate.

A better solution is to use *early-stopping* Reliable Broadcast algorithms [DRS82, PT86, TPS87]. However, such algorithms were developed for the "round model" of computation, with the unrealistic assumption that processes have *a priori* knowledge of the initial time of a broadcast. In our setting, such knowledge is not available as *any* process can initiate Distributed Bidding at *any* time. This can be overcome by using a *Firing Squad* algorithm (FSP) [CDDS85] in the request stage. This scheme requires about $(f+a)D$ time—fD for the FSP, and aD for the early-stopping Reliable Broadcasts of the bids.

Our goal is to derive algorithms that will terminate in at most $(a+c)D$ time, where c is a small constant. In the solutions outlined above, each faulty process can delay both the request stage and the bidding stage by D time each. Therefore, to achieve our goal, a process that delays the request stage must be prevented from also delaying the bidding stage. We use a *failure-detection* scheme that identifies those faulty processes that delayed the termination of the request stage, and excludes them from the bidding stage. Thus, each faulty process can only delay the algorithm in at most one of the two stages, by a total of at most D time—this gives the desired performance of at most $(a+c)D$ time. The worst-case termination time of our algorithms for the different types of failures is given in Figure 1.

crash	$min\ [(a+4)D, (f+2)D]$
send-omission	$min\ [(a+4)D, (f+2)D]$
general-omission	$min\ [(a+5)D, (f+3)D]$
arbitrary	$min\ [(a+5)D, (f+4)D]$

Figure 1: Worst-case termination time of Distributed Bidding

Our algorithms can also take advantage of the likely case that most messages are fast. For example, the algorithm tolerant of general-omission failures completes in $2d$ time (which is optimal) when no failures actually occur, and all messages take d time during the execution of the algorithm.

Distributed Bidding has many important applications, such as *Atomic Broadcast*. Informally, Atomic Broadcast requires that all processes deliver the same messages, in the same order. We use our time–efficient Distributed Bidding algorithms to derive the first known *early-delivery* Atomic Broadcast algorithms that deliver messages in time proportional to a rather than to f. Previous early–delivery Atomic Broadcast algorithms guaranteed early message delivery only in failure–free executions; but a single failure could delay message delivery by up to $O(f)$ time [GSTC90, CS87]. However, these earlier algorithms can be used in a larger class of communication networks than our algorithms.

Our early–delivery Atomic Broadcast algorithms terminate in a worst–case time of approximately $2(a + c)D$, where c is a small constant that depends on the failure model (as shown in Figure 1). This worst–case occurs in the unlikely event that the delivery of a message from a faulty process is further delayed by another message broadcast by another faulty process. Simulation results (see Appendix A) suggest that this rarely occurs, and message delivery usually completes in about $(a + c)D$ time.

5 Distributed Bidding Algorithms

Our Distributed Bidding algorithms consist of a request stage followed by a bidding stage. Some process p initiates a bid in the request stage. In the bidding stage, each process can independently disseminate its bid to all other processes. Thus, we divide the bidding stage into n concurrent, logically separate, and non-interfering instances of the same dissemination algorithm. We will restrict our attention to the dissemination of the bid of one particular process q.

For the crash and send–omission models of failures, we assume that the clocks of *all* processes that have not crashed are synchronized with each other. The corresponding Distributed Bidding algorithms tolerate any number of process failures ($f < n$). For the general–omission and arbitrary failure cases, we only assume that the clocks of *correct* processes are synchronized. The corresponding algorithms require a majority of correct processes ($2f < n$).

5.1 Crash and Send–Omission Failures

The algorithm tolerant of crash failures is a simpler version of the algorithm tolerant of send-omissions. In particular, it does not need the failure-detection mechanism that is required to deal with intermittent omission failures. This simpler algorithm is therefore omitted from this paper.

5.1.1 Description of the Algorithm for Send-Omissions

---

when p requests a bid for m_p
  $t_p := time$
  send$((wakeup, m_p, t_p), t_p)$ to all

when r first receives a message of type $((wakeup, m_p, t_p), t_s)$ from some s:
  if $r \neq p$ **then**   /* relay the *wakeup* with r's wakeup time */
    $t_r := time$
    send$((wakeup, m_p, t_p), t_r)$ to all

  /* process r starts its bidding stage */
  cobegin
    || **for all** $q \in \mathcal{P}$ **call** *determine-bid* (q)
  coend
  decide $V_r^{m_p}$

Figure 2: Request stage of Distributed Bidding algorithm tolerant of send–omissions

The algorithm tolerant of send–omission failures is given in Figures 2 and 3. The request and bidding stages of this algorithm are informally described below:

1. *Request stage*: The initiator p broadcasts a *wakeup* message with the object m_p and the current time t_p. Upon receipt of this message, each process r relays it, and starts the bidding stage by broadcasting its bid for m_p. From now on, we focus on the dissemination of process q's bid.

Note that the wakeup message may be relayed through a chain of faulty processes each one of which delays the completion of the request stage by D time.

2. *Bidding stage*: For a correct process r to decide on an outcome vector, it must decide on the bid of each process q. This decision can be reached either by receiving a copy of q's bid (directly or relayed), or by determining that q is faulty and that q's bid will never be received by any correct process. For the second case, each process must keep track of both the time that has elapsed, and the processes that failed since p initiated the bid request.

Fault-detection is a key aspect to the bidding stage. A process that failed in the request stage may have already delayed the termination of the algorithm. If this process is allowed to participate in the bidding stage, it may fail again (to properly relay q's

/* Process r executes this to determine q's bid for object m_p */
/* The bid request for m_p was initiated by process p at time t_p */
/* Process r awoke at time t_r */
Procedure *determine-bid* (q)
   if $r = q$ **then**
      $bid_q := \beta_q(m_p)$
      send$(((bid_q, m_p, t_q), t_q)$ to all
      $V_q^{m_p}[q] := \beta_q(m_p);$ **exit**
   else
      cobegin
         when r receives a message M of type $((-, m_p, -), t_s)$ from s
            if $t_s > t_r + D$ **then halt** /* r detects its failure to relay the wakeup */
            else if $t_s < t_r - D$ **then skip** /* r detects s's failure to relay the wakeup */
            else accept M
      ‖
         when r accepts $((bid_q, m_p, t_z), t_s)$ with $bid_q \neq \perp_q$
            if $t_z > t_r + D$ **then halt** /* r detects its failure to relay the wakeup */
            send$(((bid_q, m_p, t_z), t_r)$ to all
            $V_r^{m_p}[q] := bid_q;$ **exit**
      ‖
         when $time = t_p + (f + 2)D$
            $V_r^{m_p}[q] := default_q;$ **exit**
      ‖
         let k_r be such that $t_p + k_r D < t_r \leq t_p + (k_r + 1)D$
         when $time = t_p + (k_r + 1)D$ send $((\perp_q, m_p, k_r + 1), t_r)$ to all
         for $j = k_r + 2$ **to** $f + 2$ **do**
            when $time = t_p + jD$ send$((\perp_q, m_p, j), t_r)$ to all
            when $time = t_p + (j + 1)D$
               $senders := \{z | r$ accepted $((\perp_q, m_p, j), t_z)$ from z $\}$
               $faulty := faulty \cup (\mathcal{P} - senders)$
               if $|faulty| < j - 1$ **then**
                  send$(((default_q, m_p, t_r), t_r)$ to all
                  $V_r^{m_p}[q] := default_q;$ **exit**
         od
      coend

Figure 3: Bidding stage of Distributed Bidding algorithm tolerant of send–omissions

bid, for example) and thus cause an additional delay. Our algorithm prevents this by using the following fault-detection mechanism.

All messages sent by a process r are tagged with t_r, the time at which r woke up. If process r receives a message from process s such that $t_s > t_r + D$, then r halts. This is because r detected its own failure to properly relay the wakeup message to process s. However, this scheme is not sufficient. If process r receives a message from process s such that $t_r > t_s + D$, then r ignores the message. This is because r detected s's failure to properly relay the wakeup message (to r).

5.1.2 Proof of Correctness

Only the statements of the main lemmas are included. Let var_r^t denote the variable var at process r at time t.

Lemma 1 *Suppose any process r awakens at time t_r, $t_p + k_r D < t_r \leq t_p + (k_r + 1)D$.*
- *At least k_r faulty processes awoke before $t_r - D$.*
- *If r does not crash by time $t_p + (k_r + 3)D$, then $faulty_r^{t_p+(k_r+3)D}$ contains at least k_r faulty processes that awoke before $t_r - D$.*
- *If r is correct, at least k_r faulty processes that awoke before $t_r - D$ will halt by time $t_r + D$.*
- *If $r = q$, then none of the k_q faulty processes that awoke before $t_q - D$ will relay $\beta_q(m_p)$.*

Lemma 2 *If any process z sets $V_z^{m_p}[q] = default_q$ then no correct process s ever sets $V_s^{m_p}[q] = \beta_q(m_p)$.*

Theorem 3 *The algorithm given in Figures 2 and 3 is a Distributed Bidding algorithm for send–omission failures. It tolerates up to $f < n$ faulty processes, and guarantees Termination within $\min[(a + 4)D, (f + 2)D]$ time.*

5.2 General–Omission Failures

5.2.1 Description of the Algorithm

The algorithm tolerant of general–omission failures requires $n > 2f$, and is given in Figures 4 and 5.

The algorithm is similar to the one for send–omission, with one major exception—the fault–detection mechanism. In the case of send–omissions, the fault–detection worked by pairwise comparison of process wakeup times: if two processes, r and s, are awakened more than D time apart ($t_s > t_r + D$), then r failed to relay the wakeup to s. In the presence of general–omissions, this reasoning is incorrect. Process s could be the faulty process—r may have correctly relayed the wakeup and s may have failed to receive the relay.

---

when p requests a bid for m_p
  $t_p := time$
  $T_p := \{t_p\}$
  send$((wakeup, m_p, t_p), T_p)$ **to all**

when r first receives a message of type $((wakeup, m_p, t_p), T_s)$ from some s:
  if $r \neq p$ **then**   /* relay the *wakeup* with r's wakeup time */
    $t_r := time$
    $T_r := T_s \cup \{t_r\}$
    send$((wakeup, m_p, t_p), T_r)$ **to all**

  /* process r starts its bidding stage */
  cobegin
    || **for all** $q \in \mathcal{P}$ **call** *determine-bid* (q)
  coend
  decide $V_r^{m_p}$

Figure 4: Request stage of Distributed Bidding algorithm tolerant of general–omissions

Our failure–detection mechanism for general–omission failures requires a majority of correct processes. It is based on the following observation: if the first correct process to awaken is r, at time t_r, then all correct processes must awaken by $t_r + D$. Thus if a process s is correct, then its wakeup time t_s must lie in some time interval $[t, t+D]$ that includes the wakeup times of at least $n - f$ processes (those of all the correct processes). Our algorithm uses this property to detect faulty processes.

An interval $[t, t+D]$ is *good* with respect to a set of process wakeup times T if and only if it includes at least $n - f$ of the wakeup times in T. Let T_r^τ denote the set of wakeup times that r knows at time τ. Process r evaluates the following predicates:

- *plausible*$(t_s, \tau) \equiv \exists$ an interval $[t, t+D]$ which is good w.r.t. T_r^τ, and $t_s \in [t, t+D]$.

- *impossible*$(t_s, \tau) \equiv \exists$ an interval $[t, t+D]$ which is good w.r.t. T_r^τ,
  and $t_s \notin [t - D, t + 2D]$.

These two predicates are evaluated by r to determine faulty processes as follows.
  • At time $t_r + 2D$, r checks whether *plausible*$(t_r, t_r + 2D)$ holds. If not, then r halts: it detected its own failure to successfully awaken all the correct processes by time $t_r + D$.

/* Process r executes this to determine q's bid for object m_p */
/* The bid request for m_p was initiated by process p at time t_p */
/* Process r awoke at time t_r */
Procedure *determine-bid* (q)
   if $r = q$ **then**
      $bid_q := \beta_q(m_p)$
      send$((bid_q, m_p, 1), T_q)$ to all
      $V_q^{m_p}[q] := \beta_q(m_p);$ exit
   else
      cobegin
         when r receives a message M of type $((-, m_p, -), T_s)$ from s
            if $T_s \not\subseteq T_r$ **then**
               $T_r := T_s \cup T_r$
               send $((\text{wake-times}, m_p, t_p), T_r)$ to all
            if *impossible*$(t_r, time)$ **then** halt
            else if $|t_s - t_r| > D$ or *impossible*$(t_s, time)$ **then** skip
            else accept M
      || **when** $time = t_r + 2D$ and not *plausible*$(t_r, time)$ **then** halt
      || **when** r accepts $((bid_q, m_p, k), T_s)$ with $bid_q \notin \{\perp_q, default_q\}$
         if $\exists \tau \leq time$ such that *impossible*(t_q, τ) and
            $lD \leq (t_q - \tau) < (l+1)D$, for some $l \geq 0$ and
            $iD \leq (time - \tau) < (i+1)D$, for some $i \geq 0$ and
            $k < max(l+2, i+1)$ **then** skip
         else
            send$((bid_q, m_p, k+1), T_r)$ to all
            $V_r^{m_p}[q] := bid_q;$ exit
      || **when** r accepts $((default_q, m_p, -), T_s)$
         send$((default_q, m_p, -), T_r)$ to all
         $V_r^{m_p}[q] := default_q;$ exit
      || **when** $time = t_p + (f+3)D$
         $V_r^{m_p}[q] := default_q;$ exit
      || let k_r be such that $t_p + k_r D < t_r \leq t_p + (k_r + 1)D$
         when $time = t_p + (k_r + 1)D$ send $((\perp_q, m_p, k_r + 1), T_r)$ to all
         for $j = k_r + 2$ **to** $f + 3$ **do**
            when $time = t_p + jD$ send$((\perp_q, m_p, j), T_r)$ to all
            when $time = t_p + (j+1)D$
               senders := $\{z | r$ accepted $((\perp_q, m_p, j), T_z)$ from $z\}$
               faulty := *faulty* $\cup (\mathcal{P} - senders)$
               if $|faulty| < j - 2$ **then**
                  send$((default_q, m_p, -), T_r)$ to all
                  $V_r^{m_p}[q] := default_q;$ exit
         od
      coend

Figure 5: Bidding stage of Distributed Bidding algorithm tolerant of general–omissions

- If at any time τ, a correct process r determines that $impossible(t_s, \tau)$ holds for some process s, then r ignores all messages from s. In fact, we show (Lemma 4) that since $n > 2f$, s must be faulty.

- The *impossible* predicate is also used to solve another problem that does not occur in the case of send–omission failures. The wakeup of a faulty bidder q may be delayed until much after the wakeup of the first correct process. In addition to this delay, the bid of q may also be delayed through a chain of faulty processes. The failure detection mechanism that we described thus far does not prevent a faulty process from contributing to both these delays. To solve this problem, a correct process r checks whether there is a time τ such that $impossible(t_q, \tau)$. If so, r accepts q's bid only if it passes a test based on the number of processes that relayed the bid, and on the smallest time τ such that $impossible(t_q, \tau)$.

5.2.2 Proof of Correctness

Only the statements of the main lemmas are included.

Lemma 4 *If a correct process r determines that $impossible(t_z, \tau)$ holds for some process z and some time τ, then z is faulty.*

Lemma 5 *If any correct process z sets $V_z^{m_p}[q] = default_q$ because $|faulty_z| < j - 2$, then no correct process r sets $V_r^{m_p}[q] = \beta_q(m_p)$.*

Lemma 6 *If any correct process z sets $V_z^{m_p}[q] = default_q$ because it received a relay of $default_q$, then no correct process r sets $V_r^{m_p}[q] = \beta_q(m_p)$.*

Theorem 7 *The algorithm given in Figures 4 and 5 is a Distributed Bidding algorithm for general–omission failures. It tolerates up to $f \leq \lfloor \frac{n-1}{2} \rfloor$ faulty processes, and guarantees Termination within $\min[(a+5)D, (f+3)D]$ time.*

5.3 Arbitrary Failures

For systems with arbitrary failures, Distributed Bidding requires $n > 3f$ [LSP82]. Our solution is shown in Figure 6. The structure of this algorithm is different from the ones presented earlier. The algorithms for send–omission and general–omission permitted processes to begin their bidding at *different* times. In contrast, the algorithm in Figure 6 guarantees that all correct processes *that actually bid*, do so *simultaneously* at time $t_p + 3D$. This algorithm also ensures that when a correct process bids, all correct processes start to participate in the diffusion of all bids at $t_p + 3D$. This simultaneity

process p: /* p requests a bid for m_p */

$t_p := time$
send $(wakeup, m_p, t_p)$ to all

process r:

    if received $(wakeup, m_p, t_p)$ by $time = t_p + D$
        $awoke\text{-}in\text{-}time(m_p, t_p) := true$
        send (ack, m_p, t_p) to all

    if received (ack, m_p, t_p) by $time = t_p + 2D$
        send $(start, m_p, t_p)$ to all
        when $time = t_p + 3D$
            $awoke_r := awoke\text{-}in\text{-}time(m_p, t_p)$
            $bid_r := \beta_r(m_p)$
            $\mathbf{RB}(r, (m_p, t_p, bid_r, awoke_r))$ /* r broadcasts its bid for m_p */
            cobegin
                || **for all** q participate in $\mathbf{RB}(q, (m_p, t_p, bid_q, awoke_q))$
            coend
            $awaken\text{-}in\text{-}time := \{q \mid awoke_q = true\}$
            if $|awaken\text{-}in\text{-}time| \geq f+1$ **then decide** $V_r^{m_p} = <bid_1, \ldots, bid_q, \ldots, bid_n>$
            else decide NIL
            exit

    if received $(start, m_p, t_p)$ by $time = t_p + 3D$
        $awoke_r := false$
        $\mathbf{RB}(r, (m_p, t_p, \bot_r, awoke_r))$
        cobegin
            || **for all** q participate in $\mathbf{RB}(q, (m_p, t_p, bid_q, awoke_q))$
        coend

Figure 6: Distributed Bidding algorithm tolerant of arbitrary failures

permits the use of an "off-the-shelf" early-stopping Reliable Broadcast (**RB**) [DRS82, PT86,Coa88,MW88] for the diffusion of each bid.

However, if the initiator p is faulty, there could still be two correct processes r and q such that r bids for m_p and q does not. In this case, r should not decide on an outcome vector, since this vector does not include q's bid. Thus, before deciding on an outcome vector, each process r must determine that all correct processes did indeed bid. This is done as follows.

When a process r broadcasts its bid for m_p using **RB**, it also includes a flag which is *true* if r was awaken by p by time $t_p + D$. Before r decides on an outcome vector, it counts the number of bids with a *true* flag that it accepted. If r accepted at least $f+1$ such bids, it decides on the outcome vector: it knows that at least one correct process q was awake by time $t_p + D$, and q ensured that all correct processes actually bid at time $t_p + 3D$. If r accepted fewer than $f+1$ bids with a *true* flag, then it knows that the initiator p is faulty, and r decides on NIL.

The performance of our Distributed Bidding algorithm and its level of fault-tolerance depend on the particular **RB** algorithm it uses. In particular, its termination time is $3D$ greater than the termination time of the **RB** algorithm. For example, using the **RB** algorithm in [MW88], which terminates in $min[(a+2)D, (f+1)D]$, gives an overall termination time of $min[(a+5)D, (f+4)D]$. This algorithm requires $n > 6f$. The same termination time can also be achieved using the simpler **RB** algorithm in [Coa88] if $n = O(f^{1.5})$.

6 Early–Delivery Atomic Broadcast

Any process may *broadcast* any message at any time, and processes must *deliver* messages such that:

- *Validity:* If a correct process *broadcasts* m, then all correct processes eventually *deliver* m.

- *Agreement:* If a correct process *delivers* m, then all correct processes eventually *deliver* m.

- *Order:* Correct processes *deliver* messages in the same order.

Previous early–delivery Atomic Broadcast algorithms guaranteed early–delivery only in failure–free executions; however, a single failure could delay delivery by up to $O(f)$ time [GSTC90,CS87]. We use early–stopping Distributed Bidding algorithms to derive the first known early-delivery Atomic Broadcast algorithms that always deliver messages in time proportional to a rather than f. Specifically, message delivery is guaranteed within approximately $2(a + c)D$, where c is a small constant. However, our algorithms require a completely connected network. This was not necessary in [GSTC90,CS87].

Our algorithms can also take advantage of fast messages. For example, the general-omission algorithm guarantees message delivery in $2d$ time when no failures occur and all messages take d time.

Our Atomic Broadcast algorithm is based on Skeen's algorithm [Sch90]:

- *Stage 1:* The initiator p of a message m sends it to all processes.

- *Stage 2:* Each process q replies by broadcasting a *tentative* sequence number for m which must be greater than all the sequence numbers that q has seen so far. Then, q tags m with this sequence number and marks m as *pending*.

- *Stage 3:* Each process q computes the *final* sequence number for m to be the maximum of all tentative sequence numbers proposed for m. Then, q tags m with this final sequence number and marks m as *stable*.

- *Stage 4:* Each process *delivers* m when m is stable and there is no other undelivered message (either pending or stable) tagged with a lower sequence number. After delivery, m is marked as *obsolete*.

Theorem 8 *Skeen's algorithm solves Atomic Broadcast in a system with no failures.*

Unfortunately, if failures occur in Stages 1 or 2, processes may disagree on the final sequence number for m, and hence the algorithm fails. In the presence of failures, the requirements of Stages 1 and 2 constitute an instance of the Distributed Bidding problem. The tentative sequence number that q proposes for m in Stage 2 corresponds to the bid $\beta_q(m)$ that process q submits in Distributed Bidding. The specification of Distributed Bidding guarantees that if any *correct* process q bids for m in Stage 2, then either q eventually decides on an outcome vector of bids that includes the sequence numbers proposed by all the correct processes, or q eventually decides NIL.

In Stage 3, q proceeds as follows:

- If q decided on an outcome vector, then it computes a final sequence number by taking the maximum of all the bids in the vector, ignoring any default values in the vector (a default corresponds to the "bid" of a faulty process). Then, q tags m with this sequence number and marks m as *stable*. From the *Agreement* condition of Distributed Bidding, all correct processes decide on the same final sequence number.

- If q decided NIL, then it marks m as *obsolete* and never delivers m. From *Agreement*, no correct process can ever decide on an outcome vector for m, and thus no correct process ever delivers m.

Thus, one can implement Stages 1 and 2 with our *early–stopping* Distributed Bidding algorithms. This results in a family of *early–delivery* Atomic Broadcast algorithms, one for each model of failure considered in this paper. Although Distributed Bidding is invoked only once in our Atomic Broadcast algorithms, their worst–case message delivery time is exactly *twice* the time required for Distributed Bidding.

The worst case delivery time of our Atomic Broadcast is due to the following. A message is delivered only if it becomes stable *and* all messages with smaller sequence numbers are delivered. A message becomes stable at the end of the Distributed Bidding for that message. However, once a message is stable, its delivery may be delayed further: there may be another message with a lower sequence number that is still pending. This gives a total time of approximately twice the time of a Distributed Bidding algorithm. Simulation results (see Appendix A) suggest that this extreme case rarely occurs, and message delivery usually completes in about the time required for a single Distributed Bidding.

References

[CDDS85] Brian A. Coan, Danny Dolev, Cynthia Dwork, and Larry Stockmeyer. The distributed firing squad problem. In *Proceedings of the Seventeenth ACM Symposium on Theory of Computing*, pages 335–345, Providence, Rhode Island, May 1985. ACM SIGACT.

[Coa88] Brian A. Coan. Efficient agreement using fault diagnosis. In *Proceedings of the Twenty-Sixth Annual Allerton Conference on Communication, Control, and Computing*, September 1988.

[Cri89] Flaviu Cristian. Probabilistic clock synchronization. *Distributed Computing*, 3:146–158, 1989.

[CS87] Flaviu Cristian and Ray Strong. A family of early-delivery atomic broadcast protocols, June 1987. IBM Almaden Research Center, Unpublished manuscript.

[DRS82] Danny Dolev, R. Reischuk, and Ray Strong. Eventual is earlier than immediate. In *Proceedings of Twenty third Symposium on Foundations of Computer Science*, pages 196–203, November 1982.

[DS83] Danny Dolev and H. Ray Strong. Authenticated algorithms for Byzantine agreement. *SIAM Journal on Computing*, 12(4):656–666, November 1983.

[GSTC90] Ajei Gopal, Ray Strong, Sam Toueg, and Flaviu Cristian. Early-delivery atomic broadcast. In *Proceedings of Ninth ACM Symposium on Principles of Distributed Computing*, Quebec City, Canada, August 1990.

[HK89] Amir Herzberg and Shay Kutten. Efficient detection of message forwarding faults. In *Proceedings of the Eighth ACM Symposium on Principles of Distributed Computing*, Edmonton, Alberta, August 1989. ACM SIGOPS-SIGACT.

[LSP82] Leslie Lamport, R. Shostak, and M. Pease. The Byzantine generals problem. *ACM Transactions on Programming Languages and Systems*, 4(3):382–401, July 1982.

[MW88] Yoram Moses and Orly Waarts. Coordinated traversal: $(t+1)$-round Byzantine agreement in polynomial time. In *Proceedings of the Twenty-Ninth IEEE Symposium on Foundations of Computer Science*, pages 246–255, White Plains, New York, October 1988. IEEE.

[NT88] Gil Neiger and Sam Toueg. Automatically increasing the fault-tolerance of distributed systems. In *Proceedings of the Seventh ACM Symposium on Principles of Distributed Computing*, pages 248–262, Toronto, Ontario, August 1988. ACM SIGOPS-SIGACT.

[PT84] Kenneth J. Perry and Sam Toueg. An authenticated byzantine generals algorithm with early stopping. Technical Report 84-620, Cornell University, Ithaca, June 1984.

[PT86] Kenneth J. Perry and Sam Toueg. Distributed agreement in the presence of processor and communication faults. *IEEE Transactions on Software Engineering*, 12(3):477–482, March 1986.

[Sch90] Fred Schneider. Implementing fault-tolerant services using the state machine approach: A tutorial. *ACM Computing Surveys*, 1990. To Appear.

[ST87] T. K. Srikanth and Sam Toueg. Simulating authenticated broadcasts to derive simple fault-tolerant algorithms. *Distributed Computing*, 2(2):80–94, 1987.

[TPS87] Sam Toueg, Kenneth J. Perry, and T. K. Srikanth. Fast distributed agreement. *SIAM Journal on Computing*, 16(3):445–457, June 1987.

Appendix

A Simulation of Atomic Broadcast

We implemented our Atomic Broadcast algorithm for crash failures on a *BBN Butterfly GP1000*, with $n = 20$ processes.

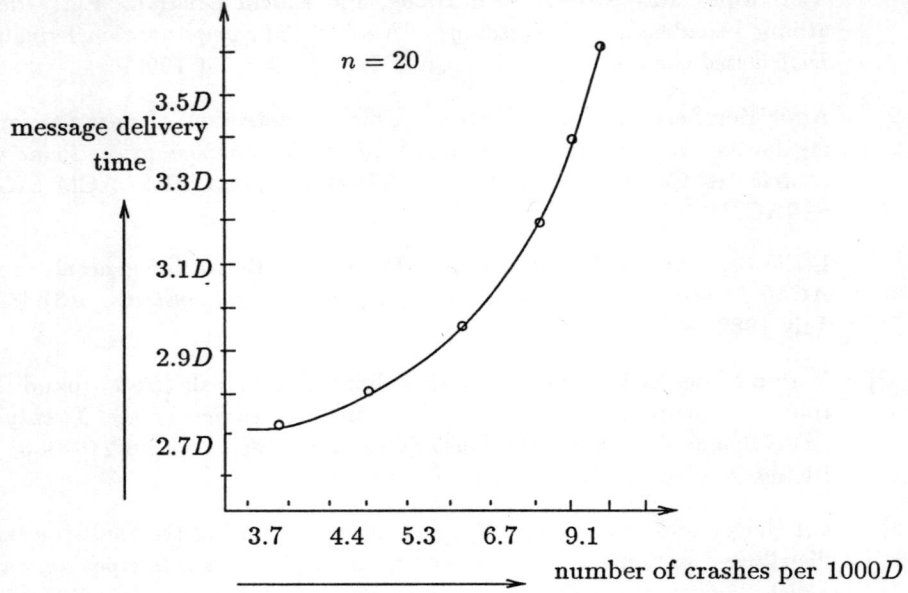

Figure 7: Average delivery time *versus* frequency of crashes per process

We measured the performance of the implementation by simulating uniformly distributed independent crash failures at each process. The average failure rates for each process ranged from 1 crash to 10 crashes every 1000 time units (where a time unit is D, the maximum link delay between two processes).

Ten processes independently generated broadcasts according to a Poisson distribution with an average rate of one every 3 time units each. Each simulation included a total of 2000 broadcasts.

The results of the simulation are shown in Figure 7. The x-axis represents the frequency of crashes per process, and the y-axis represents the average delivery time of broadcasts. It is interesting to note that even when *every* process crashes on average once every 200 time units, an unrealistically high rate of failures, Atomic Broadcasts complete in less than 3 time units on the average. Furthermore, doubling the failure rate to once every 100 time units at each process, only increases the expected message delivery time by less than one time unit.

FAST CONSENSUS IN NETWORKS OF BOUNDED DEGREE
(Extended Abstract)

Piotr Berman[†]
Department of Computer Science
The Pennsylvania State University
University Park, PA 16802, USA

Juan A. Garay[†][‡]
IBM T.J. Watson Research Center
P.O. Box 704
Yorktown Heights, NY 10598, USA

ABSTRACT

In a *Distributed Consensus* protocol all processors (of which t may be faulty) are given (binary) initial values; after exchanging messages all correct processors must agree on one of them. In this paper we focus on consensus in networks that are not completely interconnected, following the work of Dwork *et al.* [DPPU]. In such a context, complete consensus among all the correct processors is not possible and some exceptions must be allowed.

We first show how to achieve consensus in the butterfly network using $O(t + \log n \log\log n)$ one bit parallel transmission steps, while tolerating the asymptotically optimal number of faulty processors and asymptotically minimal number of exceptions. This result considerably improves upon our previous protocol, in particular it replaces the running time of $O(n \log n \log\log n)$ with an asymptotically optimal one. As in [DPPU], we can decrease the number of exceptions to $O(t)$ by using additional links, while maintaining the same running time.

The protocol is derived from a consensus protocol for complete networks that is interesting in its own right. It achieves Distributed Consensus with optimal number of processors, asymptotically optimal total bit transfer and nearly optimal number of rounds, with better constant factors than previously published results.

[†] Work partially supported by AFOSR contract 87-0400 and NSF grant CR 8805978.
[‡] This work was done while the author was at the Computer Science Departments of Bucknell University and Penn State.

1. Introduction

The *Distributed Consensus* problem (a.k.a. *Byzantine Agreement*) is one of the fundamental problems in the theory of distributed systems. There are many situations in the management of such systems with one common characteristic: a collection of processors must coordinate a decision. The problem provides perhaps the most abstract setting for the discussion of such situations, and allows for the development of elegant methods, which in turn may influence practical designs and implementations. The problem can be formally stated as follows.

Let $P = \{1, \ldots, n\}$ be a set of processors; $T \subset P$ is the set of *faulty* processors, $|T| = t$. Processors from $P - T$ are called *correct*. We assume that the correct processors initially do not know the set T. Every processor is given an initial value, 0 or 1. After the execution of a consensus protocol, the final values of the correct processors have to satisfy the following two conditions, regardless of the behavior of the faulty processors:

- *Agreement*: they are all equal.
- *Validity*: they are all equal to the initial value, if the latter is unique.

We use the simple and standard deterministic synchronous model. The computation performed by the network evolves as a series of *rounds*, during which the processors send messages, receive them and perform local computations according to the protocol (the computation time in our protocols is always polynomial).

In the context of completely interconnected networks, there are three basic aspects contributing to the quality of a protocol for Distributed Consensus:

- the ratio between n and t ($n \geq 3t+1$, [LSP]);
- the number of rounds r ($r \geq t+1$, [DS]);
- the communication complexity, alternatively given by the message length m ($m \geq 1$, obvious), or the total bit transfer B ($B > nt$, [DR]).

In the last few years there has been rapid progress towards higher quality of consensus protocols (e.g., [BD, BDDS, BG, BGP1, CW1, FM, MW]). Yet so far rather few papers dealt with the problem of achieving consensus in incomplete networks, even though such a situation is quite frequent among the existing distributed systems. One can mention the impossibility result from [D] and a quite extensive discussion in [CASD]. One should note that in the latter paper authors put various limitations on the behavior of faulty processors, e.g. the inability of forging signatures of correct processors; therefore their results are not directly comparable with ours.

In this paper we follow the work of Dwork, Peleg, Pippenger and Upfal [DPPU]. In particular we discuss the cases of the butterfly network [U] and a butterfly superimposed with a concentrator [P]. Since agreement involving all the correct processors is not possible in this setting, we must settle for an agreement with exceptions, where not all the correct processors reach agreement. Dwork *et al.* show how to simulate the transmission of a message between two processors in such a way that if none of them belongs to a set

T^+ of size $\Theta(t \log t)$ then the simulation is faithful. This makes it possible to simulate any Byzantine agreement protocol which does not rely on the privacy of the links by treating processors from T^+ as faulty.

Therefore, the number of exceptions is another aspect relevant for consensus protocols. We shall say that a protocol is X-*incomplete* if for any set of faulty processors T there exists a set $T^+ \supset T$ with at most $X+t$ elements such that all processors in $P - T^+$ reach consensus†. In the case of a complete network, $X = 0$ and $T^+ = T$; in the case of the butterfly we must have $X = \Omega(t \log t)$ [DPPU]. For some constructed bounded-degree networks, $X = O(t)$ is possible, but even there protocols with better resiliency are not known; their existence constitutes an interesting open problem.

In the context of butterfly networks, Dwork *et al.* explain how to achieve asymptotically optimal resiliency and size of T^+. However, they leave open the question of optimal running time for the simulation, which is the subject of this paper. We choose to fix the message length to 1, this way the number of parallel steps (or rounds) limits the total communication of every single processor. In particular, we prove the following two theorems, which correspond to Theorems 3 and 4 of [DPPU] (with the estimate of the running time added):

Theorem 2: The n-node butterfly network admits a t-resilient $8t \log t$-incomplete consensus algorithm that runs in time $O(t + \log n \log \log n)$, provided $36t \log t < n$.

Theorem 3: There exists a constant c and a network of degree 11 that admits a t-resilient $O(t)$-incomplete consensus algorithm that runs in time $O(t + \log n \log \log n)$, provided $ct \log t < n$.

Careful analysis shows that our protocols have lower resiliency than that of Dwork *et al.*, for which the requirement $20t \log t < n$ is sufficient. Their paper describes a transmission scheme that associates with each processor a *fan-out* set (of processors), a *fan-in* set, and for every two processors i and j, a set of disjoint paths from all processors from i's fan-out set to all processors of j's fan-in set. The transmission from i to j is performed then in three phases: i sends the message to all members of its fan-out set, then the copies of this message travel on disjoint paths to the members of the fan-in set of j, and then they are transmitted to j. Lastly, j picks the most frequently received message as the presumably correct one.

While the work of Dwork *et al.* leads to protocols with the best known resiliency and the number of exceptions, it does not show how to obtain a good running time. Their simulation of a single bit transmission between two processors thus requires $\Omega(t \log t)$ single transmissions. If we choose to simulate this way an algorithm using the minimum number of bit transmissions ($\Theta(nt)$ [BGP2, CW2]), we need $\Omega(nt^2 \log t)$ single transmissions in the butterfly, and hence $\Omega(t^2 \log t)$ parallel steps.

---

†In [DPPU] this is called X-agreement.

Thus it is relevant to investigate the ways to improve the running time, even if the resiliency is somewhat compromised. The first improvement, described in [BG], is to simulate a whole exchange round between correct processors from a complete network in the same running time as a single bit transmission. The improvement described now comes from hierarchical partitioning of the processors, which allows to perform most of the exchanges within small sets of adjacent processors.

The rest of the paper is organized as follows. In Section 2 we describe a version of a consensus protocol for complete networks that is interesting in its own right. The original version [BGP2] achieves consensus with optimal resiliency, asymptotically optimal total bit transfer and nearly optimal number of rounds (the same result was concurrently and independently obtained by Coan and Welch). The version we present here adapts particularly well to the structure of the butterfly network. In Section 3 we present the consensus protocol for the butterfly, and subsequently we discuss how to decrease the number of exceptions to $O(t)$ using additional links (following [DPPU]).

2. Consensus in Complete Networks with Optimal Bit Transfer

```
V := v_p;  (* p's initial value *)
for m := 1 to t+1 begin
                    (* Exchange 1 *)
    send(V);
    for j := 0 to 1 do
        C[j] := the number of received j's;
                    (* Exchange 2 *)
    for j := 0 to 1 do begin
        send(C[j] > n-t);
        D[j] := the number of received 1's
    end;
    V := D[1] > t;
                    (* Exchange 3 *)
    if m = p then send(V);
    if D[V] < n-t then
        V := the received message
end;
```

Fig. 1. The *Phase King* protocol: code for processor p

We first reproduce the *Phase King* protocol of [BGP] (called that way because at which phase one of the processors take charge of the computation); the optimal bit transfer is obtained by the application of the committee technique [BG] to it, and having the committees run the same protocol recursively. We assume $n > 3t$. The protocol is shown in Fig. 1; this is an asymptotically optimal consensus protocol in all cost measures: it uses messages of constant size (1 or 2 bits), and runs in $t+1$ phases, each consisting of three exchange rounds. For convenience, we present the second exchange as two one-bit exchanges. Each processor has a local variable V, and integer arrays C and D. Values 0 and 1 are identified with **false** and **true**, respectively. See [BGP1] for the proof of correctness; it relies on the fact that at least one of the "phase kings" behaves correctly.

In [BG] it is shown how to reduce the number of rounds of this type of protocol (to $t(1+1/d)$) by using "phase committees" instead of "phase kings." The committees run internally a round-optimal consensus protocol, sending the respective messages to all the processors. The pigeonhole principle assures that at least one of the committees contains a favorable proportion of correct processors. The first such committee plays the same role in assuring the agreement as the good king in *Phase King*, because it achieves consensus on a value equal to a possible message of the honest king.

The idea to achieve total optimal bit transfer is to have just two phases of *Phase King* (and two committees), while the committees internally run the same protocol recursively (i.e. they split into two subcommittees each, etc.). This way only 6 exchange rounds involve all processors, 12 rounds involve about $n/2$ processors, 24 rounds involve $n/4$ processors, etc. This results in a bit transfer of $cn^2(1+1/2+1/4+\cdots)$.

We define a hierarchy of committees of processors as follows. A committee in the hierarchy is named C_w, where $w \in \{0,1\}^*$; $C_\lambda = P$, while C_{w0} and C_{w1} form a balanced partition of C_w. A sketch of *Recursive Phase King*, the optimal bit transfer protocol, is given in Figure 2. At the top level, the processors execute $RPK(\lambda, V)$, where V contains the processor's initial binary value.

Theorem 1: Distributed consensus is achievable in $O(t)$ communication rounds, using optimal number of processors and total bit transfer $O(nt)$.

Proof: First we show that RPK achieves a valid consensus when applied to C_w such that $|C_w \cap T| < |C_w|/3$. In the case when $|C_w| \leq 3$, the claim follows trivially because C_w contains correct processors only.

In the other case, first note that either $|C_{w0} \cap T| < |C_{w0}|/3$, or $|C_{w1} \cap T| < |C_{w1}|/3$. The subcommittee with the favorable ratio plays the role of the honest king in *Phase King*, and its correctness applies.

By inspecting Fig. 2, the number of rounds can be expressed by the following recurrence (where $s = |C_w|$):

$$r(s) = \begin{cases} 1 & \text{if } s \leq 3 \\ r(\lfloor s/2 \rfloor) + r(\lceil s/2 \rceil) + 6 & \text{otherwise} \end{cases}$$

which yields $r(s) \leq 3.5s$.

Similarly, the number of bits communicated can be expressed as:

$$B(s) \leq \begin{cases} 2s & \text{if } s \leq 3 \\ B(\lfloor s/2 \rfloor) + B(\lceil s/2 \rceil) + 6.5s^2 & \text{otherwise} \end{cases}$$

```
procedure RPK(w: {0,1}*; var V: {0,1});
var C, D: array[0..1] of integer;  U: {0,1};  n_w, t_w : integer;
begin
    n_w := |C_w|;  t_w := (n_w −1) div 3;
    if |C_w| ≤ 3 then begin
        perform Phase King's Exchange 1 within C_w;
        V := C[1] > 0
    end else begin
                        (* Phase 1 *)
        perform Phase King's Exchanges 1 and 2 within C_w;
        if p ∈ C_w0 then begin
            U := V;
            RPK(w0,U);
            send U to C_w1
        end;
        if p ∈ C_w1 then
            U := the most frequently received value;
        if D[V] < n_w − t_w then
            V := U;

                        (* Phase 2 *)
        repeat Phase 1 exchanging the roles of w0 and w1
    end
end;
```

Fig. 2. *Recursive Phase King*: code for processor $p \in C_w$

For $n = 2^k$ this recurrence yields $6.5(2^{2k}+2^{2k-1}+\cdots) \leq 13n^2$. □

Remark: By stopping the recursion at committee size $c \log n$ instead of 3 (where c is a suitable constant), and making the committees run at this point an efficient round-optimal consensus protocol, Distributed Consensus can be achieved in $t+o(t)$ rounds, and same remaining parameters as in Theorem 1 (see [BGP2]).

3. Fast Consensus in Networks of Bounded Degree

The first three parts of this section deal with butterfly networks. We assume $t = O(n/\log n)$ (otherwise agreement of the majority would not be possible [D]). We use the ideas of the previous section together with communication primitives which efficiently simulate the exchange rounds on the butterfly, to obtain $O(t \log t)$-incomplete consensus in time $O(t + \log n \log \log n)$. At the end we discuss how to decrease the number of exceptions to $O(t)$ using additional links.

The general outline of our approach is as follows. We create a committee of Kt voting members (K to be established later) and $Kt \log(Kt)$ aides), to simulate *RPK* with the aides' help. Then the committee broadcasts its consensus value to all the processors. The committee's internal structure follows the structure of the butterfly.

3.1. Preliminary Definitions

A butterfly network consists of n of processors, $n = m 2^m$, each of them identified by a pair (a,u), where $a \in \{0, \ldots, m-1\}$ indicates the *rank* and $u \in \{0,1\}^m$ indicates the *column* of the processor. For convenience, we use R_i to denote the set of processors with rank i. Let e_l denote the lth unit vector. Each processor is able to exchange messages with its two successors, defined by these functions:

$$succ_0(a,u) = ((a+1) \bmod m, u),$$
$$succ_1(a,u) = ((a+1) \bmod m, u \oplus e_a).$$

We use $pred_0$ and $pred_1$ to denote the inverse functions of $succ_0$ and $succ_1$.

Within the butterfly network, we define the *committees* recursively as follows (see Fig. 3):

$$C_w = \begin{cases} \{(0,w)\} & \text{if } |w| = m \\ C_{w0} \cup C_{w1} \cup succ_0(C_{w0} \cup C_{w1}) & \text{otherwise} \end{cases}$$

Within a committee, we distinguish between the *voting members* (processors from R_0) and the committee *aids* (the rest of the processors).

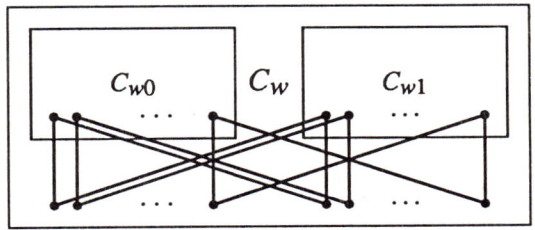

Fig. 3. The committees in the butterfly network

We choose two positive constants β and ε in such a way that

$$\beta(\varepsilon^{-1}+1) + 2\varepsilon \leq \tfrac{1}{2}$$

($\beta = 1/36$ and $\varepsilon = 1/8$ will maximize the resiliency of the protocol). Let $K = MIN\{k \in \mathbf{Z} : \beta 2^k > t\}$; we assume that $K \leq m$, in other words we assume $t < \beta 2^m$.

To gauge the faulty processors' impact on the information received by a correct processor p, we define *damage*(p) as follows:

$$damage_0(p) = 1 \text{ if } p \in T, 0 \text{ otherwise};$$
$$damage_{i+1}(p) = \tfrac{1}{2} \sum_{j=0}^{1} damage_i(succ_j(p));$$
$$damage(p) = damage_K(p).$$

We say processor p is *good* if $damage(p) < \varepsilon$ (*bad* otherwise). A committee C_w is *good* if $\sum_{p \in C_w \cap R_0} damage(p) < \beta |C_w \cap R_0|$.

Observations:

1. $\sum_{p \in R_i} damage(p) \leq t$; $\sum_{p \in P} damage(p) \leq (K+1)t$.
2. Committee $C_{0^{m-\kappa}}$ is good.
3. If committee C_w is good and $|w| < m$, then at least one of C_{w0}, C_{w1} is good.
4. If committee C_w is good, then less than $\varepsilon^{-1}\beta |C_w|$ of its voting members are bad.
5. If $|C_w \cap R_0| \leq \beta^{-1}$ and C_w is good, then $C_w \cap T = \emptyset$. □

3.2. Committee Broadcasts

Three kinds of committee broadcasts are used:

$B(C_w, C_w, ...)$: from a committee to itself (exchanges 1 and 2 from *Phase King*);

$B(C_{wb}, C_{w\bar{b}}, ...)$: from a committee to its "sister" committee (exchange 3, the king's);

$B(C_w, P, ...)$: from a committee to all processors (the final exchange).

The code for the three broadcast primitives are shown in Fig. 4. We let $|C_w| = k2^k$; $k = m - |w|$ is the number of ranks of aides in C_w. V_p contains processor p's initial value, S_p will contain the resulting sum.

Lemma 1: Assume that p, q are good voting members of C_w. Then after executing $B(C_w, C_w, V, S)$, $|S_p - S_q| < 2\varepsilon$.

Proof sketch: Let $s2^{2k}$ be the sum of numbers computed by aides from R_k at the end of the first **for** loop. In the absence of faulty processors in C_w every voting member p would compute $S_p = s$, due to the "noise" created by the faulty processors we can only assure $|S_p - s| < damage(p)$. Since p and q are good, $|S_p - S_q| \leq damage(p) + damage(q) < 2\varepsilon$. □

Lemma 2: Assume that C_w is a good committee and all of its voting members start $B(C_w, C_w, V, S)$ with $V = b$. Then after the execution, for any good voting member $p \in C_w$, $S_p < (\tfrac{1}{2} - \varepsilon)$ if $b = 0$; $S_p > (\tfrac{1}{2} + \varepsilon)$ if $b = 1$.

Proof sketch: Assume $b = 0$ (the other case is symmetric). Let $s'2^k$ be the sum of Vs of voting members and let s be defined as in the proof of Lemma 1. By Observation 4, $s' < \varepsilon^{-1}\beta$. In the absence of faulty processors in C_w we would have $s' = s$, due to the "noise" we can only assure $s < s' + \sum_{p \in C_w \cap R_0} damage(p) < s' + \beta$. Using the observation

Code for $B(C_w, C_w, V, S)$:

    voting members send V to successors;
    for $i := 1$ **to** $k-1$ **do**
        aides from R_i receive, add and send outcomes to successors;
    for $i := k$ **downto** 1 **do**
        aides from R_i receive, add and send outcomes to predecessors;
    voting members receive, add, divide by 2^{2k}, and store the outcome in S;

Code for $B(C_{wb}, C_{w\bar{b}}, V, S)$:

    voting members from C_{wb} send V to successors;
    for $i := 1$ **to** $k-1$ **do**
        aides from $C_{wb} \cap R_i$ receive, add and send outcome to successors;
    aides from R_k receive, add and send outcome to predecessors;
    for $i := k-1$ **downto** 1 **do**
        aides from $C_{w\bar{b}} \cap R_i$ receive, add and send outcome to predecessors;
    voting members from $C_{w\bar{b}}$ receive, add, divide by 2^{2k-1}, and store the outcome in S;

Code for $B(C_w, P, V, S)$:

    voting members from C_w send V to their successors;
    for $i := 1$ **to** k **do**
        aides from $C_w \cap R_i$ receive, add and send outcome to successors;
    for $i := k+1$ **to** $m-1$ **do**
        if $p \in R_i$ receives a value **then** p sends it to its successors;
    for $i := m$ **downto** 1 **do**   (* rotation *)
        processors from R_i receive from $succ_0$ and send to $pred_0$;
    all processors send the last received value to predecessors;
    for $i := 2$ **to** m **do**
        all processors receive, add and send outcome to predecessors;
    all processors receive, add, divide by 2^{k+m}, and store the outcome in S;

Fig. 4. The three committee broadcasts

from the previous proof, we get $S_p < \varepsilon^{-1}\beta + \beta + \varepsilon$, while we assumed that $\varepsilon^{-1}(\beta+1) + 2\varepsilon \leq \frac{1}{2}$. □

Lemma 3: Assume that both C_w and C_{wb} are good, and that all good voting members of C_{wb} start $B(C_{wb}, C_{w\bar{b}}, V, S)$ with $V = b$. Then for any good voting member $p \in C_{w\bar{b}}$, $S_p < (\frac{1}{2} - \varepsilon)$ if $b = 0$; $S_p > (\frac{1}{2} + \varepsilon)$ if $b = 1$.

The proof uses the same ideas as in Lemma 2.

Lemma 4: Assume that both C_w is good, and that all good voting members of C_w start $B(C_w, P, V, S)$ with $V = b$. Then for every good processor p, $S_p < (\frac{1}{2} - \varepsilon + 2\beta)$ if $b = 0$; $S_p > (\frac{1}{2} + \varepsilon - 2\beta)$ if $b = 1$.

Proof sketch: As in the previous lemmas, we need to analyze the "noise" which may affect S_p. The broadcast of C_w includes $\varepsilon^{-1}\beta$ noise of bad voting members, at the end of the second **for**-loop faulty processors add β noise, another β is added by them during the "rotation," in the final loop good processors receive ε noise from faulty processors defining their damage and β noise from the "far-away" faulty processors. $\square$

Lemma 5: $B(C_w, C_w, ...)$ and $B(C_{wb}, C_{w\bar{b}}, ...)$ use $O(k^2)$ steps, while $B(C_w, P, ...)$ uses $O(m^2)$ steps.

Proof sketch: The broadcast in Fig. 4 involves $O(k)$ rounds ($O(m)$ in the latter case), where the numbers transmitted in each round have $O(k)$ bits ($O(m)$ in the latter case).

Remark: Later we need a somewhat weaker version of Lemma 4, namely that when every good voting member p of the good committee C_w start $B(C_w, P, V, S)$ with $V_p = b$, then after the execution of the broadcast $b = S_q < \frac{1}{2}$ holds true for all good processors. Therefore in the execution of this broadcast the processors may truncate their messages to $O(\log m)$ bits, which improves the number of one-bit rounds used to $O(m \log m)$.

3.3. The Protocol

The t-resilient $O(t \log t)$-incomplete consensus protocol for the butterfly network is shown in Fig. 5. It makes use of procedure *BRPK*, a butterfly version of *RPK* that uses the communication primitives described in the previous subsection.

Lemma 6: Let C_w be a good committee and G its set of good voting members. Consider an execution of $BRPK(w, V)$. Then there exists b such that

 a) b is the initial of some $p \in G$, and
 b) b is the final value of every $q \in G$.

Proof sketch: we use induction on the committee size. When $|C_{w \cap R_0}| \leq \beta^{-1}$ then the claim follows trivially from Observation 5, as C_w contains correct processors only.

Now let us assume that all members of G started one of the two phases with $V = b$. Then after the first broadcast of the phase all members of G have $C[b] > \frac{1}{2} + \varepsilon$ and $C[\bar{b}] < \frac{1}{2} - \varepsilon$ (Lemma 2). This implies that in the subsequent pair of broadcasts members of G send unanimous values, and all of them compute $D[b] > \frac{1}{2} + \varepsilon$ and $D[\bar{b}] < \frac{1}{2} - \varepsilon$ (again Lemma 2). Thus regardless of the value of U computed in the later part of this phase all members of G will have $V = b$ at the end this phase (and at the end of Phase 2, since we can repeat the reasoning if necessary).

Note that one of the subcommittees C_{w0}, C_{w1} is good (Observation 3). Now it remains to show that all members of G have the same value of V at the end of a phase with good subcommittee. By induction, all good voting members of this subcommittee compute the same value of U, say b; Lemma 3 implies that the remaining members of G also compute $U = b$. Thus if some $p \in G$ computes $V_p = \bar{b}$ then we know that during the execution of the last **if** of the phase we have $V_p = \bar{b}$ and $D[\bar{b}]_p \geq \frac{1}{2} + \varepsilon$.

```
procedure BRPK (w: {0,1}*; var V: {0,1});
var     C, D: array[0..1] of real; U: {0,1}; W: real;
begin
    if |C_w ∩ R_01| ≤ β^{-1}K then begin
        B(C_w, C_w, V, W);
        if W < ½ then U := 0 else U := 1
    end else begin
                              (* Phase 1 *)
        B(C_w, C_w, V, C[1]);
        C[0] := 1 - C[1];
        for j := 0 to 1 do
            B(C_w, C_w, C[j] > ½+ε, D[j]);
        V := D[1] > ½-ε;
        if p ∈ C_{w0} then begin
            U := V;
            BRPK(w0, U);
            B(C_{w0}, C_{w1}, U, W)
        end;
        if p ∈ C_{w1} then
            U := W ≥ ½;
        if D[V] < ½+ε then
            V := U;

                              (* Phase 2 *)
        repeat Phase 1 exchanging the roles of w0 and w1
    end
end;
```

The butterfly consensus protocol:

```
BRPK(0^{m-K}, V);
B(C_{0^{m-κ}}, P, V, S);
V := S ≥ ½;
```

Fig. 5. BRPK and the butterfly consensus protocol

Because $D[0]$ and $D[1]$ are computed in broadcasts of C_w to itself, we know from Lemma 1 that $D[\bar{b}]_q > ½-ε$ for every q G. Lemma 2 implies that in that broadcast for at least one $q \in G$ we have $C[\bar{b}]_q > ½+ε$. Then by Lemma 1 no member of G has $C[b] > ½+ε$; as a result all members of G compute $D[b] < ½-ε$. In the case of $b=1$ this yields a contradiction, because no member of G would compute $V=1$ from $D[1] > ½-ε$. On the other hand, when $b=0$ then all members of G would compute $V=1$ from this expression, and none of them would have $U=0$, again a contradiction. □

Proof sketch of Theorem 2: The correctness of the butterfly consensus protocol follows because $C_{0^{m-κ}}$ is a good committee (Observation 2), it achieves a valid consensus (Lemma

6) and all good processors receive this consensus value correctly (Lemma 4).

To compute the number of one-bit exchange required, we define $\tau(k)$ as the number of such rounds required by $BRPK(0^{m-k}, V)$. For $k < -\log\beta$ this function is bounded by a constant, while for larger values of k it satisfies the following recurrence (due to Lemma 5):

$$\tau(k) \leq 2\tau(k-1) + ck^2$$

which implies that $\tau(k) = O(2^k)$. Consequently the time needed to execute $BRPK(0^{m-K}, V)$ is $O(2^K) = O(t)$. To finish the proof we need to apply the Remark following Lemma 5. □

3.4. Achieving Linear Number of Exceptions

Dwork *et al.* [DPPU] describe so-called *compressor* graphs which have bounded degree and which allow to execute in $O(\log n)$ a *compression* procedure. Assuming that the number of correct processors starting the compression procedure with value $V = b$ exceeds some cn, for some fixed $c > \frac{1}{2}$, only $t+1$ correct processors may fail to have $V = b$ at the end of the execution. This allows us to prove Theorem 3. (To achieve the total degree of 11, we need to decrease the resiliency by a constant factor.)

References

[B] B. Bollobás, "Random Graphs," *Combinatorics*, London Math. Society LN 52, Cambridge University Press, 1981, pp. 80-102.

[BD] A. Bar-Noy and D. Dolev, "Families of Consensus Algorithms," *Proc. 3rd Aegean Workshop on Computing*, June/July 1988, pp. 380-390.

[BDDS] A. Bar-Noy, D. Dolev, C. Dwork and H.R. Strong, "Shifting gears: changing algorithms on the fly to expedite Byzantine Agreement," *Proc. 6th PODC*, August 1987, pp. 42-51.

[BG] P. Berman and J.A. Garay, "Asymptotically Optimal Distributed Consensus," *Proc. ICALP 89*, LNCS, Vol. 372, July 1989, pp. 80-94.

[BGP1] P. Berman, J.A. Garay and K.J. Perry, "Towards Optimal Distributed Consensus," *Proc. 30th FOCS*, October 1989, pp. 410-415.

[BGP2] P. Berman, J.A. Garay and K.J. Perry, "Recursive Phase King Protocols for Distributed Consensus," PSU, CS Dept. Tech Report *CS-89-24*, August 1989.

[CASD] F. Cristian, H. Aghili, R. Strong and D. Dolev, "Atomic Broadcast: From Simple Message Diffusion to Byzantine Agreement," *Proc. 15th Inernational Symp. on Fault-Tolerant Computing*, June 1985, pp. 200-206. Revised version in IBM research report RJ5244.

[CW1] B. Coan and J. Welch, "Modular Construction of Nearly Optimal Byzantine Agreement Protocols," *Proc. 9th Annual PODC*, August 1989, pp. 295-306.

[CW2] B. Coan and J. Welch, "A Byzantine Agreement Protocol with Optimal Message Bit Complexity," *Proc. 27th Annual Allerton Conf. on Communication, Control and Computing*, 1989.

[D] D. Dolev, "The Byzantine generals strike again," *Journal of Algorithms*, Vol. 3, No. 1 (1982), pp. 14-30.

[DPPU] C. Dwork, D. Peleg, N. Pippenger and E. Upfal, "Fault Tolerance in Networks of Bounded Degree," *Proc. 18th STOC*, May 1986, pp. 370-379.

[DR] D. Dolev and R. Reischuk, "Bounds of Information Exchange for Byzantine Agreement," *JACM,* Vol. 32, No. 1, (1985), pp. 191-204.

[DS] D. Dolev and H.R. Strong, "Polynomial Algorithms for Multiple Processor Agreement," *Proc. 14th STOC*, May 1982, pp. 401-407.

[FM] P. Feldman and S. Micali, "Optimal Algorithms for Byzantine Agreement," *Proc. 20th STOC*, May 1988, pp. 148-161.

[LSP] L. Lamport, R.E. Shostak and M. Pease, "The Byzantine Generals Problem," *ACM ToPLaS,* Vol. 4, No. 3, July 1982, pp. 382-401.

[MW] Y. Moses and O. Waarts, "Coordinated Traversal: $(t+1)$-Round Byzantine Agreement in Polynomial Time," *Proc. 29th FOCS,* October 1988, pp. 246-255.

[P] N. Pippinger, "On Networks of Noisy Gates," *Proc. 26th FOCS,* October 1985, pp. 31-38.

[U] J.D. Ullman, "Computational Aspects of VLSI," Computer Science Press, 1984.

Common Knowledge and Consistent Simultaneous Coordination

Gil Neiger*
College of Computing
Georgia Institute of Technology
Atlanta, Georgia 30332-0280
gil@cc.gatech.edu

Mark R. Tuttle
DEC Cambridge Research Lab
One Kendall Square, Building 700
Cambridge, Massachusetts 02139
tuttle@crl.dec.com

Abstract

Traditional problems in distributed systems include the *Reliable Broadcast*, *Distributed Consensus*, and *Distributed Firing Squad* problems. These problems require coordination only among the processors that do not fail. In systems with benign processor failures, however, it is reasonable to require that a faulty processor's actions are consistent with those of nonfaulty processors, assuming that it performs any action at all. We consider problems requiring consistent, simultaneous coordination and analyze these problems in terms of common knowledge. (Others have performed similar analyses of traditional coordination problems [1,9].) In several failure models, we use our analysis to give round-optimal solutions. In one benign failure model, however, we show that such problems cannot be solved, even in failure-free executions.

1 Introduction

Concurrent message-passing systems provide many advantages over other kinds of systems, including efficiency and fault-tolerance. Most interesting computation in such systems requires processors to coordinate their actions in some way, but this coordination becomes especially difficult when processors can fail. Traditionally, researchers have considered problems that require only the correct processors to coordinate their actions, placing no requirements on the behavior of faulty processors. For example, in one formulation of the *Byzantine Agreement* problem (cf. [12]), each processor begins with an input bit and chooses an output bit, subject to the condition that all nonfaulty processors choose the same value as their output bit, and that this value is the value of some processor's input bit. This problem was originally proposed in the *Byzantine* failure model where faulty processors can behave in completely arbitrary ways. Consequently, only the nonfaulty processors are required to agree on their output bits, since it is impossible to force the faulty processors to do anything. Subsequently, however, a substantial body of the literature has considered this problem in systems where processors fail in relatively benign ways, such as intermittently failing to send or receive certain messages.

In the context of such benign failures, however, it seems reasonable to require that, if a faulty processors performs any action at all, then this action is *consistent* with the actions performed by the nonfaulty processors. For example, Neiger and Toueg [11] define *Uniform Agreement*, a variant of *Reliable Broadcast* in which faulty processors are forbidden to choose inconsistent values; and Gopal and Toueg [2] consider a family of related problems. It is often not only reasonable, but also critical, to consistently coordinate the behavior of *all* processors. In distributed database systems, for example, it is undesirable for sites to acquire inconsistent views of the database simply because of intermittent

---

*This author was supported in part by the National Science Foundation under grant CCR-8909663.

communication failures. Not surprisingly, however, achieving consistent coordination is more difficult than general coordination, where faulty processors are not forbidden to act inconsistently.

Problems requiring consistently coordinated actions promise to become increasingly important in the study of distributed computing. In this paper, we define the class of *consistent simultaneous choice* problems, problems in which all processors (faulty or nonfaulty) that perform an action perform the same action at the same time. We then analyze the state of knowledge that processors must attain to solve these problems and we use the results of this analysis to derive round-optimal solutions to these problems in a number of benign failure models.

The idea of analyzing the state of knowledge attained by a processor was first introduced by Halpern and Moses [5], who present a method for formally ascribing knowledge to processors and show how knowledge can be used to analyze distributed computation. In their work, the highest state of knowledge a group of processors can attain is the state of *common knowledge*: intuitively, a fact is common knowledge if everyone knows it, everyone knows that everyone knows it, and so on.

Since then, many researchers have analyzed problems in terms of the state of knowledge that processors in a system must attain to solve them. In particular, several papers have demonstrated a close relationship between common knowledge and problems requiring simultaneous coordination by the nonfaulty processors. For example, Dwork and Moses [1] consider the problem of *Simultaneous Byzantine Agreement* (a version of *Byzantine Agreement* or *Distributed Consensus* in which all correct processors choose their output bit simultaneously) in the crash failure model—a model in which processors fail by simply halting—and show that processors can reach agreement only when they have achieved common knowledge of certain facts. Using this, they derive a solution that is *optimal in all runs*: in any given context—a fixed pattern of failures and processor inputs—the round in which this solution causes processors reach agreement is at least as early as the round in which any other solution would do so. (Optimality is therefore measured in terms of rounds and not computational complexity.) Moses and Tuttle [9] extend this work to general simultaneous actions and consider systems with more general failures. They show that the coordination of any simultaneous choice requires the correct processors to achieve common knowledge. Using this observation, they too derive optimal protocols in the models they consider. However, in one of these models, the general omissions model—a model in which faulty processors may omit to send or receive messages—their optimal protocols require exponential time. In fact, using the close relationship between common knowledge and simultaneity, they are able to prove that optimally achieving such coordination in this model is inherently intractable (specifically, the problem is shown to be NP-hard).

In this paper, we perform a similar analysis of *consistent* simultaneous choice problems, but our analysis requires a generally stronger definition of common knowledge. In systems with crash and send omission failures, however, the two forms of common knowledge are equivalent (when considered with respect to the group of nonfaulty processors); the optimal protocols derived earlier [1,9] for general simultaneous choice problems also solve the corresponding consistent problems and do so optimally. In systems with general omission failures, we show that the two forms are again equivalent if fewer than half the processors in the system may fail; a minor modification to the previous protocols optimally solves consistent simultaneous choice problems in such systems. Surprisingly, in all these cases, optimal protocols for the consistent version of a problem halt as soon as optimal protocols for the original version: consistency does not delay action.

The same is not true in the general omissions model when more than half the processors may fail. In such systems, we show that the two forms of common knowledge are *not* equivalent and, furthermore, that consistent simultaneous choice problems cannot be solved. This remains true even if the definition of such problems is weakened so that processors are required to act only in failure-free executions.

The remainder of this paper is organized as follows. Section 2 defines our model of computation, and Section 3 defines consistent simultaneous choice problems. Section 4 defines a processor's state of knowledge, including common knowledge. Section 5 analyzes solutions to consistent coordination problems in terms of common knowledge and presents several optimal solutions to these problems in several variants of the omissions failure model. Section 6 shows that these problems cannot be implemented in certain systems with general omission failures. Section 7 contains some concluding remarks and mentions some open problems.

2 Model of a System

In this section, we define our model of a distributed system. Our model is closely related to others used in studying knowledge and coordination [1,5,9].

This paper considers synchronous systems of unreliable processors. Such a system consists of a finite collection $P = \{p_1, \ldots, p_n\}$ of processors, each pair of which is connected by a two-way communication link. All processors share a discrete global clock that starts at time 0 and advances in increments of one. Computation proceeds in a sequence of *rounds*, with round k taking places between time $k-1$ and time k. Each processor has a read-only *input register*, which is initialized to some value referred to as the processor's *initial input*. At time 0, each processor starts in some *initial state*, which includes its initial input. Then, in every round, the processor performs some local computation and perhaps some other actions, sends messages to a set of other processors, and receives the set of messages sent to it by other processors in that round. Each message is assumed to be tagged with the identities of the sender and intended receiver of the message, as well as the round in which it is sent. At any given time, a processor's *message history* consists of the list of messages it has received from the other processors.

A processor's *local state* (or *view*) at any given time consists of its initial input, its message history, the time on the global clock, and the processor's identity. A *global state* is a tuple $\langle s_1, \ldots, s_n, s_e \rangle$ of local states, one local state s_i for each processor p_i and one state s_e for the *environment*. Intuitively, the environment includes everything about the state of the system that cannot be deduced from the local states of the processors, such as the failure pattern and the protocol the processors follow (see below). A *run* of the system is an infinite sequence of global states. An ordered pair $\langle \rho, l \rangle$, where ρ is a run and l is a natural number, is called a *point*, and represents the state of the system after the first l rounds of ρ. Processor q's view at time l in the run ρ (at the point $\langle \rho, l \rangle$) is denoted by $v(q, \rho, l)$.

Processors in a system follow a *protocol*, which specifies what messages a processor is required to send during a round (and what other actions the processor is required to perform) as a deterministic function of its local state. In other words, a protocol is a function from a processor's local state to its next state, a list of local actions it is to perform, and a list of messages it is to send to other processors. While all processors change state and perform actions as required by the protocol, some processors may be *faulty*; such a processor may fail to send any of the messages the protocol requires it to send or fail to receive some of the messages sent to it by other processors.[1]

More precisely, a *faulty behavior* for a processor p_i is a pair of functions S_i and R_i from round numbers to sets of processors. Intuitively, $S_i(k)$ is the set of processors to which p_i fails to send a message in round k, and $R_i(k)$ is the set of processors from which p_i fails to receive a message. A *failure pattern* is a collection of faulty behaviors $\langle S_i, R_i \rangle$, one for each processor p_i. The *failure pattern* of a run is a failure pattern such that for every round k and every processor p_i

- p_i sends no messages to processors in $S_i(k)$ in round k but sends all required messages to processors not in $S_i(k)$, and

- p_i receives no messages from processors in $R_i(k)$ in round k but receives all messages sent to it by processors not in $R_i(k)$.

We assume that a run's failure pattern is encoded in the state of the environment. Since it is possible for several failure patterns to be consistent with the pattern of messages sent during a given

---

[1]This paper does not consider processors that may fail *arbitrarily* [7]; Neiger and Toueg [11] give formal definitions of a range of different failure models.

execution, it is possible for several runs to differ only in the failure pattern encoded in the state of the environment. Processor p_i is *correct* if $S_i(k) = R_i(k) = \emptyset$ for all k.

This work considers three failure models:

1. the *crash* model [3,14], in which $R_i(k)$ is empty for every k, and in which $S_i(k)$ nonempty implies $S_i(k') = P$ for every $k' > k$;[2]

2. the *send omissions* model [3,8], in which $R_i(k)$ is empty for every k, and there are no restrictions on $S_i(k)$; and

3. the *general omissions* model [9,13], in which there are no restrictions on either $R_i(k)$ or $S_i(k)$.

Thus, in the crash model, faulty processors simply halt at some point (stop sending messages); in the send omissions model, faulty processors intermittently fail to send some messages; and in the general omissions model, faulty processors intermittently fail both to send and receive messages.

Given a run ρ, if ι_i is p_i's initial input in ρ, then $\bar{\iota} = \langle \iota_1, \ldots, \iota_n \rangle$ is the *input* to ρ. A pair $\langle \pi, \bar{\iota} \rangle$, where π is a failure pattern and $\bar{\iota}$ is an input, is called an *operating environment*. Note that a run is uniquely determined by a protocol and an operating environment. Two runs of two different protocols are said to be *corresponding runs* if they have the same operating environment. The fact that an operating environment is independent of the protocol allows the comparison of different protocols according to their behavior in corresponding runs.

At several places in this paper, it will be convenient to refer to runs that differ only in some aspect of their operating environment. We say that two runs ρ and ρ' of a protocol P *differ only in* some aspect of their operating environment if and only if ρ and ρ' are the result of running P in operating environments that differ only in the given aspect. As an example, if ρ and ρ' differ only in the input to p_i, then their operating environments are identical except for p_i's input, although the actual messages sent by processors may depend on p_i's input and hence be very different in the two runs. As another example, suppose ρ and ρ' differ only in that every processor sends a message to p_i in round l (in the operating environment) of ρ'. It is possible that P does not require processors to send any messages to p_i in round l, in which case the messages actually sent in ρ and ρ' are the same; we simply mean that the operating environment itself does not keep processors from sending messages to p_i in round l of ρ'.

This work studies the behavior of protocols in the presence of a bounded number of failures (of a particular type) and a given set of possible initial inputs to the processors. It is therefore natural to identify a *system* with the set of all possible runs of a given protocol under such circumstances. Formally, a system is identified with the set of runs of a protocol $\mathcal{P}$ by n processors, at most $t \leq n-2$ of which may be faulty (in the sense of a particular failure model $\mathcal{M}$), where the initial input of each processor p_i is an element of set $\mathcal{I}$. This set of runs is denoted by the tuple $\Sigma = \langle n, t, \mathcal{P}, \mathcal{M}, \mathcal{I} \rangle$.

While a protocol may be thought of as a function of a processor's view, protocols for distributed systems are typically written for systems of arbitrarily large size. In this sense, the actions and messages required of a processor by a protocol actually depend on the number of processors in the system (and, perhaps, the bound on the number of failures) as well as the view of the processor. Therefore, a *protocol* is formally defined to be a function from n, t, and a processor's view to a list of actions the processor is required to perform, followed by a list of messages that the processor is required to send in the following round. Since each protocol is defined for systems of arbitrary size, it is natural to define a *class* of systems to be a collection of systems $\{\Sigma(n,t) \mid n \geq t+2\}$, where $\Sigma(n,t) = \langle n, t, \mathcal{P}, \mathcal{M}, \mathcal{I} \rangle$ for some fixed protocol $\mathcal{P}$, failure model $\mathcal{M}$, and input set $\mathcal{I}$.

In order to analyze systems, it is convenient to have a logical language in which we can make statements about the system. We will define such a language in Section 4. A *fact* (or *formula*) in this language will be interpreted as a property of points: a fact φ will be either true or false at a given point $\langle \rho, l \rangle$, which we denote by $\langle \rho, l \rangle \models \varphi$ or $\langle \rho, l \rangle \not\models \varphi$, respectively. Sometimes, however,

---

[2]It is common to assume, in addition, that p_i cannot perform an action after it has failed (i.e., once $S_i(k)$ is nonempty).

it is convenient to refer to facts as being about objects other than points (e.g., properties of runs). In general, a fact φ is said to be a *fact about* X if fixing X determines the truth (or falsity) of φ. For example, a fact φ is said to be a fact *about the nonfaulty processors* if fixing the identities of the nonfaulty processors determines whether nor not φ holds. That is, given any set $N \subseteq P$, either φ holds at all points $\langle \rho, l \rangle$ in which N is the set of nonfaulty processors, or at no such point. The meaning of a fact being *about the operating environment*, *about the initial states*, etc. are defined similarly.

3 Consistent Simultaneous Choices

Moses and Tuttle define a class of *simultaneous choice problems* in which nonfaulty processors are required to choose an action (such as deciding on an output bit) and perform it simultaneously. Their definition does not restrict the behavior of the faulty processors in any way. In this section, we define a class of *consistent simultaneous choice problems* (or simply *consistent choices*) in which faulty and nonfaulty processors must behave consistently: if a faulty processor performs any action at all in a run, it must perform the same action that the nonfaulty processors perform, and must perform it at the same time that the nonfaulty processors do. The definition of such a problem must tell us when each action may be performed, and describe the operating environments in which a choice among the actions must be made (that is, tell us what initial inputs and what types of processor failures are allowed).

Formally, a *simultaneous action* a_i is an action with an associated *enabling condition* ok_i, which is a fact about the input and the faulty processors. A *consistent simultaneous choice* problem $\mathcal{C}$ is determined by a set $\{a_1, \ldots, a_m\}$ of simultaneous actions (and their associated enabling conditions), together with a failure model $\mathcal{M}$ and set $\mathcal{I}$ of initial inputs. Intuitively, every run ρ of a protocol implementing $\mathcal{C}$ must satisfy the following conditions:

1. each processor performs at most one of the a_i's and does so at most once,

2. if a_i is performed by some processor at some time, then a_i is performed by all nonfaulty processors at that time,

3. if a_i is performed by some processor, then ρ satisfies ok_i, and

4. if no processors fail in ρ, then all processors perform an action.

More formally, a protocol $\mathcal{P}$ and a consistent simultaneous choice $\mathcal{C}$ determine a class of systems $\{\Sigma(n, t) \mid n \geq t + 2\}$, where $\Sigma(n, t) = \langle n, t, \mathcal{P}, \mathcal{M}, \mathcal{I} \rangle$. $\mathcal{P}$ *implements* $\mathcal{C}$ if every run of every system in the class determined by $\mathcal{P}$ and $\mathcal{C}$ satisfies the conditions 1–4 above.[3] A choice $\mathcal{C}$ is *implementable* if there is some protocol that implements it.

One example of a consistent simultaneous choice problem is a consistent, simultaneous version of the *Byzantine Agreement* problem mentioned in the introduction (cf. [1,12]). Recall that each processor begins with an input bit and chooses an output bit, subject to the condition that all output bits have the same value and that this value is the value of some processor's input bit. This problem involves a choice between two actions: a_0 corresponds to choosing 0 as the output bit, and a_1 corresponds to choosing 1. Since the value of a processor's output bit must be the value of some

---

[3]This is analogous to the definition of a simultaneous choice of Moses and Tuttle [9]. The differences are the following: their definition places restrictions only on the nonfaulty processors; their enabling conditions are derived from *pro* and *con* conditions, which are facts about the input and the existence of failures (the results of Moses and Tuttle still hold if their definition of choices is extended to allow these conditions to be facts about the input and the faulty processors); and the liveness condition given here (that processors must act in failure-free runs) is a weak version of their notion of strictness. For a further discussion of this last point, see the discussion after the definition of an optimal protocol. We sometimes refer to a simultaneous choice of Moses and Tuttle as a *general simultaneous choice*.

processor's input bit, the enabling condition ok_0 for a_0 holds if and only if some processor's input bit is 0, and ok_1 is defined analogously.

This paper concerns optimal solutions to consistent simultaneous choice problems. A protocol $\mathcal{P}$ is an *optimal* protocol for a consistent simultaneous choice $\mathcal{C}$ if (i) $\mathcal{P}$ implements $\mathcal{C}$ and (ii) given any other protocol $\mathcal{P}'$ implementing $\mathcal{C}$, for every pair of corresponding runs of the two protocols, $\mathcal{P}$ has the nonfaulty processors perform actions no later than $\mathcal{P}'$.

Note that this is a very strong definition of optimality. Whereas most definitions of optimality require only that a protocol perform as well as any other protocol in their respective worst case runs, this definition requires that a protocol do so in *every* run. It is because we are interested in this strong definition of optimality that we can make the very weak liveness condition (condition 4) in the definition of a consistent choice. This condition says that processors need only perform an action in failure-free runs. One might want to require processors to perform an action in more runs, say in all runs. Note, however, that if there is any protocol for a problem in which the nonfaulty processors perform an action in every run, then there is an *optimal* protocol in which the nonfaulty processors perform an action in every run. Consequently, because the definition of an optimal protocol is so strong and the definition of a protocol solving a consistent choice is so weak, the existence of optimal protocols (which we prove in some cases) is a very strong result, as is the impossibility of any protocol at all (which we prove in another).

4 Definitions of Knowledge

The analysis in this paper depends on a processor's knowledge at a given point of an execution. This section defines such a notion of knowledge. For the sake of this section, fix a particular system. All runs mentioned will be runs of this system, and all points will be points in such runs. The treatment here is a modification and expansion of those given by Moses and Tuttle [9] and by others [1,5].

This definition of knowledge requires a logical language that can express statements about knowledge. Recall that a *fact* (or *formula*) in this language is a property of points: a fact φ is either true or false at a given point $\langle \rho, l \rangle$, denoted $\langle \rho, l \rangle \models \varphi$ or $\langle \rho, l \rangle \not\models \varphi$, respectively. A fact is said to be *valid in the system* (for a given system) if it is true of all points in the system. A fact is said to be *valid* if it is valid in all systems. Assume the existence of an underlying logical language for representing all relevant *ground* facts—facts about the system that do not explicitly mention processors' knowledge (for example, "*the value of register x is 0*" or "*processor p_i failed in round 3*"). Formally, a ground fact φ will be identified with a set of points $\tau(\varphi)$. Intuitively, this is the set of points at which the fact holds. The truth of a ground fact φ is determined by $\langle \rho, l \rangle \models \varphi$ if and only if $\langle \rho, l \rangle \in \tau(\varphi)$. This language is then closed under the standard boolean connectives $\wedge$ and $\neg$, as well as various knowledge operators. The meaning of the boolean connectives is defined in the usual way. The meaning of the knowledge operators is the subject of the rest of this section.

The intuition underlying the standard definition of knowledge [5] is that, in any given global state, a processor considers a number of other global states to be *possible*, namely those global states in which it has the same local state; a processor knows a fact φ if and only if φ holds in all global states it considers possible. Formally, p_i considers a point $\langle \rho', l' \rangle$ to be *possible* at $\langle \rho, l \rangle$ if and only if $v(p_i, \rho, l) = v(p_i, \rho', l')$.[4] We say that p_i *knows* φ at $\langle \rho, l \rangle$, denoted $\langle \rho, l \rangle \models \mathsf{K}_i \varphi$, if and only if $\langle \rho', l' \rangle \models \varphi$ for all points $\langle \rho', l' \rangle$ that p_i considers possible at $\langle \rho, l \rangle$. Thus, a processor knows any fact that follows from the information recorded in its local state. In particular, it knows any such fact regardless of the complexity of computing that it follows from its local state.

In addition to the knowledge of individual processors, our analysis depends on knowledge among a group of processors. In particular, the state of *common knowledge* [5] plays a central role in this paper due to the close relationship between it and the simultaneous performance of an action by a group of processors (see Lemma 5). Roughly speaking, a fact φ is common knowledge to a given group if everyone in the group knows φ, everyone in the group knows that everyone knows φ, and

---

[4]Note that this implies $l = l'$ because the global clock is part of p_i's view.

so on *ad infinitum*. Given a fixed group G of processors, it is customary to define "*everyone in G knows φ,*" denoted $\mathsf{E}_G\varphi$, by

$$\mathsf{E}_G\varphi \stackrel{\text{def}}{=} \bigwedge_{p_i \in G} \mathsf{K}_i\varphi.$$

One now defines "φ *is common knowledge to G*," denoted $\mathsf{C}_G\varphi$, by

$$\mathsf{C}_G\varphi \stackrel{\text{def}}{=} \mathsf{E}_G\varphi \wedge \mathsf{E}_G\mathsf{E}_G\varphi \wedge \cdots \wedge \mathsf{E}_G^m\varphi \wedge \cdots.$$

In other words, $\langle \rho, l \rangle \models \mathsf{C}_G\varphi$ if and only if $\langle \rho, l \rangle \models \mathsf{E}_G^m\varphi$ for all $m \geq 1$, where $\mathsf{E}_G^1\varphi \stackrel{\text{def}}{=} \mathsf{E}_G\varphi$ and $\mathsf{E}_G^{m+1}\varphi \stackrel{\text{def}}{=} \mathsf{E}_G(\mathsf{E}_G^m\varphi)$.

In systems of unreliable processors, however, the groups of interest are not always fixed subsets of P like G above. In this paper, the most interesting facts will be those that are common knowledge to the set $\mathcal{N}$ of nonfaulty processors, and the value of this set changes from run to run. We use $\mathcal{N}(\rho, l)$ to denote the set of nonfaulty processors at the point $\langle \rho, l \rangle$ (and therefore in the run ρ). Moses and Tuttle argue that the generalization of common knowledge to nonconstant sets such as $\mathcal{N}$ is more subtle than simply defining

$$\mathsf{E}_\mathcal{N}\varphi \stackrel{\text{def}}{=} \bigwedge_{p_i \in \mathcal{N}} \mathsf{K}_i\varphi,$$

and they propose a generalization of $\mathsf{E}_\mathcal{N}\varphi$, which we denote by $\mathsf{F}_\mathcal{N}\varphi$ (although they continue to write $\mathsf{E}_\mathcal{N}\varphi$). They argue that one must define

$$\mathsf{F}_\mathcal{N}\varphi \stackrel{\text{def}}{=} \bigwedge_{p_i \in \mathcal{N}} \mathsf{K}_i(p_i \in \mathcal{N} \Rightarrow \varphi).$$

That is, every nonfaulty processor knows φ if and only if every nonfaulty processor knows that φ is true, given that it (the processor itself) is nonfaulty. Since it is possible for p_i to be contained in $\mathcal{N}$ without knowing this, the definition of $\mathsf{F}_\mathcal{N}\varphi$ is weaker than $\mathsf{E}_\mathcal{N}\varphi$, and the definition of common knowledge based on $\mathsf{F}_\mathcal{N}\varphi$ is therefore weaker than the definition above based on $\mathsf{E}_\mathcal{N}\varphi$. For this reason, we distinguish these two definitions of common knowledge as *weak* and *strong common knowledge*, and we denote weak and strong common knowledge of φ by $\mathsf{W}_\mathcal{N}\varphi$ and $\mathsf{S}_\mathcal{N}\varphi$, respectively. (Moses and Tuttle write $\mathsf{C}_\mathcal{N}\varphi$ for $\mathsf{W}_\mathcal{N}\varphi$.) Formally,

$$\mathsf{W}_\mathcal{N}\varphi \stackrel{\text{def}}{=} \mathsf{F}_\mathcal{N}\varphi \wedge \mathsf{F}_\mathcal{N}\mathsf{F}_\mathcal{N}\varphi \wedge \cdots \wedge \mathsf{F}_\mathcal{N}^m\varphi \wedge \cdots$$

and, as above,

$$\mathsf{S}_\mathcal{N}\varphi \stackrel{\text{def}}{=} \mathsf{E}_\mathcal{N}\varphi \wedge \mathsf{E}_\mathcal{N}\mathsf{E}_\mathcal{N}\varphi \wedge \cdots \wedge \mathsf{E}_\mathcal{N}^m\varphi \wedge \cdots.$$

Moses and Tuttle prove that weak common knowledge is a necessary and sufficient condition for the performance of simultaneous actions. In this work, we prove that strong common knowledge is a necessary and sufficient condition for the performance of *consistent* simultaneous actions. This is somewhat surprising, since it says that the solution of the harder problem depends on the simpler, less subtle definition of common knowledge. On the other hand, we will show that, in most of the cases considered by Moses and Tuttle, the two definitions of common knowledge are actually equivalent. In one case where the two definitions are different, we will show that consistent performance of simultaneous actions is impossible. Thus, this work shows precisely where the added subtlety in the definition of weak common knowledge is required for the results of Moses and Tuttle. Understanding this relationship between strong and weak common knowledge is one of the primary technical contributions of this work.

This section concludes with a few observations about properties of the two definitions of common knowledge [5]. A useful tool for thinking about $\mathsf{F}_\mathcal{N}^m\varphi$ and $\mathsf{W}_\mathcal{N}\varphi$ is a graph whose nodes are all points of the system and in which there is an edge between points $\langle \rho, l \rangle$ and $\langle \rho', l \rangle$ if and only if there is some processor $p \in \mathcal{N}(\rho, l) \cap \mathcal{N}(\rho', l)$ such that $v(p, \rho, l) = v(p, \rho', l)$. That is, there is a processor that is nonfaulty at both points and that has the same view at both points. This graph is called the *similarity graph*. An easy argument by induction on m shows that $\langle \rho, l \rangle \models \mathsf{F}_{\mathcal{N}}^m \varphi$ if and only if $\langle \rho', l \rangle \models \varphi$ holds for all points $\langle \rho', l \rangle$ at a distance m from $\langle \rho, l \rangle$ in this graph. Two points $\langle \rho, l \rangle$ and $\langle \rho', l \rangle$ are *similar*, denoted $\langle \rho, l \rangle \sim \langle \rho', l \rangle$, if they are in the same connected component of the similarity graph. It is easy to see that the following lemma holds:

Lemma 1: $\langle \rho, l \rangle \models \mathsf{W}_{\mathcal{N}} \varphi$ if and only if $\langle \rho', l \rangle \models \varphi$ for all $\langle \rho', l \rangle$ satisfying $\langle \rho, l \rangle \sim \langle \rho', l \rangle$.

Now consider a directed graph whose nodes are all points of the system and in which there is an edge from $\langle \rho, l \rangle$ to $\langle \rho', l \rangle$ if and only if there is some processor $p \in \mathcal{N}(\rho, l)$ such that $v(p, \rho, l) = v(p, \rho', l)$. This means that there is a processor that is nonfaulty at $\langle \rho, l \rangle$ (but possibly faulty at $\langle \rho', l \rangle$) and that has the same view at both points. This graph is called the *directed similarity graph*. An easy argument by induction on m shows that $\langle \rho, l \rangle \models \mathsf{E}_{\mathcal{N}}^m \varphi$ if and only if $\langle \rho', l \rangle \models \varphi$ holds for all points $\langle \rho', l \rangle$ to which there is a path from $\langle \rho, l \rangle$ of length m in this graph. Let $\langle \rho, l \rangle \rightsquigarrow \langle \rho', l \rangle$ denote the fact that $\langle \rho', l \rangle$ is reachable from $\langle \rho, l \rangle$ in the directed similarity graph. The following lemma holds for strong common knowledge:

Lemma 2: $\langle \rho, l \rangle \models \mathsf{S}_{\mathcal{N}} \varphi$ if and only if $\langle \rho', l \rangle \models \varphi$ for all $\langle \rho', l \rangle$ satisfying $\langle \rho, l \rangle \rightsquigarrow \langle \rho', l \rangle$.

Note that $\langle \rho, l \rangle \sim \langle \rho', l \rangle$ implies both $\langle \rho, l \rangle \rightsquigarrow \langle \rho', l \rangle$ and $\langle \rho', l \rangle \rightsquigarrow \langle \rho, l \rangle$; the converse does not always hold. It follows that $\mathsf{S}_{\mathcal{N}} \varphi \Rightarrow \mathsf{W}_{\mathcal{N}} \varphi$ is valid for all facts φ. (As previously noted, this also follows from the fact that $\mathsf{E}_{\mathcal{N}} \varphi \Rightarrow \mathsf{F}_{\mathcal{N}} \varphi$ is valid for all φ.)

The notions of knowledge and common knowledge defined above are closely related to modal logics [4]. Five properties of a modal operator M are

A1: the *knowledge axiom*: $\mathsf{M}\varphi \Rightarrow \varphi$;

A2: the *consequence closure axiom*: $\mathsf{M}\varphi \wedge \mathsf{M}(\varphi \Rightarrow \psi) \Rightarrow \mathsf{M}\psi$;

A3: the *positive introspection axiom*: $\mathsf{M}\varphi \Rightarrow \mathsf{MM}\varphi$;

A4: the *negative introspection axiom*: $\neg \mathsf{M}\varphi \Rightarrow \mathsf{M}\neg \mathsf{M}\varphi$; and

R1: the *rule of necessitation*: if φ is valid in the system, then $\mathsf{M}\varphi$ is valid in the system.

A modal operator M *has the properties of the modal logic S5* in a system if the axioms A1–A4 are valid in the system and R1 holds. M *has the properties of the modal logic S4* in a system if the axioms A1–A3 are valid in the system and R1 holds.

It is easy to see that the knowledge and weak common knowledge operators defined above have the properties of S5 in all systems [4]:

Theorem 3: *The operators K_i and $\mathsf{W}_{\mathcal{N}}$ have the properties of S5 in all systems.*

Strong common knowledge can only be shown to satisfy S4:

Theorem 4: *The operator $\mathsf{S}_{\mathcal{N}}$ has the properties of S4 in all systems.*

Proof: A1 holds because the relation $\rightsquigarrow$ is reflexive. A2 holds because if $\varphi \Rightarrow \psi$ and φ both hold at all points $\langle \rho', l \rangle$ such that $\langle \rho, l \rangle \rightsquigarrow \langle \rho', l \rangle$, then so does ψ. A3 holds because the relation $\rightsquigarrow$ is transitive. It is obvious that R1 holds. Thus, $\mathsf{S}_{\mathcal{N}}$ has the properties of S4. □

$\mathsf{S}_{\mathcal{N}}$ does not have the properties of S5 because the axiom A4 need not hold for $\mathsf{S}_{\mathcal{N}}$: it does not hold if the relation $\rightsquigarrow$ is not symmetric [4]. Section 6 below considers a specific system in which this occurs.

Strong and weak common knowledge also satisfy the following *fixed-point axioms*:

$$\mathsf{W}_{\mathcal{N}} \varphi \Leftrightarrow \mathsf{F}_{\mathcal{N}} \mathsf{W}_{\mathcal{N}} \varphi;$$

$$S_{\mathcal{N}}\varphi \Leftrightarrow E_{\mathcal{N}}S_{\mathcal{N}}\varphi.$$

These axioms imply that a fact's being common knowledge is "public," in the sense that a fact is common knowledge if and only if everyone knows it is common knowledge. It is these two axioms, which imply that all processors in a group "learn" of common knowledge simultaneously, that make common knowledge so important for implementing simultaneous choices. Another useful fact about common knowledge is captured by the following *induction rules*:

- if $\varphi \Rightarrow F_{\mathcal{N}}(\varphi \wedge \psi)$ is valid in the system, then $\varphi \Rightarrow W_{\mathcal{N}}\psi$ is valid in the system;
- if $\varphi \Rightarrow E_{\mathcal{N}}(\varphi \wedge \psi)$ is valid in the system, then $\varphi \Rightarrow S_{\mathcal{N}}\psi$ is valid in the system.

Intuitively, taking ψ to be φ for the moment, these rules say that if φ is "public" in the sense that everyone knows φ whenever φ holds, then φ is actually common knowledge whenever φ holds. When ψ and φ are different, these rules say that if, in addition, φ implies ψ, then ψ itself is also common knowledge.

5 Optimal Protocols

In this section, we derive optimal protocols for consistent simultaneous choice problems in the crash, send omissions, and general omissions models. Previously, Dwork and Moses [1] derived optimal protocols for *Simultaneous Byzantine Agreement* in the crash failure model, and Moses and Tuttle [9] derived optimal protocols for simultaneous choice problems in the crash, send, and general omissions failure models. The derivation of these protocols is the result of an analysis of simultaneous choice problems in terms of weak common knowledge. In this section, we show how a similar analysis of *consistent* simultaneous choice problems in terms of *strong* common knowledge leads to optimal protocols for these problems as well.

The fundamental observation leading to the earlier derivation of optimal protocols is the strong relationship between common knowledge and simultaneous actions: when a simultaneous action is performed, it is weak common knowledge that this action is being performed and, thus, that it is enabled. The following lemma shows that an analogous result holds for consistent simultaneous actions when phrased in terms of strong common knowledge. This is a stronger statement that does not hold in general for simultaneous choice problems.

Lemma 5: *Let ρ be a run of a protocol implementing a consistent simultaneous choice C. If an action a_j of C is performed by any processor at time l in ρ, then $\langle \rho, l \rangle \models S_{\mathcal{N}} ok_j$.*

Proof: Let φ be the fact "a_j is being performed by some processor." Note that $\langle \rho, l \rangle \models \varphi$. We now show that $\varphi \Rightarrow E_{\mathcal{N}}(\varphi \wedge ok_j)$ is valid in the system. It will follow by the induction rule that $\varphi \Rightarrow S_{\mathcal{N}} ok_j$ is also valid in the system and, thus, that $\langle \rho, l \rangle \models S_{\mathcal{N}} ok_j$. Let $\langle \eta, l \rangle$ be any point such that $\langle \eta, l \rangle \models \varphi$. By condition 2 of the definition of a consistent choice, all processors in $\mathcal{N}(\eta, l)$ execute a_j at time l in η, so $\langle \eta, l \rangle \models E_{\mathcal{N}}\varphi$ (every processor performing a_j certainly knows a_j is being performed). By condition 3, $\varphi \Rightarrow ok_j$ is valid in the system, so $\langle \eta, l \rangle \models E_{\mathcal{N}} ok_j$ by the consequence closure axiom A2. Thus, $\langle \eta, l \rangle \models E_{\mathcal{N}}(\varphi \wedge ok_j)$. Since $\langle \eta, l \rangle$ was chosen arbitrarily, $\varphi \Rightarrow E_{\mathcal{N}}(\varphi \wedge ok_j)$ is valid in the system. □

Lemma 5 states that no action can be performed until that action's enabling condition is strong common knowledge. Intuitively, any protocol that has processors act as soon as some enabling condition becomes strong common knowledge will have them act at least as quickly as any other protocol and should be an optimal protocol. In this sense, the derivation of optimal protocols reduces to the derivation of protocols that cause facts to become strong common knowledge as soon as possible. Both Dwork and Moses and Moses and Tuttle follow this strategy (using weak common knowledge). One class of protocols causing facts to become strong common knowledge as soon as possible are *full-information protocols*.

5.1 Full-Information Protocols

Recall that a protocol is a function specifying the actions a processor should perform and the messages that it should send as a function of n, t, and the processor's view. Thus, a protocol has two components: an *action* component and a *message* component. A protocol is said to be a *full-information protocol* [3] if its message component calls for it to send its entire view to all processors in every round. Since such a protocol requires that all processors send all the information available to them in every round, it gives each processor as much information about the operating environment as any protocol could. Consequently, if a processor cannot distinguish two operating environments during runs of a full-information protocol, then the processor cannot distinguish them during runs of any other protocol:

Lemma 6 (Moses and Tuttle): *Let τ and τ' be runs of a full-information protocol $\mathcal{F}$, and let ρ and ρ' be runs of an arbitrary protocol $\mathcal{P}$ corresponding to τ and τ', respectively. For all processors p and times l, if $v(p, \tau, l) = v(p, \tau', l)$, then $v(p, \rho, l) = v(p, \rho', l)$.*

The following corollary to Lemma 6 shows that facts about the operating environment become strong common knowledge during runs of full-information protocols at least as soon as they do during runs of any other protocol. Moses and Tuttle state a similar corollary that says that the same is true of weak common knowledge.

Corollary 7: *Let φ be a fact about the operating environment. Let τ and ρ be corresponding runs of a full-information protocol $\mathcal{F}$ and an arbitrary protocol $\mathcal{P}$, respectively. If $\langle \rho, k \rangle \models \mathsf{S}_\mathcal{N}\varphi$, then $\langle \tau, l \rangle \models \mathsf{S}_\mathcal{N}\varphi$.*

Proof: Suppose $\langle \rho, k \rangle \models \mathsf{S}_\mathcal{N}\varphi$. To prove that $\langle \tau, l \rangle \models \mathsf{S}_\mathcal{N}\varphi$, it suffices to show that $\langle \tau', l \rangle \models \varphi$ for all runs τ' of $\mathcal{F}$ such that $\langle \tau, l \rangle \leadsto \langle \tau', l \rangle$. Fix τ' and let ρ' be the corresponding run of $\mathcal{P}$. Lemma 6 and a simple inductive argument on the distance from $\langle \tau, l \rangle$ to $\langle \tau', l \rangle$ in the similarity graph show that $\langle \tau, l \rangle \leadsto \langle \tau', l \rangle$ implies $\langle \rho, k \rangle \leadsto \langle \rho', k \rangle$. Since $\langle \rho, k \rangle \models \mathsf{S}_\mathcal{N}\varphi$, we have $\langle \rho', k \rangle \models \varphi$. Since φ is a fact about the operating environment and τ' and ρ' have the same operating environment, $\langle \tau', l \rangle \models \varphi$, as desired. □

Because the enabling conditions for simultaneous actions are facts about the operating environment, the remainder of the paper concentrates on full-information protocols, which achieve common knowledge of such facts as quickly as possible.

5.2 Crash and Send Omissions

Using their version of Corollary 7 (with $\mathsf{S}_\mathcal{N}$ replaced by $\mathsf{W}_\mathcal{N}$), Moses and Tuttle derive a full-information, optimal protocol $\mathcal{F}_{MT}$ for any implementable simultaneous choice $\mathcal{C}$ (see Figure 1). The protocol is *knowledge-based* since it is programmed in a language making explicit tests for common knowledge of various facts. In this protocol, each processor broadcasts its view every round until it detects that one of the enabling actions ok_j has become weak common knowledge to the nonfaulty processors. At this point, the processor performs the action a_j, where j is the least index such that ok_j is weak common knowledge. While broadcasting views may in general require sending exponentially large messages, Moses and Tuttle show that, in any of the (benign) failure models considered here, a processor's view can be encoded as a *communication graph* whose size is polynomial in n and the current round number; thus, this protocol requires only polynomially large messages.

In order to implement $\mathcal{F}_{MT}$, it is necessary to implement the tests for weak common knowledge that appear in $\mathcal{F}_{MT}$. Because processors have to be able to test for $\mathsf{W}_\mathcal{N} ok_j$ locally, any test for $\mathsf{W}_\mathcal{N} ok_j$ must determine whether $\mathsf{W}_\mathcal{N} ok_j$ holds at a point given only the view of a processor at that point. Furthermore, this must be true of every point in every system $\Sigma(n, t)$ determined by $\mathcal{F}_{MT}$ and the simultaneous choice $\mathcal{C}$ that $\mathcal{F}_{MT}$ solves. On the other hand, because the simultaneous choice $\mathcal{C}$ restricts only the performance of actions by *nonfaulty* processors, this test need only be correct

```
repeat every round
    broadcast view to every processor
until W_N ok_j holds for some j;
j ← min{k | W_N ok_k holds};
perform a_j;
halt.
```

Figure 1: The optimal protocol $\mathcal{F}_{MT}$

when given the view of a nonfaulty processor. Formally, given any fact ψ and any set S, a *test for* ψ *for* S within a class of systems $\{\Sigma(n,t) \mid n \geq t+2\}$ is a Turing machine M that

1. accepts as input the view of any processor at any point in any system $\Sigma(n,t)$ and returns either *true* or *false*, and

2. given the view $v(p_i, \rho, l)$ of any processor p_i in S at a point $\langle \rho, l \rangle$, returns *true* if and only if $\langle \rho, l \rangle \models \psi$.

Moses and Tuttle construct tests for $W_N \varphi$ for the set $\mathcal{N}$ of nonfaulty processors, and make a rather surprising technical observation implying that, in the crash and send omissions models, even the faulty processors can use these tests: given the view of *any* processor p_i (faulty or nonfaulty) at $\langle \rho, l \rangle$, their test for $W_N \varphi$ returns *true* if and only if $\langle \rho, l \rangle \models W_N \varphi$. That is, their test is a test for $W_N \varphi$ for the set P of all processors. We refer to such a test as, simply, a *test for* $W_N \varphi$. Thus, Moses and Tuttle effectively show that

Theorem 8 (Moses and Tuttle): *If C is an implementable, consistent simultaneous choice, then $\mathcal{F}_{MT}$ is an optimal protocol for C.*

Proof: Let $\mathcal{P}$ be any protocol implementing C, and let τ and ρ be corresponding runs of $\mathcal{F}_{MT}$ and $\mathcal{P}$. First note that nonfaulty processors perform actions in τ at least as early as any processor does so in ρ: if any processor performs a_j at time l in ρ, then $S_N ok_j$ holds at $\langle \rho, l \rangle$ by Lemma 5 and also at $\langle \tau, l \rangle$ by Corollary 7; since $S_N ok_j \Rightarrow W_N ok_j$ is valid, all nonfaulty processors perform an action at time l in τ if they have not already done so. To see that $\mathcal{F}_{MT}$ actually implements C, note the following:

1. Each processor clearly performs at most one of the a_j in τ and does so at most once.

2. Suppose a_j is performed by some processor p at time l in τ. Recall that every processor is following a test M_k for $W_N ok_k$ to determine whether it should perform a_k; recall that M_k is actually a test for $W_N ok_k$ for all processors. Since p performed a_j at time l, it follows that l is the first round in which some ok_k is weak common knowledge, and that j is the least index k such that ok_k is weak common knowledge in round l. Thus, in particular, every nonfaulty processor will perform a_j at time l in τ.

3. Suppose a_j is performed by some processor p in τ. If p performs a_j at time l in τ, then $W_N ok_j$ holds at time l in τ, so τ satisfies ok_j (since ok_j is a fact about the run).

4. Suppose no processor fails in τ. Then the same is true in the corresponding run ρ of the protocol $\mathcal{P}$ implementing C, so some action must be performed at some time l in ρ. Thus, some action must be performed by the nonfaulty processors no later than time l in τ by the argument above. Since all processors are nonfaulty in τ, all processors perform this action at time l.

Thus, $\mathcal{F}_{MT}$ is an optimal protocol for $\mathcal{C}$. □

To analyze the computational complexity of $\mathcal{F}_{MT}$, one must understand the complexity the tests M_j for $W_\mathcal{N} ok_j$. To do so, one must also understand the basic complexity of determining the truth of the facts ok_j. Moses and Tuttle define the class of *practical* simultaneous choice problems to be a restriction of the class of simultaneous choice problems that, among other things, guarantees that it is possible to test in polynomial time, given the view of a processor at a point, whether this processor's view determines that ok_j must be true (or, in other words, whether this processor knows ok_j). Every natural simultaneous choice problem of which we are aware is practical, but we refer the reader to Moses and Tuttle [9] for the detailed definition. For such problems, they show that their tests for $W_\mathcal{N} ok_j$ at a point $\langle \rho, l \rangle$ run in time polynomial in n and l in both the crash and send omissions model. Thus, Moses and Tuttle effectively show that

Theorem 9 (Moses and Tuttle): *If $\mathcal{C}$ is an implementable, practical, consistent simultaneous choice, then $\mathcal{F}_{MT}$ is a polynomial-time, optimal protocol for $\mathcal{C}$.*

5.3 General Omissions with $n > 2t$

According to Lemma 5, strong common knowledge is a necessary condition for consistently performing simultaneous actions, and yet the protocol $\mathcal{F}_{MT}$ is programmed using tests for weak common knowledge. More precisely, the technical observation made by Moses and Tuttle leading to the observation that both faulty and nonfaulty processors can follow their tests for $W_\mathcal{N} \varphi$ also implies that weak common knowledge to the nonfaulty processors is equivalent to weak common knowledge to *all* processors: that is, $W_\mathcal{N} \varphi \equiv W_P \varphi$ for all facts φ. From this it follows that $W_\mathcal{N} \varphi \Rightarrow S_\mathcal{N} \varphi$, since $W_P \varphi$ clearly implies $S_\mathcal{N} \varphi$; we hence have $W_\mathcal{N} \varphi \equiv S_\mathcal{N} \varphi$, since we already observed that $S_\mathcal{N} \varphi \Rightarrow W_\mathcal{N} \varphi$.

It turns out that the same is true in the general omissions model, but only when $n > 2t$. To prove this fact, it is enough to show that the relations $\sim$ and $\leadsto$ are the same. Remember that when we make a statement like "ρ and ρ' differ only in that every processor sends a message to p_i in round l of ρ'," we mean that the operating environments of ρ and ρ' are identical except that the operating environment of ρ' does not itself keep processors from sending messages to p_i in round l.

Lemma 10: *Consider any system in the general omissions model with $n > 2t$. If $\langle \rho, l \rangle$ and $\langle \eta, l \rangle$ are two points of the system, then $\langle \rho, l \rangle \leadsto \langle \eta, l \rangle$ if and only if $\langle \rho, l \rangle \sim \langle \eta, l \rangle$.*

Proof: Since $\langle \rho, l \rangle \sim \langle \eta, l \rangle$ clearly implies $\langle \rho, l \rangle \leadsto \langle \eta, l \rangle$, we prove that $\langle \rho, l \rangle \leadsto \langle \eta, l \rangle$ implies $\langle \rho, l \rangle \sim \langle \eta, l \rangle$. It suffices to show that if $v(p, \rho, l) = v(p, \eta, l)$ for some $p \in \mathcal{N}(\rho, l)$, then $\langle \rho, l \rangle \sim \langle \eta, l \rangle$.

Suppose $v(p, \rho, l) = v(p, \eta, l)$ for some $p \in \mathcal{N}(\rho, l)$. Let $A = \mathcal{N}(\rho, l)$ and $B = \mathcal{N}(\eta, l)$. If $p \in B$, then p is nonfaulty at both points and, thus, $\langle \rho, l \rangle \sim \langle \eta, l \rangle$; suppose instead $p \in A - B$. Let $N = A \cap B$ and $F = P - (A \cup B)$; note that N is the set of processors nonfaulty in both runs and that F is the set of processors faulty in both runs. Furthermore, note that the sets $A - B$, $B - A$, N, and F partition of the set P of processors. Since all processors faulty in one or both runs are in $(A-B) \cup (B-A) \cup F$, we have $|(A - B) \cup (B - A) \cup F| \leq 2t$. Since $n > 2t$, we have $N \neq \emptyset$. Let q be a processor in N.

Let $\hat{\rho}$ and $\hat{\eta}$ be runs differing from ρ and η, respectively, only in that processors in $P - \{p\}$ receive messages from all processors in round l of $\hat{\rho}$ and $\hat{\eta}$ (and that all processors receive all messages after time l in $\hat{\rho}$ and $\hat{\eta}$). Note that A is the set of nonfaulty processors in ρ and $\hat{\rho}$ and that B is the set of nonfaulty processors in η and $\hat{\eta}$. Note also that p has the same view at time l in the four runs ρ, $\hat{\rho}$, $\hat{\eta}$, and η. It is clear that $\langle \rho, l \rangle \sim \langle \hat{\rho}, l \rangle$ since p is nonfaulty in both runs and has the same view at both points. To see that $\langle \eta, l \rangle \sim \langle \hat{\eta}, l \rangle$, let η' be a run differing from η only in that processors in $P - \{p, q\}$ receive messages from all processors in round l of η' (and assume that all processors receive all messages after time l in η'). Note that $\langle \eta, l \rangle \sim \langle \eta', l \rangle$ since q is nonfaulty in both runs and has the same view at both points. Similarly, note that $\langle \eta', l \rangle \sim \langle \hat{\eta}, l \rangle$ since there is a processor r distinct from p and q that is nonfaulty in both runs (r must exist since we always assume $n \geq t+2$) and has the same view at both points. Thus, $\langle \eta, l \rangle \sim \langle \hat{\eta}, l \rangle$.

We claim that the only missing (omitted) messages in $\hat{\eta}$ are between the sets $A - B$ and $B - A$ and between the sets P and F. To prove this, it suffices to show that there are no missing messages among processors in A or among processors in B. First, since $B = \mathcal{N}(\hat{\eta}, l)$ is the set of nonfaulty processors in $\hat{\eta}$, there are certainly no missing messages between processors in B in $\hat{\eta}$. Second, since $A = \mathcal{N}(\hat{\rho}, l)$ is the set of nonfaulty processors in $\hat{\rho}$, there are no missing messages between processors in A in $\hat{\rho}$. In particular, p's view at $\langle \hat{\rho}, l \rangle$ records the fact that there are no missing messages between processors in A for the first $l - 1$ rounds of $\hat{\rho}$ and that there are no missing messages from processors in A to p in round l. Since p has the same view at $\langle \hat{\eta}, l \rangle$, the same is true of $\hat{\eta}$. Since, furthermore, all processors in $P - \{p\}$ receive messages from all processors in round l of $\hat{\eta}$ (and in particular from processors in A), there are no missing messages between processors in A in $\hat{\eta}$.

Since the only missing messages in $\hat{\eta}$ are between the sets $A - B$ and $B - A$ and between the sets P and F, it is consistent with this failure pattern that the faulty processors in $\hat{\eta}$ are the processors in $P - A = (B - A) \cup F$ (and not the processors in $P - B = (A - B) \cup F$ as is actually the case in $\hat{\eta}$). Let $\hat{\eta}'$ be the run identical to $\hat{\eta}$, except that $\mathcal{N}(\hat{\eta}', l) = A$ instead of $\mathcal{N}(\hat{\eta}, l) = B$. Since q is nonfaulty in both runs $\hat{\eta}$ and $\hat{\eta}'$—note that $q \in N = A \cap B = \mathcal{N}(\hat{\eta}, l) \cap \mathcal{N}(\hat{\eta}', l)$—and certainly has the same view at both points, $\langle \hat{\eta}, l \rangle \sim \langle \hat{\eta}', l \rangle$. Note, furthermore, that p is nonfaulty in both $\hat{\rho}$ and $\hat{\eta}'$ since $p \in A$, and that p also has the same view at $\langle \hat{\eta}', l \rangle$ and $\langle \hat{\rho}, l \rangle$, so $\langle \hat{\eta}', l \rangle \sim \langle \hat{\rho}, l \rangle$. By the transitivity of the similarity relation, therefore, we have $\langle \rho, l \rangle \sim \langle \eta, l \rangle$. □

It follows from Lemma 10 that strong and weak common knowledge are equivalent with respect to the nonfaulty processors:

Theorem 11: *Consider any system in the general omissions model with $n > 2t$. The formula $W_\mathcal{N}\varphi \Leftrightarrow S_\mathcal{N}\varphi$ is valid in the system for every fact φ.*

Proof: It suffices to show that $W_\mathcal{N}\varphi \Rightarrow S_\mathcal{N}\varphi$ is valid in the system, since we already noted in Section 4 that $S_\mathcal{N}\varphi \Rightarrow W_\mathcal{N}\varphi$ is valid. Suppose $\langle \rho, l \rangle \models W_\mathcal{N}\varphi$. Lemma 2 implies that to prove that $\langle \rho, l \rangle \models S_\mathcal{N}\varphi$ we need only show that $\langle \rho', l \rangle \models \varphi$ for all $\langle \rho', l \rangle$ such that $\langle \rho, l \rangle \leadsto \langle \rho', l \rangle$. Since $\langle \rho, l \rangle \leadsto \langle \rho', l \rangle$ implies $\langle \rho, l \rangle \sim \langle \rho', l \rangle$ by Lemma 10, and since $\langle \rho, l \rangle \models W_\mathcal{N}\varphi$ implies $\langle \rho', l \rangle \models \varphi$ for all $\langle \rho', l \rangle$ such that $\langle \rho, l \rangle \sim \langle \rho', l \rangle$ by Lemma 1, we have $\langle \rho', l \rangle \models \varphi$ for all $\langle \rho', l \rangle$ such that $\langle \rho, l \rangle \leadsto \langle \rho', l \rangle$. Thus, $\langle \rho, l \rangle \models S_\mathcal{N}\varphi$. □

Moses and Tuttle prove that $\mathcal{F}_{MT}$ is an optimal protocol for simultaneous choice problems in the general omissions model, and they show how to implement tests for $W_\mathcal{N}\varphi$ for the nonfaulty processors in polynomial space, although they are not able to do so in polynomial time. The fact that weak and strong common knowledge are equivalent (when $n > 2t$) suggests that $\mathcal{F}_{MT}$ might also be an optimal protocol for *consistent* simultaneous choice problems in the general omissions model, as well as in the crash and send omissions models. But while they show that the nonfaulty processors can test for common knowledge of ok_j, it is not immediately clear that faulty processors can do the same. This raises two possible problem: (i) that the nonfaulty processors will know ok_1 and ok_2 are common knowledge while the faulty processors know only that ok_2 is common knowledge, resulting in the nonfaulty processors performing a_1 and the faulty processors performing a_2, and (ii) that the nonfaulty processors will know ok_1 is common knowledge at time k while the faulty processors will not know ok_1 is common knowledge until time $k+1$, resulting in the nonfaulty processors performing a_1 at time k and the faulty ones at time $k + 1$. Fortunately, we can use Lemma 10 again to prove that every processor—faulty or nonfaulty—knows the truth of $S_\mathcal{N}\varphi$, provided it does not know it is faulty (that is, its local state does not prove it must be faulty):

Lemma 12: *Consider any system in the general omissions model with $n > 2t$. The formula $\neg K_i(p_i \text{ is faulty}) \Rightarrow K_i(S_\mathcal{N}\varphi) \vee K_i(\neg S_\mathcal{N}\varphi)$ is valid in the system for every processor p_i and every fact φ.*

Proof: Suppose $\langle \rho, k \rangle \models \neg K_i(p_i \text{ is faulty})$. We prove that $\langle \rho, k \rangle \models S_\mathcal{N}\varphi$ implies $\langle \rho, k \rangle \models K_i(S_\mathcal{N}\varphi)$, and an analogous argument proves that $\langle \rho, k \rangle \models \neg S_\mathcal{N}\varphi$ implies $\langle \rho, k \rangle \models K_i(\neg S_\mathcal{N}\varphi)$.

```
nonfaulty ← true
repeat every round
    broadcast view to every processor;
    if K_i(p_i is faulty) then nonfaulty ← false
until nonfaulty and S_N ok_j holds for some j;
j ← min{k | S_N ok_k holds};
perform a_j;
halt.
```

Figure 2: The optimal protocol $\mathcal{F}'_{MT}$

To show that $\langle \rho, k \rangle \models \mathsf{K}_i(\mathsf{S}_\mathcal{N}\varphi)$, it is enough to show that $\langle \rho', k \rangle \models \mathsf{S}_\mathcal{N}\varphi$ for every point $\langle \rho', k \rangle$ such that p_i has the same view at $\langle \rho, k \rangle$ and $\langle \rho', k \rangle$. Given such a point $\langle \rho', k \rangle$, by Lemma 2, it is enough to show that $\langle \rho'', k \rangle \models \varphi$ for every point $\langle \rho'', k \rangle$ such that $\langle \rho', k \rangle \leadsto \langle \rho'', k \rangle$. Fix such a point $\langle \rho'', k \rangle$. Since $\langle \rho, k \rangle \models \neg \mathsf{K}_i(p_i \text{ is faulty})$, there exists a point $\langle \eta, k \rangle$ such that p_i is nonfaulty in η and p_i has the same view at both $\langle \rho, k \rangle$ and $\langle \eta, k \rangle$, and hence also at $\langle \rho', k \rangle$ and $\langle \eta, k \rangle$; in other words, $\langle \eta, k \rangle \leadsto \langle \rho, k \rangle$ and $\langle \eta, k \rangle \leadsto \langle \rho', k \rangle$. By Lemma 10, $\langle \eta, k \rangle \leadsto \langle \rho, k \rangle$ implies $\langle \eta, k \rangle \sim \langle \rho, k \rangle$, which implies $\langle \rho, k \rangle \leadsto \langle \eta, k \rangle$. It follows that

$$\langle \rho, k \rangle \leadsto \langle \eta, k \rangle \leadsto \langle \rho', k \rangle \leadsto \langle \rho'', k \rangle,$$

and hence, by Lemma 2, that $\langle \rho'', k \rangle \models \varphi$ since $\langle \rho, k \rangle \models \mathsf{S}_\mathcal{N}\varphi$. □

One consequence of this result is that any test for $\mathsf{S}_\mathcal{N}\varphi$ for the set of nonfaulty processors is almost a test for $\mathsf{S}_\mathcal{N}\varphi$ for the set of *all* processors: such a test correctly computes whether $\mathsf{S}_\mathcal{N}\varphi$ holds when applied to the view of *any* processor—faulty or nonfaulty—provided the processor's view does not *prove* that it must be faulty.

Corollary 13: *Consider any system in the general omissions model with $n > 2t$. Let M_φ be a test for $\mathsf{S}_\mathcal{N}\varphi$ for the nonfaulty processors. If $\langle \rho, k \rangle \models \neg \mathsf{K}_i(p_i \text{ is faulty})$, then M_φ on input $v(p_i, \rho, k)$ returns true if and only if $\langle \rho, k \rangle \models \mathsf{S}_\mathcal{N}\varphi$.*

Proof: Since $\langle \rho, k \rangle \models \neg \mathsf{K}_i(p_i \text{ is faulty})$, there is a point $\langle \eta, k \rangle$ such that p_i is nonfaulty at $\langle \eta, k \rangle$ and has the same view at $\langle \rho, k \rangle$ and $\langle \eta, k \rangle$. By Lemma 12, processor p_i knows the truth of $\mathsf{S}_\mathcal{N}\varphi$ at $\langle \rho, k \rangle$, so $\mathsf{S}_\mathcal{N}\varphi$ holds at $\langle \rho, k \rangle$ if and only if $\mathsf{S}_\mathcal{N}\varphi$ holds at $\langle \eta, k \rangle$. Since the test M_φ must return the same answer when given p_i's view at these two points as input (the views are the same), and since M_φ returns true at $\langle \eta, k \rangle$ if and only if $\langle \eta, k \rangle \models \mathsf{S}_\mathcal{N}\varphi$ (M_φ is a test for the nonfaulty processors), it follows that M_φ returns true at $\langle \rho, k \rangle$ if and only if $\langle \rho, k \rangle \models \mathsf{S}_\mathcal{N}\varphi$. □

In particular, the polynomial-space tests for $\mathsf{S}_\mathcal{N} ok_j$ for the nonfaulty processors given by Moses and Tuttle in the general omissions model are accurate tests for $\mathsf{S}_\mathcal{N} ok_j$ when given the view of any processor whose state does not prove that it is faulty. Notice that a processor p_i knows it is faulty (its view proves it is faulty) if and only if p_i is faulty in every failure pattern consistent with the messages recorded in p_i's view as missing, and this can also be checked in polynomial-space. Thus, the protocol $\mathcal{F}'_{MT}$ given in Figure 2 is an optimal protocol for a consistent choice running in polynomial space.

Theorem 14: *If C is an implementable, consistent simultaneous choice in the general omissions model with $n > 2t$, then $\mathcal{F}'_{MT}$ is an optimal protocol for C. Furthermore, if C is practical, then $\mathcal{F}'_{MT}$ runs in polynomial space.*

Proof: Let $\mathcal{P}$ be any protocol implementing $\mathcal{C}$, and let τ and ρ be corresponding runs of $\mathcal{F}'_{MT}$ and $\mathcal{P}$. Remember that every processor is following a test M_k for $S_\mathcal{N} ok_k$ to determine whether it should perform a_k, where M_k is a test for $S_\mathcal{N} ok_k$ that returns true at a point if and only if $S_\mathcal{N} ok_k$ holds at that point, given as input the view of a processor that does not know it is faulty. Furthermore, remember that a processor performs an action only if it does not know it is faulty (and hence only when these tests M_k are correct). First note that nonfaulty processors perform actions in τ at least as early as any processor does so in ρ: if any processor performs a_j at time l in ρ, then $S_\mathcal{N} ok_j$ holds at $\langle \rho, l \rangle$ by Lemma 5 and also at $\langle \tau, l \rangle$ by Corollary 7, so all nonfaulty processors (who certainly don't know they are faulty) perform an action at time l in τ if they have not already done so. To see that $\mathcal{F}'_{MT}$ actually implements $\mathcal{C}$, note the following:

1. Each processor clearly performs at most one of the a_j in τ and does so at most once.

2. Suppose a_j is performed by some processor p at time l in τ. Since p performed a_j at time l, it does not know it is faulty at time l or at any earlier time, so the tests M_k are correct through time l. It follows that l is the first round in which some ok_k is strong common knowledge, and that j is the least index k such that ok_k is strong common knowledge in round l. Thus, in particular, every nonfaulty processor will perform a_j at time l in τ.

3. Suppose a_j is performed by some processor p in τ. If p performs a_j at time l in τ, then $S_\mathcal{N} ok_j$ holds at time l in τ, so τ satisfies ok_j (since ok_j is a fact about the run).

4. Suppose no processor fails in τ. Then the same is true in the corresponding run ρ of the protocol $\mathcal{P}$ implementing $\mathcal{C}$, so some action must be performed at some time l in ρ. Thus, some action must be performed by the nonfaulty processors no later than time l in τ by the argument above. Since all processors are nonfaulty in τ, all processors perform this action at time l.

Thus, $\mathcal{F}'_{MT}$ is an optimal protocol for $\mathcal{C}$. □

Moses and Tuttle prove that testing for weak common knowledge (and thus for strong common knowledge, since the definitions are equivalent) is NP-hard in the general omissions model when $n > 2t$. Consequently, assuming $P \neq NP$, the protocol $\mathcal{F}'_{MT}$ cannot be implemented in polynomial time in this model. In fact, Moses and Tuttle prove that any optimal protocol for general simultaneous choice problems requires processors to perform NP-hard computations, and, thus, that optimal protocols are inherently intractable. Given the equivalence of weak and strong common knowledge, a reduction similar to that given by Moses and Tuttle shows that the same is true of optimal protocols for consistent simultaneous choice problems.

6 An Impossibility Result

While weak and strong common knowledge are equivalent in the general omissions model when $n > 2t$, they are *not* equivalent when $n \leq 2t$. It is not hard to show that some fact about the initial state must become weak common knowledge by time $t+1$ at the latest in any run [1,5,9]. This section shows that, in many runs, such facts *never* become strong common knowledge; this is true even in runs in which no failures occur. This implies that consistent simultaneous choice problems cannot be implemented in these systems.

This distinction between general and consistent simultaneous choices comes about because of the fact that, when $n \leq 2t$, the system can become partitioned into two sets of at most t processors such that processors within the same set communicate with no trouble, but processors in different sets never communicate. Since it is consistent with this failure pattern that either set of processors is the set of faulty processors, no processor can know whether or not it is faulty. In the context of general simultaneous choices, this is not a problem, since the behavior of the faulty processors is unimportant, so each set can simply behave as if it is the set of nonfaulty processors. In the context of *consistent* choices, however, this behavior is critical: because the processors may be isolated from important information and not know whether or not they are faulty, the correct processors can become "paralyzed" and unable to act.

To make this precise, we say that two sets A and B *partition* the set of processors if A and B are two nonempty, disjoint sets of processors such that $A \cup B = P$ and $|A|, |B| \leq t$. Given a failure-free run ρ, let ρ_k be the run identical to ρ except that, for every round $l > k$, no processor in B sends or receives a message to or from any processor in A in round l of ρ_k; note that A is the set of nonfaulty processors in ρ_k. We say that ρ_k *the result of partitioning ρ into A and B from time k*. The following lemma says that $\langle \rho, l \rangle \leadsto \langle \rho_k, l \rangle$ for every $k \geq 0$, and in particular for $k = 0$.

Lemma 15: *Consider any system in the general omissions model with $n \leq 2t$, and let ρ be any failure-free run of this system. Suppose A and B partition the set of processors, and suppose ρ_k is the result of partitioning ρ into A and B from time k. Then $\langle \rho, l \rangle \leadsto \langle \rho_k, l \rangle$ and $\langle \rho_k, l \rangle \leadsto \langle \rho, l \rangle$ for every $k \geq 0$.*

Proof: The proof is by reverse induction on k. For the base case of $k = l$, the result is trivially true since every processor has the same view at time l in ρ and ρ_l. In this case, we actually have $\langle \rho, l \rangle \sim \langle \rho_l, l \rangle$.

For $k < l$, suppose the inductive hypothesis holds for $k + 1$; that is, $\langle \rho, l \rangle \leadsto \langle \rho_{k+1}, l \rangle$ and $\langle \rho_{k+1}, l \rangle \leadsto \langle \rho, l \rangle$. Let η_1 be a run differing from ρ_{k+1} only in that no processor in B receives a message for any processors in A in round k of η_1. Note that A is the set of nonfaulty processors in both runs and that every nonfaulty processor has the same view at time l in both runs, so $\langle \rho_{k+1}, l \rangle \leadsto \langle \eta_1, l \rangle$ and $\langle \eta_1, l \rangle \leadsto \langle \rho_{k+1}, l \rangle$. Since the only missing messages in η_1 are between processors in A and processors in B, and since both A and B are of size at most t, it is consistent with the failure pattern in η_1 that either A or B is the set of faulty processors. Let η_2 be a run identical to η_1, except that now A is the set of faulty processors and B is the set of nonfaulty processors. Since every nonfaulty processor in η_1 has the same view at time l in the two runs, $\langle \eta_1, l \rangle \leadsto \langle \eta_2, l \rangle$. Similarly, since every nonfaulty processor in η_2 has the same view at time l in the two runs, $\langle \eta_2, l \rangle \leadsto \langle \eta_1, l \rangle$.[5] Now let η_3 be a run differing from η_2 only in that no (faulty) processor in A receives a message from any processor in B in round k of η_3. Because the set of nonfaulty processors is the same in η_2 and η_3, and because every nonfaulty processor has the same view at time l in both runs, we have $\langle \eta_2, l \rangle \leadsto \langle \eta_3, l \rangle$ and $\langle \eta_3, l \rangle \leadsto \langle \eta_2, l \rangle$. Note that η_3 is the result of partitioning ρ into B and A from time k, while the desired ρ_k is the result of partitioning ρ into A and B from time k; that is, the only difference between the two runs in that A is the set of nonfaulty processors in ρ_k, while B is the set of nonfaulty processors in η_3. On the other hand, since every processor (faulty or nonfaulty) has the same view at time l in both runs, it is clear that $\langle \eta_3, l \rangle \leadsto \langle \rho_k, l \rangle$ and $\langle \rho_k, l \rangle \leadsto \langle \eta_3, l \rangle$. It follows that $\langle \rho, l \rangle \leadsto \langle \rho_k, l \rangle$ and $\langle \rho_k, l \rangle \leadsto \langle \rho, l \rangle$, as desired. □

Using Lemma 15, we can now show that, even in failure-free runs, no fact about the input and the faulty processors—and thus no enabling condition ok_j—can become strong common knowledge.

Lemma 16: *Consider any system in the general omissions model with $n \leq 2t$. Let φ be a fact about the input and the faulty processors that is not valid in the system. If $\langle \rho, l \rangle$ is any point of any failure-free run ρ of the system, then $\langle \rho, l \rangle \not\models S_\mathcal{N} \varphi$.*

Proof: Let $\vec{i}$ be the input to ρ. Since φ is a fact about the input and the faulty processors that is not valid in the system, there is an input vector $\vec{i}' \in \mathcal{I}^n$ and a set $N' \subseteq P$ such that $\langle \rho', l \rangle \not\models \varphi$ for any run ρ' with input $\vec{i}'$ and with N' as its set of nonfaulty processors. Two input vectors $\vec{i}_a$ and

---
[5]Note that we do not have $\langle \eta_1, l \rangle \sim \langle \eta_2, l \rangle$, since there is no processor that is nonfaulty in both runs. For this reason, the proof does not apply when $\leadsto$ is replaced by $\sim$ and, thus, the consequences of this lemma apply only to strong common knowledge and not to weak common knowledge.

$\bar{\iota}_b$ are said to be *adjacent*—denoted $\bar{\iota}_a \leftrightarrow \bar{\iota}_b$—if they differ on the initial state of only one processor; that is, for some processor p_i, we have $\iota_{j,a} = \iota_{j,b}$ for all $j \neq i$. It is not hard to see that for some $m \geq 0$ there are vectors $\bar{\iota}_0, \bar{\iota}_1, \ldots, \bar{\iota}_m$ such that $\bar{\iota} = \bar{\iota}_0 \leftrightarrow \bar{\iota}_1 \leftrightarrow \cdots \leftrightarrow \bar{\iota}_m = \bar{\iota}'$. For every j, let ρ_j be the failure-free run with input $\bar{\iota}_j$.

To complete the proof, it is enough to show that $\langle \rho, l \rangle \rightsquigarrow \langle \rho_j, l \rangle$ for every $j \geq 0$. In particular, suppose $\langle \rho, l \rangle \rightsquigarrow \langle \rho_m, l \rangle$. Let ρ' be the run differing from ρ_m only in that no processor in $P - N'$ receives any message in round l of ρ'. Since $\bar{\iota}'$ is the input to ρ' and N' is the set of nonfaulty processors in ρ', we have $\langle \rho', l \rangle \not\models \varphi$. Furthermore, since processors in N' have the same view at time l in both runs, we have $\langle \rho_m, l \rangle \rightsquigarrow \langle \rho', l \rangle$. It follows that $\langle \rho, l \rangle \rightsquigarrow \langle \rho', l \rangle$ and yet $\langle \rho', l \rangle \not\models \varphi$; thus, $\langle \rho, l \rangle \not\models \mathsf{S}_{\mathcal{N}} \varphi$ by Lemma 2.

We proceed by induction on j to show that $\langle \rho, l \rangle \rightsquigarrow \langle \rho_j, l \rangle$ for every $j \geq 0$. For $j = 0$, we are done since $\rho = \rho_0$. For $j > 0$, suppose the inductive hypothesis holds for $j-1$; that is, $\langle \rho, l \rangle \rightsquigarrow \langle \rho_{j-1}, l \rangle$. By definition, ρ_{j-1} and ρ_j differ only in the input to some processor p. Consider any partition of P into sets A and B with $p \in B$, and let η_{j-1} and η_j be the result of partitioning ρ_{j-1} and ρ_j, respectively, into A and B from time 0. By Lemma 15, we have $\langle \rho_{j-1}, l \rangle \rightsquigarrow \langle \eta_{j-1}, l \rangle$ and $\langle \eta_j, l \rangle \rightsquigarrow \langle \rho_j, l \rangle$. Because η_{j-1} and η_j differ only in the input to p, and because there is no message from $p \in B$ to any processor in A in either run, all (nonfaulty) processors in A have the same view at time l in both runs, and we have $\langle \eta_{j-1}, l \rangle \rightsquigarrow \langle \eta_j, l \rangle$. It follows that $\langle \rho, l \rangle \rightsquigarrow \langle \rho_j, l \rangle$, as desired. □

It follows from Lemma 16 that there is no solution to any consistent simultaneous choice in this model:

Theorem 17: *In the general omissions model with $n \leq 2t$, no consistent simultaneous choice is implementable.*

Proof: Suppose by way of contradiction that some protocol $\mathcal{P}$ implements some consistent simultaneous choice $\mathcal{C}$ in the general omissions model with $n \leq 2t$. Let ρ be a failure-free run of $\mathcal{P}$. Since $\mathcal{P}$ implements $\mathcal{C}$, all processors must perform some action a_j at some time l in ρ, and $\langle \rho, l \rangle \models \mathsf{S}_{\mathcal{N}} ok_j$ by Lemma 5. But Lemma 16 says that this is impossible, since ok_j is a fact about the input and faulty processors. It follows that $\mathcal{P}$ does not implement $\mathcal{C}$ after all. □

Theorem 17 states that consistent simultaneous choices cannot be implemented in systems with general omission failures in which half or more of the processors can fail. This is in marked contrast to the results of Section 5.3, which show that such choices can be implemented in systems in which fewer than half of the processors can fail.

7 Conclusions

In this paper, we have studied optimal solutions to consistent simultaneous choice problems in a variety of simple failure models. Consistent simultaneous choice problems differ from more general simultaneous choice problems in that faulty processors must perform an action (if they perform any action at all) that is consistent and simultaneous with the actions performed by the nonfaulty processors. This additional requirement seems a natural one to make in systems with benign failures. In this paper, we have presented a complete study of optimal solutions to such problems in a number of simple failure models. In the crash and send omissions failure models, optimal protocols for simultaneous choice problems derived elsewhere [1,9] are also optimal protocols for consistent simultaneous choice problems and run in polynomial time. We have shown that, in the general omissions failure model, when fewer than half the processors are faulty (that is, when $n > 2t$), a simple modification of these optimal protocols are optimal protocols in this model as well. These protocols require exponential time, but any optimal protocol in this model is inherently intractable (it requires processors to perform NP-hard computations). Furthermore, in each of these models, the optimal protocols for the consistent version of a simultaneous choice problem halt at precisely the same time as the optimal protocols for the original version. Thus, we can obtain consistency in these models at no extra cost. Finally, we have shown that, in the general omissions failure model, when half or more of the processors may fail (that is, when $n \leq 2t$), there is no solution to any consistent simultaneous choice problem: processors cannot consistently coordinate actions even in failure-free runs.

One of the technical contributions of this work is exploring the relationship between the definition of common knowledge used by Moses and Tuttle [9] and the original definition proposed by Halpern and Moses [5]. Moses and Tuttle define weak common knowledge, and Halpern and Moses essentially define strong common knowledge (although their definition is formulated in terms of fixed sets and not nonconstant sets such as the set of nonfaulty processors). Because we have shown that the two definitions are equivalent in most of the cases considered by Moses and Tuttle, we have shown that their work could have been simplified slightly by using the less subtle definition of common knowledge originally proposed by Halpern and Moses. On the other hand, where we have shown the two definitions are different (in the general omissions model with $n \leq 2t$), consistent simultaneous choice problems cannot be solved, whereas Moses and Tuttle have shown that the original simultaneous choice problems *can* be solved. In this sense, our work shows precisely where the subtlety in the definition of strong common knowledge is required by the work of Moses and Tuttle; it also shows that the difference between strong and weak common knowledge is at the heart of the difference between the consistent and original versions of simultaneous choice problems.

In this work we have analyzed problems requiring simultaneous coordination of actions, but non-simultaneous coordination is also of interest. For example, Halpern, Moses, and Waarts [6] have considered the *Eventual Byzantine Agreement* problem, a problem in which correct processors must agree on the value of their output bits but need not choose these output bits at the same time. They show that solving such problems requires a variant of common knowledge called *continual common knowledge* and give a two-step method for transforming any solution to this problem into a round-optimal solution (optimal in the sense that no other solution outperforms it in all operating environments). Since they study this problem in two of the benign failure models considered in our paper, it is again interesting to consider consistent formulations of this problem. Using a definition of continual common knowledge strengthened in a way similar to the way we have defined strong common knowledge, Neiger [10] has generalized their work to a general class of nonsimultaneous choice problems requiring consistent coordination. In particular, he has generalized their transformation and so can transform any solution to a consistent coordination problem into an optimal solution. The number of steps in this transformation depends on the number of actions from which the processors must choose.

There still remain a few open problems to consider. First, the precise complexity of solutions to consistent simultaneous choice problems in the general omissions model when $n > 2t$ is still an open question: all we have established is that it is somewhere between NP and PSPACE. As noted by Moses and Tuttle [9], the precise complexity of solutions to *general* simultaneous choice problems in this model is also an open problem. Furthermore, the complexity of solutions to general problems is still open even when $n \leq 2t$, although we have established that solutions to *consistent* problems are impossible. Finally, we conjecture that the impossibility result of Section 6 applies to the nonsimultaneous *consistent coordination* problems mentioned above [10].

Acknowledgements

We thank Yoram Moses for enlightening discussions on the relationship between this work and the work of Moses and Tuttle and for pointing out a minor error in an earlier version of $\mathcal{F}'_{MT}$.

References

[1] Cynthia Dwork and Yoram Moses. Knowledge and common knowledge in a Byzantine environment: Crash failures. *Information and Computation*, 88(2):156–186, October 1990.

[2] Ajei Gopal and Sam Toueg. Reliable broadcast in synchronous and asynchronous environments (preliminary version). In J.-C. Bermond and M. Raynal, editors, *Proceedings of the Third International Workshop on Distributed Algorithms*, volume 392 of *Lecture Notes on Computer Science*, pages 110–123. Springer-Verlag, September 1989.

[3] Vassos Hadzilacos. *Issues of Fault Tolerance in Concurrent Computations*. Ph.D. dissertation, Harvard University, June 1984. Department of Computer Science Technical Report 11-84.

[4] Joseph Y. Halpern and Yoram Moses. A guide to the modal logic of knowledge and belief. In *Proceedings of the Ninth International Joint Conference on Artificial Intelligence*, pages 480–490. Morgan-Kaufmann, August 1985.

[5] Joseph Y. Halpern and Yoram Moses. Knowledge and common knowledge in a distributed environment. *Journal of the ACM*, 37(3):549–587, July 1990.

[6] Joseph Y. Halpern, Yoram Moses, and Orli Waarts. A characterization of eventual Byzantine agreement. In *Proceedings of the Ninth ACM Symposium on Principles of Distributed Computing*, pages 333–346, August 1990.

[7] Leslie Lamport, Robert Shostak, and Marshall Pease. The Byzantine generals problem. *ACM Transactions on Programming Languages and Systems*, 4(3):382–401, July 1982.

[8] C. Mohan, R. Strong, and S. Finkelstein. Methods for distributed transaction commit and recovery using Byzantine agreement within clusters of processors. In *Proceedings of the Second ACM Symposium on Principles of Distributed Computing*, pages 89–103, August 1983.

[9] Yoram Moses and Mark R. Tuttle. Programming simultaneous actions using common knowledge. *Algorithmica*, 3(1):121–169, 1988.

[10] Gil Neiger. Using knowledge to achieve consistent coordination in distributed systems. In preparation, July 1990.

[11] Gil Neiger and Sam Toueg. Automatically increasing the fault-tolerance of distributed algorithms. *Journal of Algorithms*, 11(3):374–419, September 1990.

[12] M. Pease, R. Shostak, and L. Lamport. Reaching agreement in the presence of faults. *Journal of the ACM*, 27(2):228–234, April 1980.

[13] Kenneth J. Perry and Sam Toueg. Distributed agreement in the presence of processor and communication faults. *IEEE Transactions on Software Engineering*, 12(3):477–482, March 1986.

[14] Richard D. Schlichting and Fred B. Schneider. Fail-stop processors: an approach to designing fault-tolerant computing systems. *ACM Transactions on Computer Systems*, 1(3):222–238, August 1983.

Agreement on the Group Membership in Synchronous Distributed Systems

Rogério de Lemos   Paul D. Ezhilchelvan

Computing Laboratory
University of Newcastle upon Tyne, NE1 7RU, UK.

Abstract
When a group of processors in a distributed system cooperate with each other on processing of a common task, it is often necessary for the non-faulty processors to have a mutually consistent knowledge of the set of processors that can be considered to be non-faulty. The set of non-faulty processors in the group - known as the *group membership* - will change for example when a processor crashes or when a crashed processor, after restart, joins the group. These changes should be known by all non-faulty processors as quickly as possible within a known bounded time interval. We present an algorithm by which non-faulty processors of a group of bounded size will be able to maintain a consistent and timely knowledge of the group membership. Processors in the group are assumed to execute the algorithm in a synchronous manner and at periodic intervals or cycles of some fixed length. In an execution of the proposed algorithm, every non-faulty processor knows of any processor failure within at most two cycles following the cycle in which the failure occurred, and a restarted processor can join the group in two cycles. At most less than half the number of processors are assumed to fail in any three consecutive cycles.
Keywords: group membership, distributed algorithms, broadcast networks, fault-tolerance.

1. Introduction

We consider a group of potentially faulty processors which communicate with each other only by message passing, and cooperate on processing of a common task. The set of non-faulty, or correctly functioning, processors - called the *group membership* - can change with time. The changes can be due to processor failures or due to failed processors being repaired and restarted. Non-faulty processors should be made aware of changes in the group membership promptly and should maintain a mutually consistent knowledge of the group membership. Having a consistent knowledge of the group membership is essential for non-faulty processors in the group to cooperate in processing a common task. Since a change in the group membership cannot be detected and known instantaneously by all non-faulty processors, the solution is to have a known and bounded interval within which a change in the group membership can be observed by all non-faulty processors.

The problem of maintaining a consistent and timely knowledge of the group membership among non-faulty processors within the context of a synchronous distributed system was first considered in /Cristian 88/, and was later studied in the context of TDMA based distributed real-time systems in /Kopetz 89/ /Ezhilchelvan 90/. In this paper, we study the group membership problem in the context of a synchronous distributed system where the processors are connected by a reliable broadcast network which guarantees message delivery in a bounded time interval; we present an algorithm by which non-faulty processors can maintain a consistent and timely knowledge of the group membership in the presence of processor failures and joins; it is assumed that less than half the number of processors in the group membership can fail in an interval of some bounded length. A faulty processor is assumed to suffer omission failures in sending and receiving broadcast messages and also to suffer 'crash' failures (halting to function). The group membership algorithm presented here, upon execution by processors in the group, will guarantee the following:

1. at any given time, non-faulty processors have *identical* knowledge of the group membership;

2. a processor failure or join will be known to all non-faulty processors in the group within a *known and bounded* interval following the processor failure or join;

3. a failed but not crashed processor will know of its send/receive omission failure in a *known and bounded* time interval.

The rest of the paper is organized as follows. The next section describes the chosen system architecture and failure assumptions involved in the design of the group membership algorithm. In section 3, the group membership algorithm is presented. Correctness of the algorithm is discussed in section 4 and section 5 concludes the paper.

2. System Architecture and Failure Assumptions

In the design of the algorithm, the processor group is assumed to be of a bounded size, where the maximum number of processors is fixed to be n, $n \geq 3$. The processors communicate with each other only by message passing and are connected by a broadcast network. The group membership algorithm takes advantage of some of the properties of the broadcast network which, if failure free, guarantees that messages transmitted are delivered at all destinations without any message corruption, and within a bounded interval of length δ, $\delta > 0$, where δ is measured according to any processor clock. All the processor's clocks are assumed to be kept synchronized within a known bound ε. Every broadcast message is (time) stamped, while being sent, with the reading of the local clock. Thus, if the broadcast network is failure free and the sending processor is functioning properly, then a broadcast message with timestamp T_S will be delivered to all destinations by $T_S + \delta + \varepsilon$ - according to any processor clock. We will also assume that broadcast messages are sequentially numbered at the source so that a receiving processor can detect the absence or loss of missing messages by monitoring the sequence numbers of messages received. The sequence numbering of messages at every processor is assumed to commence with a predetermined and known number.

Every processor in the group is assumed to have a process, called the *Received-Message-Monitor*, which monitors the sequence numbers of messages received from every other processor in order to detect the absence of lost messages. Note that the loss of a message or messages from a processor can be detected only by receiving a higher numbered message from that processor. Since broadcast messages are sequentially numbered before being sent, a lost message cannot have a timestamp that is greater than the timestamp of the message whose reception led to the detection of that message loss. Only for the purpose of the membership algorithm, a lost message will be considered to have the same timestamp as that of the message whose reception led to the detection of that message loss. For every message loss detected, the *Received-Message-Monitor* process will record the sequence number and the 'supposed' timestamp of the lost message. In addition, the process, after being started, will also record the sequence number of the very first message it receives from each processor in the group.

We will use the term 'cycle' to refer to the time interval where the processors in the group update their view of the group membership in a synchronous manner and at periodic intervals of length π, $\pi \geq \delta + \varepsilon$. The cycles are sequentially numbered and, throughout this paper, passage of time intervals will also be referred to in terms of these cycles. The first cycle will be said to begin or the zeroth cycle will be said to have ended when the processor group is initially formed, say, by T_0 according to the clock of any processor in the group. $T_i = T_0 + i\pi$ will denote the end of the ith cycle, $i \geq 0$. At clock time T_0, every processor in the group is considered to be functioning with an implication that an already crashed processor, if any, will be counted to have failed after the group is formed.

In the following, three major assumptions considered in the design of the algorithm are stated. They are concerned with the types of processor failures and the maximum number of failures allowed in a given interval.

Assumption A1

The broadcast network is assumed to be reliable and only processors can fail. The following three categories of processor failures are considered:

(i) *send failure*: a processor suffers a send failure, when it fails to send to the network the message it attempts to broadcast.

(ii) *receive failure*: a processor suffers a receive failure, when it fails to receive a message delivered to it by the broadcast network.

(iii) *crash failure*: a processor crashes, when it halts to function.

Since the network is assumed to be reliable, if there is a failure in interprocessor communication, then that must be due to processor failure. When a processor suffers a send failure, the message it attempted to broadcast will not be received by any other processor. If a processor detects the absence of a message from another processor, it cannot conclude whether it has suffered a receive failure or the other processor has suffered a send or a crash failure. Send/receive failures can occur in a processor when the message buffering and processing capacity of that processor is exceeded or due to transient failures in the processor's communication interface. We assume that these failures are restricted (by the use of message checksums or possibly digital signatures) to result in the affected messages being detected and ignored (as opposed to messages being undetectably or maliciously corrupted and delivered with incorrect contents to the application processes). A reliable broadcast network can be realized by having redundant communication channels and/or by redundant message passing on a given channel. Broadcast protocols which guarantee bounded message delivery time and atomicity properties in the presence of network failures and processor failures of the types considered here do exist in the literature /Babaoglu 85/ /Cristian 85/ /Cristian 90/. A crash failure is also called halting failure /Birman 87/, fail-silent failure /Powell 88/, and, with subtle differences, fail-stop failure /Schlichting 83/.

At any given time, a processor will be referred to as non-faulty, if it has not failed until that time ever since it started functioning as a member of the group. The maximum number of processors in the group which can fail during an interval of given length is assumed in the next assumption.

Assumption A2

The maximum number of processors that fail during any three consecutive cycles will be less than half the total number of non-faulty processors in the group at the beginning of the three cycle period, and the restarted processors that joined the group membership during that period.

If N_i is the number of non-faulty processors in the group by the end of cycle i, and J_i is the number of restarted processors that joined the group during cycle i, then, by A2, for any $i \geq 1$, $N_i \geq \lceil (N_{i-3} + J_{i-2} + J_{i-1} + J_i + 1)/2 \rceil$. If failed processors are not restarted, $N_i \geq \lceil (N_{i-3} + 1)/2 \rceil$ for any $i \geq 1$, where, for k, $-2 \leq k \leq -1$, $N_k = N_0$ (the number of non-faulty processors at the start of the first cycle) and $J_{k+1} = 0$. Since every processor in the group at the end of the zeroth cycle is considered to be functioning, $N_0 = n$.

Assumption A3

A processor that attempts to join the group will not suffer receive failure until it joins the group.

After joining the group, a processor can fail subject to assumption A2. We now proceed to present the algorithm.

3. The Group Membership Algorithm

The group membership algorithm can be viewed to perform two distinct operations: the localization of failed processors in the group and the restoration into the group of processors which are repaired or restarted. For the description of the algorithm we consider the processor group to be $\{P_1, P_2, ..., P_n\}$, with P_a, $1 \leq a \leq n$, being processors in the group. Each processor maintains a vector, called the Membership Status Vector and denoted as the MSV, which contains the processor's view of the membership status of all processors (including itself) in the group. Each ath entry, $1 \leq a \leq n$, in the MSV will indicate the membership status of processor P_a and the following notations will be used to indicate the status: the entry '1' for a processor will indicate that the processor is considered to be in the group membership, the entry '0' for a processor will indicate that the processor is removed from, and hence is not in, the group membership, and the entry 'a' for a processor will indicate that the processor is considered to be in the group membership but some of its recent messages are observed to have been lost. A processor's observation of loss of messages from another processor could be due to its receive failures or the other processor's send failures or both.

A processor's MSV is a vector of '1's at the start of the first cycle (i.e., when the group is initially formed) and is updated and renewed at every cycle through the execution of the membership algorithm: at clock time $(\delta+\varepsilon)$ before the end of every cycle, processors update their MSV's to account for any absence of messages detected during the cycle and exchange the updated MSV's with each other and compute a new MSV based on their own MSV and the MSV's received during the exchange. The computation of the new MSV will indicate any changes in the group membership and every member processor will consider the new MSV as the MSV for the next cycle. In the following presentation of the algorithm for a processor, say P_a, it is assumed that each statement of the algorithm can be executed by a processor in zero time (this will call for an appropriate increase in δ).

3.1. The Main Body of the Algorithm

3.1.1. The Algorithm

```
       task Membership (MSV: array [1..n] of char; Processor-Status: Status);
          var
             Om, m: integer; a, b: integer;
             T: Clock-Time; L:List of message-sequence-numbers;

                subtask Diagnose (MSV: array [1..n] of char; Om,m: integer);
                        ......
                subtask Restoration (MSV: array [1..n] of char; Om,m: integer);
                        ......
          begin
             a:= MyOrder-Number;
1a)        if Processor-Status = initialization then
                Start(Received-Message-Monitor); Fill-with-1(MSV);
                m := n;               /* m, the no. of '1's in the MSV, is initialised to n
                Om := n;              /* Om is the old value of m and is initialised to n
                T := T₀ + π;          /* T initialised to the end of first cycle
                Processor-Status := member;
1b)        fi;
2a)        if Processor-Status = restart then
                Restoration (MSV, Om, m);
                Processor-Status := member;
2b)        fi;
3a)        repeat
                wait Clock.Get = T-(δ+ε);
                   do
                      for b:=1 to n
                         do
                            if (b ≠ a and MSV[b] ≠'0') then
                               List-Absent-Messages(b, T-π-2(δ+ε), T-2(δ+ε), L)
```

```
                              if non-empty(L) then
                                   MSV[b] := 'a';
                              fi;
                         fi;
                    od;
               Send-for-Broadcast (MSV);
               wait Clock.Get = T;
                    Diagnose (MSV, Om, m);
               T := T + π;
          od;
3b)  until (Processor-Status = failed or m < 3);
     if (m < 3) then
          Raise-Exception (no-quorum) fi;
     endtask {Membership};
```

3.1.2. Explanation

The *Membership* task has two parameters which are also global variables within the host processor (P_a). The parameter *Processor-Status* indicates the status of the processor with respect to the group membership: it is set to *initialization*, while the group is being formed; it is set to *member*, if the processor considers itself to be in the group membership; it is *failed*, if the processor, due to its failure, removes itself from the group membership; and, it can also be *restart* when the processor is being restarted.

When the *Membership* task is invoked the *Processor-Status* can either be set to *initialization* or *restart*. When *Processor-Status* is set to *initialization* the first block of statements (from 1a to 1b) are executed: the process *Received-Message-Monitor*, which mainly detects and records any message loss, is started; the MSV is initialised to a vector of '1's by *Fill-with-1()* operation; the integer m, the number of '1's in the MSV, is set to n - the maximum size of the group and the integer Om, the old value of m or the number of '1's in the MSV prior to the current one, is also initialised to n; the variable T is set to $T_0 + π$, and the *Processor-Status* to *member*. When *Processor-Status* is set to *restart* the second block of statements (from 2a to 2b) are executed: the *Restoration* subtask is executed to prepare the processor to join the group by computing ₃ first MSV; and the *Processor-Status* is set to *member*. After the execution of either of these blocks the periodic execution of the *Diagnose* subtask commences at the third block of statements (from 3a to 3b). Before the execution of the *Diagnose* subtask, the MSV is updated to account for any absence of message(s) whose loss have been detected by receiving message(s) with timestamp(s) T_s, T-π-$2(δ+ε) < T_s ≤ T$-$2(δ+ε)$:

The procedure, *List-Absent-Messages*(b, T-π-$2(δ+ε)$, T-$2(δ+ε)$, L), upon invocation, will return in L the list of sequence numbers of P_b's messages that were recorded (by the process *Received-Message-Monitor*) to be absent with the supposed timestamp in the range (T-π-$2(δ+ε)$, T-$2(δ+ε)$]. The list L will be empty, if no message is detected to be absent. Recall that a message with timestamp T_s is guaranteed to be delivered before local clock time $T_s+(δ+ε)$. Therefore, in the executions of the algorithm at a given clock time, the values of L returned for a given P_b will be identical for any two non-faulty processors in the group. For every P_b, $1 ≤ b ≤ n$ and $b ≠ a$, if P_a gets a non-empty L for P_b, it makes the bth entry in its MSV to be 'a'. (Actually, the entry 'a' will be suffixed with the message sequence number(s) in L. If P_a's entry for P_b is already an 'a', the new suffix will be appended with the old one. For the sake of simplicity in the presentation, the suffix for 'a's are dropped throughout the paper and will be recalled when necessary).

The updated MSV is then sent to be broadcast to other processors in the group. The message containing the updated MSV is broadcast, like any other application message, with a timestamp and an appropriate sequence number.

At time T, the *Diagnose* subtask is executed to obtain the new membership of the group. During the execution of the subtask *Diagnose*, if a processor diagnoses itself to have failed, the *Processor-Status* is set to *failed* which will halt the current execution of the *Membership* task. The execution will also be stopped if m is less than three in which case an exception is raised to indicate the absence of quorum to have a processor group.

Referring to the figure 1 below, the repetitive execution of the third block of statements can be seen in detail. (In the figure, n is assumed to be much larger than $(\delta+\varepsilon)$.) At $T_i-(\delta+\varepsilon)$,

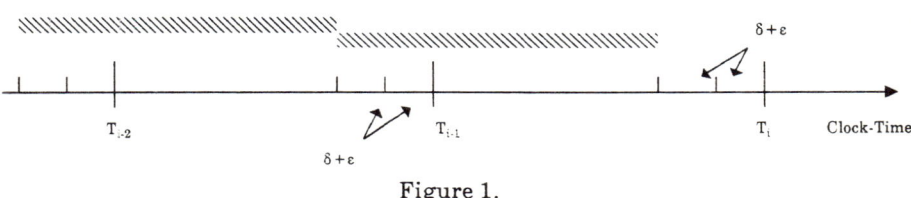

Figure 1.

$i \geq 1$, a processor with *Processor-Status* = member will update its MSV computed at T_{i-1}. This update will take into account of missing messages from every other processor in the timestamp range $(T_{i-1}-2(\delta+\varepsilon), T_i-2(\delta+\varepsilon)]$ - shown in the figure by the shaded band. (In case of $i=1$, the lower bound in the range will be T_0.) The updated MSV is broadcast to other processors and a new MSV is computed, at clock time T_i, based on the MSV's received.

3.2. The Diagnose of Faulty Processors

In this subsection, we present the algorithm which enables a processor to localize failed processors in the group, through the exchange of MSV's between the members of the group. The algorithm (for a processor P_a) is presented below and, in any execution, the following conditions will be met: (1) MSV's used by any two non-faulty processors in a given cycle will be identical; (2) a processor failure will be known to every non-faulty processor in a known and bounded time interval; (3) a failed but not crashed processor will know of its failure in a known and bounded time interval.

3.2.1. The Algorithm

```
    subtask Diagnose (MSV: array[1..n] of char; Om, m: integer);
      var
        NewMSV, RecMSV: array [1..n] of char;
        a, b, c: integer;
        Nm: integer;
```

```
            begin
                a := MyOrder-Number;
                Nm := 0;                            /*Nm is the number of '1's in the new MSV
                for b := 1 to n
                    do
                        c := b;
                        if (b ≠ a) then
1a)                         if (MSV-Received-from (P_b)) then
                                RecMSV := MSV from P_b;
                                if MSV ≠ RecMSV then
                                    NewMSV[c] := '0';
                                else
                                    NewMSV[c] := '1';
                                    Nm := Nm + 1;
                                fi;
                            else
                                if MSV[c] ≠ '1' then
                                    NewMSV[c] := '0';
                                else
                                    NewMSV[c] := 'a';
                                fi;
                            fi;
                        else
                            NewMSV[c] := '1';  Nm := Nm + 1;    /* entry for '1' itself
                        fi;
1b)                 od;
2a)             if Nm ≥ ⌈(Om+1)/2⌉ then
                    MSV := NewMSV; Om := m; m := Nm;
                else
                    Processor-Status := failed;
                    Fill-with-0(MSV);
2b)             fi;
            endsubtask {Diagnose};
```

3.2.2. Explanation

The MSV is continually renewed by repeated executions of the subtask *Diagnose*. In the first block of statements (from 1a to 1b), the *NewMSV* - the MSV to be used for the next cycle - is being formed with each of its entries being derived from the MSV's, or their absence, from the corresponding processors: if the MSV broadcast by a processor, say P_b, is received by P_a, then the boolean *MSV-Received-from(P_b)* will return true. If the received MSV, the *RecMSV*, from say P_b is the same as the MSV, then the entry for P_b in the *NewMSV* will be '1'; the entry will be '0', if the *RecMSV* is different. Two MSV's will considered to be the same, if they have identical entries at all respective positions. (Two 'a's will be identical if their suffixes are identical.)

While the boolean *MSV-Received-from(P_b)* is not true, the entry for P_b in the *NewMSV* will be 'a', if the entry for P_b in the MSV is '1'; it will be '0', if the entry for P_b in the MSV is 'a' (indicating that some of P_b's messages have already been observed to be missing) or '0' (indicating P_b is already considered to be removed from the group membership). (When 'a' is entered for P_b due to *MSV-Received-from(P_b)* being false, the 'a' will be suffixed with "msv", since it is not possible, at this time, to determine the sequence number of P_b's missing message that contains its updated MSV.)

In the second block, if Nm - the number of '1's in the *NewMSV* - is at least the majority in Om, MSV is set to *NewMSV*; if the *NewMSV* does not have enough number of '1's, then the *Processor-Status* is set to *failed* which will bring the execution to a halt.

3.3. The Restoration of Faulty Processors

In this subsection, we present the algorithm which enables a restarted processor to join the group. A restarted processor will be said to join the group in cycle i, when its entry is known to all non-faulty processors in the group since the beginning of that cycle. A restarted processor can be a new or repaired processor or a former member processor that restarts itself in the hope that its failure will not occur again. It does not suffer any receive failures until it joins the group (assumption A3) and is assumed to have its clock in the required ε-synchronism. A restarted processor, which may decide to join the group at any time, will be referred to as an incoming processor until it actually joins the group. (An incoming processor may suffer a send failure and, as a result, it may have to make a few attempts before it eventually joins the group.)

When an incoming processor attempts to join the group, processors which consider themselves to be in the group membership, will be broadcasting and renewing their MSV's. An incoming processor restores its view on the existing group membership by obtaining an MSV for itself and by broadcasting that MSV to processors in the group to declare its intention to join the group. The activities carried out by an incoming processor to compute an MSV are described here by referring to the figure 1: Suppose that an incoming processor is to declare its joining intentions at T_i-$(\delta+\varepsilon)$, for $i \geq 2$. In order to be able to declare its joining intensions at its clock time T_i-$(\delta+\varepsilon)$, the incoming processor must start its *Received-Message-Monitor* process at least as early as T_{i-2}-$2(\delta+\varepsilon)$-ε according to its clock so that messages received with timestamp as small as T_{i-2}-$2(\delta+\varepsilon)$ can have their sequence numbers monitored. At T_{i-1}, it computes an MSV - called its first MSV - based on the MSV's it received and by using an algorithm, called the *Restoration* subtask, described below. This algorithm is designed to provide, upon execution, an incoming processor with an MSV that is the same as the one computed by a non-faulty processor at T_{i-1}, provided that the incoming processor does not suffer receive failures.

After having computed its first MSV, an incoming processor proceeds to the next cycle (i.e. the ith cycle) and at clock time T_i-$(\delta+\varepsilon)$ it behaves as if it were already in the group (see 3a-3b in the presentation of the *Membership* task): it updates its first MSV to account for any missing message from every given processor in the timestamp range $(T_{i-1}$-$2(\delta+\varepsilon), T_i$-$2(\delta+\varepsilon)]$; it broadcasts its updated MSV and receives other processors' MSV's; and, at the end of the ith cycle it computes its second MSV in the manner described in the *Diagnose* subtask. If the number of '1's in its second MSV is as large as required, it considers itself to have joined the group.

When an incoming processor broadcasts its first updated MSV, it declares its attempts to join the group. Since it suffers no receive failures, every non-faulty processor in the group finds the incoming processor's MSV to be the same as its own and considers the incoming processor to have become another member in the system by replacing '0' in its MSV by '1'; on the other hand, if the incoming processor fails to broadcast its first updated MSV, in the subsequent execution of the *diagnose* subtask it will find every non-faulty processor's MSV having a '0' (not '1', as it has) for itself. An incoming processor that broadcasts its first updated MSV at T_i-$(\delta+\varepsilon)$, will join the group in cycle (i+1). Thus, for an incoming processor to join the group at T_i, it must have started its *Received-Message-Monitor* process at its clock time no later than T_{i-2}-$2(\delta+\varepsilon)$-ε. In the following, we discuss and present the *Restoration* subtask by which an incoming processor computes its first MSV.

3.3.1. Design Background

Let P_a be an incoming processor which started monitoring the group communication messages as early as T_{i-2}-$2(\delta+\varepsilon)$-ε according to its clock. At its clock time T_{i-1}, P_a would have received the MSV's broadcast by member processors by the (near) end of (i-2)th cycle and (i-1)th cycle. Let the MSV's received by P_a be classfied as the *old* MSV's - the ones broadcast by the end of (i-2)th cycle, and the *recent* MSV's - the ones broadcast by the end of (i-1)th cycle. (The distinction of MSV's is possible for P_a due to different message timestamps). Out of these MSV's and what has been recorded by its *Received-Message-Monitor* process, P_a should compute its first MSV at the end of cycle i-1 which will have to be the same as the MSV computed by any non-faulty processor in the group. To do so, it first constructs a matrix, called the Restoration Syndrome Matrix and denoted as the RSM, such that the bth row, $1 \leq b \leq n$ and $b \neq a$, of the RSM is the recent MSV from P_b or a row of 'a's if P_a has not received a recent MSV from P_b, and the ath row is a row of '0's. The RSM, so formed, will be modified and the information recorded by P_a's *Received-Message-Monitor* process about messages missing in the timestamp range (T_{i-2}-$2(\delta+\varepsilon)$, T_{i-1}-$2(\delta+\varepsilon)$] will be used for the modification.

Recall that the *Received-Message-Monitor* process of P_a also records the sequence number of the very first broadcast message received from every processor. Let S_c be the first message which P_a received from P_c. Any failure(s) of P_c in sending message(s) with sequence number(s) larger than S_c will be eventually detected by P_a, so long as P_c does not crash before detection. However, the send failures of P_c, if any, that occurred before the message with sequence number S_c was sent, will not be detected by P_c. Also, if a processor P_b, $b \neq c$, reports in its MSV the absence of P_c's message with a sequence number smaller than S_c, then P_a, unlike a non-faulty member processor, cannot determine by itself whether P_c suffered a send failure or P_b suffered a receive failure. (Recall that the entry 'a' in a processor's MSV is suffixed with the sequence number of messages found to be missing by that processor.) However, if the recent MSV of P_b reports of the absence of P_c's message with a sequence number larger than S_c, then P_c can ascertain whether it is due to P_c's send failure or P_b's receive failure. Thus, among the processor failures diagnosed by non-faulty member processors at their clock time T_{i-1}, it is easy for P_a to detect the failures that occurred after P_a received the first messages from the concerned processors, while it is not so straight forward to detect the ones that occurred earlier on.

The modification of RSM by P_a - carried out in order to compute the MSV computed by non-faulty member processors - will proceed in two stages. The first stage of modification is to make RSM reflect the failures that occurred after P_a received a message from the concerned processors and, the second stage is to make it reflect the failures occurred earlier to P_a receiving the first messages from the concerned processors. In the first stage, the following is carried out: for every P_b, $1 \leq b \leq n$ and $b \neq a$, if an old MSV from P_b has not been received, then the bth row of RSM is made into a row of '0's. This means that P_b has crashed or suffered a failure to send its MSV at its clock time $T_{i-2}-(\delta+\varepsilon)$ and that the bth row is made into a row of '0's to give '0' for P_b in the first MSV being computed by P_a. If P_b, on the other hand, has sent its old MSV, then for every c, $1 \leq c \leq n$, $c \neq b$ and $c \neq a$, RSM[b][c] is verified for being *inconsistent* with what is recorded by the *Received-Message-Monitor* process. Note that RSM[b][c] will represent P_b's view about P_c. RSM[b][c] is said to be *inconsistent*, if the following is true: RSM[b][c] reports the absence of a message whose sequence number is larger than S_c and is not recorded by P_a's *Received-Message-Monitor* process (i.e., RSM[b][c] is 'a' with suffix containing a sequence number that is larger than S_c and is not recorded by *Received-Message-Monitor* process); or, RSM[b][c] is '1' while P_a has detected the absence of a message with the supposed timestamp in the range $(T_{i-2}-2(\delta+\varepsilon), T_{i-1}-2(\delta+\varepsilon)]$, It is possible for P_b not to observe P_c's send failure and thereby to have RSM[b][c]='1', when P_b fails to receive every one of P_c's messages that were broadcast following the send failure. If RSM[b][c] is *inconsistent* with what P_a's *Received-Message-Monitor* process has recorded for P_c's messages, then P_a will replace the bth row into a row of '0's.

Having thus completed the first stage, P_a proceeds to the next stage by constructing a vector, called Summation Vector and denoted as SV, such that SV[c], $1 \leq c \leq n$, is the number of '1's in the cth column of matrix RSM. Recall that RSM[b][c] will represent the view of P_b about P_c. If for some c, $1 \leq c \leq n$, SV[c]=1, it will (be later shown to) imply that only RSM[c][c] is 1 in the cth column (and that only P_c views itself to be correct). This means that the send failure of P_c has been detected by every non-faulty member processor in the group and therefore this failure must have happened before P_a received its first message from P_c. If SV[c]=0, either P_c is not in the group or every processor has an 'a' for P_c in its recent MSV and P_c itself has failed to send its MSV in cycle i-1. If SV[c]>1, then it implies that P_c broadcast its MSV in cycle i-2 and there exists at least one other processor in the group which has agreed with that MSV and detects no absence of messages from P_c at $T_{i-1}-(\delta+\varepsilon)$. This means that only processors such as P_c for which SV[c]>1, can be considered by P_a to be eligible to be in the group membership at the end of cycle i-1.

However, among these 'eligible' processors, some may have suffered a failure to send their recent MSV in cycle i-1 or a receive failure that could not be detected by P_a in the first stage of modification. Such processors will be identified by P_a in the following manner: if no recent MSV has been received (identified by a row of 'a's in the RSM) from an eligible processor, then that processor will be considered to have failed to send its recent MSV. To identify eligible processors that suffered a receive failure that was not detected in the first stage of modification, P_a will inspect every eligible processor's row in the RSM. If a row has 'a' for another eligible processor and is not a row of 'a's, then that row is replaced by a row of '0's.

After these modifications, the diagonal elements of the RSM will be chosen to form the respective entries in the required MSV for P_a. The algorithm is presented below with provisions for treating the first MSV's broadcast by other incoming processors, if any.

3.3.2. The Algorithm

```
    subtask Restoration (MSV: array [1..n] of char; Om, m: integer);
      var
        Flag-Join: boolean;
        a, b, c: integer;
        RefMSV: array [1..n] of char;
        RSM: array [1..n] of array [1..n] of char;
        SV: array [1..n] of integer;
      begin
        a := MyOrder-Number; Flag-Join := false;
        Start(Received-Message-Monitor);
        Estimate-Joining-Time (T);
        wait clock.get = T-π
1a)     for b := 1 to n                                  /* RSM formed
          do  if (b ≠ a) then
                   if (recent-MSV-Received-from (Pb)) then
                        RSM[b] := recent MSV from Pb;
                   else Fill-Row-with-a(b);
                   fi;
              else Fill-Row-with-0(b);
              fi;
1b)       od;
2a)     for b := 1 to n                                  /* Stage-1 modification
          do  if (Old-MSV-Absent-from (Pb)) then
                   Fill-Row-with-0(b)
              else for c := 1 to n
                        do   if (inconsistent(RSM[b][c])) then
                                 Fill-Row-with-0(b) fi;
                        od
              fi;
2b)       od
3a)     Calculate-SV;                                    /* SV computed
        for c := 1 to n
          do  if (SV[c] = 1) then
                   RSM[c][c] := '0'
              fi;
              if (SV[c] > 1                              /*Pc is eligible and has
                  and Suffered-Receive-Failure(c)        /*an 'a' for another
                  and RSM[c][c] ≠ 'a') then              /*eligible processor and
                   RSM[c][c] := '0'                      /*has sent its recent
              fi;                                        /*MSV, so...
              if RSM[c][c] = '0' and RSM[c] ≠ Row-of-0s then
                                                         /* broadcast from another
                   Flag-Join := true; SV[c] := -1;  /* incoming node identified
              fi;
              MSV[c] := RSM[c][c];  /* form MSV with diagonal elements of RSM
3b)       od;
```

4a)   c := 1;
        while MSV[c] ≠ '1' **do** c := c + 1 **od**;   /*look for a member,P_c, in the new MSV
        RefMSV := RSM[c];
        Om = Number-of-'1's-in-RefMSV;
        if Flag-Join = true **then**
            for c := 1 **to** n
                do   **if** SV[c] = -1 **and** RSM[c] = RefMSV
                        then MSV[c] := '1' **fi**;
                od;
        fi;
4b)   m := Number-of-'1's-in-MSV;
        endtask {Restoration};

3.3.3. Explanation

A processor starts the *Received-Message-Monitor* process which, in addition to recording the sequence number and the supposed timestamp of every missing message, records, where possible, the sequence number of the very first message received from every processor in the group. It computes, as T, the earliest time at which it can join the group. If T_m is its clock time at which it started the *Received-Message-Monitor* process, then T will be chosen such that $T = T_e + 2\pi$, where T_e is the smallest value of its clock time at which a cycle has been observed to have ended and $T_e - T_m \geq 2(\delta + \varepsilon) + \varepsilon$. In other words, T is computed such that the sequence numbers of received messages can be monitored at least from $T-2\pi-2(\delta+\varepsilon)-\varepsilon$ and T denotes the end of a cycle. (The MSV's received in the intervals $[T-2\pi-(\delta+\varepsilon)-\varepsilon, T-2\pi]$, $[T-\pi-(\delta+\varepsilon)-\varepsilon, T-\pi]$ will be respectively called old and recent.) At clock time $T-\pi$, the matrix RSM is formed in the first block of statements (statements 1a to 1b). The instructions *Fill-Row-with-0(b)* and *Fill-Row-with-a(b)*, upon execution, will have the bth row of the RSM made into a row of '0's and a row of 'a's respectively. The RSM subsequently enters modification in the second block (statements 2a to 2b).

In the third block, every processor P_c for which SV[c] > 1, is verified for having suffered a receive failure that could not be detected by the *Received-Message-Monitor*. RSM[c][c] = 'a' will imply that recent MSV from P_c was not received and P_c is to get an entry 'a' in the first MSV. The boolean *Suffered-Receive-Failure(c)* will become true, if RSM[c][c] ≠ 'a' and the cth row has an 'a' for another eligible processor i.e., for some d, d ≠ c, RSM[c][d] = 'a' and SV[d] > 1. If RSM[c][c] = '0' and the cth row is not a row of '0's, P_c is marked to be an incoming processor broadcasting its first MSV by setting SV[c] = -1. (The MSV of a member processor will have '1' for itself.). The diagonal elements of the RSM form the MSV. In the fourth block, *RefMSV* is taken to be the row in the RSM of a processor whose entry in the newly computed MSV is '1'. The entries for incoming processors, if any, are decided by comparing their rows in the RSM with *RefMSV*. (The entry for an incoming processor will always be '1' due to A3. Yet this step is included here, as A3 can be, and will be, removed in the future version of the paper). Om and m are computed to be the number of '1's in the *RefMSV* and the final form of the new MSV respectively.

4. Correctness of the Algorithm

In showing the correctness of the algorithm, we make two assumptions just to simplify the presentation of the correctness arguments: no processor will broadcast more than one message with the same timestamp; and, a processor takes zero time to execute the instructions of the algorithm. By the second assumption, a message containing an MSV that was updated at clock time $T_i-(\delta+\varepsilon)$ will be broadcast with the timestamp $T_i-(\delta+\varepsilon)$. The first assumption will enable us to deal with messages only in terms of message timestamps and therefore to avoid the explicit use of message sequence numbers in presenting the correctness arguments.

The following notations will be used: a cycle i, $i \geq 1$, which terminates for a processor at its clock time T_i, will be denoted as C_i; the membership status vector, MSV, computed by a processor at the end of C_i, $i \geq 1$, will be denoted as the MSV_i of that processor and m_i will denote the number of '1's in the MSV_i of a processor. A processor's MSV_i, after being updated at clock time $T_{i+1} - (\delta+\varepsilon)$, will be called the updated MSV_i of that processor. N_i will denote the total number of processors that remained non-faulty upto the end of cycle C_i. A processor will be said to have remained non-faulty until the end of C_i, if that processor, ever since it joined the group, has suffered no send failures until it broadcast its updated MSV_{i-1}'s at its clock time $T_i-(\delta+\varepsilon)$ and no receive failure in receiving messages broadcast with timestamp T_s, $T_s \leq T_i-(\delta+\varepsilon)$, and does not crash until its clock time T_i. $MSV_0, m_0,$ and N_0 will denote the respective quantities at the start of the first cycle. By the algorithm, MSV_0 of a functioning processor is to be initialised to a vector of '1's and m_0 to n - the number of processors in the group. Recall that all processors in the group at the start of the first cycle are considered to be functioning (with an implication that any processor which has already crashed will be considered to fail in the first cycle i.e. before the first execution of the algorithm). Thus, $N_0 = n$. When the group is initially formed, every processor in the group will initialise its Om - denoted here as m_{i-1} - to n. Finally, a note on terminology: we will say that a processor suffered a send (or receive) failure in a cycle, say C_{i-2}, $i > 2$, if it suffers a failure in sending (or receiving) a message with timestamp T_s, $T_{i-3}-(\delta+\varepsilon) < T_s \leq T_{i-2}-(\delta+\varepsilon)$. A processor will be said to have failed in cycle C_{i-2}, if it suffered a send or receive failure in that cycle or it crashed at its clock time T, $T_{i-3} < T \leq T_{i-2}$. (See that such a processor won't be in N_{i-2}).

Before presenting the correctness of the algorithm, it is useful to make the following observation: Suppose that a processor suffers a send failure in cycle C_{i-2}. This failure can result in a discontinuity in its message sequence numbering being observed by another processor only when the second processor receives a message with higher sequence number from the failed processor. Thus, a non-faulty processor may not indicate in its MSV a loss of message from a processor that suffered a send failure in C_{i-2}, until its clock time $T_{i-1}-(\delta+\varepsilon)$ before which time it was to receive the updated MSV_{i-3} from the failed processor. Similarly, if a processor suffers a receive failure in cycle C_{i-2}, it may not indicate in its MSV, until its clock time $T_{i-1}-(\delta+\varepsilon)$, a loss of message from the processor whose message it failed to receive. This means that (barring an exceptional case discussed in the proof for lemma 2) there can be an interval of length at most 2π between the occurrence of a failure and the start of agreement over the observation of message loss caused by that failure.

The correctness of the algorithm is discussed by indicating, through a series of lemmas, that the following conditions are met in any execution of the algorithm: (B1) MSV_i of any two non-faulty processors will be identical; (B2) a processor failure will be known to a non-faulty processor in two cycles following the cycle during which the failure happened; (B3) a failed but functioning processor will know of its failure no later than the next cycle. Due to space limitations, many lemmas are not formally proved but are merely stated. First, we will consider the algorithm without processor restarts.

4.1. Algorithm without Processor Restarts

When failed processors are not restarted, assumption A2 will imply that $N_i \geq \lceil (N_{i-3}+1)/2 \rceil$ for every i, $i \geq 1$, where $N_k = N_0$ for any $k < 0$. The first lemma is about meeting B1.

Lemma 1

The membership status vectors, MSV_i's, $i \geq 1$, of non-faulty processors will be identical in any execution of the algorithm.

Proof

By the algorithm MSV_0's are initialised to a vector of '1's. Due to bounded message delivery time, bounded clock synchronisation, and the send failure assumption of A1, what is (not) received by one non-faulty processor by its clock time $T_i-(\delta+\varepsilon)$ is also (not) received by any other non-faulty processor by that time according to its clock. Hence the lemma.

Remark: A non-faulty processor will have '1' in its MSV_i and updated MSV_i, $i \geq 1$, for every other non-faulty processor in the group. The following lemmas are stated for a single non-faulty processor, say P_a, $1 \leq a \leq n$, since at the end of every C_i, $i \geq 1$, MSV_i's are identical for all non-faulty processors. The next two lemmas are concerned with meeting B2.

Lemma 2

A processor, say P_b, $1 \leq b \leq n$ and $b \neq a$, that fails during C_{i-2}, $(i-2) \geq 1$, will be diagnosed by any non-faulty processor P_a no later than the end of C_i, given that P_a has $m_j \geq \lceil (m_{j-2}+1)/2 \rceil$, at the end of every C_j, $1 \leq j \leq i$, during the execution of the algorithm.

Proof

The lemma can be seen to be true from the algorithm: when P_b suffers a send failure, P_a will have '0' for P_b in its MSV_{i-2} or in its MSV_{i-1} depending on whether P_a is able to observe a discontinuity in the sequence numbering of P_b's messages at its clock time $T_{i-2}-(\delta+\varepsilon)$ or $T_{i-1}-(\delta+\varepsilon)$ respectively. When P_b suffers a receive failure in receiving a message from P_c, $c \neq b$, it will have an 'a' for P_c in its updated MSV_{i-3} or in its updated MSV_{i-2} depending on whether it is able to observe a discontinuity (caused by its own receive failure) in the sequence numbering of P_c's messages at its clock time $T_{i-2}-(\delta+\varepsilon)$ or $T_{i-1}-(\delta+\varepsilon)$ respectively; If P_b does not suffer send failures in broadcasting its updated MSV's, P_a will have '0' for P_b in its MSV_{i-2} or in its MSV_{i-1} depending on whether P_b had an 'a' for P_c in its updated MSV_{i-3} or in its updated MSV_{i-2} respectively. Suppose that P_b suffers a receive failure and does not observe any discontinuity in message sequence numbering at its clock time $T_{i-2}-(\delta+\varepsilon)$. If P_b crashes in the next cycle or fails to send its updated MSV_{i-2} at $T_{i-1}-(\delta+\varepsilon)$, P_a can observe only P_b's send failure and can remove P_b only at the end of C_i (This is the exceptional case mentioned before). If P_b crashes after broadcasting its updated MSV_{i-3}, then P_a will have 'a' for P_b in its MSV_{i-1} and will remove P_b at its clock time T_i. Hence the lemma.

Remark: Lemma 2 implies that P_a will meet B2, if it has $m_i \geq \lceil(m_{i-2}+1)/2\rceil$ at the end of every cycle ever since the execution started. Lemma 3 shows that it is possible by assumption A2. First, a corollary is stated based on the proof for lemma 2.

Corollary 1

If a processor P_b, $1 \leq b \leq n$ and $b \neq a$, fails during C_{i-2}, $(i-2) \geq 1$, then a non-faulty processor P_a will have either '0' or 'a' as an entry for P_b in its MSV_{i-1} and will have $m_{i-1} \leq N_{i-2}$, provided P_a has $m_j \geq \lceil(m_{j-2}+1)/2\rceil$, at the end of every C_j, $1 \leq j \leq i$, during the execution of the algorithm.

Proof

From the proof given for lemma 2, it can be seen that P_a's entry for P_b in its MSV_{i-1} will be 'a' in the exceptional case and when P_b crashes after broadcasting its updated MSV_{i-3} that is the same as P_c's updated MSV_{i-3}; the entry will be '0' in all other cases. Thus any processor that failed in C_j, $j \leq i-2$, will not have an entry of '1' in P_a's MSV_{i-1}. Therefore, $m_{i-1}=N_{i-2}$ will be true for P_a, if no processor counted N_{i-2} does not fail in C_{i-1}. But when processors that were counted in N_{i-2} suffer failures in C_{i-1}, they may reduce the number of '1's in P_a's MSV_{i-1}. Thus, P_a can have $m_{i-1} \leq N_{i-2}$.

Lemma 3

At the end of any cycle C_i, $i \geq 1$, a non-faulty processor P_a would have had $m_j \geq \lceil(m_{j-2}+1)/2\rceil$, for every j, $1 \leq j \leq i$, if $N_j \geq \lceil(N_{j-3}+1)/2\rceil$ and $N_j \geq 3$, where $N_k = N_0$ for any $k < 0$ (Assumption A2).

Proof

As failed processors are not restarted, $N_{i-1} \geq N_i$, for all $i \geq 1$; since P_a has '1' in its MSV_i for any other non-faulty processor, P_a will have $m_i \geq N_i$ for all i, $i \geq 1$; also, $N_0 = n$ (by assumption); $m_0 = n$ and $m_{-1} = n$ (by algorithm). This means, for P_a:

$m_i \geq N_i \geq \lceil(N_0+1)/2\rceil = \lceil(n+1)/2\rceil \geq \lceil(m_{i-2}+1)/2\rceil$, for $i = 1, 2$ and 3.

The rest of the proof is by induction. Assume that the lemma is true for some i, $i \geq 3$. From the corollary stated above, P_a will have $m_{i-1} \leq N_{i-2}$. At the end of C_{i+1}, P_a will have $m_{i+1} \geq N_{i+1}$; by A2, $N_{i+1} \geq \lceil(N_{i-2}+1)/2\rceil$. Hence, for P_a, $m_{i+1} \geq N_{i+1} \geq \lceil(N_{i-2}+1)/2\rceil \geq \lceil(m_{i-1}+1)/2\rceil$. Thus, the lemma is true by induction.

Lemma 4

A processor P_b which failed in C_{i-1}, $i-1 \geq 1$, and did not crash until the end of C_i, will diagnose its failure no later than the end of C_i, if, for every j, $1 \leq j \leq i$, $N_j \geq \lceil (N_{j-3}+1)/2 \rceil$ and $N_j \geq 3$, where $N_k = N_0$ for any $k < 0$.

Proof

Since P_b fails only in C_{i-1}, it can be treated like a non-faulty processor (at least) until it computes its MSV_{i-2} at its clock time T_{i-2}. Therefore, it will have $m_{i-2} \leq N_{i-3}$ and $m_{i-3} \leq N_{i-4}$ (as per corollary 1). Because of its failure in C_{i-1}, either its updated MSV_{i-2} or its updated MSV_{i-1} will be different from the respective updated MSV's of a non-faulty processor. Let S_j, $i-1 \leq j \leq i$, denote the number of '1's in P_b's MSV_{j-2} which turned to either '0' or 'a' in P_b's MSV_j. Suppose that the updated MSV_{i-2} of P_b is different from the updated MSV_{i-2} of a non-faulty processor. This means that in the execution of the algorithm at the end of C_{i-1}, P_b will have '0' in its MSV_{i-1} for every non-faulty processor's MSV it receives. Thus, the number of '1's in P_b's MSV_{i-2} which turn to either '0' and 'a' in P_b's MSV_{i-1} will be at least as large as N_{i-1}.

$S_{i-1} \geq N_{i-1} \geq \lceil (N_{i-4}+1)/2 \rceil \geq \lceil (m_{i-3}+1)/2 \rceil$. Hence, $m_{i-1} \geq \lceil (m_{i-3}+1)/2 \rceil$ will not be true for P_b.

If P_b's updated MSV_{i-2} is not different from the MSV_{i-2} of a non-faulty processor, then its updated MSV_{i-1} will be different from that of a non-faulty processor due to its failure in C_{i-1}. This means that if P_b has $m_{i-1} \geq \lceil (m_{i-3}+1)/2 \rceil$, then it will have, at its clock time T_i,

$S_i \geq N_i \geq \lceil (N_{i-3}+1)/2 \rceil \geq \lceil (m_{i-2}+1)/2 \rceil$. Therefore, it will not have $m_i \geq \lceil (m_{i-2}+1)/2 \rceil$. Hence the lemma.

4.2. Algorithm with Processor Restarts

Correctness of the Restoration Subtask

The restoration subtask is correct when an incoming processor, say P_a, which computes, at the end of cycle, say C_{i-1}, a vector that is the same as the MSV_{i-1} of a non-faulty processor in the group, provided P_a has received all broadcasts with timestamp in $(T_{i-2}-2(\delta+\varepsilon), T_{i-1}-2(\delta+\varepsilon)]$ during cycle C_{i-1}. Due to A3, P_a will not suffer any receive failure until it joins the group. The subtask can be seen to be correct through much of the arguments used in presenting the design background. Essential to these arguments is the lemma that is stated and proved below.

Lemma 5

In an execution of the restoration subtask by processor P_a in cycle C_{i-1}, $SV[c]=1$ will imply that P_c, $c \neq a$, has been observed to have suffered a send failure before $T_{i-1}-(\delta+\varepsilon)$.

Proof

$SV[c]=1$ means that there is only one '1' in the cth column of P_a's matrix RSM. If that entry is in c^{th} row, then it will imply that, in P_a's view, only the processor P_c, but no other processor, at clock time $T_{i-1}-(\delta+\varepsilon)$, considers P_c to be failure free. This means that P_c had suffered a send failure before $T_{i-1}-(\delta+\varepsilon)$. Let us assume that the entry '1' is in the dth row, $d \neq c$ and $d \neq a$.

By this assumption, P_c did not fail to send its updated MSV_{i-3}, because P_d has received P_c's updated MSV_{i-3} and has agreed with it, while no non-faulty processor has agreed with the updated MSV broadcast by P_c. This means that the processor P_d must have found every MSV it received from a non-faulty processor in C_{i-2} to be in disagreement with its own MSV_{i-3}. Such a processor, by the arguments in the proof for Lemma 4, would have removed itself from the group at the end of C_{i-2} and would not have broadcast anything in C_{i-1}. Therefore the assumption cannot be correct. Hence the lemma.

Correctness of the Membership Algorithm

The membership algorithm will be correct, if conditions B1 to B3 of previous section are met in any execution. In the new context where restarted processors attempt to join the group, lemmas 1 and 2 remain true as before. J_i, $i \geq 3$, is the number of restarted processors which joined the group at the end of cycle C_{i-1} by broadcasting the updated MSV_{i-2} of a non-faulty processor in the group and J_i will be counted in a non-faulty member processor's m_{i-1}. Therefore, in corollary 1, $m_{i-1} \leq N_{i-2}$ will change to $m_{i-1} - J_i \leq N_{i-2} + J_{i-1}$. Using the inequality $m_{i-1} \leq N_{i-2} + J_{i-1} + J_i$ in the place of $m_{i-1} \leq N_{i-2}$ and based on the general version of A2, lemmas 3 and 4 can be argued to be correct in the new context in the same manner adopted in the previous subsection. In the following, we illustrate some aspects of these arguments to indicate the reader the correctness of lemmas 3 and 4 in the new context.

Suppose that, in an execution of the membership algorithm, a non-faulty member processor has $m_j \geq \lceil (m_{j-2}+1)/2 \rceil$ true for all j, $1 \leq j \leq i$. At its clock time T_{i+1}, the non-faulty member processor will have $m_{i+1} - J_{i+1} \geq N_{i+1} \geq \lceil (N_{i-2}+J_{i-1}+J_i+J_{i+1}+1)/2 \rceil \geq \lceil (m_{i-1}+J_{i+1}+1)/2 \rceil \geq \lceil (m_{i-1}+1)/2 \rceil$. Consider the correctness argument provided for lemma 4 in section 3. When restarted processors are attempting to join the group, P_b can be seen, by A2, to have either for $j=i-1$ or $j=1$: $S_j \geq N_j \geq \lceil (N_{j-3}+J_{j-2}+J_{j-1}+J_j+1)/2 \rceil \geq \lceil (m_{j-2}+J_j+1)/2 \rceil \geq \lceil (m_{j-2}+1)/2 \rceil$. Thus the lemmas 3 and 4 will be correct in the new context.

5. Concluding Remarks

A group membership algorithm has been presented for a reliable broadcast network which guarantees bounded message delivery time. The working of the algorithm is illustrated by means of examples in section 4. The two basic operations of the algorithm are the localization of faulty processors, and the restoration into the group of processors that are repaired and restarted. In any execution of this algorithm non-faulty processors know of a failure at the end of the next cycle except in the case where a faulty processor suffers a send failure in the next cycle. In the exceptional case, one complete cycle following the send failure will be needed for the failure to be known. Thus, a failure occurred at T is guaranteed to be known to all non-faulty member processor simultaneously and no later than $T+3\pi+(\delta+\varepsilon)$, where π is the cycle length, δ the bound on message delivery time, and ε the bound on clock difference. Restarted processors are assumed not to suffer receive failures and can join the group at any time and independent of each other.

The group membership algorithm can also be viewed as a means of achieving message ordering with respect to processor failures. The broadcast network properties guarantee that messages can be ordered identically and in the sent order at all non-faulty destinations. This ordering however will not reflect missing messages due to send failures at the source. Execution of the group membership algorihm will achieve this - an identical sent ordering that reflects source failures. Thus, if a processor attemps to broadcast messages m_1, m_2, and m_3 (in that order) and fails to broadcast m_2, then the non-faulty processors can see this, using the group membership algorithm, as m_1, send-failure, and m_3. They can remove the sending processor from the group considering m_1 as the last message 'received' from the sender and thus maintaining an identical and source level fifo ordering among received messages and source failures. Alternatively, non-faulty processors may attempt, as best as they can, to retrieve the missing message and to keep the failed processor in the group. But such attempts will violate the property of bounded message delivery time and make the message communication asynchronous. Asynchronous broadcast protocols such as /Birman 87/, /Peterson 89/, /Melliar-Smith 90/ attempt to retrieve a missing message (at the expence of a slow performance). It is our future plan to study, in the context of synchronous message communication, various types of message ordering (notably causal) with respect to processor failures in a group.

6. Acknowledgments

This work was supported in part by a grant from the U.K. Science and Engineering Research Council and from the European Community ESPRIT-II research programme Project P2252 (DELTA-4). The first author would like to acknowledge the financial support of CAPES/Brazil.

References

/Birman 87/   Birman, K.; Joseph, T. "Reliable Communication in the Presence of Failures". ACM Transactions on Computer Systems, Vol. 5, No 1. February 1985. pp 47-76.

/Cristian 85/   Cristian, F.; Aghili, H.; Strong, R.; Dolev, D. "Atomic Broadcast: From Simple Message Diffusion to Byzantine Agreement". Proceedings 15th International Symposium on Fault-Tolerant COmputing. Ann Arbor, MI. June 1985. pp 200-206.

/Cristian 88/   Cristian, F. "Agreeing on who is Present and who is Absent in a Synchronous Distributed System". 18th International Symposium on Fault-Tolerant Computing. Tokyo, Japan. June 1988. pp 206-211.

/Cristian 90/   Cristian, F. "Synchronous Atomic Broadcast for Redundant Broadcast Channels". IBM Research Report RJ7203. April 1990.

/Ezhilchelvan 90/ Ezhilchelvan, P.D.; Lemos, R. "A Robust Group Membership Algorithm for Distributed Real-Time Systems". Proceedings of the 11th Real-Time Systems Symposium. Orlando, Florida. December 1990.

/Kopetz 89/ Kopetz, H.; Grunsteidl, G.; Reisinger, J. "Fault-Tolerant Membership Service in a Distributed Real-Time System". Int. Conference on Dependable Computing for Critical Applications. Santa Barbara, CA. August, 1989. pp 167-174.

/Melliar-Smith 90/ Melliar-Smith, P.M.; Moser, L.M.; Agarwala. "Broadcast Protocols for Distributed Systems". IEEE Transactions on Parallel and Distributed Systems Vol.1, No 1. January 1990. pp 17-25.

/Navaratnam 88/ Navaratnam, S.; Chanson, S.; Neufeld, G. "Reliable Group Communication in Distributed Systems". Proc 8th International Conference on Distributed Computing Systems. June, 1988. pp 439-446.

/Peterson 89/ Peterson, L.; Buchholz, N.C.; Schlichting, R.D. "Preserving and Using Context Information in Interprocess Communication". ACM TOCS Vol. 7, No. 3. August 1989. pp 217-246,.

/Powell 88/ Powell, D. et al. "The Delta-4 Approach to Dependability in Open Distributed Computing Systems. 18th International Symposium on Fault-Tolerant Computing. Tokyo, Japan. June 1988. pp 83-93.

/Schlichting 83/ Schlichting, R.D.; Schneider, F.B. "Fail-Stop Processors: An Approach to Design Fault-Tolerant Computing Systems". ACM Transactions on Computer Systems, Vol 1, No 3. August 1983. pp 222-234.

TIGHT BOUNDS ON THE ROUND COMPLEXITY OF DISTRIBUTED 1-SOLVABLE TASKS[1]

(Extended Abstract)

Ofer Biran, Shlomo Moran and Shmuel Zaks
Department of Computer Science
Technion, Haifa, Israel 32000

ABSTRACT

A distributed task T is 1-solvable if there exists a protocol that solves it in the presence of (at most) one crash failure. A precise characterization of the 1-solvable tasks was given in [BMZ]. In this paper we determine the number of rounds of communication that are required, in the worst case, by a protocol which 1-solves a given 1-solvable task T for n processors. We define the radius $R(T)$ of T, and show that if $R(T)$ is finite, then this number is $\Theta(\log_n R(T))$; more precisely, we give a lower bound of $\log_{(n-1)} R(T)$, and an upper bound of $2+\lceil \log_{(n-1)} R(T) \rceil$. The upper bound implies, for example, that each of the following tasks: renaming, order preserving renaming ([ABDKPR]) and binary monotone consensus [BMZ] can be solved in the presence of one fault in 3 rounds of communications. All previous protocols that 1-solved these tasks required $\Omega(n)$ rounds. The result is also generalized to tasks whose radii are not bounded, e.g., the approximate consensus and its variants [DLPSW, BMZ].

---

[1] This research was supported in part by Technion V.P.R. Funds - Wellner Research Fund and Loewengart Research Fund, and by the Foundation for Research in Electronics, Computers and Communications, administrated by the Israel Academy of Sciences and Humanities.

1. INTRODUCTION

An asynchronous distributed network consists of a set of processors, connected by communication lines, through which they communicate in order to accomplish a certain task; the time delay on the communication lines is finite, but unbounded and unpredictable. In this paper we study the case when at most one processor is faulty, which means that all of its messages are not delivered from some point on (fail-stop failure). It was shown in [FLP] that it is impossible to achieve a distributed consensus for this case. This result was extended in several directions. In [DDS] the features of asynchrony that yield the result of [FLP] and related results were analyzed. In [DLPSW] it was shown that approximate consensus, in which all processors must agree on values that are arbitrarily close to one another, is possible in the presence of few faulty processors. In [ABDKPR] few other problems were shown to be solvable in the presence of faulty processors. In [MW] a class of tasks was shown not to be solvable in the presence of one faulty processor (not 1-solvable). In [BMZ] we provided a complete characterization of the 1-solvable tasks.

Let T be an 1-solvable task. In this paper we analyze the *round complexity* of T, which is the number of communication rounds that are required, in the worst case, by any protocol that 1-solves T, and provide optimal bounds (up to a constant *additive* factor) for this number. This measure attempts to capture the notion of *time complexity* for asynchronous, fault tolerant protocols. We first consider *bounded* tasks, which are tasks that can be 1-solved by protocols that require at most a constant number of rounds in all possible executions (e.g., the renaming tasks and the strong binary monotone consensus task [ABDKPR, BMZ]). Then we generalize our results for unbounded tasks (like the approximate consensus and its variants [DLPSW, BMZ]).

The outline of our proof is as follows: For a distributed task T, let X_T be the set of possible input vectors for T. First we show, by using the result in [BMZ], that if T is 1-solvable, then there is a set R_T of *radius functions* related to T, where each radius function ρ is a mapping $\rho : X_T \to N$, which maps each input vector $\vec{x}$ on a positive integer $\rho(\vec{x})$. We use this set to define $R(T)$, the radius of the task T, as $R(T) = \min_{\rho \in R_T} \max_{\vec{x} \in X_T} \rho(\vec{x})$.

In proving our bounds, we first consider only tasks T for which $R(T)$ is finite, and show that these are exactly the bounded tasks. In fact, we show that if $R(T)$ is finite then the round complexity of T is $\Theta(\log_n R(T))$; more precisely, we give a lower bound of $\log_{(n-1)} R(T)$, and an upper bound of $2 + \lceil \log_{(n-1)} R(T) \rceil$. We then extend the results to arbitrary task T. In the general case, the round complexity of T is not a constant, but a function of the input vector. Since there is no natural total order on these functions, we cannot define the optimal round complexity of T, but only define the set of *minimal* round complexity functions of T, in the natural partial ordering of functions. This set is defined by a correspondence to the set of minimal radius functions in R_T.

The upper bound implies, for example, that each of the following tasks: renaming with $n+1$ new names, order preserving renaming with $2n-1$ new names ([ABDKPR]), and strong binary monotone consensus [BMZ] can be solved in the presence of one fault in three rounds of communications. All previous protocols that 1-solved these tasks required $\Omega(n)$ rounds. For the case where $R(T)$ is infinite, we extend the optimal bounds of [Fe] for the approximate consensus: In particular, we show that similar bounds hold for variants of the approximate consensus that were studied in [BMZ], which are considerably harder than the (original) approximate consensus.

The rest of the paper is organized as follows: In Section 2 we provide the preliminary definitions. In Section 3 we define standard protocols and round complexity. In Section 4 we define the radius of a task. The lower and upper bounds for bounded tasks are presented in Sections 5 and 6. In Section 7 we generalize our results for arbitrary tasks and in the Appendix we bring some applications.

2. PRELIMINARY DEFINITIONS AND NOTATIONS

2.1 Asynchronous Systems

An *asynchronous distributed system* is composed of a set $V = \{P_1, P_2, \ldots, P_n\}$ of n *processors* ($n \geq 3$), each having a unique *identity*. We assume that the identities of the processors are mutually known, and w.l.o.g. that the identity of P_i is i. Our results are applicable also to the model in which the identities are not mutually known (or absent, provided that the inputs are distinct). The processors are connected by *communication links*, and they communicate by exchanging messages along them. Messages arrive with no error in a finite but unbounded and unpredictable time; however, one of the processors might be faulty, in which case messages might not have these properties (the exact definition is given in the sequel).

2.2 Decision Tasks

Definition: Let X and D be sets of *input values* and *decision values*, respectively. A *distributed decision task* T is a function

$$T: X_T \to 2^{D^n} - \{\emptyset\},$$

where $X_T \subseteq X^n$. X_T is called the *input set* of the task T. The *decision set* of the task T is the union of the sets $T(\vec{x})$ over all $\vec{x} \in X_T$. Each vector $\vec{x} = (x_1, x_2, \ldots, x_n) \in X_T$ is called an *input vector*, and it represents the initial assignment of the *input value* $x_i \in X$ to processor P_i, for $i = 1, 2, \ldots, n$. Each vector $\vec{d} = (d_1, d_2, \ldots, d_n) \in D_T$ is called a *decision vector*, and it represents the assignment of a *decision value* $d_i \in D$ to processor P_i, for $i = 1, 2, \ldots, n$.

Thus, a decision task T maps each input vector to a non-empty set of allowable decision vectors. We assume that all tasks T discussed in this paper are *computable*, in the sense that the set $\{(\vec{x}, \vec{d}) : \vec{x} \in X_T \text{ and } \vec{d} \in T(\vec{x})\}$ is recursive.

Examples:

(1) **Consensus [FLP]:** A consensus task is any task T where $X_T = X^n$ for an arbitrary set X, and such that $T(\vec{x}) \subseteq \{(0,0,...,0),(1,1,...,1)\}$ for every input vector $\vec{x} \in X_T$. Let $\vec{0}$ denote the vector $(0,0,...,0)$, and $\vec{1}$ denote the vector $(1,1,...,1)$. A *strong* consensus task is a consensus task T, in which there exist two input vectors $\vec{u}$ and $\vec{v}$ such that $T(\vec{u}) = \{\vec{0}\}$ and $T(\vec{v}) = \{\vec{1}\}$. The main result in [FLP] implies that a strong consensus task is not 1-solvable.

(2) **Strong Binary Monotone Consensus [BMZ]:** This is probably the strongest variant of the consensus task which is 1-solvable. To simplify the definition, assume that n is even: The input is an integer vector $\vec{x} = (x_1, \cdots, x_n)$, and $T(\vec{x})$ consists of all vectors $\vec{d} = (d_1, \cdots, d_n)$ where each d_i is one of the two medians of the multiset $\{x_1, \cdots, x_n\}$, and $d_i \leq d_{i+1}$ (the "strong" stands for the fact that the two values must be the medians).

(3) **Renaming [ABDKPR]:** This task is defined for a given integer K, where $K \geq n$. The input set X_T is the set of all vectors $(x_1, \cdots, x_n)$ of distinct integers. For a given input $\vec{x}$, $T(\vec{x})$ is the set of all integer vectors $\vec{d} = (d_1, \cdots, d_n)$ satisfying $1 \leq d_i \leq K$ and such that for each i, j, $d_i \neq d_j$. In order to prevent trivial solutions in which P_i always decides on i, this task assumes a model in which the processors identities are not known.

(4) **Order Preserving Renaming (OPR) [ABDKPR]:** This task is similar to the renaming task, with the additional requirement that for each i, j, $x_i < x_j$ implies $d_i < d_j$.

(5) **Approximate Consensus [DLPSW]:** This task is defined for any given $\varepsilon > 0$. The input set X_T is Q^n, where Q is the set of rational numbers, and for a given input $\vec{x} = (x_1, \cdots, x_n)$, $T(\vec{x})$ is the set of all vectors $\vec{d} = (d_1, \cdots, d_n)$ satisfying $|d_i - d_j| \leq \varepsilon$ and $m \leq d_i \leq M$ ($1 \leq i, j \leq n$), where $m = \min\{x_1, \cdots, x_n\}$ and $M = \max\{x_1, \cdots, x_n\}$.

(6) **Strong Binary Monotone Approximate Consensus [BMZ]:** This is a harder variant of the approximate consensus task which is still 1-solvable. To simplify the definition, assume that n is even: The input is the same as for the approximate consensus. For an input $\vec{x} = (x_1, \cdots, x_n)$, $T(\vec{x})$ consists of all vectors $\vec{d} = (d_1, \cdots, d_n)$ satisfying: $\vec{d}$ has at most two distinct entries, which lie between the two medians of the multiset $\{x_1, \cdots, x_n\}$, and $d_i \leq d_{i+1} \leq d_i + \varepsilon$.

2.3. Protocols and Executions

A *protocol* for a given network is a set of n programs, each associated with a single processor in the network. Each such program contains operations of sending a message to a neighbor, receiving a message and processing information in the local memory.

If the network is initialized with the input vector $\vec{x} \in X^n$ (i.e., the value x_i is assigned to processor P_i), and if each processor executes its own program in the protocol α, then the sequence of operations performed by the processors is called an *execution of α on input $\vec{x}$*. (We assume that no two operations occur simultaneously; otherwise, we order them arbitrarily. For more formal definitions see, e.g., [KMZ].)

Definition: A vector $\vec{d} = (d_1, d_2, \cdots, d_n)$ is an *output (decision) vector of α on input $\vec{x}$* iff there is an execution of α on $\vec{x}$ in which processor P_i *decides* on d_i, by writing d_i in a write-once register.

2.4. Faults and 1-Solvability

Definition: A processor P is *faulty* in an execution e if all the messages sent by P during e from some point on are never received (a *fail-stop* failure; see, e.g., [FLP]. Also known as *crash* failure; see, e.g., [NT]).

Definition: A protocol α *1-solves* a task T if for every execution of α on input $\vec{x} \in X_T$ in which at most one processor is faulty, the following two conditions hold:

(1)   All the non-faulty processors eventually decide.

(2)   If no processor is faulty in the execution, then the output vector belongs to $T(\vec{x})$.

When such a protocol α exists we say that the task T is *1-solvable*.

The definition above does not require the processors to halt after reaching a decision. However, in the case of a single failure, it is not hard to see that a processor that learns that $n-1$ processors had already decided may halt. Hence, in this case, reaching a decision by all non-faulty processors is sufficient to guarantee halting. For this reason, in this paper we shall restrict the discussion to protocols in which the processors are guaranteed to halt in every possible execution. (Note that in the case of $t > 1$ crash failures, there exist tasks which can be t-solved only by protocols that do not guarantee termination, e.g, the renaming tasks [ABDKPR]. For more on the termination requirement for multiple failures see [TKM]).

3. STANDARD PROTOCOLS AND ROUND COMPLEXITY

In this paper we bound the number of communication rounds that are required by protocols that 1-solve a given task. This number attempts to capture the notion of *time complexity* for asynchronous, fault tolerant protocols, in which every processor is guaranteed to halt. We model an arbitrary t-resilient protocol that works in rounds of communications by the notion of *standard protocol*. The definitions and discussion below are restricted to the case $t = 1$.

3.1. Standard Protocols

A protocol that 1-solves a task T is *standard protocol* if it works in rounds of communications, as follows. In each round a processor broadcasts a message (which includes the round number), which is a function of its state, to all the processors (including itself), and waits until it receives $n-1$ messages of this round (or less if it heard on processors that had already halted). During this period of waiting, it might receive messages from different rounds. Those of higher rounds are saved until the processor itself reachs these rounds. Messages of previous rounds (might be one such message per each previous round) - called *late messages* - are gathered with the $n-1$ of this round to form a set M. Then the processor computes its next state, which is a function of M and its previous state. The state of a processor includes its write-once register.

Our notion of standard protocol is similar to the one used in [Fe]. It can be shown that this notion is general enough for the sake of lower bounds, by using *full information* protocols [Fe, FL].

Formally, the standard protocol for P_k:

```
r ← 0
state ← INIT_k
while state <> HALT do
        r ← r+1
        BROADCAST (r, MESSAGE_FUNCTION_k(state))
        WAIT until you RECEIVE n−1− [# of halted processors] messages of the form (r, *)
        M ← {m | a message (r′, m), r′≤r was received in the above WAIT}
        state ← STATE_FUNCTION_k(state, M)
end
```

3.2. Round Complexity

Definition: Let T be a task and α a standard protocol that 1-solves T. The *round complexity of α on input* $\vec{x}$, denoted $rc_\alpha(\vec{x})$, is the maximum round number, over all executions of α on input $\vec{x}$, that a correct processor reaches.

The *round complexity* of α, denoted $rc_\alpha(T)$ is defined by: $rc_\alpha(T) = \max_{\vec{x} \in X_T} rc_\alpha(\vec{x})$.

The *round complexity* $rc(T)$ of a task T is defined by: $rc(T) = \min \{rc_\alpha(T) \mid \alpha \text{ 1-solves } T\}$.

Note that $rc(T)$ may be infinite; This is the case only when the input set X_T is infinite, and for any protocol α that 1-solves T and for any constant C, there is an input $\vec{x}$ such that $rc_\alpha(\vec{x}) > C$.

Definition: A 1-solvable task T is *bounded* iff $rc(T)$ is finite, and is *unbounded* otherwise.

We will first present results for bounded tasks, and then extend them to results which are applicable for unbounded tasks as well.

4. COVERING FUNCTIONS AND RADII OF TASKS

We first give some basic definitions from [BMZ] which are needed for this paper.

4.1 Adjacency graphs, partial vectors, covering vectors and i-anchors

Definition: Let $S \subseteq A^n$, for a given set A. Two vectors $\vec{s}_1, \vec{s}_2 \in S$ are *adjacent* if they differ in exactly one entry. The *adjacency graph of* S, $G(S) = (S, E_s)$, is an undirected graph, where $(\vec{s}_1, \vec{s}_2) \in E_s$ iff $\vec{s}_1$ and $\vec{s}_2$ are adjacent. For a task T and an input vector $\vec{x}$ for T, $G(T(\vec{x}))$ is the *decision graph of* $\vec{x}$.

Definition: A *partial vector* is a vector in which one of the entries is not specified; this entry is denoted by '*'. For a vector $\vec{s} = (s_1, \cdots, s_n)$, $\vec{s}^i$ denotes the partial vector obtained by assigning $*$ to the i-th entry of $\vec{s}$, i.e., $\vec{s}^i = (s_1, \cdots, s_{i-1}, *, s_{i+1}, \cdots, s_n)$. $\vec{s}$ is called an *extension* of $\vec{s}^i$.

Definition: Let $\vec{x}^i$ be a partial input vector and $\vec{d}^i$ a partial decision vector of a task T. We say that $\vec{d}^i$ is a *covering vector* for $\vec{x}^i$ if for each extension of $\vec{x}^i$ to an input vector $\vec{x} \in X_T$, there is an extension of $\vec{d}^i$ to a decision vector $\vec{d} \in T(\vec{x})$.

Definition: A vector $\vec{d}$ is an *i-anchor* of an input vector $\vec{x}$ if $\vec{d} \in T(\vec{x})$ and $\vec{d}^i$ is a covering vector for $\vec{x}^i$.

Example: consider the *OPR* task for $n=3$ processors and $K=5$. For the partial input vector $\vec{x}^2 = (10, *, 30)$ there is a unique covering vector $\vec{d}^2 = (2, *, 4)$, and the input vector $\vec{x} = (10, 20, 30)$ has a unique 2-anchor $\vec{d} = (2, 3, 4)$. In the *OPR* task with $n=3$ and $K=6$ there are three covering vectors for $\vec{x}^2$: $(2,*,4)$, $(2,*,5)$, and $(3,*,5)$. Thus, $\vec{x}$ has four 2-anchors: $(2,3,4)$, $(2,3,5)$, $(2,4,5)$ and $(3,4,5)$.

4.2 Covering functions and radii of tasks

Definition: A *covering function* for a given task T is a function that maps each partial input vector to a corresponding covering vector for it.

Definition: Let T be a task, CF a covering function for T, and $\vec{x} \in X_T$ an input vector. An *anchors tree for $\vec{x}$ based on CF* is a tree in $G(T(\vec{x}))$ that, for each i ($1 \leq i \leq n$), includes an i-anchor which is an extension of $CF(\vec{x}^i)$.

We now reformulate Theorem 3 of [BMZ] to a form suitable to our discussion:

Theorem [BMZ]: A task T is 1-solvable if and only if there exists a covering function CF for T, s.t. for each input vector $\vec{x} \in X_T$, there is an anchors tree for $\vec{x}$ based on CF. □

A covering function satisfying the condition of Theorem [BMZ] is termed a *solving covering function for T*.

We are now ready to define the concept of the *radius* of a task T, which plays an essential role in this paper. All the definitions below refer to an arbitrary 1-solvable task T.

Definition: Let CF be a solving covering function for T, and $\vec{x}$ an input vector in X_T. $\rho_{CF}(\vec{x})$ is the minimum possible radius of an anchors tree for $\vec{x}$ based on CF.

By the above definition, each solving covering function CF defines a *radius function* $\rho_{CF}: X_T \to N$. The set of all radius functions for T is denoted by $\mathbf{R}_T$. That is,
$$\mathbf{R}_T = \{\rho_{CF}: CF \text{ is a solving covering function for } T\}.$$
$R(T)$, the radius of the task T, is defined by:

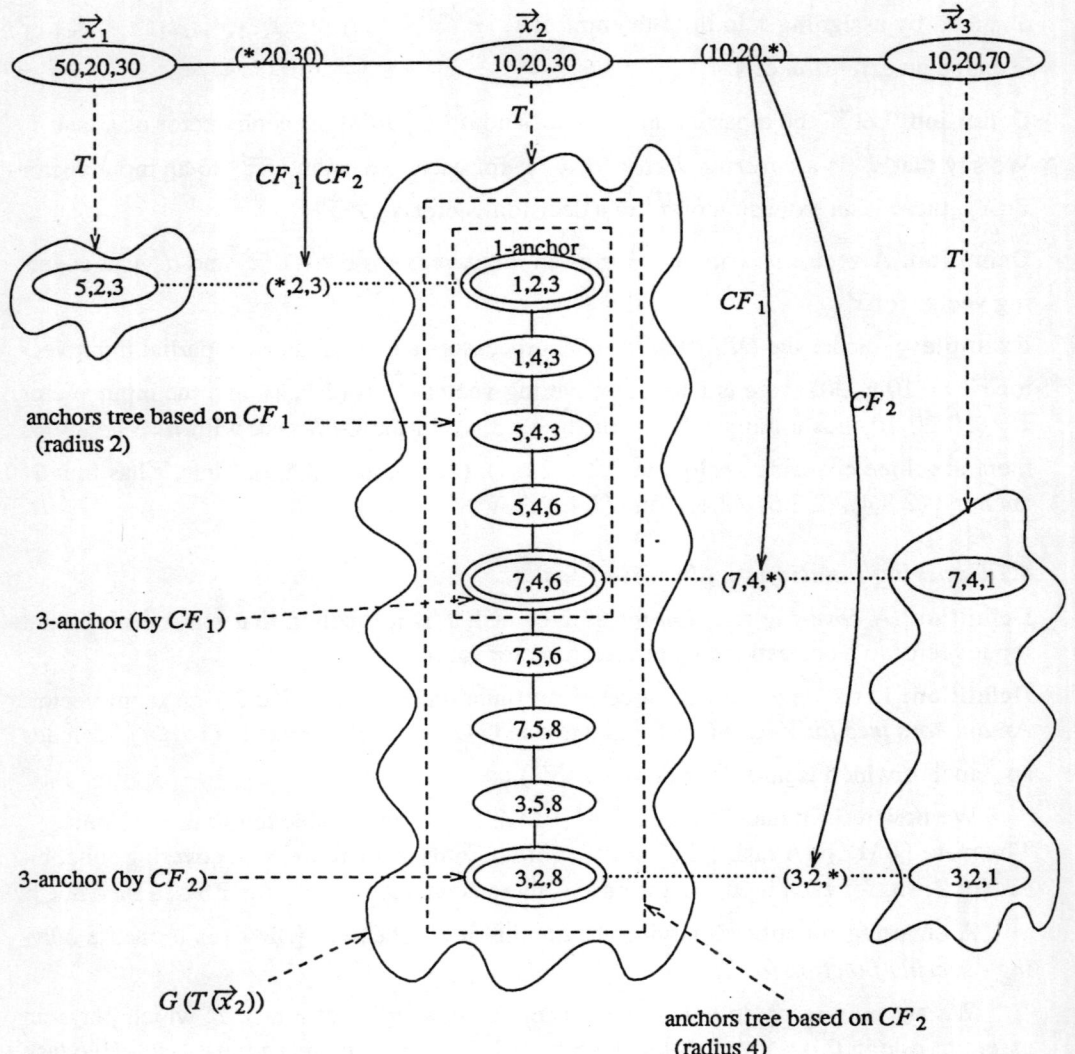

Figure 1: A task T with $R(T)=2$ $(=\rho_{CF_1}(\vec{x}_2))$

$$R(T) = \min_{\rho_{CF} \in R_T} \max_{\vec{x} \in X_T} \rho_{CF}(\vec{x}).$$

Note that $R(T)$ may be infinite; This is the case only when the input set X_T is infinite, and for any radius function ρ_{CF} in R_T and for any constant C, there is an input $\vec{x}$ such that $\rho_{CF}(\vec{x}) > C$. As we shall show, $R(T)$ is finite iff T is a bounded task.

A covering function CF, and the corresponding radius function ρ_{CF}, are *optimal* for a bounded task T if $\max_{\vec{x} \in X_T} \rho_{CF}(\vec{x}) = R(T)$.

Example: Consider the following task T, in which X_T composed of only 3 input vectors:
$\vec{x}_1 = (50,20,30), \vec{x}_2 = (10,20,30)$ and $\vec{x}_3 = (10,20,70)$.
$T(\vec{x}_1) = \{ (5,2,3) \}$,
$T(\vec{x}_2) = \{ (1,2,3),(1,4,3),(5,4,3),(5,4,6),(7,4,6), (7,5,6),(7,5,8),(3,5,8),(3,2,8) \}$ and
$T(\vec{x}_3) = \{ (7,4,1),(3,2,1) \}$.
Now, in choosing an optimal covering function for T, the only partial input vectors that should be considered are those which might be extended to more than one input vector (if $\vec{x}^i$ might be extended to a unique input vector $\vec{x}$, then any vector $\vec{d}$ in $T(\vec{x})$ is an i-anchor of $\vec{x}$, so the need to select an i-anchor does not impose any constrain on the anchors tree). Thus we consider only (*,20,30) and (10,20,*), so the only anchors that constrain the anchors tree are the 1-anchor and the 3-anchor. From the decision graphs (see Figure 1), clearly $\vec{x}_2$ is dominate for $R(T)$, since any anchors tree of the others two is composed of a single vertex. There are only 2 covering functions (which are different in the 2 key partial vectors):
$CF_1((*,20,30)) = (*,2,3), CF_1((10,20,*)) = (7,4,*)$ and
$CF_2((*,20,30)) = (*,2,3), CF_2((10,20,*)) = (3,2,*)$.
In the anchors tree based on CF_1 (in $G(T(\vec{x}_2))$) the 1-anchor is (1,2,3), the 3-anchor is (7,4,6), and thus the radius is 2 (a line, with center (5,4,3)). In anchors tree based on CF_2 the 3-anchor is (3,2,8), and the radius is 4. So CF_1 is the optimal covering function, and $R(T) = 2$.

More examples appear in the Appendix.

5. LOWER BOUND

The following proposition shows that for proving lower bounds, it suffices to consider standard protocols that do not use late messages, since the use of late messages cannot improve the round complexity.

Proposition 1: Let α be a standard protocol which 1-solves a task T. Then the protocol α', which is identical to α except that it does not use late messages (i.e. , - the set M in line 7 of the standard protocol is built only of round r messages - the late messages are ignored) also 1-solves T, and for every input $\vec{x}$, $rc_{\alpha'}(\vec{x}) \leq rc_{\alpha}(\vec{x})$.

Proof: For each execution (which is determined by the messages scheduling) of α' on input $\vec{x}$ consider a similar execution, in which all late messages are delivered only after $rc_\alpha(\vec{x})$ rounds. α' clearly works the same in both executions (namely, the states of the processors after each round are the same). But up to round $rc_\alpha(\vec{x})$ in the second execution, α' works exactly like α, and thus is guaranteed to halt with legal output vector before round $rc_\alpha(\vec{x}) + 1$. $\square$

Theorem 1: Let T be a bounded task. Then its round complexity $rc(T)$ satisfies
$$rc(T) \geq \log_{(n-1)} R(T).$$

Sketch of proof: The theorem is proved by the next two lemmas. First we need the following definition:

Definition: The *extended task* of a task T, denoted T^E, is a task with the same input set as T, but each processor P_k has to decide on a partial vector $\vec{d}^i$ s.t. $i \neq k$ (instead on a single decision value), such that:

(1) The partial vectors are consistent (i.e., every two non-'*' j'th entries in two partial vectors are equal).

(2) The complete vector $\vec{d}$ composed of the partial vectors (note that for each entry there is at least one partial vector with a value different than '*' in this entry) belongs to $T(\vec{x})$.

Lemma 1: If the round complexity of T is s, then the round complexity of T^E is at most $s+1$.

Proof: Let α be a protocol which 1-solves T, and $rc_\alpha(T) = s$. We construct a protocol α' which 1-solves T^E, and $rc_{\alpha'}(T^E) = s+1$. α' is the same as α, with one additional round, in which each processor sends the decision value it would have decided on in α, wait for $n-1$ such values, and decides on the partial vector composed of these values. $\square$

In view of of Lemma 1, in order to prove Theorem 1 it suffices to show that if the round complexity of T^E is s, then $R(T)$, the radius of T, is at most $(n-1)^{s-1}$. For this, we shall show that a standard protocol α that does not use late messages and 1-solves T^E in s rounds implies a solving covering function for T, CF_α, such that $\rho_{CF_\alpha}(\vec{x}) \leq (n-1)^{s-1}$ for every input vector $\vec{x}$. Note that an execution of α is completely determined by the lists of $n-1$ processors from which each processor receives messages in each round. We need one more definition:

Definition: e is an *r-rounds execution* of a standard protocol A if e is the first r rounds of an execution of A. e is an *r-rounds i-sleeping execution* if during e, no processor P_j, $j \neq i$, ever receives a message from P_i.

We now define CF_α. For a given $\vec{x}$ and i, $CF_\alpha(\vec{x}^i)$ is the partial vector $\vec{d}^i$ output by each of the $n-1$ processors (excluding P_i) in an s-rounds i-sleeping execution of α^2 on input $\vec{x}$. (The validity of this definition follows from the fact that α 1-solves T^E.)

Lemma 2: Let $1 \leq i < j \leq n$ and let $\vec{x}$ be an input vector. Then for $r=1, \cdots, s$, there exists a sequence of $D = (n-1)^{r-1}$ r-rounds executions of α, $e_1, \cdots, e_D$, and a sequence of $D+1$ processors $Q_0, \cdots, Q_D$, s.t. $Q_i \in \{P_1, \cdots, P_n\}$, $Q_0 \neq P_i$, $Q_D \neq P_j$, satisfying the following:

a. The history of processor Q_0 in e_1 is the same as its history in an r-rounds i-sleeping execution on input $\vec{x}$, and the history of processor Q_D in e_D is the same as its history in an r-rounds j-sleeping execution on input $\vec{x}$.

b. For $k=1, \cdots, D-1$, processor Q_k has the same history in executions e_k and e_{k+1}.

Sketch of proof: The proof is based on an inductive construction, starting with $r=D=1$, where the base of the induction is an 1-round execution e_1, in which Q_0 receives $n-1$ messages from all processors except P_i, Q_1 receives $n-1$ messages from all processors except P_j, and the set of messages received by the other processors is arbitrary. In the induction step we replace each $r-1$-rounds execution by $n-1$ r-rounds executions, and appropriately extend the sequence of processors. This construction uses ideas similar to ones appearing in [Fe], and will be given in details in the full paper. □

We now use Lemma 2 to show that $R(T) \leq D = (n-1)^{s-1}$. For this, apply Lemma 2 for $r=s$. Then each execution e_k defines an output vector $\vec{d_k} \in T(\vec{x})$, composed of the partial vectors output by the processors. Moreover, (a) of the lemma implies that $\vec{d_1}$ and $\vec{d_D}$ are i-anchor and j-anchor of $\vec{x}$, resp., and (b) of the lemma implies that for every k, $\vec{d_k}$ and $\vec{d_{k+1}}$ are either the same vector or are adjacent. Thus, $(\vec{d_1}, \cdots, \vec{d_D})$ is a path of length at most $D-1$ from an i-anchor to a j-anchor of $\vec{x}$. Since this holds for every i and j, $\rho_{CF_\alpha}(\vec{x}) < D$. Since $\vec{x}$ is arbitrary, we have that $R(T) \leq D$. This completes the proof of Theorem 1. □

6. UPPER BOUND

6.1 The protocol

Theorem 2: The round complexity of a bounded task T is at most $2 + \lceil \log_{(n-1)} R(T) \rceil$.

Proof: We present a protocol that 1-solves T, and whose round complexity is $2 + \lceil \log_{(n-1)} R(T) \rceil$. The protocol is an improvement of the protocol in [BMZ], whose round complexity is $2 + R(T)$ ($2+2R(T)$ if the number of processors, n, is 3). Like the protocol in [BMZ], this protocol is based on a given solving covering function CF. Informally, this protocol differs from the one in [BMZ] in two ways. First, in each execution of this protocol all the vectors that may be suggested by the processors belong to a single path in the anchors tree of CF, while the protocol in [BMZ] may use a larger

---
[2] We assume here that the partial vector decided on in an i-sleeping execution of α by a processor P_j ($j \neq i$) is a $\vec{d}^i$. We can make this assumption since the protocol α' in the proof of Lemma 1 is actually such a protocol.

portion of that tree. Second, the convergence to two adjacent vertices on that path is done by an averaging process, similar to the one used in approximate consensus protocols, and not in the step by step fashion of the protocol in [BMZ].

Let CF be an optimal solving covering function of T (i.e., $R(T) = \max_{\vec{x} \in X_T} \rho_{CF}(\vec{x})$). By the computability of T, it follows that there is an algorithm $TREE$ that on input $\vec{x}$, outputs a minimum radius anchors tree $TREE(\vec{x})$ based on CF, with a center $ROOT(\vec{x})$ as its root. Our protocol assumes that each processor has a copy of the algorithms CF and $TREE$ above.

The general outline of the algorithm is as follows: In the first two stages each processor P_k is trying to find out the input vector $\vec{x}$. For this, it first broadcasts its input value and receives $n-1$ input values (including its own), which determine a partial input vector $\vec{x}^j$ (note that $j \neq k$). Then it broadcasts $\vec{x}^j$ and waits for $n-1$ such partial vectors. At this point, there are two kinds of processors: those who know only partial input vector $\vec{x}^j$ (it is the same $\vec{x}^j$ for all these processors !), and those who know the complete input vector $\vec{x}$.

Now, the processors perform a simple averaging approximate consensus, for $\lceil \log_{n-1} R(T) \rceil$ rounds, with two kinds of initial values: those who know $\vec{x}^j$ start with zero, and those who know $\vec{x}$ start with $R(T)$. During these rounds, each of the processors appends to its messages whatever it knows from the two things - $\vec{x}$ and j. After these rounds, each processor will have a value v in $[0, R(T)]$ s.t. the maximum difference between the values is 1. If v is equal to zero (in this case P_k still knows only $\vec{x}^j$) then P_k decides on $CF(\vec{x}^j)$ (deciding on a (partial) output vector $(d_1, \cdots, d_k, \cdots, d_n)$ means, in particular, that d_k is the decision value of P_k). Otherwise P_k knows $\vec{x}$ (and thus can compute $TREE(\vec{x})$; actually, it will only have to compute $ROOT(\vec{x})$, or the path in $TREE(\vec{x})$ from the j-anchor to $ROOT(\vec{x})$). If v is equal to $R(T)$, then P_k decides on $ROOT(\vec{x})$. Otherwise, P_k knows $\vec{x}$ and j. Then, it normalizes the value v to the length of the path from the j-anchor to $ROOT(\vec{x})$ (which is less or equal $R(T)$), and decides on the p-th (the normalized value) vector on this path. Since the difference between the v values was at most 1, this ensures that each non-faulty processor will decide on one out of two adjacent vertices (vectors) (this guarantees that the actual output vector is one of these two vectors, and hence it is in $T(\vec{x})$).

The protocol for P_k:
A. BROADCAST x_k and WAIT until you RECEIVE $n-1$ stage-A messages
B. you know $\vec{x}^j$. BROADCAST $\vec{x}^j$ and WAIT until you RECEIVE $n-1$ stage-B messages

C. {approximate consensus stage} if you know only $\vec{x}^j$ then $v \leftarrow 0$ else $v \leftarrow R(T)$
   for $r = 1$ to $\lceil \log_{(n-1)} R(T) \rceil$ do
           $info \leftarrow \vec{x}$ and / or j (whatever you know of the two)
           BROADCAST $(r, info, v)$ and WAIT until you RECEIVE $n-1$ messages of round r
           $v \leftarrow$ the average of the $n-1$ v's received in this round
   end

D. if $v = 0$ (you know only $\vec{x}^j$) then DECIDE $CF(\vec{x}^j)$
   else if $v = R(T)$ (you know only $\vec{x}$) then DECIDE $ROOT(\vec{x})$
   else (you know $\vec{x}$ and j) do
           Let l be the length of the path in $TREE(\vec{x})$ between the j-anchor and $ROOT(\vec{x})$
           $p \leftarrow \lfloor v l / R(T) \rfloor$
           DECIDE on the p'th vector of the path in $TREE(\vec{x})$ between the j-anchor and $ROOT(\vec{x})$
           (the j-anchor is number 0 in the path, and $ROOT(\vec{x})$ number l)
   end
   HALT

6.2 Correctness proof

In the correctness proof we will show that all non-faulty processors decide on two adjacent vectors in the path between a j-anchor to $ROOT(\vec{x})$. This j is determined if some processors know only $\vec{x}^j$ after stage B (such $\vec{x}^j$ is unique since $n-1$ is a majority). If there are no such processors, then they all decide on $ROOT(\vec{x})$.

After $\lceil \log_{(n-1)} R(T) \rceil$ rounds of approximate consensus in stage C the difference between the v's will be at most 1 (since it is reduced at least by a factor of $n-1$ each round). If no processor finish it with $v = R(T)$ or $v = 0$ then clearly the maximum difference between the p's is 1 since $l \leq R(T)$. If some processor finished with $v = R(T)$ (and decided on $ROOT(\vec{x})$) then all the v's are in the range $[R(T)-1, R(T)]$, and the minimum possible p is $l-1$ (the number of the vector adjust to $ROOT(\vec{x})$). The argument for $v = 0$ is similar (deciding on $CF(\vec{x}^j)$ is exactly like deciding on the j-anchor). □

7. GENERALIZATION

In this section we generalize our results to hold for arbitrary tasks. In the general case, the round complexity of a protocol that 1-solves a (possibly unbounded) task T is not a constant, but a function from the set of input vectors X_T to the positive integers.

Definition: Let T be a 1-solvable task. A function $f : X_T \rightarrow N$ is a *round complexity function of T* if there exists a protocol α that 1-solves T, and for each $\vec{x} \in X_T$ $rc_\alpha(\vec{x}) \leq f(\vec{x})$ ($rc_\alpha(\vec{x})$ is defined in Section 3.2).

Since in general there is no natural total order on such functions, we cannot define the optimal round complexity of a task T, but only define the set of *minimal* round complexity functions of T, in the natural partial ordering of functions, as follows:

Definition: Let f and g be two functions defined on the same domain X. Then f is *smaller* than g if $f \neq g$ and for all $x \in X$, $f(x) \leq g(x)$. A function g is *minimal* in a set of function **F** if there is no $f \in$ **F** such that f is smaller than g.

We define the set of minimal round complexity functions of a task T by a correspondence to the set of minimal radius functions in R_T: we show that for each minimal radius function ρ_{CF} in R_T there corresponds a minimal round complexity function which is $\theta(\log_n \rho_{CF})$, and these are the only minimal round complexity functions of T.

Theorem 1u: Let T be a task and ρ_{CF} be a minimal radius function in R_T. Then, there is no round complexity function of T which is smaller than $\log_{(n-1)} \rho_{CF}$.

Proof: Similar to the proof of Theorem 1, and using the minimality of ρ_{CF}. □

We now extend the upper bound to hold also for unbounded 1-solvable tasks. In fact, the next theorem show that for each radius function ρ_{CF} in R_T there is a protocol whose round complexity for each input $\vec{x}$ is $O(\log_n \rho_{CF}(\vec{x}))$.

Theorem 2u: Let ρ_{CF} be a radius function for a task T. Then, $3 + \lceil \log_{(n-1)} \rho_{CF} \rceil$ is a round complexity function of T.

Proof: We only need few minor changes in the protocol of Section 6: First, all occurrences of $R(T)$ are replaced by $\rho_{CF}(\vec{x})$. Now, the problem is that processors that at the beginning of stage C know only $\vec{x}^j$, cannot compute $\lceil \log_{(n-1)} \rho_{CF}(\vec{x}) \rceil$ - the number of approximate consensus rounds. To solve this problem, we add an initialization round in stage C (this idea is borrowed from [DLPSW]) in which a processor that receives a message with $v=0$ sets its own v to 0, and a processor that all the $n-1$ v values it receives are 0 (and thus still knows only $\vec{x}^j$), broadcasts a "FINISH" message, and exits stage C. A processor that receives in the next rounds a "FINISH" message, sets its v to 0, broadcasts a "FINISH" message and exits stage C. Thus, if some processor broadcasts "FINISH" message in the initialization round, then all processors set their v to 0, and it follows that all the v's will be zero after stage C. The rest of the correctness proof is similar to the one in Section 6. □

REFERENCES

[ABDKPR]   C. Attiya, A. Bar-Noy, D. Dolev, D. Koller, D. Peleg, R. Reischuk, *Achievable cases in an asynchronous environment*, **Proc. of the 28th FOCS**, October 1987, pp. 337-346.

[BMZ]   O. Biran, S. Moran and S. Zaks, *A combinatorial characterization of the distributed tasks which are solvable in the presence of one faulty processor*, **Proc. of the 7th PODC**, 1988, pp. 263-273.

[DDS]   D. Dolev, C. Dwork and L. Stockmeyer, *On the minimal synchronism needed for distributed consensus*, **Journal of the ACM**, Vol. 34 no. 1, pp. 77-97.

[DLPSW]   D. Dolev, N. A. Lynch, S. Pinter, E. Stark and W. Weihl, *Reaching approximate agreement in the presence of faults*, **Journal of the ACM**, Vol. 33 no. 3 (1986), pp. 499-516.

[Fe]   A. D. Fekete, *Asynchronous Approximate Agreement*, **Proc. of the 6th PODC**, 1987, pp. 64-76.

[FL]   G. N. Frederickson and N. A. Lynch, *Electing a leader in a synchronous ring* **Journal of the ACM**, Vol. 34 No. 1 (1987), pp. 98-115.

[FLP]   M. J. Fischer, N. A. Lynch and M. S. Paterson, *Impossibility of distributed consensus with one faulty process*, **Journal of the ACM**, Vol. 32 No. 2 (1985), pp. 373-382.

[KMZ]   E. Korach, S. Moran and S. Zaks, *Tight lower and upper bounds for some distributed algorithms for a complete network of processors*, **Proc. of the 3rd PODC**, pp. 199-207.

[MW]   S. Moran and Y. Wolfstahl, *Extended impossibility results for asynchronous complete networks*, **Information Processing Letters**, 26, 1987, pp. 145-151.

[NT]   G. Neiger and S. Toueg, *Automatically increasing the Fault-tolerance of distributed systems*, **Proc. of the 7th PODC**, pp. 248-262.

[TKM]   G. Taubenfeld, S. Katz and S. Moran, *Initial failures in distributed computations*, to appear in **Journal of Parallel and Distributed Computing**.

APPENDIX: APPLICATIONS

We present here new optimal bounds on the round complexity of the 1-solvable tasks mentioned in the paper. The first three examples deal with bounded tasks, and provide upper bounds of 3 rounds for the tasks involved (it can be shown that 2 rounds are not enough). All previous protocols that 1-solved these tasks required $\Omega(n)$ rounds. The bounds are proved by presenting a covering function CF for each task T which prove that $R(T) \le n-1$ (and hence $\log_{n-1} R(T) \le 1$). Actually, each of the covering functions presented will be optimal. The last example deal with the strong binary monotone approximate consensus, and provide a bound of $4 + \log_{n-1}(\frac{d-c}{\varepsilon})$, where d and c are the two medians of the numbers of the input vector. This is approximately the same bound that is proved optimal in [Fe] for the task of approximate consensus, which seems to be considerably simpler than the strong binary monotone approximate consensus. (We note, however, that the bounds in [Fe] apply to multiple failures.)

The formal definitions of the tasks discussed below are given in Section 2.2.

(1) **Binary Monotone Consensus:** Let $\vec{x}^i = (x_1, \cdots, x_{i-1}, *, x_{i+1}, \cdots, x_n)$ be a partial input vector for this task. Again, we assume for simplicity that n is even. In this case there is a unique possible covering function CF, defined by $CF(\vec{x}^i) = (c, \cdots, c)$, where c is the median of the multiset $\{x_1, \cdots, x_{i-1}, x_{i+1}, \cdots, x_n\}$
We now describe anchors trees based on CF. For a given input vector $\vec{x}$, let c and d be the two medians of the multiset $\{x_1, \cdots, x_n\}$. If $c = d$ then the anchors tree consists of the single vertex $(c, \cdots, c)$. Otherwise, it consists of the path $[(c, \cdots c, d), (c, \cdots, c, d, d) \cdots, (c, d, \cdots, d)]$. In the first case the radius of the tree is 0, and in the second is $\frac{n}{2} - 1$. It can be shown that this anchor tree is of minimum possible radius, and hence $R(T) = \frac{n}{2} - 1$.

(2) **Renaming with $n+1$ new names:** In this task the input to each processor is its id, and the id's are not mutually known. Such a task cannot be modeled as a function from input vectors to output vectors, since there is no fixed order among the processes. Instead, it is modeled as a function between input sets to allowed output sets [BMZ]. By adapting the definitions for this model, as done in [BMZ], we get a that $R(T) \le n-1$.

(3) **Order Preserving Renaming with $2n-1$ new names:** This task is order invariant, i.e: $T(\vec{x})$ depends only on the relative order among the entries of $\vec{x}$. CF is also order invariant, and we describe $CF(\vec{x}^i)$ only for the case that the entries in $\vec{x}$ are monotone increasing (i.e., $x_i < x_{i+1}$). The adaptation of the definition to other order types is straight forward. In this case, $CF(\vec{x}^i) = (2, 4, \cdots, 2i-2, *, 2i, \ldots, 2n-2)$. A suitable anchors tree of such $\vec{x}$ is the path of length $2n-2$ (and hence of radius $n-1$) starting at $(1, 2, 4, \cdots, 2n-2)$ and ending in $(2, 4, \cdots, 2n-2, 2n-1)$, that passes via all the i-anchors. (e.g., for $n=3$ this path is $[(1,2,4), (1,3,4), (2,3,4), (2,3,5), (2,4,5)]$).

(4) **Binary Monotone Approximate Consensus (for a given ε):** The input is the same as for the binary monotone consensus. The (unique) covering function CF is the same as for the binary monotone consensus. The minimal radius anchors tree based on CF is also similar to the one for the binary consensus, but this time ε must be taken into account:
For a given input vector $\vec{x}$, let c and d be the two medians of the multiset $\{x_1, \cdots, x_n\}$. Assume for simplicity the ε divides $d-c$. If $c = d$ then the anchors tree consists of the single vertex $(c, \cdots, c)$. Otherwise, it consists of the path $[(c, \cdots, c, c+\varepsilon), (c, \cdots, c, c+\varepsilon, c+\varepsilon),$

$\cdots$, $(c,c+\varepsilon,\cdots,c+\varepsilon)$, $(c+\varepsilon,\cdots,c+\varepsilon)$, $\cdots$, $(d-\varepsilon,\cdots,d-\varepsilon,d-\varepsilon)$, $\cdots$, $(d,\cdots,d-\varepsilon)]$. In the first case the radius of the tree is 0, and in the second is $n(\frac{d-c}{\varepsilon})$. Thus, the upper bound provided by our results (for unbounded tasks) is at most $5 + \log_{n-1}(\frac{d-c}{\varepsilon})$.

A Time-Randomness Tradeoff
for Communication Complexity [1]

Rudolf Fleischer[2]   Hermann Jung[3]   Kurt Mehlhorn[2]

ABSTRACT   We present a tight tradeoff between the expected communication complexity $\bar{C}$ and the number R of random bits used by any Las Vegas protocol (for a two-processor system) for the list-disjointness function of two lists of n numbers of n bits each. This function evaluates to 1 if and only if the two lists correspond in at least one position. We show a $\log(n^2/\bar{C})$ lower bound on the number of random bits used by any Las Vegas protocol, $\Omega(n) \leq \bar{C} \leq O(n^2)$. We also show that expected communication complexity $\bar{C}$, $\Omega(n \log n) \leq \bar{C} \leq O(n^2)$, can be achieved using no more than $(1 + o(1)) \log(n^2/\bar{C})$ random bits.

1. Introduction

The use of randomness has led, for a variety of problems, to algorithms which are more efficient than the best known deterministic algorithms. This suggests the following question : How much does a single random bit actually help?

One approach ([KPU]) to this question is to treat randomness as a resource and to analyze the amount of randomness needed to obtain a certain performance. If successful, this approach leads to tradeoffs between randomness and other complexity measures such as computation time and memory requirement. In this paper we prove such a tradeoff in the context of communication complexity. Previously a tradeoff between randomness and computation time was shown for oblivious routing in computer networks ([KPU]), and between randomness and memory for caching algorithms ([RS]).

Recently [CG] investigated tradeoffs between randomness and communication complexity for Monte Carlo protocols, i.e. protocols which are allowed to give wrong answers with some probability less than half. Their results are related to ours in the following sense: Roughly speaking, if you truncate a Las Vegas protocol after $\bar{C}$ bits you get a Monte Carlo protocol with same complexity. Hence lower bounds for Monte Carlo protocols also imply lower bounds for Las Vegas protocols (but not vice versa !).

The communication complexity of Boolean functions is a complexity measure corresponding to the amount of information transfer necessary to compute the function (see [F87],[F89],[MS],[LS88],[HR88],[AUY],[PS],[Y79]). Let $f : X_1 \times X_2 \to \{0,1\}$ be a Boolean function. Assume that P_1 and P_2 are two processors such that P_i knows

---

[1] This work was supported by the DFG, SFB 124, TP B2, VLSI-Entwurf und Parallelität and ESPRIT P3075 ALCOM

[2] Department of Computer Science, University of Saarland, 6600 Saarbrücken, Germany

[3] Department of Computer Science, Humboldt University, O-1086 Berlin, Germany

argument $x_i \in X_i$, $i = 1, 2$. The processors now exchange messages according to some probabilistic protocol. The protocol terminates when both processors have correctly determined the function value $f(x_1, x_2)$. This type of protocol is called Las Vegas protocol (uses randomness but always computes the correct answer). The following quantities measure the performance of a protocol :

Randomness R, i.e. the (maximal) number of random bits used on any input (x_1, x_2).

Communication Complexity C, i.e. a bound on the number of bits exchanged by the processors.

Failure Probability Q_C, i.e. the probability that the algorithm exchanges more than C bits.

Expected Communication Complexity $\bar{C}$, i.e. the (maximal) expected number of bits exchanged for any input (x_1, x_2).

Another measure for randomness is the entropy of a random source ([KY]). We remark that our lower bounds in chapter 4 (as well as the explicit upper bounds) hold as well for this measure instead of R.

We prove a tradeoff between randomness, communication complexity and failure probability for the list-disjointness function. Let $X_1 = X_2 = X^n$ where $X = \{0,1\}^n$. For $(x_1, \ldots, x_n), (y_1, \ldots, y_n) \in X^n$ define $LD((x_1, \ldots, x_n), (y_1, \ldots, y_n)) := 1$ if $\exists j : x_j = y_j$ and 0 otherwise. Note that the inputs to both processors consist of n^2 bits each. It is known ([MS]) that any deterministic algorithm (i.e. $R = 0$) for LD must exchange n^2 bits in the worst case (i.e. $\bar{C} \geq n^2$, and if $C = o(n^2)$ then $Q > 0$) and that there is a probabilistic algorithm with $\bar{C} = O(n)$ ([F87]). It is also known that this quadratic gap is the maximal possible one ([AUY]).

We will show in this paper :

Theorem 1 *(Lower bound)* Let $0 \leq r \leq \log n$ [a].

    a) Any protocol for LD that exchanges less than $C = n^2/2^r$ bits with probability $1 - Q$, $0 < Q < 1$, requires randomness $R \geq r - \log Q = \log(n^2/(CQ))$.

    b) Any protocol for LD that exchanges an expected number of $\bar{C} = n^2/2^r$ bits requires randomness $R \geq r = \log(n^2/\bar{C})$.

Theorem 2 *(Nonconstructive Upper Bound)* For $0 \leq r \leq \log n - \log \log n - 5$ a protocol exists which uses $R = r + \log r + 4$ random bits and exchanges an expected number of $\bar{C} \leq n^2/2^{r-2}$ bits. In other words, $R \leq (1 + o(1)) \cdot \log(n^2/\bar{C})$.

---

[a] *log* denotes logarithm of base 2, whereas *ln* denotes the natural logarithm of base *e*

Theorems 1 and 2 together imply that one additional random bit reduces the expected communication complexity by factor $\frac{1}{2}$. The proof of Theorem 2 is non-constructive, i.e. involves a counting argument. We also have an explicit construction, which does not give quite as good bounds however.

Theorem 3 *(Explicit Upper Bound)   For $0 \leq r \leq \log n - \log\log n - O(1)$ a simple protocol with randomness $R = 2r$ and expected communication complexity $\bar{C} \leq n^2/2^{r-2}$ exists. In other words, $R \leq (2 + o(1)) \cdot \log(n^2/\bar{C})$.*

2. The Old Algorithm

Let $x_1, \ldots, x_n, y_1, \ldots, y_n$ be n-bit integers (i.e. $0 \leq x_i, y_i < 2^n$) and

$$LD((x_1, \ldots, x_n), (y_1, \ldots, y_n)) := \begin{cases} 1 & \text{if } \exists j : x_j = y_j \\ 0 & \text{if } \forall j : x_j \neq y_j \end{cases}.$$

Fürer's time-optimal Las Vegas algorithm ([F87]) for LD uses $\Omega(n)$ random bits, which is far away from the lower bound stated in section 4. We will use the main idea of his algorithm to prove our much better results, so we first present Fürer's algorithm.

W.l.o.g. assume $n = 2^k$ for $k \in \mathbb{N}$. Let $\mathcal{P}_i := \{p \text{ prime} \mid 2^{2^{i-1}} \leq p < 2^{2^i}\}$, i.e. all $p \in \mathcal{P}_i$ have binary length between $2^{i-1} + 1$ and 2^i.

Algorithm A :

(1) P_1 chooses for $i = 4, \ldots, k-1$ a prime p_i from $\mathcal{P}_i$ at random with uniform distribution and sends it to P_2. After that P_1 computes the sequence $x_j^k := x_j$, $x_j^i := x_j^{i+1} \bmod p_i$ for all j, $1 \leq j \leq n$, and i, $k \geq i \geq 4$. P_2 does the same for the y_j.

(2) P_1 continues as follows :

```
j := 0;
repeat
    j := j + 1;
    i := 3;
    repeat
        i := i + 1;
        P₁ sends xⁱⱼ to P₂;
    until xⁱⱼ ≠ yⁱⱼ or i = k;
until xⁱⱼ = yⁱⱼ or j = n;
if xⁱⱼ = yⁱⱼ then f((x₁,...,xₙ),(y₁,...,yₙ)) := 1
else f((x₁,...,xₙ),(y₁,...,yₙ)) := 0;
```

Observe that $x_j^i \neq y_j^i$ implies $x_j \neq y_j$. So the inner loop checks $x_j \stackrel{?}{=} y_j$ correctly. For the analysis of the expected communication complexity of Algorithm A and the other algorithms we need the following Lemmas which generalize [F87].

Lemma 2.1 $|\mathcal{P}_i| \geq 2^{2^i - i - 3}$ for $i \geq 4$.

Proof: We know by the prime number theorem for $x \geq 12$ ([G72],[HW]):

$$\frac{1}{3} \cdot \frac{x}{\log x} \leq \frac{1}{3} \cdot \frac{x}{\ln x} \leq \text{\# primes less or equal } x \leq 16 \cdot \frac{x}{\ln x} \leq 24 \cdot \frac{x}{\log x}.$$

Therefore $|\mathcal{P}_i| \geq \frac{1}{3} \cdot \frac{2^{2^i}}{2^i} - 24 \cdot \frac{2^{2^{i-1}}}{2^{i-1}} = \frac{2^{2^i} - 2^{2^{i-1}+7} - 2^{2^{i-1}+4}}{3 \cdot 2^i} \geq 2^{2^i - i - 3}$. □

Lemma 2.2 If $z_1 \neq z_2$ are m-bit integers and $\mathcal{P}$ is a set of primes with binary length at least l, then there are at most $\lfloor \frac{m}{l} \rfloor$ different primes $p \in \mathcal{P}$ with $z_1 \equiv z_2$ modulo each of them.

Proof: Otherwise we would have $z_1 \equiv z_2$ modulo the product of these primes. But because this product has more than m bits, this would imply $z_1 = z_2$. □

Analysis of Algorithm A:

In step (1) P_1 uses (if $k \geq 5$) at least $r := \sum_{i=4}^{k-1} \log |\mathcal{P}_i| \geq \frac{n}{4}$ random bits (but no more than n) and also sends these bits to P_2.

In step (2) first assume $x_j \neq y_j$. Given $x_j^{i+1} \neq y_j^{i+1}$, the relative probability for $x_j^i = y_j^i$ is at most $\frac{3}{2^{2^i-i-3}}$ by the Lemmas above. In this case at most $\sum_{l=4}^{i+1} 2^l \leq 2^{i+2}$ bits are sent by P_1 to P_2 until $x_j \neq y_j$ is recognized. Therefore the expected communication complexity of the inner loop is $O(1)$ if $x_j \neq y_j$.

If $x_j = y_j$, then P_1 sends $\sum_{l=4}^{k} 2^l \leq 2n$ bits at most, but this can only happen once because the algorithm stops afterwards. So we have proven:

Theorem 2.3 *Algorithm A has expected communication complexity $\bar{C} = O(n)$ and uses between $\frac{n}{4}$ and n random bits.*

3. Improved Algorithms

It is easy to reduce the number of random bits in Algorithm A as follows :

Algorithm B :

Run Algorithm A, but do not use primes with length more than $2\log n$, i.e. restrict i to the range $4 \leq i \leq 1 + \log\log n$.

Theorem 3.1 *Algorithm B has expected communication complexity $\bar{C} = O(n)$ and randomness $2\log n \leq R \leq 3\log n$.*

Proof : Substitute $k' = \log\log n + 2$ for $k = \log n$ in the analysis of Algorithm A. Further observe that the relative probability for $x_j \equiv y_j \bmod p_{k'-1}$ if $x_j \neq y_j$ is less than $\frac{n}{\log n}/2^{2^{\log\log n+1}-\log\log n-4} = \frac{2^4}{n}$ by Lemmas 2.1 and 2.2 (because $p_{k'-1}$ has length $\log n + 1$ at least). Therefore the expected number of bits exchanged for checking $x_j \neq y_j$ is still $O(1)$. □

But our main goal is a tradeoff between the number of random bits and the expected communication complexity. We will first present an easy algorithm which is optimal up to factor two for the number r of random bits used (if this number is not too big).

Algorithm C :

Let S be any set of 4^r primes of length $n/2^r$ each. S is known to both P_1 and P_2.

(1) P_1 chooses a prime p from S at random and tells P_2 which one was chosen.

(2) P_1 sends $x_j \bmod p$ to P_2 for all j, and if $x_j \equiv y_j \bmod p$, then it also sends x_j. If $x_j = y_j$, then the algorithms halts.

Theorem 3.2 *Algorithm C can be applied if $r \leq \log n - \log\log n - O(1)$. It uses $2r$ random bits and has expected communication complexity $\bar{C} \leq n^2/2^{r-2}$.*

Proof : Note first that the required set S of primes exists provided that $r \leq \log n - \log\log n - O(1)$. This is an easy consequence of Lemma 2.1. Observe next that for $p \in S$

$$\text{prob}(x_j \equiv y_j \bmod p \text{ if } x_j \neq y_j) \leq \frac{n}{\frac{n}{2^r}} \cdot \frac{1}{4^r} = \frac{1}{2^r}$$

by Lemma 2.2. Therefore an expected number of $\frac{n}{2^r} + \frac{1}{2^r} \cdot n + 2$ bits has to be exchanged to check $x_j \neq y_j$. If $x_j = y_j$, then n more bits are sent from P_1 to P_2. □

Algorithm C uses twice as many random bits as the lower bound requires (Theorem 4.2). We will now show that $\bar{C} = O(n^2/2^r)$ can already be obtained with $r + o(r)$ random bits; this is optimal up to lower order terms. Unfortunately, we have no explicit construction of such an algorithm. We can only prove its existence by a counting argument. The algorithm is a refinement of Algorithm A. The main idea is to restrict the choice of random primes to small subsets of $\mathcal{P}_i$. Also, because we aim at communication complexity $O(n^2/2^r)$, we need the modulo cascade only from $\mathcal{P}_{k-1}$ to $\mathcal{P}_{k-r}$ (any $p \in \mathcal{P}_{k-r}$ has $2^{k-r} = n/2^r$ bits at most, so we can afford to send x_j^{k-r}).

Def. 3.3 Let $\mathcal{P}$ be the set of all sequences $s = (p_{k-1}, \ldots, p_{k-r})$ with primes $p_j \in \mathcal{P}_j$, and be (x, y) a pair of numbers with $x \neq y$.
(1) A sequence $s \in \mathcal{P}$ is called i-bad for (x, y) if
$$(\ldots (x \bmod p_{k-1}) \ldots) \bmod p_i = (\ldots (y \bmod p_{k-1}) \ldots) \bmod p_i \, .$$
(2) A subset $\mathcal{L} \subseteq \mathcal{P}$ is called bad for (x,y) if an i exists such that at least $Q_i \cdot L$ sequences of $\mathcal{L}$ are i-bad for (x,y); here $L := |\mathcal{L}|$ and $Q_i := \frac{n}{r \cdot 2^{r+i+1}} \leq \frac{1}{2}$.

Theorem 3.4 If we restrict Algorithm A to sequences s of an $\mathcal{L} \subseteq \mathcal{P}$ which are good for all (x,y), $x \neq y$ and $0 \leq x, y < 2^n$, we get expected communication complexity $\bar{C} \leq \dfrac{n^2}{2^{r-1}} + (r+2) \cdot n + \log L$ and use only $\log L$ random bits.

Proof : $\log L$ bits are required to transmit the choice of the sequence s. For fixed $x_j \neq y_j$ and all i there are at most $\lfloor Q_i \cdot L \rfloor$ sequences of $\mathcal{L}$ which are i-bad for (x_j, y_j), i.e. $x_j^i = y_j^i$. Hence P_1 sends an expected number of $\dfrac{n}{2^r} + \sum_{i=k-r}^{k-1} \dfrac{\lfloor Q_i \cdot L \rfloor}{L} \cdot 2^{i+1} \leq \dfrac{n}{2^{r-1}}$ bits to check $x_j \neq y_j$ (and P_2 sends at most r one-bit answers).

If $x_j = y_j$, then at most $2n$ bits are exchanged. □

Theorem 3.5 If $r \leq \log n - \log \log n - 5$, then an $\mathcal{L} \subseteq \mathcal{P}$, $|\mathcal{L}| = r \cdot 2^{r+4}$, exists which is good for all (x,y), $x \neq y$ and $0 \leq x, y < 2^n$.

Cor. 3.6 If $r \leq \log n - \log \log n - 5$, then there exists an algorithm for LD with expected communication complexity $\bar{C} \leq n^2/2^{r-2}$ which uses only $r + \log r + 4$ random bits.

To prove Theorem 5 we first need two technical Lemmas.

Lemma 3.7 For $i \geq 4$ and all (x,y) with $x \neq y, 0 \leq x, y < 2^n$, is
$$b_i(x,y) := \#\,i\text{-bad sequences for } (x,y) \leq 6 \cdot \frac{|\mathcal{P}|}{|\mathcal{P}_i|} \, .$$

Proof : We have by Lemma 2.2

$$b_{k-1}(x,y) \leq 3 \cdot \frac{|\mathcal{P}|}{|\mathcal{P}_{k-1}|}$$

and $\quad b_l(x,y) \leq b_{l+1}(z) + |\mathcal{P}_{k-1}| \cdots |\mathcal{P}_{l+1}| \cdot 3 \cdot |\mathcal{P}_{l-1}| \cdots |\mathcal{P}_{k-r}|$.

Because of $|\mathcal{P}_{l+1}| \geq 2 \cdot |\mathcal{P}_l|$ (Lemma 2.1) we conclude :

$$b_i(x,y) \leq 3 \cdot \left(\frac{|\mathcal{P}|}{|\mathcal{P}_{k-1}|} + \cdots + \frac{|\mathcal{P}|}{|\mathcal{P}_i|} \right) \leq 6 \cdot \frac{|\mathcal{P}|}{|\mathcal{P}_i|} \,. \qquad \square$$

Lemma 3.8 If $\log n \geq i \geq \log n - r \geq 5$, $r \leq \log n - \log\log n - 5$ and $1 \leq c \leq \min\{2^6, n^2\}$, then

$$2^{2n} \cdot \left(2^4 \cdot \left(\frac{c}{Q_i \cdot |\mathcal{P}_i|} \right)^{Q_i} \right)^L \leq \frac{1}{2r} \,.$$

Proof : (1) Since $i \leq \log n$ and $c \leq n^2$ we have

$$\begin{aligned}
i \;\geq\; \log n - r \;&\geq\; \log(2\log n + 2\log n) + 2 \\
&\geq\; \log\bigl(\log n + (\log n - \log\log n - 5) + 4 \\
&\quad + \log(\log n - \log\log n - 5) + \log c\bigr) + 2 \\
&\geq\; \log(\log n + r + \log r + \log c + 4) + 2 \\
&\geq\; \log(\underbrace{2i - \log n + r + \log r + \log c + 4}_{=:d}) + 2 \,.
\end{aligned}$$

Hence $2^i - d \geq 2^i - 2^{i-2}$.

(2) Let $a := \dfrac{1}{2^i \cdot Q_i} = \dfrac{r \cdot 2^{r+1}}{n} \leq \dfrac{\log n \cdot 2^{\log n - \log\log n - 4}}{n} = \dfrac{1}{16}$.

Then for $c \leq 2^6$ and $i \geq 5$:

$$\begin{aligned}
|\mathcal{P}_i| \;\geq\; 2^{2^i - i - 3} \;&>\; 2^{2^{i-1} + i + \log c + \log a} \\
&\geq\; 2^{8 a 2^i} \cdot c \cdot a \cdot 2^i \\
&=\; c \cdot 2^{\frac{8}{Q_i}} \cdot \frac{1}{Q_i}
\end{aligned}$$

$$\implies \quad Q_i \cdot \log \frac{Q_i \cdot |\mathcal{P}_i|}{c} > 8 \,.$$

(3) It follows:
$$\frac{2n + \log 2r}{Q_i \cdot \log \frac{Q_i \cdot |\mathcal{P}_i|}{c} - 4} \leq \frac{2n + 1 + \log \log n}{\frac{n}{r \cdot 2^{r+i+1}} \cdot \log \left(\frac{n}{r \cdot 2^{r+i+1}} \cdot \frac{2^{2^i - i - 3}}{c} \right) - 4}$$

$$\leq \frac{4n \cdot r \cdot 2^{r+i+1}}{n \cdot (2^i - d) - r \cdot 2^{r+i+3}}$$

$$\stackrel{(1)}{\leq} \frac{4n \cdot r \cdot 2^{r+i+1}}{n \cdot (2^i - 2^{i-2}) - a \cdot n \cdot 2^{i+2}}$$

$$\stackrel{(2)}{\leq} \frac{r \cdot 2^{r+i+3}}{2^i - 2^{i-2} - 2^{i-2}}$$

$$= r \cdot 2^{r+4}$$

$$= L$$

$$\stackrel{(2)}{\Longrightarrow} \quad L \cdot \left(Q_i \cdot \log \frac{Q_i \cdot |\mathcal{P}_i|}{c} - 4 \right) \geq 2n + \log 2r$$

$$\Longrightarrow \quad 2n + 4L + L \cdot Q_i \cdot \log \frac{c}{Q_i \cdot |\mathcal{P}_i|} \leq \log \frac{1}{2r} \,. \qquad \square$$

Proof of Theorem 3.5:

We show that the fraction of $\mathcal{L}$'s which are bad for some pair (x, y), $x \neq y$ and $0 \leq x, y < 2^n$, is less than $\frac{1}{2}$.

For a fixed (x, y) an $\mathcal{L} \subseteq \mathcal{P}$ is bad for (x, y) if there is an i, $k - r \leq i \leq k - 1$, such that at least $\lceil Q_i \cdot L \rceil = Q_i \cdot L$ (because $n = 2^k$) sequences of $\mathcal{L}$ belong to the at most $6 \cdot \frac{|\mathcal{P}|}{|\mathcal{P}_i|} =: c \cdot \frac{|\mathcal{P}|}{|\mathcal{P}_i|}$ sequences which are i-bad for (x, y) (Lemma 3.7). Hence

$$\frac{\#\text{bad } \mathcal{L}}{\#\mathcal{L}} \leq \frac{2^{2n} \cdot \sum_{i=k-r}^{k-1} \binom{c \cdot \frac{|\mathcal{P}|}{|\mathcal{P}_i|}}{Q_i \cdot L} \cdot \binom{|\mathcal{P}|}{(1-Q_i) \cdot L}}{\binom{|\mathcal{P}|}{L}}$$

$$\leq 2^{2n} \cdot \sum_{i=k-r}^{k-1} \left(\frac{c \cdot |\mathcal{P}| \cdot e}{|\mathcal{P}_i| \cdot Q_i \cdot L} \right)^{Q_i \cdot L} \cdot \left(\frac{|\mathcal{P}| \cdot e}{(1 - Q_i) \cdot L} \right)^{(1-Q_i) \cdot L} \cdot \left(\frac{L}{|\mathcal{P}| - L} \right)^L$$

(because of Sterling's Approximation $\binom{n}{m} \leq \left(\frac{n \cdot e}{m} \right)^m$)

$$= 2^{2n} \cdot \sum_{i=k-r}^{k-1} \left(\left(\frac{c \cdot |\mathcal{P}| \cdot e}{|\mathcal{P}_i| \cdot Q_i \cdot L} \cdot \frac{(1 - Q_i) \cdot L}{|\mathcal{P}| \cdot e} \right)^{Q_i} \cdot \frac{|\mathcal{P}| \cdot e}{(1 - Q_i) \cdot L} \cdot \frac{L}{|\mathcal{P}| - L} \right)^L$$

$$\leq 2^{2n} \cdot \sum_{i=k-r}^{k-1} \left(\left(\frac{c \cdot (1-Q_i)}{|\mathcal{P}_i| \cdot Q_i} \right)^{Q_i} \cdot \frac{2e}{1-Q_i} \right)^L \qquad (|\mathcal{P}| \geq 2L)$$

$$\leq \sum_{i=k-r}^{k-1} 2^{2n} \cdot \left(\left(\frac{c}{|\mathcal{P}_i| \cdot Q_i} \right)^{Q_i} \cdot 2^4 \right)^L \qquad (Q_i \leq \frac{1}{2})$$

$$\leq \frac{1}{2} \qquad \text{(by Lemma 3.8)}$$

, i.e. at least half of all $\mathcal{L} \subseteq \mathcal{P}$, $|\mathcal{L}| = r \cdot 2^{r+4}$, are good for all (x,y), $x \neq y$ and $0 \leq x, y < 2^n$. $\square$

4. The Lower Bound

Our proof of the lower bound is based on the lower bound proof for deterministic communication complexity as given in [MS] and generalized in [F89].

Let F be the functional matrix of some function $f : X \times Y \to \{0,1\}$, i.e. $F[x,y] := f(x,y)$. It is well known that any probabilistic algorithm $\mathcal{A}$ can be transformed into an equivalent algorithm $\mathcal{A}'$ which first does some random step and afterwards runs deterministically. If we have r random bits, this means that we first choose a deterministic algorithm out of a set of 2^r algorithms, which then solves the problem.

Lemma 4.1 For any set of 2^r deterministic algorithms for f there exists an input (x,y) such that all of these 2^r algorithms need at least communication complexity $\frac{\log rank(F)}{2^r}$ on that input. Here $rank(F)$ denotes the rank of the matrix F over the field of two elements (or any other field).

Proof: Let $\{A_1, \ldots, A_{2^r}\}$ be a set of deterministic algorithms for f. We know that $\log rank(F)$ is a lower bound for the deterministic communication complexity of f ([MS],[F89]). Assume w.l.o.g. that algorithms $A_1, \ldots, A_a$ start with processor P_1 sending the first bit (and $A_{a+1}, \ldots, A_{2^r}$ with P_2). Then the set of all functions $h : \{1, \ldots, a\} \to \{0,1\}$ induces a disjunct partition of the input values X (of processor P_1) as follows :

$$X_h := \{x \in X \mid \text{Algorithm } A_i \text{ sends } h(i) \text{ as its first bit}, 1 \leq i \leq a\}$$

, i.e. X_h is a maximal subset of X with the property that each algorithm does the same for all $x \in X_h$. Because there are only 2^a different functions h there must be one function h_1 such that F restricted to X_{h_1} has at least rank $\frac{rank(F)}{2^a}$.

In the same way we can find a subset Y_{h_2} of the input values Y of processor P_2 (considering the algorithms $A_{a+1}, \ldots, A_{2^r}$ which have P_2 sending the first bit) such that the resulting functional matrix has at least rank $\frac{rank(F)}{2^a \cdot 2^{2^r-a}} = \frac{rank(F)}{2^{2^r}}$.

As long as the matrix has rank > 1 the computation cannot stop (because not both processors can know the correct result). Hence we can iterate this procedure until $\left(2^{2^r}\right)^C \geq rank(F)$, but then all algorithms had to communicate C bits. □

From this follows immediately

Theorem 4.2 *Any probabilistic algorithm for f using r random bits has expected communication complexity $\bar{C} \geq \frac{\log rank(F)}{2^r}$.* □

An alternative proof of this Theorem is by simulating all sequences of random bits by a deterministic protocol (see [CG]).

As suggested by [KPU], Lemma 4.1 can also be used to derive a relationship between communication complexity, failure probability and randomness.

Theorem 4.3 *Any protocol for f that exchanges less than $C = \frac{\log rank(F)}{2^r}$ bits with probability $1 - Q$, $0 < Q < 1$, requires randomness*
$$R \geq r - \log Q = \log \frac{\log rank(F)}{CQ}.$$

Proof : If the probabilistic algorithm exchanges on every input with probability $1 - Q$ less than C bits, then the sum of the probabilities given to any set of 2^r deterministic algorithms must be bounded by Q (this follows directly from Lemma 4.1). But then there must be at least $\frac{2^r}{Q}$ deterministic algorithms that are selected with positive probability. □

Cor. 4.4 *Let $0 \leq r \leq \log n$.*

a) *Any protocol for LD that exchanges less than $C = n^2/2^r$ bits with probability $1 - Q$, $0 < Q < 1$, requires randomness $R \geq r - \log Q = \log(n^2/(CQ))$.*

b) Any protocol for LD that exchanges an expected number of $\bar{C} = n^2/2^r$ bits requires randomness $R \geq r = \log(n^2/\bar{C})$.

Proof: [MS] have shown $rank(F) = 2^{n^2}$. □

5. Further Research

(1) We should try to close the gap between the upper and lower bound (Cor. 3.6 and Cor. 4.4).

(2) The upper bound (Theorem 3.5) applies only for $r \leq \log n - \log\log n - 5$. We should also find efficient algorithms for $\log n - \log\log n - 4 \leq r \leq \log n$.

(3) [HR88] exhaustively studied k-round protocols (i.e. the processors are only allowed to send a restricted number of messages of arbitrary length). It should be possible to prove similar results for this kind of communication complexity.

References

[AUY] A.V. Aho, J.D. Ullman, M. Yannakakis
"On notions of information transfer in VLSI circuits"
Proc. 15th ACM STOC 1983, 133–139

[CG] R. Canetti, O. Goldreich
"Bounds on tradeoffs between randomness and communication complexity"
to appear in FOCS 1990

[F87] M. Fürer
"The power of randomness for communication complexity"
Proc. 19th ACM STOC 1987, 178–181

[F89] R. Fleischer
"Communication complexity of multi processor systems"
IPL **30** (1989), 57-65

[G72] K.-B. Gundlach
"Einführung in die Zahlentheorie
BI Bd. 772, 1972

[H86] B. Halstenberg
"Zweiprozessorkommunikationskomplexität"
Diplomarbeit
University of Bielefeld 1986

[HR87] B. Halstenberg, R. Reischuk
"Relations between communication complexity classes"
Proc. 3rd Structure Conference 1988

[HR88]   B. Halstenberg, R. Reischuk
"On different modes of communication"
Proc. 20th ACM STOC 1988, 162–172

[HW]   Hardy & Wright
"An Introduction to the Theory of Numbers"
5th edition, Oxford 1979

[KPU]   D. Krizanc, D. Peleg, E. Upfal
"A time-randomness tradeoff for oblivious routing"
Proc. 20th ACM STOC 1988, 93–102

[KY]   D.E. Knuth, A.C Yao
"The complexity of nonuniform random number generation"
in *Algorithms and Complexity*, Ed. J.E. Traub, Academic Press
N.Y., 1976, 357–428

[LS81] R.J. Lipton, R. Sedgewick
"Lower bounds for VLSI"
Proc. 13th ACM STOC 1981, 300–307

[LS88] L. Lovász, M. Saks
"Lattices, möbius functions and communication complexity"
Proc. 29th IEEE FOCS 1988, 81–90

[MS]   K. Mehlhorn, E.M. Schmidt
"Las Vegas is better than determinism in VLSI and distributed
    computing"
Proc. 14th ACM STOC 1982, 330–337

[PS]   C.H. Papadimitriou, M. Sipser
"Communication complexity"
Proc. 14th ACM STOC 1982, 196–200
Journal of Computer and System Sciences **28**, 1984, 260–269

[RS]   P. Raghavan, M. Snir
"Memory versus randomization in on-line algorithms"
ICALP 1989, 687–703

[Y77]   A.C. Yao
"Probabilistic computations: toward a unified measure of complexity"
Proc. 18th IEEE FOCS 1977, 222–227

[Y79]   A.C. Yao
"Some complexity questions related to distributive computing"
Proc. 11th ACM STOC 1979, 209–213

[Y83]   A.C. Yao
"Lower bounds by probabilistic arguments"
Proc. 24th IEEE FOCS 1983, 420–428

BOUNDS ON THE COSTS OF REGISTER IMPLEMENTATIONS

SOMA CHAUDHURI
Department of Computer Science and Engineering
University of Washington
Seattle, Washington

JENNIFER WELCH
Department of Computer Science
University of North Carolina
Chapel Hill, North Carolina

Abstract

A fundamental aspect of any concurrent system is how processes communicate with each other. Ultimately, all communication involves concurrent reads and writes of shared memory cells, or *registers*. The stronger the guarantees provided by a register, the more useful it is to the user, but the harder it may be to implement in practice. Thus it is of interest to determine which types of registers can implement which other types of registers. The types of registers studied in this paper are safe vs. regular, 1-reader vs. n-readers, and binary vs. k-ary. Algorithms for various implementations have been previously developed. These have, for the most part, concentrated on the relative computability between different types of registers. In contrast, this paper studies the relative complexity of such algorithms, by considering the costs incurred when implementing one type of register (the *logical* register) with registers of another type (*physical* registers). The cost measures considered are the number of physical registers and the number of reads and writes on the physical registers required to implement the logical register. Bounds on the number of physical operations can be easily converted to provide time bounds for the logical operations. Tight bounds are obtained on the cost measures in many cases, and interesting trade-offs between the cost measures are identified. The lower bounds are shown using information-theoretic techniques. Two new algorithms are presented that improve on the costs of previously known algorithms: the *hypercube* algorithm implements a k-ary safe register out of binary safe registers, requiring only one physical write per logical write; and the *tree* algorithm implements a k-ary regular register out of binary regular registers, requiring only $\log k$ physical operations per logical operation. Both algorithms use novel combinatorial techniques.

1 Introduction

A fundamental aspect of any concurrent system is how processes communicate with each other. Ultimately, all communication involves concurrent accesses to shared memory cells, or registers. The stronger the guarantees provided by the shared memory, the more useful it is to the user, but the harder it may be to implement in practice. Thus it is of interest to determine which types of registers can implement which other types. Many such implementations are known [Blo87, BP87, Lam86, LTV90, NW87, Pet83, SAG87, Tro89, Vid88, VA86].

The contribution of this paper is to study the *costs* of implementing one type of register (the *logical* register) out of registers of another type (the *physical* registers). Cost measures considered are the number of physical registers, and the number of operations on the physical registers used to perform the operations of the implemented register. Bounds on the number of physical operations can be used to obtain time bounds for the logical operations in terms of the time taken by the physical operations.

A *register* is a shared variable or memory cell that supports concurrent reading and writing by a collection of processing entities. The operations of reading and writing are not instantaneous; instead, they have duration in time, from a starting point to an ending point. Although each entity accessing a register is assumed to issue operations sequentially, operations on behalf of different entities can overlap in time.

A variety of types of registers can be defined, differing in several dimensions, including the number of concurrent readers supported, the number of concurrent writers supported, the number of values the register can take on, and the strength of the consistency guarantees provided in the presence of concurrent operations. Throughout this paper we assume there is only one writer and a fixed number of readers, leaving two parameters of interest: the number of values and the consistency guarantees. We distinguish between binary registers and k-ary registers, for $k > 2$. (A k-ary register can take on k different values.)

Lamport [Lam86] defines three kinds of consistency guarantees, called safe, regular, and atomic. Roughly speaking, a read of a *safe* register always returns the most recent value written to the register, unless the read overlaps with a write, in which case any legal value of the register can be returned. A read of a *regular* register always returns the most recent value written, unless the read overlaps one or more writes, in which case it returns either the old value or one of the values written by an overlapping write. An *atomic* register provides the illusion, via the values returned by read operations, that each operation happens at a single instant in time within its range, *i.e.*, that the operations are totally ordered. In this paper, we only consider safe and regular registers.

The types of registers defined form a hierarchy of stronger and weaker definitions. For example, an n-reader, k-ary, regular register, for $n > 1$ and $k > 2$, can "implement" a 1-reader, binary, safe register *a fortiori*, simply because the former has a stronger definition than the latter. Lamport [Lam86] describes implementations among safe and regular one-writer registers (as well as atomic), showing that in many cases weaker register types can implement stronger register types.

We study the costs incurred by implementations between register types. Let M, R, and W be the minima, over all implementations between two particular types of registers, of the number of physical registers, the maximum number of physical operations in a logical read, and the maximum number of physical operations in a logical write, respectively. In particular, our algorithms will involve no physical reads in a logical write and no physical writes in a logical read. Our lower bound results give bounds on the number of physical reads per logical read, and the number of physical writes per logical write. These are stronger results than just giving bounds on the number of physical operations per logical action.

For implementing a k-ary safe register out of binary safe registers, we show tight bounds of $R = \lceil \log k \rceil$, $W = 1$, and $M = \lceil \log k \rceil$. The upper bound of 1 on W is obtained from a new algorithm, which we call the *hypercube* algorithm. The best previous upper bound on W was $\lceil \log k \rceil$ [Lam86]. These three optimal bounds are not obtained simultaneously in a single algorithm, and in fact, we show some non-trivial trade-offs between the three cost measures.

For implementing a k-ary regular register out of binary regular registers, we show the tight bound that $R = \lceil \log k \rceil$, and the bounds $1 \leq W \leq \lceil \log k \rceil$, and $\max\{\lceil \log k \rceil + 1, 2(\log k) - \log \log k - 2\} \leq M \leq \min\{k - 1, n(3 \log k + 68)\}$, where n is the number of readers of the logical register. The upper bounds on R and W are simultaneously achieved by a new algorithm, which we call the *tree* algorithm. We also present some lower bounds on R and M that follow if we restrict attention to implementations that use only a small constant number of physical writes per logical write.

Our results for binary to k-ary implementations are summarized in Tables 1 and 2. Table 1 gives the bounds when all algorithms are considered. Table 2 gives the bounds when certain classes of algorithms are considered, as specified by the column labeled S—namely, 1-write algorithms, c-write algorithms, and $\lceil \log k \rceil$-register algorithms.

All of the lower bounds mentioned above are new. Little previous work has been done concerning lower bounds or trade-offs for register implementations. One such previous result is in [Lam86], where it is shown that in any implementation of an atomic register using regular registers, a read of the logical register must involve a write to a physical register. Tromp [Tro89] uses this result to show that three binary safe registers are necessary to construct a binary atomic register.

In Section 2 we present our model and some results that are true for all implementations. The bulk of the paper concerns implementing k-ary registers out of binary registers. Section 3 considers safe registers and Section 4 considers regular registers. Our conclusions are in Section 5. Due to lack of space, some proofs have been sketched or omitted. The complete proofs can be found in [CW90].

	Safe		Regular	
	lower	upper	lower	upper
R	$\lceil \log k \rceil$	$\lceil \log k \rceil$	$\lceil \log k \rceil$	$\lceil \log k \rceil$
W	1	1	1	$\lceil \log k \rceil$
M	$\lceil \log k \rceil$	$\lceil \log k \rceil$	$\max\{\lceil \log k \rceil + 1,$ $\lceil 2 \log k - \log \log k \rceil - 2\}$	$\min\{k-1,$ $n(3 \log k + 68)\}$

Table 1: Independent Bounds for Binary to k-ary Algorithms

S			Safe		Regular	
			lower	upper	lower	upper
$\{A \mid W_A = 1\}$		R_S	$k-1$	$2^{\lceil \log k \rceil} - 1$	$k-1$	∞
		M_S	$k-1$ or k^*	$2^{\lceil \log k \rceil} - 1$	k	∞
$\{A \mid W_A = c\}$		R_S	$(c!k/2)^{1/c}$	$c - 2 + \lceil k/2^{c-2} \rceil$	$(c!k/2)^{1/c}$	∞
		M_S	$(c!k/2)^{1/c}$	$c - 2 + \lceil k/2^{c-2} \rceil$	$(c!k/2)^{1/c}$	∞
$\{A \mid M_A = \lceil \log k \rceil\}$		W_S	$\lceil \log k \rceil$	$\lceil \log k \rceil$	∞	∞

Table 2: Trade-Off Results for Binary to k-ary Algorithms

2 Preliminaries

In this section, we give formal definitions for the types of registers that we will study (n-reader, k-ary, safe and regular), describe the rules we impose on implementing one type of register with another, and define the cost measures we will use. Then we present some definitions and lemmas that are true for implementations between any types of registers.

2.1 Model

We model system components using a state machine whose state transitions are labeled with **actions**. If there is a transition from a state labeled with an action, then that action is **enabled** in that state. The state machine is deterministic in that every transition from a particular state is labeled with a different action. An **execution** of an automaton is an alternating sequence of states and actions, beginning with an initial state, in which each action is enabled in the previous state and each state change correctly reflects the transition relation for the intervening action. A **schedule** of an automaton is the sequence of actions extracted from an execution.

We model an n-reader, k-ary safe (or regular) register by an automaton X as follows. Let V be the value set of the register with $|V| = k$ and initial value $v_0 \in V$. Let N be a set of size n identifying the n readers. The actions of X are $\{\text{read}(i) : i \in N\} \cup \{\text{write}(v) : v \in V\} \cup \{\text{return}(i,v) : i \in N, v \in V\} \cup \{\text{ack}\}$. These are the start and finish of the read and write operations on the register: a read is terminated by a return and a write by an ack; the i parameter of a read identifies the particular reader.

We restrict the register automaton X to have the following property. Every schedule α of X is **well-formed**, meaning that for all $i \in N$, the restriction of α to reads and returns for i consists of alternating reads and returns, beginning with a read, and the restriction of α to writes and acks consists of alternating writes and acks, beginning with a write. (This models the sequential nature of the individual processing entities that access the register.) Given a sequence α, each read(i) instance and the following return(i,v) instance constitute an **operation**, and the same for each write(v) instance and the following ack. The two members of the same operation **match** each other. Every schedule α contains at most one unmatched write and at most one unmatched read for each i; these are called **pending in** α.

We also require that operations can be initiated at any time, as long as well-formedness is not violated. For every schedule α of X, let s_α be the state resulting from that schedule. If no write is pending in α,

---
[†]$k-1$ if k is a power of 2, k otherwise

then write(v) is enabled in s_α, for all v. Similarly, for all $i \in N$, if no read(i) is pending in α, then read(i) is enabled in s_α. We call this the **free initiation** property.

The automaton X must return correct values for reads, where the notion of correct depends on whether the register is safe or regular. Consider any (completed) read operation in any schedule. If no write operation overlaps this read operation, then the read must return the value of the most recent preceding write; if there is no preceding write, then the read must return the initial value v_0. Suppose a write does overlap the read. If the register is safe, then the read can return any value in V. If the register is regular, then the read must return either the value of the most recent preceding write (or v_0 if there is no such write) or some value written by an overlapping write.

The preceding intuitive discussion is now formalized. We define two kinds of **possible values** on finite sequences α, denoted $PV^{\text{safe}}(\alpha)$ and $PV^{\text{regular}}(\alpha)$. Suppose there is no write action in α. Then $PV^{\text{safe}}(\alpha) = PV^{\text{regular}}(\alpha) = \{v_0\}$. Suppose $\alpha = \alpha_1 \text{write}(v) \alpha_2$, where there is no write action in α_2. If there is an ack action in α_2 (i.e., no write is pending), then $PV^{\text{safe}}(\alpha) = PV^{\text{regular}}(\alpha) = \{v\}$. If there is no ack action in α_2 (i.e., a write is pending), then $PV^{\text{safe}}(\alpha) = V$ and $PV^{\text{regular}}(\alpha) = \{v\} \cup PV^{\text{regular}}(\alpha_1)$. When the superscript "safe" or "regular" on PV is clear from context, it will be dropped. We require that for any schedule α of X, for every read operation in α, the value returned is in $PV(\alpha')$, where α' is some prefix of α that ends after the operation starts and before it finishes.

Finally, we require that the register be **wait-free**: for every finite schedule α of X, if operation $\mathcal{O}$ is pending in α, then there is a schedule $\alpha\pi$, where π is a single action, such that $\mathcal{O}$ is not pending in $\alpha\pi$. This condition states that at any point in an execution at which an operation is pending, it is possible to complete the operation without waiting for any other operation to start or complete.

We now define the "rules" for implementing one type of register, the **logical** register, out of registers of another type, the **physical** registers. The **type** of a register specifies the number of readers supported, the number of values it can take on, and whether it is safe or regular. The building blocks for an implementation are physical registers, read processes, and write processes. There is one write process (since we are only considering 1-writer registers); the number of read processes is the number of readers to be supported by the logical register. Each read or write process is an automaton that communicates with the outside world via (logical) READ, WRITE, RETURN, and ACK actions (the actions of a logical register) and with the physical registers via (physical) read, write, return and ack actions. (Logical actions are denoted by upper-case, physical actions by lower-case.) The read and write processes cannot communicate directly with each other. Also, in any schedule, no physical operation is pending unless a logical operation is pending, and at most one physical operation is pending at any point.

We proceed more formally. Assume particular logical and physical register types with associated value sets V and V' respectively. Let m be the number of physical registers.

A **read process** is an automaton RP_i, $i \in N$, that has the actions READ(i) and RETURN(i, v), $v \in V$ (by which it communicates with the outside world), the actions read$_j(i)$ and return$_j(i, v')$, $v' \in V'$ (by which it reads registers, where j ranges over the physical registers read), and the actions write$_j(v')$, $v' \in V'$, and ack$_j$ (by which it writes registers, where j ranges over the physical registers written).

A **write process** is an automaton WP that has the actions WRITE(v), $v \in V$, and ACK (by which it communicates with the outside world), the actions read$_j(0)$ and return$_j(0, v')$, $v' \in V'$ (by which it reads registers, where j ranges over the physical registers read), and the actions write$_j(v')$, $v' \in V'$, and ack$_j$ (by which it writes registers, where j ranges over the physical registers written).

We restrict each read process automaton RP_i to have the following property. Exactly one group of actions is enabled in each state, where each action is in its own group, except that for each j, the set of actions $\{\text{return}_j(i, v') : v' \in V'\}$ forms a group. A READ(i) transition leads to a state in which either a read$_j$, a write$_j$, or a RETURN(i, v) is enabled. A RETURN(i, v) transition leads to a state in which READ(i) is enabled. A read$_j$ transition leads to a state in which the return$_j$ group is enabled. A write$_j$ transition leads to a state in which ack$_j$ is enabled. A return$_j$ or ack$_j$ transition leads to a state in which read$_l$, write$_l$, or RETURN(i, v) is enabled, for some $l \in N$ and some $v \in V$.

The write process has similar restrictions except that READ(i) is replaced with the group $\{\text{WRITE}(v) : v \in V\}$, all the RETURN($i, v$)'s are replaced with ACK, and the variable i in the physical action names is replaced with 0.

We now describe formally how to **compose** n read processes (RP_1 to RP_n), one write process (WP), and m physical registers (X_1 to X_m), in such a way as to produce another automaton A. First, we require that the read$_j$, return$_j$, write$_j$, and ack$_j$ actions of the read and write processes "match up" with the actions of the physical registers, *i.e.*, for each action π of a physical register, there is exactly one read or write process for which π is a (physical) action, and for each (physical) action π of a read or write process, there exists exactly one physical register for which π is an action. Note that for all j, the actions with subscript j are actions of register X_j. Therefore each logical action is the action of exactly one read or write process, while each physical action is the action of one read or write process and one register. For any register X_l, exactly one read or write process has the actions $write_l$ and ack_l, *i.e.*, there is a sole writer to the register.

The state set of the composition A is the cross product of the state sets of the component automata; thus each state of A is an $(n + m + 1)$-tuple. The actions of A are the READ, WRITE, RETURN and ACK actions (the "logical" actions) and the $read_j$, $return_j$, $write_j$, and ack_j actions of the physical registers (the "physical" actions). Finally, we describe the transition function of A. Suppose s is a state of A. We say that the logical action π is enabled in s if it is enabled in the state in s of the unique read or write process for which π is an action. We say that the physical action π is enabled in s if it is enabled in the states in s of both the read or write process and the register for which π is an action. The transition consists of each component automaton for which π is an action performing the action concurrently, while the remaining components do nothing.

We need some notation to distinguish the possible values of different physical registers as well as the logical register. Let α be a schedule of A. For any physical register X in the composition, let $PV_X(\alpha)$ be equal to $PV(\beta)$, where β is the restriction of α to actions of X. Let $PV_A(\alpha)$ be equal to $PV(\beta)$, where β is the restriction of α to the logical actions of A.

A **register implementation algorithm** (or simply **algorithm** for short) is a composition A of n read processes (RP_1 to RP_n), one write process (WP), and some number of physical registers (X_1 to X_m) such that the composition is a logical register. This means that the schedules of A, when restricted to the logical actions, satisfy the conditions for a register of the logical type. These conditions are (1) well-formedness and free initiation, which follow from the restrictions on read and write processes, (2) that logical READs RETURN values that are correct according to the possible values of the logical register, which must be ensured by the code of the read and write processes, and (3) the wait-free property, which also must be ensured by the code of the read and write processes. We are interested in a particular class of implementations, called **wait-free**, with the property that each read or write process can always complete a pending logical operation through a bounded number of its own actions. Formally, for every finite schedule α of A, if operation $\mathcal{O}$ is pending in α, then there is a constant l_α such that for every sequence β where $\alpha\beta$ is a schedule of A, if β consists of at least l_α actions of $\mathcal{O}$'s read/write process, then $\mathcal{O}$ is not pending in $\alpha\beta$.

We now define the cost measures.

Consider two register types, physical and logical, and let A be an algorithm for a physical-to-logical register implementation. Let M_A be the number of physical registers used in A, let R_A be the maximum number of physical operations performed during any logical READ in any execution of A, and let W_A be the maximum number of physical operations performed during any logical WRITE in any execution of A. Given a set S of physical-to-logical register implementations, let M_S be the minimum of M_A over all $A \in S$, R_S be the minimum of R_A over all $A \in S$, and W_S be the minimum of W_A over all $A \in S$. Finally, let $M = M_S$, $R = R_S$, and $W = W_S$, where S is the set of all physical-to-logical register implementations (for these two types). (The physical and logical register types are implicit parameters to M, R, and W.)

In the rest of this paper, we derive upper and lower bounds on M, R, and W, and trade-offs between them, for different physical and logical register types.

These bounds on R and W can be converted into time bounds for performing logical operations as follows. Suppose we know bounds R_l, R_u, W_l, and W_u such that $R_l \leq R \leq R_u$ and $W_l \leq W \leq W_u$. Let r be an upper bound on the time to read a physical register and let w be an upper bound on the time to write a physical register. Let s be an upper bound on the time for a read or write process to perform an action once it becomes enabled. Our upper bounds on R and W come from algorithms, all of which have the property that no logical READ involves a physical write and no logical WRITE involves a physical read. Since we assume that all physical operations are enclosed within logical operations and that only one physical operation can be pending at a time, we deduce that an upper bound on the worst case time to perform a READ of a logical register that is implemented with physical registers is $R_u(r + s) + s$. Similarly, an upper bound on the worst case time to perform a WRITE of a logical register that is implemented with physical registers is $W_u(w + s) + s$. Our lower bounds on R and W do not assume that logical READs do not involve physical writes, or that logical WRITEs do not involve physical reads, and thus they imply analogous lower bounds on the worst case times.

2.2 General Results

Fix any two physical and logical register types. Given a finite schedule σ of an algorithm A, let the **configuration** of σ be the tuple of sets of possible values of the physical registers at the end of the schedule, *i.e.*, if X_i is the i-th physical register, then the i-th element of the configuration is $PV_{X_i}(\sigma)$. A configuration is **stable** if each element of the tuple is a singleton set. Thus it can be represented as $x_1 \ldots x_m$, where x_i is the possible value of register X_i for all i. The **initial configuration** is the (stable) configuration of the empty schedule, consisting of the initial value of each physical register.

Let $\mathcal{WO}$ (for "write-only") be the set of all schedules of A in which only WP takes steps and no physical write is pending. Let $\mathcal{S} = \{C : C \text{ is the configuration of some } \sigma \in \mathcal{WO}\}$. It is easy to see that all configurations in $\mathcal{S}$ are stable.

Let $\mathcal{WOC}$ (for "write-only, completed") be the set of all schedules of A in which only WP takes steps and no *logical* WRITE is pending. Let $\mathcal{T} = \{C : C \text{ is the configuration of some } \sigma \in \mathcal{WOC}\}$. It is easy to see that $\mathcal{T} \subseteq \mathcal{S}$. Every configuration in $\mathcal{T}$ is defined to be a **terminal** configuration.

For each $i \in N$, define $L_i : \mathcal{S} \to V$ as follows. Let $C \in \mathcal{S}$ and $\sigma \in \mathcal{WO}$ such that C is the configuration of σ. Then $L_i(C) = v$, where

$$\sigma \text{ READ}(i) \; \alpha \text{ RETURN}(i, v)$$

is a schedule of A such that α consists solely of actions of RP_i and contains no RETURN. That is, L_i is the logical value returned by RP_i when RP_i starts in its local initial state and the physical registers have the values specified in C. The next lemma shows that L_i is well-defined, *i.e.*, that the current configuration (values of the physical registers) and nothing else determines the value of the logical register (as perceived by RP_i). We omit the proof due to brevity.

Lemma 1 *For any algorithm A, the function L_i is well-defined for all i.*

The next lemma states that under certain circumstances, each L_i is equal to the possible value of the logical register.

Lemma 2 *For any algorithm A, if σ is in $\mathcal{WOC}$ with configuration C, then $PV_A(\sigma) = \{L_i(C)\}$ for all i.*

Proof: Since σ is in $\mathcal{WOC}$, we know that $\sigma = \text{WRITE}(v_1) \; \alpha_1 \text{ ACK} \ldots \text{WRITE}(v_l) \; \alpha_l \text{ ACK}$ for some $v_1, \ldots, v_l$, where α_i, for all i, consists of physical actions by the write process. It is easy to see that the logical possible value of σ is $\{v_l\}$. By Lemma 1 and the safe or regular property, $L_i(C) = v_l$.
□

Define $L : \mathcal{T} \to V$ to be $L(C) = L_i(C)$ for any i. By the previous lemma, L is well-defined. It is easy to see that for each $v \in V$, there is a $C \in \mathcal{T}$ such that $L(C) = v$.

The next lemma gives some straightforward lower bounds on R, W, and M.

Lemma 3 $R \geq 1$, $W \geq 1$, and $M \geq 1$.

3 k-ary Safe Register From Binary Safe Registers

We consider the problem of implementing an n-reader, k-ary, safe register out of n-reader, binary, safe registers, for any $n \geq 1$, where $k > 1$. Subsection 3.1 is devoted to proving tight, independent bounds on R, W and M. In Subsection 3.2, we present an algorithm A such that $W_A = 1$. We also show some nice combinatorial properties related to one-write algorithms. Subsection 3.3 discusses algorithms which allow c physical accesses per logical WRITE. We also give some additional trade-offs between the cost measures.

We show that the independent bounds are not achievable simultaneously. We use some interesting combinatorial techniques in developing these results. In particular, we show that there are underlying combinatorial questions that we are required to answer to obtain upper and lower bounds for these problems. Let the value set of the logical register be $V = \{0, \ldots, k-1\}$.

3.1 Independent Bounds

Theorem 4 *The implementation of an n-reader, k-ary, safe register by n-reader, binary, safe registers gives the following independent bounds: $R = \lceil \log k \rceil$, $W = 1$, and $M = \lceil \log k \rceil$.*

Proof: The upper bounds on R and M follow from the binary representation algorithm in [Lam86] described below. The upper bound on W follows from our hypercube algorithm presented in Section 3.2. The lower bound on W follows from Lemma 3.

We now show the lower bound on M. Choose any algorithm A. For each $v \in V$, let C_v be an element of T such that $L(C_v) = v$. By Lemma 1, if $v \neq w$, then $C_v \neq C_w$. Since there are k distinct C_v's and each is a bit string of the same length, the length of each bit string must be at least $\lceil \log k \rceil$. Thus $M_A \geq \lceil \log k \rceil$. Since A was chosen arbitrarily, $M \geq \lceil \log k \rceil$.

We now show the lower bound on R. For each $v \in V$, there is a schedule σ_v of A of the form

$$\text{WRITE}(v)\ \alpha_v\ \text{ACK READ}(1)\ \beta_v\ \text{RETURN}(1, v),$$

where α_v consists solely of actions of WP and contains no ACK, and β_v consists solely of actions of RP_1 and contains no RETURN.

By the definition of read processes, for all distinct v and w, $\beta_v \neq \beta_w$ and the maximal common prefix of β_v and β_w is immediately followed by a return(0) action from some physical register X in β_v and by a return(1) action from X in β_w (or vice versa). I.e., RP_1 does the same thing in β_v and β_w until it reads a different value. Let γ_v be the sequence of physical values read in β_v, for all v.

Thus, if $v \neq w$, then the sequence γ_v of physical values read in β_v is not equal to the sequence γ_w of physical values read in β_w. There are k distinct sequences of physical values corresponding to the γ_v's, i.e., k binary strings. Thus at least one string, say that corresponding to γ_v, must have length at least $\lceil \log k \rceil$, implying that β_v contains at least $\lceil \log k \rceil$ physical reads.

Thus $R_A \geq \lceil \log k \rceil$. Since A was chosen arbitrarily, $R \geq \lceil \log k \rceil$. □

The **binary representation algorithm** in [Lam86] implements an n-reader, k-ary, safe register out of $\lceil \log k \rceil$ n-reader, binary, safe registers. The write process writes the binary representation of the logical value into the physical registers. Each read process reads all the physical registers and returns the logical value whose binary representation was read, as long as the value is less than k. Otherwise, it returns any value less than k. This algorithm implies that $R \leq \lceil \log k \rceil$, $W \leq \lceil \log k \rceil$, and $M \leq \lceil \log k \rceil$. By Theorem 4, the number of registers and number of physical reads in the binary representation algorithm are both optimal.

The **unary representation algorithm** presented next shows that $W \leq 2$. There are $k - 1$ physical registers, $X_1, \ldots, X_v$. Logical value 0 is represented when all registers are 0. Logical value $v \neq 0$ is represented when X_v is 1 and the other registers are 0. Each read process reads registers X_1, X_2, etc., in order, until reading a 1, and RETURNs logical value v, where X_v is the register that returned 1. To WRITE logical value v, the write process writes 0 to X_w, where w is the old value of the logical register, and writes 1 to X_v.

In the next subsection, we will present an algorithm which brings down the number of physical writes per logical WRITE to 1.

3.2 One-Write Algorithms

In this subsection, we discuss the class of one-write algorithms. We show that their existence depends on satisfying a combinatorial coloring property of hypercubes.

We now describe our new **hypercube algorithm** (see Figure 1), which shows that $W \leq 1$. For now, assume that k is a power of 2. Later we will show how to remove this restriction.

We notice an interesting relationship between the correctness of the hypercube algorithm and coloring the nodes of a $(k-1)$-dimensional hypercube with k colors such that each node has a neighbor with each of the $k - 1$ colors other than its own. The following definition and lemmas formalize this idea. (Nodes are labeled with $(k-1)$-bit strings, the colors are elements of V, and the function is the coloring.)

A function g is said to have the **rainbow-coloring property** if $g : \{0,1\}^{k-1} \to V$ such that for all $x \in \{0,1\}^{k-1}$, and for all $v \in V$, if $v \neq g(x)$, then there exists $y \in \{0,1\}^{k-1}$ such that $v = g(y)$ and x and y differ in exactly one bit.

Physical Registers: $X_1, \ldots, X_{k-1}$, initially $X_j = 1$ iff $j = v_0$, for all $j \in \{1, \ldots, k-1\}$
Read Process RP_i, $1 \leq i \leq n$: variables $x_1, \ldots, x_{k-1}$
    READ(i):
        for $j := 1$ to $k-1$ do $x_j := $ read X_j endfor
    RETURN($i, f(x_1 \ldots x_{k-1})$)
Write process WP: variables $x_1, \ldots, x_{k-1}$, initially $x_j = 1$ iff $j = v_0$, for all $j \in \{1, \ldots, k-1\}$;
        old, initially $old = v_0$
    WRITE(v):
        write $\overline{x_j}$ to X_j, where $bin(j) = bin(v) \oplus bin(old)$
        $old := v$
        $x_j := \overline{x_j}$
    ACK

Figure 1: Hypercube Algorithm

We define a function $f : \{0,1\}^{k-1} \to V$ *with the rainbow-coloring property* for use in the algorithm. For positive integer $i < k$, let $bin(i)$ be the binary representation of i in $\log k$ bits. For $x \in \{0,1\}^{k-1}$, let x_i be the ith bit of x, i.e., $x = x_1 x_2 \ldots x_{k-1}$. For all $x \in \{0,1\}^{k-1}$, we define $f(x)$ to be the element of V whose binary representation is:

$$x_1 \circ bin(1) \oplus x_2 \circ bin(2) \oplus \cdots \oplus x_{k-1} \circ bin(k-1),$$

where $\oplus$ represents exclusive-or and $\circ$ represents multiplication. (We define multiplication of $bin(i)$ by the bit 0 to be the zero-vector of length $\log k$, and multiplication of $bin(i)$ by the bit 1 to be $bin(i)$.) Since each x_i is either 0 or 1 and each $bin(i)$ consists of $\log k$ bits, this expression consists of $\log k$ bits and thus represents a value in the range 0 to $k-1$, i.e., a value in V.

The intuition behind the coloring function f is that we want to go from a $(k-1)$-bit string, the label of a node in the hypercube, to a $(\log k)$-bit string, indicating one of k colors. Given a node with label x, the color assigned is the one whose binary representation is equal to the exclusive-or of the set of $bin(i)$, for all i such that $x_i = 1$. Note that, if two nodes x and y differ in the single bit i, then $f(x) \oplus f(y) = bin(i)$. So, given the color of a node x, we can derive the color of any adjacent node y in a consistent manner.

Lemma 5 below proves that if a function exists with the rainbow-coloring property, the hypercube algorithm correctly implements a k-ary safe register using binary safe registers such that each logical WRITE requires one physical write. Lemma 6 proves that our simple and elegant function f *does* satisfy the rainbow-coloring property.

Lemma 5 *If function f has the rainbow-coloring property, then the hypercube algorithm is correct.*

Proof Sketch: The heart of the argument is showing that every READ RETURNs a correct value. Fix a READ(i) operation in some schedule. Clearly, if a WRITE operation overlaps the READ(i), then the RETURNed value, which can be arbitrary, is correct.

Suppose no WRITE operation overlaps the READ(i) operation. If no WRITE operation precedes the READ(i), clearly the READ(i) RETURNs the initial value. Suppose some WRITE precedes the READ(i). Then the schedule is of the form

$$\alpha \text{ WRITE}(u) \ \beta \ \text{ACK} \ \gamma \ \text{READ}(i) \ \delta \ \text{RETURN}(i, v)$$

for some u, where β contains no ACK, γ contains no WRITE, and δ contains no RETURN($i, *$) or WRITE.

We must show $v = u$. Let C be the configuration read by the READ(i). Then $v = f(C)$. Clearly C is also the configuration immediately after the ACK. Let D be the configuration immediately after α.

We need to prove that $f(C) = u$ (more precisely, that $f(C) = bin(u)$). To do this, we show that the write process behaves correctly. Note that at the end of schedule α, the internal variables $x_1, \ldots, x_{k-1}$ of WP correspond to the values of the physical registers in configuration D. This is clear from the fact that only WP writes into the physical registers, and it always modifies its internal variables to match the values of the physical registers.

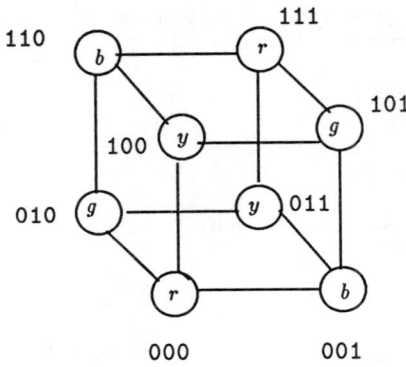

Figure 2: An Example Illustrating the Hypercube Algorithm

Let $w \in V$ be such that $f(D) = bin(w)$. By the algorithm, C and D differ only in bit j, where $bin(j) = bin(u) \oplus bin(w)$. Therefore, by the definition of the coloring function f, $f(C) \oplus f(D) = bin(j)$. By substitution, $f(C) \oplus bin(w) = bin(u) \oplus bin(w)$, which implies that $f(C) = bin(u)$.

□

Lemma 6 *The function f defined for the hypercube algorithm (when k is a power of 2) has the rainbow-coloring property.*

Proof: The following two facts together show that f has the rainbow-coloring property.

- For all $x, y \in \{0, 1\}^{k-1}$ which differ in exactly one bit, $f(x) \neq f(y)$.

- For all $x, y, z \in \{0, 1\}^{k-1}$ such that $y \neq z$ and y and z both differ from x in exactly one bit, $f(y) \neq f(z)$.

We prove the first fact. Let x and y differ in bit i. Then $f(x) \oplus f(y) = bin(i)$. Since $bin(i) \neq 0^{\log k}$, this implies that $f(x) \neq f(y)$. The second fact can be proved similarly. Let x and y differ in bit i, and let x and z differ in bit j. Then y and z differ in exactly two bits, bits i and j. Then $f(y) \oplus f(z) = bin(i) \oplus bin(j)$. Since $i \neq j$, $bin(i) \neq bin(j)$ and, therefore $f(y) \neq f(z)$.

□

Figure 2 illustrates how our algorithm works in the simple case where $k = 4$. Our hypercube is then a 3-dimensional cube, whose vertices can be colored with 4 colors, r, b, g and y. Note that the coloring satisfies the rainbow-coloring property.

Combining Lemmas 5 and 6 shows that the hypercube algorithm is a one-write algorithm (using $k-1$ registers) if k is a power of 2. To obtain a one-write algorithm for values of k that are not powers of 2, we modify the power-of-2 hypercube algorithm for $m - 1$ physical registers, where $m = 2^{\lceil \log k \rceil}$, *i.e.*, m is the smallest power of 2 larger than k. The modification is to change the RETURN statement to be RETURN($\min\{k-1, f(x_1 \ldots x_{m-1})\}$). This implementation of a k-ary register by binary registers will not cause the binary registers to take on all possible 2^{m-1} values, *i.e.*, no stable configuration of the algorithm will be mapped to a value that is out of the range of the logical register. However, a slow read process, which overlaps a number of writes, might (spuriously) observe a configuration corresponding to a value larger than $k-1$, thus necessitating the modification. Thus we have shown the following theorem.

Theorem 7 *The hypercube algorithm correctly implements a k-ary safe register using binary safe registers.*

The following theorem summarizes our results for the class of 1-write algorithms.

Theorem 8 *Let S be the set of algorithms A such that $W_A \leq 1$. Then*

- $k - 1 \leq R_S \leq 2^{\lceil \log k \rceil} - 1$,
- $M_S = k - 1$, *if k is a power of 2, and*
- $k \leq M_S \leq 2^{\lceil \log k \rceil} - 1$, *if k is not a power of 2.*

Proof: All the upper bounds follow from the hypercube algorithm. The rest of the proof concerns the lower bounds.

Choose an algorithm $A \in S$. Let C_{v_0} be the initial configuration. Then $L(C_{v_0}) = v_0$. For all $v \neq v_0$, let C_v be the configuration of a schedule in $\mathcal{WOC}$ of the form

$$\text{WRITE}(v) \; \alpha_v \; \text{ACK},$$

where α_v contains no ACK. Lemma 2 implies that $L(C_v) = v$. Lemma 1 implies that for all $v \neq v_0$, $C_v \neq C_{v_0}$. Since α_v only contains one physical write, C_{v_0} and C_v differ in a single bit, say that for physical register X_v. Lemma 1 implies that for all distinct v and w (not equal to v_0), $C_v \neq C_w$. Thus C_{v_0} differs from each C_v in a different bit, i.e., $X_v \neq X_w$.

Since there are $k - 1$ choices for $v \neq v_0$, there are at least $k - 1$ physical registers. Since A was chosen arbitrarily, $M_S \geq k - 1$. The improved lower bound of k for M_S when k is not a power of 2 follows from Lemmas 9 and 10 below.

To show $R_S \geq k - 1$, we assume, for contradiction, that $R_A < k - 1$. Consider the schedule

$$\text{READ}(1) \; \beta \; \text{RETURN}(1, v_0),$$

where β consists solely of actions of RP_1 and contains no RETURN. β contains a sequence of less than $k - 1$ physical reads. Let X_v (as defined above) be one of the physical registers not read in β; note that $v \neq v_0$. Since C_{v_0} differs from C_v in the value of register X_v and nowhere else, an easy induction on the length of β shows that

$$\text{WRITE}(v) \; \alpha_v \; \text{ACK} \; \text{READ}(1) \; \beta \; \text{RETURN}(1, v_0)$$

is a schedule of A, violating the safe condition since $v \neq v_0$. We therefore have a contradiction, implying $R_A \geq k - 1$. Since A was chosen arbitrarily, it follows that $R_S \geq k - 1$. □

We now consider the number of registers when k is not a power of 2. Lemma 9, which is the converse of Lemma 5, shows that the existence of a function with the rainbow-coloring property is necessary for the existence of a one-write algorithm using $k - 1$ registers. Lemma 10, which is the converse of Lemma 6, shows that when k is not a power of 2, no function with the rainbow-coloring property can exist. Together, these two lemmas imply that if k is not a power of 2, then any one-write algorithm must use more than $k - 1$ registers.

Lemma 9 *If there is an algorithm A with $W_A = 1$ and $M_A = k - 1$, then there exists a function with the rainbow-coloring property.*

Proof Sketch: The proof consists of verifying that L has the rainbow-coloring property. □

Lemma 10 *If k is not a power of 2, then there is no function with the rainbow-coloring property.*

Proof: Assume in contradiction that there is a function f with the rainbow-coloring property. Choose any color, say blue, and let b be the number of nodes colored blue by f. Let B be the set of edges in the hypercube that have one endpoint colored blue and one endpoint not colored blue. Since each non-blue node is adjacent to exactly one blue node and there are $2^{k-1} - b$ non-blue nodes, $|B|$ must be $2^{k-1} - b$. However, since each blue node is adjacent to $k - 1$ non-blue nodes and there are b blue nodes, $|B|$ must be $b(k - 1)$. Therefore, $2^{k-1} - b = b(k - 1)$, which implies that $2^{k-1} = kb$. This implies that k divides 2^{k-1}, contradicting the fact that k is not a power of 2. □

In this subsection, we showed that the existence of a one-write implementation of a k-ary safe register was based on solving an underlying combinatorial problem. Specifically, a one-write algorithm using $k-1$ physical registers exists if and only if we can color a $(k-1)$-dimensional hypercube with k colors such that each node has a neighbor with every color other than its own. We can generalize this to any number of physical registers as follows. A one-write algorithm using m registers exists if and only if we can *partially color* an m-dimensional hypercube with k colors, such that each *colored* node has a neighbor with every color other than its own. By a partial coloring, we mean a coloring where not all nodes of the graph need to be colored.

Lemmas 6 and 10 imply that there is rainbow-coloring of the $(k-1)$-dimensional hypercube if and only if k is a power of 2. Kant and van Leeuwen [KvL90] have independently shown the same result. Their proof uses notions from coding theory and is based on showing a correspondence between 1-error-correcting codes and these colorings. They applied this result to the file distribution problem.

3.3 c-write Algorithms and Trade-off Results

As we showed for 1-write algorithms, the problem of implementing c-write algorithms can also be shown to have a corresponding parallel in a combinatorial problem. Here, we are interested in a partial coloring of the m-dimensional hypercube such that for each colored node, there exists a node of every other color *within a distance of c* from this node.

This combinatorial characterization also helps us obtain lower bounds for M and R. For example, we know that for a one-write algorithm to exist which uses m physical registers, there must be a configuration C_0 which differs from $k-1$ different configurations C_v in exactly one bit. Since each configuration is represented in m bits, this says that there is a binary string of length m which differs from $k-1$ different strings of the same length in exactly one bit. To satisfy this combinatorial property, we require that $m \geq k-1$. This sequence of reasoning was implicit in the proof of Theorem 8.

Along similar lines, for a c-write algorithm to exist which uses m registers, there must exist a binary string of length m which differs from $k-1$ different strings of the same length in at most c bits.

We formalize this property in Lemma 11 and Theorems 13 and 14 below. These theorems give lower bounds on M and R for c-write algorithms, i.e., algorithms that use a small bounded number of physical writes per logical WRITE.

The next result is Theorem 15, which gives trade-offs on W versus R and M. An application of this result is to give upper bounds on M and R for c-write algorithms.

The final result in this subsection, Theorem 16, states that if no more than $\lceil \log k \rceil$ registers are used, then some WRITE must write at least $\lceil \log k \rceil$ physical registers.

Lemma 11 *Given any binary string x of length m, if there are at least k distinct strings of length m which differ from x in at most c bits, where $c \leq (\log k)/3$, then $m \geq (c!k/2)^{1/c}$.*

Proof: Let x be a string of length m. The number of distinct strings of length m which differ from x in at most c bits is

$$\binom{m}{0} + \binom{m}{1} + \binom{m}{2} + \cdots + \binom{m}{c}$$

Since we know that there are at least k such distinct strings, we have the following inequality.

$$\sum_{i=0}^{c} \binom{m}{i} \geq k$$

We obtain the following upper bound on $\binom{m}{i}$, for all i.

$$\binom{m}{i} = \frac{\overbrace{m(m-1)(m-2)\cdots(m-i+1)}^{i \text{ factors}}}{i!} \leq \frac{m^i}{i!}$$

To get an upper bound on the entire summation, we need the following claim, which is taken from [Tya88]. First, we introduce some notation. Let $s_{m,j}$ denote $\sum_{i=0}^{j} \binom{m}{i}$. Let $b_{m,i}$ denote $\binom{m}{i}$.

Claim 12 *If $1 \leq j \leq m/3$, then $s_{m,j} \leq 2b_{m,j}$.*

Proof Sketch: We compute a lower bound for $b_{m,j}/b_{m,j-1} = \frac{m-j+1}{j}$. Note that $\frac{m-j+1}{j}$ is larger than 2 for $j \leq m/3$. Therefore, for $j \leq m/3$, $b_{m,j}/b_{m,j-1} > 2$. The remaining proof is by induction on j and uses this lower bound.

End of Claim

The above claim holds for $j = c$ since we know that $m \geq \log k$ (it takes $\log k$ bits to represent k distinct values), and this implies that $c \leq m/3$. Now, using the above claim and our previous upper bound for $\binom{m}{i}$, we have

$$\sum_{i=0}^{c} \binom{m}{i} \leq 2 \binom{m}{c} \leq \frac{2m^c}{c!}$$

So, $k \leq 2m^c/c!$ and by manipulating this inequality, we get the result $m \geq (c!k/2)^{1/c}$. □

Theorem 13 *For all algorithms A, if $W_A = c$, where $c \leq (\log k)/3$, then $M_A \geq (c!k/2)^{1/c}$.*

Proof: Given an algorithm A such that $W_A = c$, where $c \leq (\log k)/3$, let C_{v_0} be the initial configuration. Then $L(C_{v_0}) = v_0$. For all $v \neq v_0$, the schedule σ_v of the form WRITE(v) α_v ACK yields the terminal configuration C_v. Since each WRITE can initiate at most c physical writes, each C_v differs in at most c bits from C_{v_0}.

Since there are k values v, there must be at least k terminal configurations C_v differing in at most c bits from C_{v_0}. The number of registers used in the algorithm is M_A. Each terminal configuration is therefore a binary string of length M_A. Therefore, there are at least k strings of length M_A which differ in at most c bits from C_{v_0}. Since, $c \leq (\log k)/3$, Lemma 11 applies, and we have the result $M_A \geq (c!k/2)^{1/c}$. □

Theorem 14 *For any algorithm A, if $W_A = c$, where $c \leq (\log k)/3$, then $R_A \geq (c!k/2)^{1/c}$.*

Proof: For any algorithm A, where $W_A \leq c$, consider the following schedules, for all v,

WRITE(v) α_v ACK READ(1) β_v RETURN(1, v),

where α_v and β_v contain only physical actions. We claim that for some v, β_v initiates at least $(c!k/2)^{1/c}$ physical reads. We prove this by contradiction.

Suppose, for every v, β_v initiates at most p physical reads where $p < (c!k/2)^{1/c}$. Let ρ_v be the sequence of values read, in order, on accessing any given register *for the first time* in β_v. Note that we don't include values obtained from registers which have been read before or been written before in β_v. Clearly, $|\rho_v| \leq p$.

We define the string δ_v, for every v, as follows. For every bit in ρ_v, if the value is the *same* as the initial value of the register read, place the bit 0 in δ_v. If the value is *different* from the initial value of the register read, place the bit 1 in δ_v. Since α_v contains at most c writes, δ_v can contain at most c 1's. Also, each δ_v is distinct. (Otherwise, if for some v, v' such that $v \neq v'$, $\delta_v = \delta_{v'}$, then a READ in both cases would RETURN the same value, which would be a contradiction.) Therefore, $\{\delta_v | v \in V\}$ is a set of k distinct strings of length at most p which differ from the zero-vector in at most c bits. Since $p < (c!k/2)^{1/c}$, this contradicts Lemma 11.

Therefore, for some v, β_v initiates at least $(c!k/2)^{1/c}$ physical reads. This gives our lower bound for R_A. □

The binary representation algorithm yields an upper bound of $\lceil \log k \rceil$ for R, W and M. The unary representation algorithm brings down the upper bound for W to 2, while pushing up the bounds for R and M to $\Omega(k)$. This suggests a trade-off between these measures. We can construct a class of algorithms, by borrowing from both algorithms mentioned above, which have bounds on R_A and M_A varying from $\Theta(\log k)$ to $\Theta(k)$ and bounds on W_A varying from $\Theta(\log k)$ to $\Theta(1)$.

Theorem 15 *For any m, $1 \leq m \leq k$, there is an algorithm A such that $R_A = M_A = \lceil \log m \rceil + \lceil k/2^{\lceil \log m \rceil} \rceil$, and $W_A = \lceil \log m \rceil + 2$.*

Proof: We implement our k-ary register by combining an a-ary register and a b-ary register as follows. Let a be the smallest power of 2 which is at least as large as m, i.e., $a = 2^{\lceil \log m \rceil}$. Let $b = \lceil k/a \rceil$. We implement an a-ary register by the *binary* representation method, and a b-ary register by the *unary* representation method. Both these methods have been described earlier. Let the values represented by the a-ary register be in $A = \{1, \ldots, a\}$ and the values represented by the b-ary register be in $B = \{1, \ldots, b\}$. We obtain an ab-ary register by combining these two registers, where the ab values represented are in $A \times B$. Note that $ab \geq k$, so we have our k-ary register. The bounds follow by inspection.

□

The preceding theorem helps us to derive upper bounds for M_S and R_S, where S is the class of c-write algorithms. Choose $m = 2^{c-2}$. Since $c \leq \log k$, it follows that $m \leq k$ and Theorem 15 applies. Therefore, there exists an algorithm A such that

- $W_A = c$,
- $R_A = c - 2 + \lceil k/2^{c-2} \rceil$, and
- $M_A = c - 2 + \lceil k/2^{c-2} \rceil$.

We thus have the corresponding upper bounds for R_S and M_S, where S is the class of c-write algorithms. Clearly, the upper bounds obtained earlier for the class of 1-write algorithms also hold for c-write algorithms. These new bounds surpass the earlier bounds when $c \geq 3$.

The next theorem states that if an algorithm uses only $\lceil \log k \rceil$ physical registers, then some logical WRITE must use at least $\lceil \log k \rceil$ physical writes.

Theorem 16 *For any algorithm A, if $M_A \leq \lceil \log k \rceil$, then $W_A \geq \lceil \log k \rceil$.*

Proof: Let A be an algorithm with $M_A = \lceil \log k \rceil$. (We have already shown M_A cannot be smaller.) Since the physical registers are binary, $|\mathcal{T}| \leq 2^{\lceil \log k \rceil}$. Recall that for all $v \in V$, there is an $x \in \mathcal{T}$ with $L(x) = v$.

Let U be the subset of $\mathcal{T}$ such that x is in U if and only if there is no $y \neq x$ in $\mathcal{T}$ such that $L(y) = L(x)$. Thus for each configuration x in U, x is the only terminal configuration which has the logical value $L(x)$.

Claim 17 *There is an $x \in U$ such that $\overline{x} \in \mathcal{T}$. ($\overline{x}$ is the binary string that differs from x in every bit.)*

> **Proof:** Suppose there is no such x. Let $|U| = l$. Each element of U corresponds to a distinct element of V, accounting for l elements of V. The remaining $k - l$ elements of V are represented among the configurations of $\mathcal{T}$ that are not in U and are not the inverse of an element of U. There are at most $2^{\lceil \log k \rceil} - 2l$ of these configurations. There are at least two of these configurations for each remaining element of V. Thus $2^{\lceil \log k \rceil} - 2l \geq 2(k - l)$, which gives a contradiction.
>
> **End of Claim**

Choose $x \in U$ such that $\overline{x} \in \mathcal{T}$. Let σ be a schedule in $\mathcal{WOC}$ with configuration $\overline{x}$. Suppose $L(x) = v$. Then there is a schedule τ in $\mathcal{WOC}$ of the form

$$\sigma \text{ WRITE}(v) \, \alpha \text{ ACK,}$$

where α contains no ACK. The configuration of τ must be x since $x \in U$. Thus α contains at least $\lceil \log k \rceil$ writes, and $W_A \geq \lceil \log k \rceil$.

□

4 k-ary Regular Register From Binary Regular Registers

We now shift our attention to regular registers. We would like to implement n-reader, k-ary, regular registers using n-reader, binary, regular registers. (Binary safe registers could be used just as well as binary regular registers, since Lamport showed that a binary regular register can be implemented with a binary safe register using one write per WRITE and one read per READ [Lam86].)

As with safe registers, the problem of implementing k-ary regular registers can also be shown to have a parallel in a combinatorial problem. If there exists an algorithm to implement a k-ary regular register which uses m binary registers, then there is a partial k-coloring of an m-dimensional hypercube with the following restriction. For each colored vertex v, let c be its color. Then, for each color c_i such that $c_i \neq c$ (there are $k-1$ such colors), there exists a path in the hypercube from v to some vertex v_i with color c_i all of whose intermediate vertices are colored c.

This characterization takes care of a slow WRITE which overlaps a number of READs. The path corresponds to the intermediate configurations reached during a WRITE. It makes sure that whatever value is RETURNed by a READ which sees an intermediate configuration preserves the regular property of registers. Note, however, that while this restriction is necessary for an algorithm, it is not sufficient. This is because the restriction doesn't take care of the problem of a slow READ overlapping a number of WRITEs, as we will show later. In particular, our hypercube algorithm for safe registers satisfies this characterization, but cannot be used to implement a regular register. Therefore, this characterization may help us get a lower bound for this problem, but not an upper bound.

Subsection 4.1 shows our independent bounds on R, W, and M. Subsection 4.2 contains our trade-off results. As before, we let $V = \{0, \ldots, k-1\}$.

4.1 Independent Bounds

The following theorem establishes the independent bounds achieved for this problem.

Theorem 18 *The implementation of an n-reader, k-ary, regular register by n-reader, binary, regular registers gives the following independent bounds:*

- $R = \lceil \log k \rceil$,

- $1 \leq W \leq \lceil \log k \rceil$, and

- $\max\{\lceil \log k \rceil + 1, \lceil 2 \log k - \log \log k \rceil - 2\} \leq M \leq \min\{k-1, n(3 \log k + 68)\}$.

Proof: The lower bound for R follows directly from the same result for safe registers. The lower bound for W follows from Lemma 3. The lower bound for M is shown in Lemmas 21 and 22 below.

The upper bounds on R and W appear simultaneously in the tree algorithm, presented below. However, this algorithm uses $k-1$ physical registers. Lamport [Lam86] describes a complex composition of implementations to achieve an algorithm using $n(3 \log k + 68)$ 1-reader physical registers (recall that n is the number of readers for the logical register). It is unknown whether a better result, for example without the factor of n, is possible by taking advantage of the additional power when the physical registers are n-reader.

□

Note that our hypercube algorithm, which we used to implement a k-ary safe register from binary safe registers, cannot be used to implement a k-ary regular register. The reason for this is as follows. In case of a slow READ which overlaps a number of WRITEs, the physical reads initiated by the READ may return a set of register values which do not represent a configuration achieved by the automaton during the course of the READ. Thus a logical value may be RETURNed which doesn't correspond to a value written by an overlapping or last preceding WRITE. A stronger result, stating that no 1-write algorithm using $k-1$ registers can implement a k-ary regular register from binary regular registers, is proven in Theorem 23.

We now present our new **tree algorithm**, which gives the bounds of $R \leq \lceil \log k \rceil$, $W \leq \lceil \log k \rceil$, and $M \leq k-1$. Let T be a binary tree with k leaves, $k-1$ internal nodes, and height $\lceil \log k \rceil$. (It can be shown that such a tree exists for all k.) The algorithm uses $k-1$ physical registers; each register corresponds to an internal node of T, thus giving the bound on M. Each k-ary value corresponds to a leaf of T.

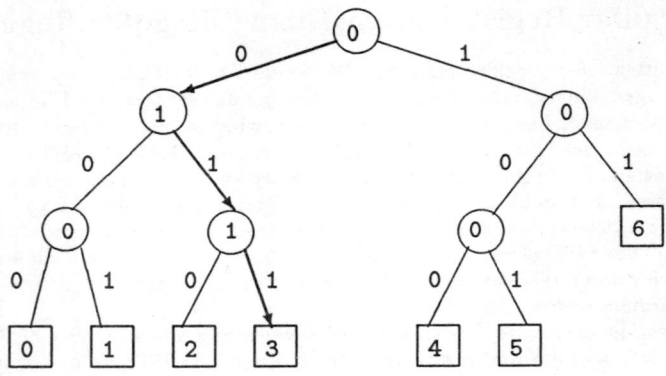

Figure 3: An Example Illustrating the Tree Algorithm

Let v_0 be the initial value of the logical register. The initial values of the physical registers are those that would result from starting with all 0's in the physical registers and then executing a single WRITE(v_0) operation as described below.

A WRITE(v) operation writes the set of physical registers on the path from the leaf labeled v to the root, in order from leaf to root. If the leaf is the left (resp., right) child of its parent, then 0 (resp., 1) is written into the register corresponding to the parent of the leaf. If the node z corresponding to the previous register written is the left (resp., right) child of its parent, then 0 (resp., 1) is written into the register corresponding to the parent of z. Since the number of internal nodes on the path is at most $\lceil \log k \rceil$, the bound on W follows.

A READ operation reads the set of physical registers on the path from the root to some leaf, in order from root to leaf, where the path is determined dynamically as follows. Let z be the node corresponding to the previous register read. If the previous register read returned 0 (resp., 1), then let y be the node corresponding to the left (resp., right) child of z. If y is not a leaf, then the next register to be read is the one corresponding to y. If y is a leaf, then RETURN the k-ary value labeling y. Since the number of internal nodes on the path is at most $\lceil \log k \rceil$, the bound on R follows.

Figure 3 illustrates a 7-ary register with value 3. The path marked on the tree corresponds to the physical registers read by a logical READ operation.

In order to prove the correctness of the tree algorithm, we need some definitions and a lemma. We define a physical read r to **reflect** a physical write w, in a given schedule, if r and w access the same physical register, and either (1) w completely precedes r, or (2) w and r overlap and r returns the value that w writes. We say that a logical READ R **notices** a logical WRITE W if there exists a physical register s such that R contains a physical read that reflects a physical write contained in W.

Lemma 19 *Given any schedule of the tree algorithm, and any READ R in the schedule, R RETURNs the value written by the last WRITE W that R notices (note that there is a total order among the WRITE operations). If no such WRITE exists, R RETURNs the initial value.*

Proof: Let R be a READ in some schedule. Suppose R notices no WRITEs. Then every physical read r initiated by R returns the initial value of the physical register read. Therefore, R RETURNs the initial value of the logical register.

Otherwise, R notices some WRITEs. Let W be the last WRITE that R notices. Let s be the last register read by R such that R's read from s reflects W's write to s. Clearly, R reads the value b written by W into s. Otherwise, there is a later WRITE W_1 such that W_1 writes s and R notices W_1, which contradicts the fact that W is the last WRITE that R notices.

Without loss of generality, let $b = 0$. (The argument for $b = 1$ is identical by replacing "left" in the following discussion with "right".)

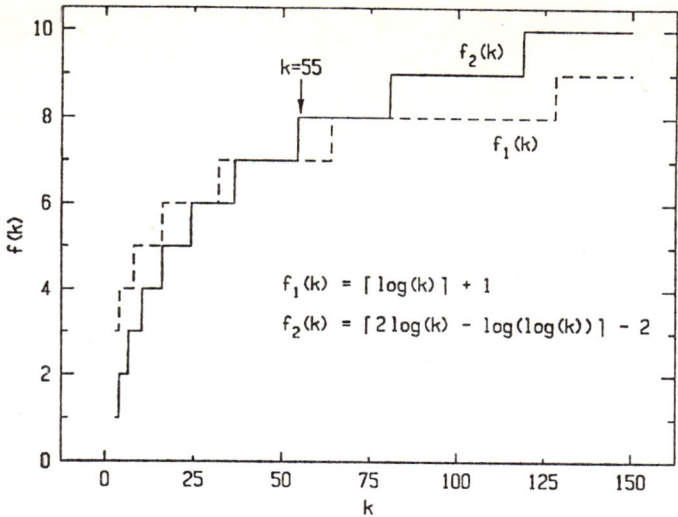

Figure 4: The Lower Bound for m at Different Values of k

We claim that s is the last register read by R. Suppose not. Then, R next reads the register t corresponding to the left son of s. Since W wrote b in register s, it must have earlier written to register t. This contradicts the definition of s.

Now, the left son of s must be a leaf node. Let v be the label of this leaf node. Clearly, v is RETURNed by R. Since W writes b into s, the logical value written by W is v.

□

Theorem 20 *The tree algorithm implements a k-ary regular register using binary regular registers.*

Proof: We need to argue that our logical k-ary register behaves correctly; *i.e.*, given that our algorithm is implemented using regular binary physical registers, it actually implements a regular k-ary register. Clearly the algorithm has the wait-free property.

Given any schedule, and any READ R in that schedule, we need to prove that R RETURNs the value of one of the WRITE operations it overlaps with or the last preceding WRITE W_1 (or the initial value, in the case that no WRITE completely precedes R). We consider two cases.
Case 1: R notices no WRITEs.
Since R reads the root node, and any WRITE must write into the root node, it follows that no WRITE completely precedes R. By Lemma 19, R RETURNs the initial value, and this satisfies regularity.
Case 2: R notices some WRITEs.
Let W_1 be the last WRITE that R notices. By Lemma 19, R RETURNs the value written by W_1. We show that W_1 either overlaps with R or is the last WRITE preceding R. This would satisfy regularity.

Clearly, W_1 cannot completely follow R, since, by the definition of *notice*, W_1 writes into some physical register which is subsequently read by R. The only other case to consider is that W_1 precedes another WRITE W_2, which completely precedes R. Since W_1 is the last WRITE that R notices, R does not notice W_2. Since W_2 completely precedes R, R must read the root node after W_2 writes into it, which implies that R does notice W_2. This gives a contradiction. Therefore, W_1 either overlaps with R or is the last WRITE preceding R.

□

We present our lower bounds for M below. Both of the bounds we obtain are significant for different values of k. Figure 4 illustrates which bound is better for particular values of k.

Lemma 21 $M \geq \lceil \log k \rceil + 1$.

Proof Sketch: Choose any algorithm A. We assume, for contradiction, that $M_A = \lceil \log k \rceil$. Note that the lower bound for M of $\lceil \log k \rceil$, proved for safe registers, holds here as well. For all $v \in V$, let C_v be the stable configuration resulting from the schedule consisting of the complete operation WRITE(v).

Choose $v \in V$. For each $w \in V$, $w \neq v$, consider the schedule consisting of the complete operation WRITE(v) followed by the complete operation WRITE(w). Let C_{vw} be the stable configuration resulting from this schedule.

The automaton goes through a sequence α of configurations from C_v to C_{vw} during the second WRITE in this schedule. By Lemma 2, $L_1(C_{vw}) = w$ and $L_1(C_v) = v$. Since $w \neq v$ and L_1 is a function by Lemma 1, $C_{vw} \neq C_v$. Thus, there is a stable configuration in α that is different from C_v. Let D_{vw} be the first such configuration. D_{vw} and C_v differ in a single bit, i.e., in the value of a single register.

Since there are only $\lceil \log k \rceil$ bits in each configuration, there are only $\lceil \log k \rceil$ configurations which differ in a single bit from C_v. Since there are $k-1$ values in V different from v, there exist distinct w and u in V such that $D_{vw} = D_{vu}$. Call this configuration D_v. By regularity, $L_1(D_{vw}) \in \{v, w\}$ and $L_1(D_{vu}) \in \{v, u\}$. Thus $L_1(D_v) = v$.

Since $L_1(C_v) = v$, all the C_v's are distinct. Since $L_1(D_v) = v$, all the D_v's are distinct. It is easy to see that $C_v \neq D_w$ for all v and w. Thus there are at least $2k$ distinct stable configurations, requiring at least $\lceil \log k \rceil + 1$ registers. Therefore, we have a contradiction.

□

Lemma 22 $M \geq \lceil 2 \log k - \log \log k \rceil - 2$.

Proof Sketch: Choose any algorithm A. Let m be the number of registers used in the algorithm. We claim that for any two k-ary values v and w, there exist a pair of configurations D_v and D_w which differ in exactly one bit such that $L_1(D_v) = v$ and $L_1(D_w) = w$. Suppose this is not true for some v and w. Then, consider the schedule consisting of the complete operation WRITE(v) followed by the complete operation WRITE(w). The sequence of configurations reached during the second WRITE includes a configuration C such that $L_1(C)$ is neither v nor w. Thus a READ at this point would RETURN neither v nor w, contradicting regularity.

Let c_v be the number of configurations C in $\mathcal{S}$ such that $L_1(C) = v$, for each k-ary value v. Let $c = \min\{c_x | x \in V\}$, and let $v \in V$ be such that $c = c_v$. For each value w such that $w \neq v$, there are configurations D_v and D_w in $\mathcal{S}$ which differ in exactly one bit such that $L_1(D_v) = v$ and $L_1(D_w) = w$. Since each configuration C, such that $L_1(C) = v$, has m neighbors, and there are $k-1$ values w, it follows that $cm \geq k-1$. Since there are k different values and at most 2^m possible configurations, $ck \leq 2^m$. These two inequalities imply $(k-1)/m \leq 2^m/k$. By manipulating this inequality, we obtain our result.

□

4.2 Trade-Offs

We have the following lower bounds for R and M relating to one-write algorithms. In particular, we show that any one-write algorithm for this problem would require at least k registers. In other words, our hypercube algorithm for safe registers does not work for regular registers.

Theorem 23 For all algorithms A, if $W_A = 1$ then $R_A \geq k-1$ and $M_A \geq k$.

Proof: The lower bound for R_A follows from the same result for safe registers. By using a similar argument, we can actually make the additional claim that every READ reads at least $k-1$ distinct physical registers. We use this claim in the following proof of the bound for M_A.

To show $M_A \geq k$, suppose in contradiction that a one-write algorithm A exists which uses $k-1$ registers. Then Lemma 9 carries over from the safe case, implying that the function L has the rainbow-coloring property. Let C_0 be the initial configuration; clearly, $L(C_0) = v_0$. Consider the following schedule α:

$$\text{READ}(1) \; \delta \; \text{RETURN}(1, v_0)$$

where δ consists only of physical actions taken by RP_1. We claim that δ does not contain any physical write.

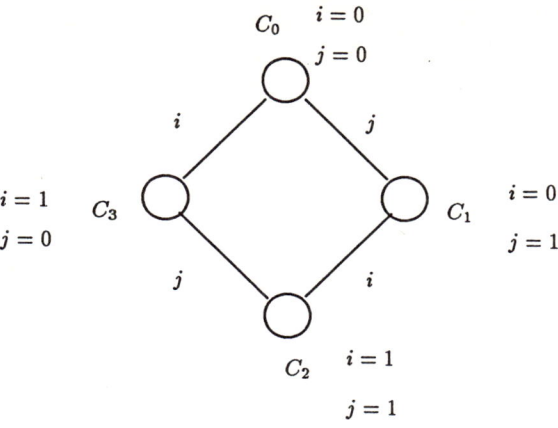

Figure 5: Relationship Between the Four Configurations

Claim 24 *The sequence of actions δ does not contain a physical write.*

Proof Sketch: Suppose δ does contain a physical write of some physical register i. We construct a new schedule where we insert the complete WRITE(v) operation which modifies register i *immediately before* the physical write in δ. If RP_1 performs another READ immediately after completing the first one in both schedules, it cannot distinguish between the two schedules since the logical WRITE has been obliterated by the physical write. Thus, in the new schedule, the second READ will RETURN v_0 as in the old schedule, while it should RETURN v. This violates regularity.

End of Claim

Now, we continue with the proof of the theorem. Pick two distinct registers (call them registers i and j) which are read in schedule α.

We define C_1 to be the stable configuration which differs from C_0 in position j, C_3 to be the stable configuration which differs from C_0 in position i, and C_2 to be the stable configuration which differs from C_0 in positions i and j. For all $l \in \{1, 2, 3\}$, C_l is a terminal configuration. Let $L(C_l) = v_l$. It is easy to verify that v_0, v_1, v_2 and v_3 are distinct values in V. Suppose, without loss of generality, that the initial value of both registers i and j is 0. Figure 5 illustrates the relation between the four configurations defined. Adjacent configurations differ in a single bit. The label on the edge between two configurations corresponds to the particular bit in which they differ.

Now, consider the sequences of actions, specified in Table 3, which can be applied at a configuration C_{start} and results in the configuration C_{finish}.

We claim that if we have a schedule σ with the configuration C_{start} and no pending WRITE, we can concatenate an appropriate sequence of actions β (from Table 3) to σ to obtain the schedule σ' with the configuration C_{finish}. The sequence β is a single logical WRITE which consists of a single physical write (and possibly some physical reads)—thus none of the γ_{ab}'s contain any physical writes. It is easy to see that each β exists.

We create a new schedule α' by taking α and inserting certain sequences at certain points, according to the following rules. First, we insert β_{01} before READ(1), resulting in configuration C_1. Then, before each read$_j$ of RP_1, if the configuration is C_1, we insert $\beta_{12}\beta_{23}$, resulting in configuration C_3. Before each read$_i$ of RP_1, if the configuration is C_3, we insert $\beta_{32}\beta_{21}$, resulting in configuration C_1. To see that α' is a schedule, it is sufficient to observe that the only time the configuration changes within the schedule is when a sequence β_{ab} is inserted. This follows from the fact, proven in Claim 24, that α contains no

C_{start}	sequence β	C_{finish}
C_0	$\beta_{01} = \text{WRITE}(v_1)\ \gamma_{01}\ \text{write}_j(1)\gamma'_{01}\ \text{ACK}$	C_1
C_1	$\beta_{12} = \text{WRITE}(v_2)\ \gamma_{12}\ \text{write}_i(1)\gamma'_{12}\ \text{ACK}$	C_2
C_2	$\beta_{23} = \text{WRITE}(v_3)\ \gamma_{23}\ \text{write}_j(0)\gamma'_{23}\ \text{ACK}$	C_3
C_3	$\beta_{32} = \text{WRITE}(v_2)\ \gamma_{32}\ \text{write}_i(1)\gamma'_{32}\ \text{ACK}$	C_2
C_2	$\beta_{21} = \text{WRITE}(v_1)\ \gamma_{21}\ \text{write}_i(0)\gamma'_{21}\ \text{ACK}$	C_1

Table 3: Sequences for Proof of Theorem 23

physical writes. In particular, inserting β_{01} changes the configuration to C_1, inserting $\beta_{12}\beta_{23}$ changes the configuration to C_3, and inserting $\beta_{32}\beta_{21}$ changes the configuration to C_1. We can prove, by a simple induction, that the configuration reached by any prefix of schedule α' up to a read_i by RP_1 is always C_1. Similarly, the configuration reached by any prefix of schedule α' up to a read_j by RP_1 is always C_3. Therefore, read_i and read_j always return the value 0. It follows that v_0 is the value RETURNed by the READ(1) in the schedule α'. Since, to satisfy regularity, the READ should RETURN v_1, v_2 or v_3, we have a contradiction.

□

We conclude this section with a trade-off result relating to a constant number of writes. This follows from the identical result derived in the safe case.

Theorem 25 *For all algorithms A, if $W_A = c$, where $c \leq (\log k)/3$, then $M_A \geq (c!k/2)^{1/c}$ and $R_A \geq (c!k/2)^{1/c}$.*

5 Conclusion

We have demonstrated upper and lower bounds on the number of physical registers, the number of physical reads in a logical read, and the number of physical writes in a logical write, for a variety of register implementations. In many cases, our bounds are tight. Some of our upper bounds follow from two new algorithms that we present, one for implementing a k-ary safe register out of binary safe registers, and another for implementing a k-ary regular register out of binary regular registers. We also presented several interesting trade-offs between these cost measures, for implementing k-ary registers out of binary registers. The bounds on the number of physical operations can be converted into bounds on the time to perform the logical operations, in terms of the time for the physical operations.

Our technical report [CW90], besides containing full proofs, also shows tight bounds for implementing n-reader safe or regular registers out of 1-reader safe or regular registers (namely, $R = 1$, $W = n$, and $M = n$), and for implementing binary regular registers out of binary safe registers (namely, $R = W = M = 1$).

Future work includes finding such bounds for more algorithms, in particular, those involving atomic registers and multi-writer registers. We also do not yet have tight bounds on W and M for implementing k-ary regular registers out of binary regular registers. It would be interesting to see if better bounds are possible in some cases than those obtained by composing the algorithms we have. A final question is what difference does it make, if any, if clocks are available to the read and write processes?

6 Acknowledgments

We would like to thank Brian Coan for the proof of Lemma 10. Akhilesh Tyagi suggested the combinatorial techniques required in proving Theorems 13 and 14. He also helped to obtain the proof of correctness for our tree algorithm. George Welch kindly prepared the graphical display of Figure 4. We are grateful to Hagit Attiya, Paul Beame, Richard Ladner, and Martin Tompa for many helpful comments on an earlier version of this paper which served to improve some of the results and presentation. We would also like to thank Mike Saks for some helpful conversations. The notes [LG89] from the Distributed Computing class taught by Nancy Lynch at MIT served as a useful survey and tutorial of earlier work.

References

[Blo87] Bard Bloom. Constructing Two-Writer Atomic Registers. In *Proceedings of the Sixth Annual ACM SIGACT-SIGOPS Symposium on Principles of Distributed Computing*, pages 249–259, August 1987.

[BP87] James E. Burns and Gary L. Peterson. Constructing Multi-Reader Atomic Values from Non-Atomic Values. In *Proceedings of the Sixth Annual ACM SIGACT-SIGOPS Symposium on Principles of Distributed Computing*, pages 222–231, August 1987.

[CW90] Soma Chaudhuri and Jennifer L. Welch. Bounds on the Costs of Register Implementations. Technical Report TR90-025, University of North Carolina, Chapel Hill, June 1990.

[KvL90] Goos Kant and Jan van Leeuwen. The File Distribution Problem for Processor Networks. In *Lecture Notes in Computer Science 447: Proceedings of the Second Scandinavian Workshop on Algorithm Theory*, pages 48–59. Springer-Verlag, July 1990.

[Lam86] Leslie Lamport. On Interprocess Communication. *Distributed Computing*, 1(1):86–101, 1986.

[LG89] Nancy A. Lynch and Kenneth J. Goldman. Distributed Algorithms: Lecture Notes for 6.852. Research Seminar Series MIT/LCS/RSS 5, Massachusetts Institute of Technology, May 1989.

[LTV90] Ming Li, John Tromp, and Paul M. B. Vitanyi. How to Share Concurrent Wait-Free Variables. submitted for publication, June 1990.

[NW87] Richard Newman-Wolfe. A Protocol for Wait-Free, Atomic, Multi-Reader Shared Variables. In *Proceedings of the Sixth Annual ACM SIGACT-SIGOPS Symposium on Principles of Distributed Computing*, pages 232–248, August 1987.

[Pet83] Gary Peterson. Concurrent Reading While Writing. *ACM Transactions on Programming Languages and Systems*, 5(1):46–55, 1983.

[SAG87] Ambuj K. Singh, James H. Anderson, and Mohamed G. Gouda. The Elusive Atomic Register Revisited. In *Proceedings of the Sixth Annual ACM SIGACT-SIGOPS Symposium on Principles of Distributed Computing*, pages 206–221, August 1987.

[Tro89] J. T. Tromp. How to Construct an Atomic Variable. Technical Report CS-R8939, Centre for Mathematics and Computer Science, Amsterdam, October 1989.

[Tya88] A. Tyagi. *The Role of Energy in VLSI Computations*. PhD thesis, Department of Computer Science, University of Washington, Seattle, 1988. *Available as UWCS Technical Report Number 88-06-05*.

[VA86] Paul M. B. Vitanyi and Baruch Awerbuch. Atomic Shared Register Access by Asynchronous Hardware. In *Proceedings of the Twenty-seventh Annual IEEE Symposium on Foundations of Computer Science*, pages 233–243, October 1986.

[Vid88] K. Vidyasankar. Converting Lamport's Regular Register to Atomic Register. *Information Processing Letters*, 28:287–290, 1988.

A Bounded First-In, First-Enabled Solution to the l-Exclusion Problem

Yehuda Afek*   Danny Dolev[†]   Eli Gafni[‡]

Michael Merritt[§]   Nir Shavit[¶]

Abstract

This paper presents a solution to the *first-in, first out* l-exclusion problem of [FLBB79]. Unlike the solution in [FLBB79], this solution is achieved without the use of powerful read-modify-write synchronization primitives, and it requires only bounded size shared memory. Moreover, this solution has the extra property of being *first-in, first-enabled*, a property which subsumes *first-in, first-out*. Use of the *concurrent time-stamp system* of [DS89] is key in solving the problem within bounded size shared memory.

1 Introduction

Consider a system of n asynchronous processes that communicate only by reading from and writing to bounded atomic registers. The program of each process contains a piece of code called the *critical section*. The l-exclusion problem is to guarantee that the system does not enter a global state in which more than l processes are executing their critical section [FLBB79].

To illustrate the problem, imagine that each process controls some device which from time to time needs to enter a mode of high electrical power consumption. The main circuit breaker can withstand at most l devices at high electrical power consumption. By allowing each process to switch its device on only when it is in its critical section, an l-exclusion solution will protect the circuit breaker from burning out.

The l-exclusion problem was first introduced and solved by Fischer, Lynch, Burns, and Borodin in [FLBB79]. The problem is an extension of mutual exclusion (where $l = 1$), a classic problem in

---

*AT&T Bell Laboratories and Tel-Aviv University.

[†]IBM Almaden Research Center and Hebrew University Jerusalem.

[‡]Tel-Aviv University and University of California, Los Angeles, Supported by NSF Presidential Young Investigator Award under grant DCR84-51396 & matching funds from XEROX Co. under grant W881111.

[§]AT&T Bell Laboratories.

[¶]IBM Almaden Research Center and Stanford University. Parts of this research were conducted while the author was at the Hebrew Univerity, visiting AT&T Bell Laboratories and MIT.

concurrency control [Dij65, Lam74]. The ability of a solution to withstand the slow-down or even the crash of processes (up to $l-1$ of them), as well as the absence of collaboration of processes not requesting a resource, are inherent to the problem. (Solutions to the more difficult *l-assignment* problem [ABND+87] must assign each process in the critical section to a distinct "slot", in addition to maintaining the *l*-exclusion property.)

Lamport observed [Lam86] that a first-in, first-out mutual exclusion algorithm can be constructed by preceding a mutual exclusion mechanism with an independent fifo-queue. The possibility of faults precludes the straightforward use of any such first-in, first-out mechanism–the first process to enter the mechanism may fail, preventing progress by those following [FLBB79, DGS88]. To resolve this dilemma, the notion of process enabling was introduced: a process is *enabled* to enter the critical section when sufficiently many local steps of that process will carry it into the critical section, independently of the actions of other processes. First-in, first-enabled was introduced as the natural fairness condition for *l*-exclusion, generalizing the first-in, first-out condition for mutual-exclusion.

The only formerly known first-in, first-enabled solution to the *l*-exclusion problem, due to [FLBB79, FLBB89], was defined and solved based on the use of a strong *read-modify-write* primitive[1]. Rudolph, in [Rud81], solved the problem assuming the slightly weaker *fetch-and-add* primitive. (Herlihy has shown that atomic read-modify-write and fetch-and-add primitives cannot be implemented from shared memory without introducing waiting [H88].) In [Pet81, Pet88, DGS88], bounded solutions were presented using only shared memory, but they implement a weaker form of n^2-*fairness*. (A process may be overtaken by as many as n^2 later processes. An appendix of [DS89] describes an earlier version of the algorithm of this paper, and erroneously claims it to be first-in, first-enabled. This earlier version implements $\sqrt{n}$-fairness.)

Though bounded memory *first-in, first-out* solutions to the mutual exclusion problem were presented in [Kat78, Lam86], it was not known how to generalize these techniques to the case of *l*-exclusion. It seems that in order to achieve first-in, first-enabled fairness, a process must be able to deduce the relative order among other processes, not only between itself and others, and this must be done in the face of concurrent state changes by the other processes. It has remained open whether a bounded solution having a level of fairness higher than n^2-*waiting* was achievable, without resorting to the use of communication primitives of greater power than shared memory.

This paper presents a first-in, first-enabled solution to the *l*-Exclusion problem, using only shared memory communication primitives. The solution requires only bounded size shared memory, and is achieved without the use of strong synchronization primitives. The definitions and proofs are all natural generalizations of mutual exclusion. The protocol makes use of a *concurrent time-stamp system*, an unbounded version of which was implicitly used in [Lam74]. A bounded, wait-free implementation of a concurrent time-stamp system from shared memory is presented in [DS89].

---

[1] The power of such operations is in the assumption that processes do not fail during such an operation.

2 The l-Exclusion Problem

Consider a system of n asynchronous processes communicating via shared memory consisting of *single-writer, multi-reader* atomic registers. The program of every process consists of two distinguished sections: a *remainder section* and a *critical section*. Each process alternates between executing its remainder and its critical section as follows:

> **Process i:**
> **repeat forever**
>   *remainder_section$_i$*
>   *critical_section$_i$*
> **end repeat;**

The l-exclusion problem is to guarantee that the system does not enter a global state in which more than l processes are executing their critical section [FLBB79]. For an application of such a problem imagine that each process P_i controls some device D_i which from time to time needs to enter a mode of high electrical power consumption. The main circuit breaker can withstand at most l devices at high electrical power consumption. By allowing each process to switch its device on only when it is in its critical section, an l-exclusion solution will protect the circuit breaker from burning out.

To coordinate the entrance to the critical section an *entry* and an *exit* code are added to the program of each process as follows:

> **Process i:**
> **repeat forever**
>   *remainder_section$_i$*
>   *entry$_i$*
>   *critical_section$_i$*
>   *exit$_i$*
> **end repeat;**

The following properties are required from any solution to the problem:

l-Exclusion: No more than l processes concurrently execute their critical sections.

If a process i takes only finitely many steps in a given run of the system, or enters the critical section and never leaves, we say i is *faulty* in the run.

l-Deadlock Avoidance: If fewer than l processes are faulty in a run, any process that is not faulty and leaves the remainder region later re-enters it.

Lockout Freedom: In an infinite run of the system if less than l processes are eventually forever outside the remainder region, and infinitely many steps are taken by process i, then process i enters its critical section infinitely often.

(In this paper we count as steps of i both internal actions of process i and external actions of i by which it communicates with shared data structures.)

The fairness property of lockout freedom can be strengthened in the following way. The definition of the entry section is refined to consist of two parts, a *doorway* and a *waiting_room*. An execution of the doorway consists of a bounded number of shared memory operations while an execution of the waiting_room may consist of an unbounded number of such operations, as a process is busy waiting there for room in the critical_section to become available.

> **Process i:**
> **repeat forever**
> > *remainder_section$_i$;*
> > *doorway$_i$;*
> > *waiting_room$_i$;*
> > *critical_section$_i$;*
> > *exit$_i$*;
>
> **end repeat;**

The strengthened fairness property is:

First-Come, First-Served: If process i finishes executing its doorway before process j begins executing its doorway, then i executes its critical section before j does.

As noted by [FLBB79], the first-come, first-served property and the l-deadlock avoidance property cannot be mutually satisfied when l is greater than 1. Because there is room for l processes in the critical section, one must allow later processes to utilize space in the critical section even if earlier processes (i.e. ones with higher priority) are slow. One cannot assure that the later processes will execute their critical sections after the earlier ones, but rather that later processes "keep space open" for earlier ones. Thus, for the general l exclusion problem, this condition must be weakened. Rather than requiring the earliest process to be serviced first, we require instead that it be *enabled* first.

Definition 2.1 *Process i in a global state S is* **enabled** *(to enter its critical section), if there exists $k \geq 0$ such that in any execution from S in which i executes k operations and local steps, i has* **entered** *critical_section$_i$.*

Note that an enabled process remains enabled until it leaves the critical section. Thus, the fairness property of lockout freedom can be strengthened to the following property:

First-In, First-Enabled: Let α be a finite execution ending in a state in which process i is in the waiting room and j is in the critical section. If i last left the doorway before j last entered it, then i is enabled to enter its critical section.

Note that for $l = 1$, where l-*exclusion* reduces to *mutual exclusion*, the first-in, first-enabled property implies the first-come, first-served property.

```
procedure Label_i;
  LabelBegin_i;
    (∀j ∈ {1,...,n}||) Temp_i[j] := Read_i(Label[j])
    Write_i(Label[i] := max_{j∈{1..n}}(Temp_i[j] + 1));
  LabelReturn_i;

procedure Scan_i;
  ScanBegin_i;
    (∀j ∈ {1,...,n}||) Temp_i[j] := Read_i(Label[j])
  ScanReturn_i(order(Temp_i));
```

Figure 1: *UCTSS*: an unbounded concurrent time-stamp system implementation

3 The Solution

The algorithm presented in Figure 2 is a first-in, first-enabled solution to the l-exclusion problem using a concurrent time-stamp system. A concurrent time-stamp system is an object shared by the processes which interact with it via two operations, Label and Scan. Intuitively, in the Label operation a process atomically reserves for itself a time-stamp that is larger than all the time-stamps previously reserved. The Scan operation returns a permutation of the processes, such that the jth location in the permutation is the rank of a time-stamp reserved by process j among a set of times-stamps, where each was reserved for the corresponding process at some time during the Scan.

A bounded implementation of a concurrent time-stamp system from shared memory is presented in [DS89], justifying its use as a primitive in our algorithm.

3.1 Concurrent Time-Stamp Systems

This section presents a particular, unbounded implementation of a concurrent time-stamp system, *UCTSS*, and uses its behavior to specify the safety properties required of arbitrary implementations. The system *UCTSS* consists of an array of n atomic registers [L86b], $Label[1,...,n]$, each $Label[i]$ written by process P_i and read by all. Each process also maintains a local array $Temp_i[1,...,n]$ of integer variables. The function $order(Temp_i)$ returns the permutation $(s_i,...,s_n)$, where s_k is the rank of the pair $(Temp_i[k], k)$ in the lexicographically ordered set $\{(Temp_i[1], 1),...,(Temp_i[n], n)\}$. (We denote this ranking by $\prec$.) Each process P_i can perform two types of operations defined by the Scan$_i$ and Label$_i$ subroutines of Figure 1.

We assume that no process invokes more than one of these subroutines concurrently. The notation $(\forall j \in \{1,...,n\}||) S_j$ indicates that the S_j statements may be executed in any concurrent or sequential order.

3.1.1 Behavioral specification of concurrent time-stamp systems

Different scenarios of invocations of these subroutines by the processes give rise to a set of possible sequences of LabelBegin, LabelReturn, ScanBegin and ScanReturn events, which we denote $Behavior(UCTSS)$.

The substitution of different bodies for the $Label_i$ and $Scan_i$ subroutines to produce an alternative system X, will similarly give rise to alternative sets of sequences of LabelBegin, LabelReturn, ScanBegin and ScanReturn events, which we denote $Behavior(X)$.

Definition 3.1 *A concurrent time-stamp system for n processes is any system X which provides a pair of subroutines, $Label_i$ and $Scan_i$, for each process P_i, whose syntactic interface is identical to the $Label_i$ and $Scan_i$ subroutines above, and such that $Behavior(X) \subset Behavior(UCTSS)$.*

An implementation of a concurrent time-stamp system from shared memory is presented in [DS89], where both the size of the memory and the number of reads and writes per Scan or Label operation are bounded. The existence of this bounded implementation of a concurrent time-stamp system means that the Scan and Label operations may be treated as primitives in designing bounded concurrent algorithms. At the same time, since any concurrent time-stamp system is behaviorally indistinguishable from $UCTSS$, one can reason about the correctness of systems employing a time-stamp system in terms of this particular unbounded implementation. (The internal state of $UCTSS$ acts as an auxiliary variable in such proofs.)

Specifically, use of this bounded implementation in the algorithm of the next section, provides the desired bounded solution to the l-exclusion problem, in that the shared memory size is bounded and the number of shared memory operations in the doorway and exit regions are also bounded. However, the correctness proofs are based solely on the much simpler specification derived from $UCTSS$.

3.2 The l-Exclusion Algorithm

In the code of process P_i as described in Figure 2, Lines 2, 3, and 4 are the doorway part of the code, lines 5, 6, 7, 8, and 9 are the waiting_room, and lines 11, and 12 are the exit code. When entering the doorway a process first raises a flag by setting x_i to true, then it reserves a time-stamp by performing a label operation, and finally records in S_i all the processes whose flag is set (i.e., that were not in the remainder) before it left the doorway. In the waiting_room a processes waits until the number of processes with a smaller time-stamp that are not in the remainder now and that were not in the remainder when it went through the doorway, is smaller than l. A process i in the waiting_room need not consider the time-stamp of processes that were in the remainder when i left the doorway, since such processes might temporarily have a smaller time-stamp, but by the time they next pass through the doorway, their time-stamp will be larger than that of i. After leaving the critical_section a process takes down the flag x_i, signaling that it went back to the remainder. Finally, it performs an additional $Label_i$ operation, so that when it resets x_i in its next pass through the doorway, the time-stamp it holds will not be too early.

```
repeat forever
    1.    remainder section_i

    2.    x_i := true;                                            /* enter doorway */
    3.    Label_i;
    4.    S_i := {j|x_j = true};                                  /* leave doorway */

    5.    repeat                                                  /* enter waiting_room */
    6.        (∀j ∈ S_i||) y_j := x_j;
    7.        l̄_i := Scan_i ;
    8.        Test_i := {j ∈ S_i| y_j = true ∧ l̄_i[j] ≺ l̄_i[i]} ;
    9.    until (|Test_i| < l);                                   /* leave waiting_room */

   10.    critical section_i

   11.    x_i := false;                                           /* enter exit region */
   12.    Label_i;                                                /* leave exit region */
end repeat;
```

Figure 2: A *First-In, First-Enabled* l-exclusion algorithm

4 Correctness proof

We begin by stating several rather straightforward properties of the Label and Scan operations that will be used in the proofs below.

Lemma 4.1 *The following are properties of concurrent time-stamp systems:*

- *If i finishes a Label operation before j begins one, then any Scan performed entirely after both labeling operations, and entirely before any subsequent labeling operations by i and j, returns $\bar{\ell}[i] \prec \bar{\ell}[j]$.*

- *If a Scan_i operation returns $\bar{\ell}_i[i] \prec \bar{\ell}_i[j]$, then every Scan_i operation thereafter returns $\bar{\ell}_i[i] \prec \bar{\ell}_i[j]$, until the next Label_i operation. (To be used in Lemma 4.3 and 4.4.)*

- *Suppose a Label_i operation entirely precedes a Label_j and a Scan_j operation, where the Scan_j returned $\bar{\ell}_j[j] \prec \bar{\ell}_j[k]$. Then if i later performs a Scan_i operation, without having executed a new Label_i operation, then the Scan_i operation returns $\bar{\ell}_i[i] \prec \bar{\ell}_i[k]$. (To be used in Lemma 4.5.)*

Proof The reader can verify that these are properties of the specific implementation $UCTSS$, and hence of general concurrent time-stamp systems, in particular the bounded implementation of [DS89]. □

Lemma 4.2 *The code in Figure 2 satisfies the l-exclusion property.*

Proof Since any concurrent time-stamp system is an implementation of the particular system $UCTSS$, it suffices to consider runs of the l-exclusion algorithm utilyzing $UCTSS$ as the implementation of the Label and Scan operations. By way of contradiction, assume that in some reachable state s there is a set C of more than l processes in the critical section. Let m be the process in C whose label in the concurrent time-stamp system in state s is the largest. Thus the last label operation (line 3) of each of the other processes in C could not have started after m finished its Label$_m$ operation in line 3. Hence, each of the processes in C had finished its last execution of line 2 when m last started line 4, and when m executed line 4 it must have seen $x_j = true$ for every process j in C. Furthermore, since in $UCTSS$ the labels read by a sequence of Label operations form a monotonically non-decreasing sequence, the label of each process in C was smaller than the label of m when m executed its last Scan$_m$ operation. Finally, the value of x_j is $true$ between j's leaving line 2 and j beginning line 10, after the critical section. Hence, every element of C will be included in the set $Test_m$. But $|C| \geq l$, contradicting the assumption that m entered the critical section. □

The next lemma argues that in consecutive executions of the until loop by process i, the size of $Test_i$ is monotonically non-increasing.

Lemma 4.3 *Let T^k and T^{k+1} be the values of $Test_i$ computed in two successive iterations of the waiting_room loop. If $j \in T^{k+1}$, then $j \in T^k$.*

Proof Assume that $j \in T^{k+1}$. Then $j \in S_i$, and in this pass, the value of x_j is $true$ and $\bar{\ell}_i[j] \prec \bar{\ell}_i[i]$. Suppose that $j \notin T^k$. Then in this previous pass through the loop, i observed either x_j as $false$ or $\bar{\ell}_i[i] \prec \bar{\ell}_i[j]$. First consider the case that x_j was observed to be $false$. Then process i read x_j at least three times; once $true$ (in the most recent execution of line 4, since $j \in S_i$), then $false$ and then $true$ again. In between these three observations of x_j, i did not perform any label operation. The semantics of the atomic variable x_j imply that a write by j of $x_j := false$ must be serialized between the first two reads of x_j by i, and another write by j of $x_j := true$ must be serialized between the last two reads by i. But between these two writes by j a label operation by j occurred; that is, j must have finished a label operation before i read $x_j = true$ the final time, and this label operation of j started after i left the doorway, hence after i's last label operation. By the first property in Lemma 4.1, $\bar{\ell}_i[i] \prec \bar{\ell}_i[j]$ in subsequent Scan$_i$'s, a contradiction. It only remains to show that if $\bar{\ell}_i[i] \prec \bar{\ell}_i[j]$ in the kth pass through the loop, then this ordering will continue to be observed until i executes another label operation. But this follows from the second part of Lemma 4.1. □

Lemma 4.4 *The code in Figure 2 satisfies the l-deadlock avoidance property.*

Proof Assume that in some infinite run there exists a set, $Stuck$, of $k \geq 1$ non-faulty processes that leave the remainder region and never re-enter it. Since these processes are non-faulty, they take an infinite number of steps outside the critical section, and so fail the waiting-room loop test infinitely many times. Then each process i in $Stuck$ performed a final Label_i operation before entering the loop, and (in $UCTSS$) has a final value of Label_i. Let i be the process in this set with the (lexicographically) smallest final value of (Label_i, i).

Since all the processes in $Stuck$ eventually finish their final Label operations, eventually the Scan_i's always see them ordered after i, $\bar{\ell}_i[i] \prec \bar{\ell}_i[j]$, and so the processes in $Stuck$ eventually do not contribute to the failure of i's loop test.

By the first part of Lemma 4.1, if any other process j runs a Label_j operation that strictly follows i's last Label_i operation, then every Scan_i that in turn follows this Label_j operation will observe $\bar{\ell}_i[i] \prec \bar{\ell}_i[j]$ and hence thereafter j will not contribute to the failure of i's loop tests. Call the set of such j the $LateLabels$.

Finally, denote by $Remainder$ the set of processes j that eventually enter the remainder section and never leave. Eventually, i will always observe $x_j = \mathit{false}$, and so the processes in $Remainder$ will eventually stop contributing to the failure of i's tests.

Since i continues to fail the loop test, it follows that there are at least l processes that are in neither $Stuck$, $LateLabels$ nor $Remainder$. Then these processes either stop taking steps or take infinitely many steps in the critical section. □

Lemma 4.5 *The code in Figure 2 satisfies the first-in, first-enabled property.*

Proof Let α be a finite system run ending in a state in which process i is in the waiting room and j is in the critical section, and suppose that i last left the doorway before j last entered it.

Either i is enabled to enter the critical section, or every run extending α by $2n+2$ operations of i (n reads and a Scan_i in each pass through the loop) cause i to execute a pass through the waiting room loop that takes place completely after j's entrance to the critical section. (That is, i may finish one pass through the loop that was concurrent with j's entrance to the critical section, and then make another complete pass.) If the concurrent time-stamp system implementation is wait-free (as are $UCTSS$ and the bounded implementation in [DS89]), such extensions exist, regardless of the behavior of other processes. Suppose i is not enabled to enter the critical section, and choose any such extension of α. Let T_i be the value of $Test_i$ computed by i in this last pass through the loop, and let T_j be the value of $Test_j$ computed by j in its last pass through the loop. Since j passed the test and i did not, it follows that there is a process k in T_i that is not in T_j.

Following arguments similar to the proof of Lemma 4.3, if k is in $Test_i$, x_k must have had the value $true$ continuously since the last execution of line 4 by i. Thus k is in both S_j and S_i. Since k is not in T_j, it must be that $\bar{\ell}_j[j] \prec \bar{\ell}_j[k]$. Then by the last part of Lemma 4.1, $\bar{\ell}_i[i] \prec \bar{\ell}_i[k]$, which would exclude k from T_i, a contradiction. □

References

[ABND+87] H. Attiya, A. Bar-Noy, D. Dolev, D. Koller, D. Peleg, and R. Reischuk. Achievable cases in an asynchronous environment. In *Proc. of the 28th IEEE Annual Symp. on Foundation of Computer Science*, pages 337–346, October 1987.

[AG88] J. H. Anderson, and M. G. Gouda, The virtue of patience: concurrent programming with and without waiting. unpublished manuscript, Dept. of Computer Science, Austin, Texas, January 1988.

[DGS88] D. Dolev, E. Gafni, and N. Shavit. Towards a non-atomic era: ℓ-exclusion as a test case. In *Proceedings of the 20th Annual ACM Symposium on Theory of Computing*. ACM SIGACT, ACM, 1988.

[Dij65] E.W. Dijkstra. Solution of a problem in concurrent programming control. *Communications Of The ACM*, 8:165, 1965.

[DS89] D. Dolev and N. Shavit. Bounded concurrent time-stamp systems are constructible. In *Proceedings of the 21st Annual ACM Symposium on Theory of Computing, Seattle, Washington*, pages 454–465. ACM SIGACT, ACM, 1989.

[FLBB79] M. Fischer, N. Lynch, J. Burns, and A. Borodin. Resource allocation with immunity to limited process failure. In *Proceedings of 20th FOCS*, pages 234–254, October 1979.

[FLBB89] M. Fischer, N. Lynch, J. Burns, and A. Borodin. Distributed fifo allocation of identical resources using small shared space. *ACM Transactions on Programming Languages and Systems*, 11(1):90–114, January 1989.

[H88] M. P. Herlihy, Wait free implementations of concurrent objects, *Proc. 7th ACM Symp. on Principles of Distributed Computing*, 1988, pp. 276–290.

[Kat78] H. Katseff. A new solution to the critical section problem. In *Proceedings of the 10th Annual ACM Symposium on Theory of Computing*, pages 86–88. ACM, 1978.

[Lam74] L. Lamport. A new solution of *dijkstra's* concurrent programming problem. *Communications of the ACM*, 78(8):453–455, 1974.

[L86a] L. Lamport, On interprocess communication. Part I: Basic formalism. *Distributed Computing 1, 2* 1986, 77–85.

[L86b] L. Lamport, On interprocess communication. Part II: Algorithms. *Distributed Computing 1, 2* 1986, pp. 86–101.

[Lam86] Leslie Lamport. The mutual exclusion problem.part ii: Statement and solutions. *J. ACM*, 33(2):327–348, 1986.

[Pet81] G. L. Peterson. Myths about the mutual exclusion problem. *Information Processing Letters*, 12(3):115–116, 1981.

[P83] G. L. Peterson, Concurrent reading while writing. *ACM Transactions on Programming Languages and Systems*, Vol. 5, No. 1 (January 1983), pp. 46–55.

[Pet88] G. Peterson. personal communication. unpublished, 1988.

[Rud81] Larry Rudolph. *Software Structures for Ultra-Parallel Computing*. PhD thesis, New York University, 1981.

LIST OF PARTICIPANTS

4th Int. Workshop on Distributed Algorithms

Agrawal, D.
Auletta, V.
Babaoglu, O.
Beauquier, J.
Biran, O.
Black, R.
Bonuccelli, M.
Budhiraja, N.
Butelle, F.
Chandra, T.D.
Charron-Bost, B.
Cori, R.
Dwork, C.
Even, S.
Ezhilchelvan, P.D.
Feng, A.
Fleischer, R.
Gafni, E.
Garay, J.A.
Gastin, P.
Grønning, P.
Israeli, A.
Itai, A.
Janardan, R.
Jayanti, P.
Kutten, S.
Lavallée, I.
Lavault, C.
Lüling, R.
Mantzaris, S.
Mattern, F.

Moran, S.
Nanni, U.
Negro, A.
Neiger, G.
Ojala, L.
Pagli, L.
Pani, G.
Park, J.
Persiano, G.
Pezzoli, G.
Priese, L.
Rajsbaum, S.
Sandoz, A.
Santoro, N.
Scarano, V.
Schwarz, R.
Skyum, S.
Spirakis, P.
Stockmeyer, L.
Tan, R.
Toueg, S.
van Haaften, P.
van Leeuwen, J.
Villain, V.
Vitányi, P.M.B.
Vito, L.P.
Vito, M.
Welch, J.L.
Yung, M.
Zaks, S.
Zito, M.

AUTHOR INDEX

Afek, Y. 15, 422
Agrawal, D. 245
Beauquier, J. 57
Berman, P. 263, 321
Biran, O. 373
Budhiraja, N. 304
Chandra, T.D. 289
Chaudhuri, S. 402
Cheng, S. Wing 133
Cidon, I. 169, 185
Dolev, D. 422
Dwork, C. 213
El Abbadi, A. 245
Ezhilchelvan, P.D. 353
Fleischer, R. 390
Gafni, E. 422
Garay, J.A. 321
Gastin, P. 57
Gopal, A. 304
Gopal, I. 185
Grønning, P. 151
Hagihara, K. 122
Israeli, A. 1
Itai, A. 29
Jalfon, M. 1
Janardan, R. 133
Jayanti, P. 277
Jung, H. 390
Kröger, B. 90
Kutten, S. 15, 169, 185
Lavallée, I. 41
Lavault, C. 41
Lemos, R. de 353
Løvengreen, H.H. 151
Lüling, R. 90
Mansour, Y. 169
Masuzawa, T. 122
Mehlhorn, K. 390
Merritt, M. 422
Monien, B. 90
Moran, S. 373
Neiger, G. 334
Nielsen, T.Q. 151
Obradovic, M. 263
Ofek, Y. 192
Park, J. 122
Peleg, D. 71, 169
Rajsbaum, S. 102
Sandoz, A. 228
Schiper, A. 228
Shavit, N. 422
Sidi, M. 102
Tokura, N. 122
Toueg, S. 277, 289, 304
Tuttle, M.R. 334
Villain, V. 57
Vornberger, O. 90
Welch, J.L. 402
Yung, M. 15, 192
Zaks, S. 373